Landslide Hazards, Risks, and Disasters

T0305845

Hazards and Disasters Series

Landslide Hazards, Risks, and Disasters

Series Editor

John F. Shroder
Emeritus Professor of Geography and Geology
Department of Geography and Geology
University of Nebraska at Omaha
Omaha, NE 68182

Volume Editor

Tim Davies
Geological Sciences
University of Canterbury
Christchurch, New Zealand

ELSEVIER

AMSTERDAM BOSTON HEIDELBERG LONDON NEW YORK OXFORD
PARIS SAN DIEGO SAN FRANCISCO SINGAPORE SYDNEY TOKYO

Elsevier
Radarweg 29, PO Box 211, 1000 AE Amsterdam, Netherlands
The Boulevard, Langford Lane, Kidlington, Oxford OX5 1GB, UK
225 Wyman Street, Waltham, MA 02451, USA

Notices

Knowledge and best practice in this field are constantly changing. As new research and experience broaden our
understanding, changes in research methods, professional practices, or medical treatment may become
necessary.

Practitioners and researchers must always rely on their own experience and knowledge in evaluating and using
any information, methods, compounds, or experiments described herein. In using such information or methods
they should be mindful of their own safety and the safety of others, including parties for whom they have a
professional responsibility.

To the fullest extent of the law, neither the Publisher nor the authors, contributors, or editors, assume any liability
for any injury and/or damage to persons or property as a matter of products liability, negligence or otherwise, or
from any use or operation of any methods, products, instructions, or ideas contained in the material herein.

Library of Congress Cataloging-in-Publication Data
Landslide hazards, risks, and disasters / volume editor, Tim Davies.
    pages cm. -- (Hazards and disasters series)
  ISBN 978-0-12-396452-6 (hardback)
1. Landslides--Risk assessment. 2. Landslides--Prevention. 3. Landslide hazard analysis. I. Davies,
Timothy R. H., editor of compilation.
  QE599.A2L363 2014
  363.34'9--dc23
                   2014035690

British Library Cataloguing in Publication Data
A catalogue record for this book is available from the British Library

ISBN: 978-0-12-396452-6

For information on all Elsevier publications visit
our web site at http://store.elsevier.com

Working together
to grow libraries in
developing countries

ELSEVIER | Book Aid International

www.elsevier.com • www.bookaid.org

Contents

1. Landslide Hazards, Risks, and Disasters: Introduction
Tim Davies

2. Landslide Causes and Triggers
Samuel T. McColl

3. Mass Movement in Bedrock
Marc-André Brideau and Nicholas J. Roberts

11. Remote Sensing of Landslide Motion with Emphasis
     on Satellite Multitemporal Interferometry
     Applications: An Overview
     Janusz Wasowski and Fabio Bovenga

12. Small Landslides—Frequent, Costly, and Manageable
     Elisabeth T. Bowman

13. Analysis Tools for Mass Movement Assessment
     Stefano Utili and Giovanni B. Crosta

Contributors

Fabio Bovenga, National Research Council, CNR-ISSIA, Italy

Elisabeth T. Bowman, Department of Civil and Structural Engineering, University of Sheffield, South Yorkshire, UK

Marc-André Brideau, Simon Fraser University, Burnaby, BC, Canada

John J. Clague, Centre for Natural Hazard Research, Simon Fraser University, Burnaby, BC, Canada

Giovanni B. Crosta, Università degli Studi di Milano-Bicocca, Italy

Tim Davies, Geological Sciences, University of Canterbury, Christchurch, New Zealand

Audray Delcamp, Department of Geography, Faculty of Sciences, Vrije Universiteit Brussel, Brussel, Belgium

Philip Deline, EDYTEM Lab, Université de Savoie, CNRS, Le Bourget-du-Lac, France

Kenneth Hewitt, Geography and Environmental Studies, Wilfrid Laurier University, Waterloo, Ontario, Canada

Oliver Korup, Institute of Earth and Environmental Science, University of Potsdam, Potsdam, Germany

Samuel T. McColl, Physical Geography Group, Institute of Agriculture and Environment, Massey University, New Zealand

Bill Murphy, Leeds University, UK

Natalya Reznichenko, Department of Geography, Durham University, Durham, UK

Nicholas J. Roberts, Simon Fraser University, Burnaby, BC, Canada

Dan Shugar, Department of Geography, University of Victoria, British Columbia, Canada

Rosanna Sosio, Università degli Studi di Milano-Bicocca, Italy

Stefano Utili, School of Engineering, University of Warwick, Coventry, UK

Benjamin van Wyk de Vries, Laboratoire Magmas et Volcans, Univeristé Blaise Pascal, Clermont-Ferrand, France

Gonghui Wang, Disaster Prevention Research Institute, Kyoto University, Kyoto, Japan

Jeff Warburton, Durham University, UK

Janusz Wasowski, National Research Council, CNR-IRPI, Italy

General hazards, risks, and disasters: Hazards are processes that produce danger to human life and infrastructure. Risks are the potential or possibilities that something bad will happen because of the hazards. Disasters are that quite unpleasant result of the hazard occurrence that caused destruction of lives and infrastructure. Hazards, risks, and disasters have been coming under increasing strong scientific scrutiny in recent decades as a result of a combination of numerous unfortunate factors, many of which are quite out of control as a result of human actions. At the top of the list of exacerbating factors to any hazard, of course, is the tragic exponential population growth that is clearly not possible to maintain indefinitely on a finite Earth. As our planet is covered ever more with humans, any natural or human-caused (unnatural?) hazardous process is increasingly likely to adversely impact life and construction systems. The volumes on hazards, risks, and disasters that we present here are thus an attempt to increase the understanding about how to best deal with these problems, even while we all recognize the inherent difficulties of even slowing down the rates of such processes as other compounding situations spiral on out of control, such as exploding population growth and rampant environmental degradation.

Some natural hazardous processes such as volcanoes and earthquakes that emanate from deep within the Earth's interior are in no way affected by human actions, but a number of others are closely related to factors affected or controlled by humanity, even if however unwitting. Chief among these, of course, are climate-controlling factors, and no small measure of these can be exacerbated by the now obvious ongoing climate change at hand (Hay, 2013). Pervasive range and forest fires caused by human-enhanced or induced droughts and fuel loadings, mega-flooding into sprawling urban complexes on floodplains and coastal cities, biological threats from locust plagues, and other ecological disasters gone awry; all of these and many others are but a small part of the potentials for catastrophic risk that loom at many different scales, from the local to planet girdling.

In fact, the denial of possible planet-wide catastrophic risk (Rees, 2013) as exaggerated jeremiads in media landscapes saturated with sensational science stories and end-of-the-world Hollywood productions is perhaps quite understandable, even if simplistically short-sighted. The "end-of-days" tropes promoted by the shaggy-minded prophets of doom have been with us for centuries, mainly because of Biblical verses written in the early Iron-Age during remarkably pacific times of only limited environmental change. Nowadays however, the Armageddon enthusiasts appear to want the worst to validate their death desires and prove their holy books. Unfortunately we are all entering times when just a few individuals could actually trigger societal breakdown by error or terror, if Mother Nature does not do it for us first. Thus we enter contemporaneous times of considerable peril that present needs for close attention.

These volumes we address here about hazards, risks, and disasters are not exhaustive dissertations about all the dangerous possibilities faced by the ever-burgeoning human populations, but they do address the more common natural perils that people face, even while we leave aside (for now) the thinking about higher-level existential threats from such things as bio- or cybertechnologies, artificial intelligence gone awry, ecological collapse, or runaway climate catastrophes.

In contemplating existential risk (Rossbacher, 2013) we have lately come to realize that the new existentialist philosophy is no longer the old sense of disorientation or confusion at the apparently meaninglessness or hopelessly absurd worlds of the past, but instead an increasing realization that serious changes by humans appear to be afoot that even threaten all life on the planet (Kolbert, 2014; Newitz, 2013). In the geological times of the Late Cretaceous, an asteroid collision with Earth wiped out the dinosaurs and much other life; at the present time by contrast, humanity itself appears to be the asteroid.

Misanthropic viewpoints aside, however, an increased understanding of all levels and types of the more common natural hazards would seem a useful endeavor to enhance knowledge accessibility, even while we attempt to figure out how to extract ourselves and other life from the perils produced by the strong climate change so obviously underway. Our intent in these volumes is to show the latest good thinking about the more common endogenetic and exogenetic processes and their roles as threats to everyday human existence. In this fashion, the chapter authors and volume editors have undertaken to show you overviews and more focused assessments of many of the chief obvious threats at hand that have been repeatedly shown on screen and print media in recent years. As this century develops, we may come to wish that these examples of hazards, risks, and disasters are not somehow eclipsed by truly existential threats of a more pervasive nature. The future always hangs in the balance of opposing forces; the ever-lurking, but mindless threats from an implacable nature, or heedless bureaucracies countered only sometimes in small ways by the clumsy and often feeble attempts by individual humans to improve our little lots in life. Only through improved education and understanding will any of us have a chance against such strong odds; perhaps these volumes will add some small measure of assistance in this regard.

Landslide hazards, risks, and disasters: The chapters in this volume on landslides provide the latest thinking on understanding these ever-so-common phenomena that seem to threaten societies almost everywhere in the world that slopes exist. In fact, as one of the most pervasive processes at work almost everywhere on the Earth's surface, slope failure is characteristically the one process that most people appear to ignore, all too commonly to their detriment.

Downslope movements of rock and soil debris and earth by falling, toppling, sliding, and flowing with great variations of velocities and water contents, whether liquid water, solid ice present such a plethora of movement mechanics, that mass-movement specialists have struggled to understand and protect against landslides for many decades. As one of those natural hazard specialists, it has indeed been a pleasure for me to see that continued new thinking on the phenomena has progressed so well; in fact, even though landslide processes are so common and so hard to avoid entirely, the new observations and understandings have led to a more intelligent siting of structures, as well as better engineering designs to cope with instability. As humans continue to expand their populations and move into regions with ever-increasing hazard potentials for slope stability, the observations and understandings presented in these chapters can help in detailing the means to avoid mass-movement problems.

Volume editor Tim Davies is from New Zealand, where the combinations of strong relief, active tectonic forces, high seismicity, strong climatic controlling factors, and many weak lithologies produce exceptionally active slope failure and mass movement at practically every turn. Dr Davies has given us a comprehensive volume that not only addresses the great variety of landslides in the world, but that also addresses a number of examples so that the reader will receive a good education on what has happened, what can happen, and how to best understand the many subtle variations in slope-failure phenomena so that avoidance of the dangerous or damaging processes can be effected best. The variety of chapters presented about a diverse and difficult hazard are sufficiently authoritative and new that even those quite accustomed to the rich literature of landslides will still find much that is new and interesting here.

John (Jack) Shroder
Editor-in-Chief
July 14, 2014

REFERENCES

Hay, W.W., 2013. Experimenting on a Small Planet: A Scholarly Entertainment. Springer-Verlag, Berlin, 983 p.
Kolbert, E., 2014. The Sixth Extinction: An Unnatural History. Henry Holt & Company, NY, 319 p.
Newitz, A., 2013. Scatter, Adapt, and Remember. Doubleday, NY, 305 p.
Rees, M., 2013. Denial of catastrophic risks. Science 339 (6124), 1123.
Rossbacher, L.A., October 2013. Contemplating existential risk. Earth, Geologic Column 58 (10), 64.

This volume is one of a nine-volume set, each of which deals with the hazards, risks, and disasters associated with specific types of landscape process. The other volumes treat hydro-meteorological, volcanic, landslides, earthquakes, sea and ocean processes, snow and ice processes, wildfires, and biological and environmental processes; whereas the final volume presents a cross-disciplinary overview of natural hazards and disasters in society.

The present volume focuses on the hazards, risks, and disasters associated with landslides of various types (e.g., bedrock, volcanic, peat-slide, rock-ice, landslide-dam) and from a variety of perspectives (e.g., paleoslope failures, susceptibility, remote sensing). The purpose is to make available to readers, both students and professionals, a set of discussions by a range of experts on topics receiving current attention in the general area of landslide hazards, risks, and disasters. Authors from a range of backgrounds have contributed, with a number of contributions from younger researchers who have a refreshingly modern approach to landslide problems.

Acknowledging the recent publication of an excellent book on *Landslides: Types, Mechanisms and Modeling* edited by John Clague and Doug Stead, this volume has tried not to duplicate or overlap unnecessarily with the contents of that book. By contrast with the Clague and Stead book, the focus herein is to develop improved understanding of, and hence ability to foresee and thus make expectable, "the effects of landslides on society." To that end some basic concepts are required, particularly where these may not be readily available elsewhere; but throughout, the aim is to further our understanding of landslides and the ways in which they affect humankind and its assets. Hence the "Hazard–Risk–Disaster" sequence, wherein the *hazard* is essentially a natural landslide process (perhaps quantified by its probability of occurrence); *risk* is the product of the probability and consequence of interaction between a landslide and a societal asset; and a *disaster* is the event that results from this interaction (usually inferred to be of major impact with respect to the part of society affected).

The introductory Chapter 1 is intended to provide a general context for the more specialized chapters that follow, by outlining some overarching concepts of hazards, risks, and disasters. It also introduces some developing concepts and ideas in a range of topics that may point the way to improvements in our attempts to reduce the damage, disruption, and deaths caused by landslides in the future.

In Chapter 2 McColl outlines the slope preconditioning, preparation, and triggering processes required to initiate a landslide that may have the potential to cause a disaster, depending on whether assets are likely to be affected by it. Brideau and Roberts describe the conditions and processes that lead to failure of rock slopes in Chapter 3, and discuss the strategies available to manage potential rock-slope instabilities; while in Chapter 4 Murphy focuses specifically on the processes by which earthquakes can destabilize slopes leading to coseismic landsliding, emphasizing the importance of vertical accelerations and topographic amplification, and sounding a cautionary note with respect to the reliability of presently available analysis techniques. In Chapter 5 Van Wyk De Vries and Delcamp outline the factors and processes that cause large-scale collapse of volcanic edifices, and the geometry of the resulting debris avalanche deposits; they also mention the range of secondary hazards resulting from such phenomena. In the following chapter Warburton describes the state of the art in understanding of peat landslides, emphasizing the remaining uncertainties that cause hazard assessment of these events to remain problematic at present.

In Chapter 7 Sosio describes the comparatively recently recognized class of events known as "rock-snow-ice" avalanches, which are represented by the Huascaran events of 1962 and 1970 in Peru, and the 2002 Kolka-Karmadon event in North Ossetia. This chapter complements that of Deline et al. (Chapter 9) describing the processes and effects of rock avalanches that travel over, and deposit on, glaciers, and both exemplify the reality that natural events rarely behave in such a way that they can be satisfactorily described within the confines of a single specialization.

Korup and Wang in Chapter 8 extend previous knowledge of landslide dam hazards and risks to the occurrence of multiple episodes of such events, as have been documented from recent severe earthquakes and storms. They emphasize the additional challenges posed to emergency response and recovery by multiple landslide-damming episodes. In Chapter 10 Clague discusses the nature and accuracy of information on hazards and risks that can be derived from study of the deposits of past landslides, discussing in particular the data bias resulting from unequal preservation of deposits of different ages and its application to risk analysis.

Wasowski and Bovenga provide in Chapter 11 a review of presently available methods for utilizing satellite-based remote sensing to detect and monitor, and outline the risk management and research opportunities offered by these rapidly advancing technologies.

In Chapter 12 Bowman addresses the comparatively neglected topic of assessment and management of the risks associated with the smaller but correspondingly more frequent landslides that make up most of the event population, focusing on the widespread impacts of these "minor" events on society, and on the techniques available and required for their economic mitigation. Finally Utili and Crosta (Chapter 13) outline and critically assess the methods currently available for assessing the stability of slopes, in the context of the technologies now available to facilitate these assessments, emphasizing the continuing necessity for adequate data on which to base even sophisticated analyses.

It is evident from many of these contributions that, even after a century or more of work, and with all the benefits of modern technologies, serious deficiencies remain in our ability to understand, and therefore to avoid or mitigate, future landslide disasters. Clearly landslide research needs to continue into the future, and we can confidently anticipate improvements in knowledge and even understanding of landslide processes. However, in order that these improvements are reflected in reduction in landslide disasters, this effort must be complemented with significantly increased involvement of landslide experts in the development of strategies that will allow society and communities to plan sustainable futures that are resilient to both anticipated and unexpected landslide events.

Tim Davies
Volume Editor

Chapter 1

Landslide Hazards, Risks, and Disasters: Introduction

Tim Davies

Geological Sciences, University of Canterbury, Christchurch, New Zealand

ABSTRACT

Consideration of the nature and occurrence of landslide hazards leads to perspectives on the dominant role of landsliding in the geomorphology of active orogens, and consequently the major role that landsliding plays in determining hazard- and riskscapes both within and downstream of these areas. Discussion of landslide risks suggests that probabilistic analyses are only likely to be reliable in planning location-specific landslide risk management strategies for small, frequent events, and the potential for identifying sites of future landslides—both rainfall generated and coseismic—is examined. Finally the role of landslides in triggering consequential hazards, such as tsunami, river flooding, and debris flows, is emphasized.

1.1 INTRODUCTION

Landslides are a ubiquitous phenomenon on a planet that, like Earth, is tectonically active. However society is generally inclined to view them as exceptional events that occur very infrequently, and usually elsewhere, and their inevitable impacts on society worldwide and over extra-human timescales have hitherto been considered rarely, if at all, in societal planning. In the last decade, however, global landslide occurrence and impact has been better documented (e.g. Petley, 2012), and the seriousness of this hazard underlined. Interestingly, climate change is shown to have much less influence on the number of landslide fatalities than population growth (Petley, 2010), and if this trend continues, landslide fatalities will continue to increase. There is thus good reason to examine the role of landslides as threats to society, and to seek for avenues whereby this threat can be reduced in the future.

Landslide Hazards, Risks, and Disasters. http://dx.doi.org/10.1016/B978-0-12-396452-6.00001-X

1.2 UNDERSTANDING LANDSLIDE HAZARDS

Landslide hazards are, in essence, landslides which have the potential to affect society detrimentally. One may debate whether or not all landslides constitute hazards to society, and whether in principle any landslide anywhere is a potential hazard if there is any possibility that humankind is now or will at some time in the future make itself vulnerable to the effects of that landslide thus generating a risk. There are very few if any places on Earth where this possibility is zero, thus to a fair approximation all terrestrial landslides can be considered to be hazards. So, on that basis, are landslides on the Moon, but we have to draw a line somewhere...

Landslides are a crucial component of Earth's geological cycle, in which tectonic plate motion causes parts of the crust to be continuously uplifted above a base level; they are then continuously eroded down again by gravity and gravity-driven water flow toward base level. Landslides represent the directly gravity-driven component of erosion, and they occur in sizes ranging from individual rocks falling to whole mountainsides collapsing. There is increasing evidence, by way of magnitude–frequency data, that larger landslides deliver more sediment to river systems over time than do smaller ones, so that large, infrequent events dominate the sediment supply spectrum (Korup and Clague, 2009); and, since the majority of river-transported sediment originates in slope failures, this emphasizes the significance of landsliding in geomorphology—including fluvial geomorphology—especially in and adjacent to active orogens. Even a mountain range such as the Southern Alps of New Zealand, which was heavily glaciated prior to 18 ka, today shows little evidence of any erosion process other than mass movement (Figure 1.1).

Thus, in steep, active terrain, the progress of geology requires that land-slides will continue to occur on hillslopes in the future; and the increasing presence of people and their assets on, in the vicinity of and downstream of these hillslopes, means that landslide-generated disasters will inevitably occur—and to an increasing extent—in the future.

An interesting fact discussed at some length by Korup and Clague (2009) is the fairly consistent variation of probability of occurrence for landslides of different sizes. These, irrespective of type and trigger, appear to follow a common distribution for larger events (Figure 1.2), suggesting that there is some factor constraining the frequency of occurrence of landslides of various sizes. This obviously has relevance for assessing hazards and risks from landslides. Recently, research into complex systems has shown that distributions of this type are very common for such systems in many contexts (geomorphic, societal, financial, biological, ecological, etc.); and has in

FIGURE 1.1   The Southern Alps, New Zealand from the west. All of the landforms in the upper part of the picture are mass movement related, with a complete absence of glacial landforms (in spite of a small glacier being visible). The prominent high terraces in the center of the picture are glacial and/or tectonic in origin, while the sediment being reworked by the river at bottom is largely landslide derived.

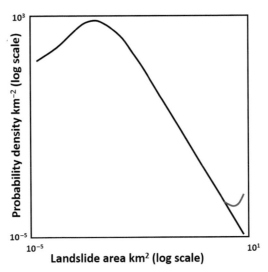

FIGURE 1.2   Landslide area—probability density distribution. (*After Malamud et al., (2004).*) The red curve indicates "dragon-king" events (see text).

addition identified that the very largest events can depart significantly from this distribution. These mega-events, known as "dragon-king" events (Sornette, 2009), occur much more frequently than the distribution would suggest (indicated by the red line in Figure 1.2), and appear to reflect the fact that these events occupy a large proportion of the space available to them—thus such an event has as its environment the system boundaries, which is not the case for smaller events whose environment usually does not include these boundaries. In landslide terms this is equivalent to the probability distribution of landslides from a given hillslope being limited because the physical extent of the hillslope limits the volume of the largest landslides that can occur. The tendency for still larger events to occur is constrained by the system boundaries, so that these larger events are in fact manifested as smaller (but still very large) events, which thus acquire correspondingly higher (but still small) frequencies. This higher than expected frequency of the largest events is clearly a concern in anticipating future landslide disasters, as these events can give rise to the largest disasters; and, although they occur very rarely in a given place, they will inevitably occur and can occur at any time, so treating them as a low priority is not sensible. Hence the largest events occur more frequently than the probability function for smaller events would suggest.

To better deal with landslide hazards, it is first necessary to know where landslides are likely to occur; and, second, how big they will be. These two steps can lead to an *event scenario* for the hazard. Estimating a probability for an event of given size and location is much more difficult, and is in fact of lesser value from a disaster reduction perspective, because we are interested predominantly in reducing the impacts of the *next* disaster that will affect a locality, and probabilities give no reliable information about when that will occur or how big it will be, even in an ideal world with perfect magnitude—frequency information. Thus a landslide susceptibility map, together with an event scenario, provides local people and their governments with realistic information about what can happen there at any time; this can then be used, in conjunction with information on the location of societal assets, to develop a *societal consequence scenario* that forms the basis for designing hazard avoidance and/or damage reduction (assuming that preventing a major landslide is an unrealistic ambition) at all scales from personal to societywide and for thinking about disaster recovery frameworks.

The tricky bit in this train of logic lies in choosing the magnitude of the event scenario. A case can be made for selecting the worst-case scenario, on the basis that a community that has thought through how to cope with this scenario can also cope with anything smaller (although here it needs to be recognized that calculated worst-case scenarios—"maximum credible events"—have recently been dramatically exceeded in a number of earthquake disasters); however, this is likely to be criticized as scare mongering, on the grounds that the maximum possible event occurs incredibly rarely. Thus a somewhat less catastrophic scenario is likely to gain broader acceptance, although here it is easy to set off down the slippery slope of associating probabilities with scenarios. Recent geo-disasters have been spoken of by scientists as having return periods of tens of thousands of years, so it is quite clear that a useful disaster event scenario needs to be substantially worse than the commonly used "100-year" event. People who experience major disasters learn that statistical improbability does not prevent rare events from happening at any time.

Developing a consequence scenario from an event scenario depends on having (or developing) information about where vulnerable assets are located, and where they will be located in the future. A distinct benefit of consequence scenarios is that if (as is often the case) the landslide location (event scenario) is uncertain, then the areal extent of the consequence scenario can be expanded accordingly to accommodate this uncertainty. Foreseeing landslide effects is clearly difficult; one needs to know the volume of the landslide and its location to predict the extent of its deposit and what assets and people it will affect. As the basis for an event scenario, the specific volume and location can be assumed on the basis of credible science, but the implications of this specific scenario then need to be extended across the full area that any landslide can possibly affect. Again it may be tempting to plan regionally for the most likely (highest probability) event; but there is a need to be aware that, although this event is indeed the most likely, its probability is nevertheless small, and *the probability that something different will happen is very much larger*. Thus planning specifically for the most likely event is not rational.

A potentially confusing factor in identifying and quantifying landslide hazards is the geomorphic similarity between moraine deposits and landslide deposits. In recent years many deposits previously classified as moraines (some of which have indeed been used as paleoclimatic indicators) have been reinterpreted as deposits of large landslides (e.g., Hewitt, 1999; McColl and Davies, 2011; Reznichenko et al., 2012). Obviously the hazard implications of such reinterpretation are significant; moraines indicate zero likelihood of repeat events until glaciers advance again, and even then the threat is minor due to the slow speed of glacial advance, but landslide deposits indicate the potential for further landslides at any time in the future, with potentially much more serious consequences.

Predicting the locations of future landslides is obviously difficult, but modern technologies allow at least the relative susceptibilities of different locations to landsliding to be estimated. For example, Kritikos et al. (in review) analyzed the spatial distribution of landslides caused by the Northridge and Wenchuan earthquakes with respect to ground shaking, topography, distance from active faults, slope gradient, and slope position, and were able on this basis to explain with >90% accuracy the relative spatial distribution of landslides resulting from the Chichi earthquake in Taiwan. This explanation used no data from Taiwan landslides, so the technique is applicable to regions with no previous landslide data; similar analyses for rainfall-generated landslides should be feasible. While providing only relative information, this technique nevertheless is useful for scenario-based planning using relative vulnerabilities of, for example, highways and utilities, so that preparation for future landslide events can take place.

It is relevant to note here that relatively little focus is to be found, in this book, on climate change as a factor in future landslide hazards, risks, or disasters. This is because, of the variables that affect the impact of landslides on society, the best quantified in the past, and thus the best predicted in the future, is the increase in global population and the degree of encroachment of society into landslide-prone territory. Recent data show unsurprisingly that landslides kill people mainly where people live in landslide-prone areas (Petley, 2012), and it follows that more people will be killed in future as a result of population increase in these areas. The effects of climate change on landsliding are in comparison poorly understood, poorly defined, and in some cases equivocal. A similar effect occurs with landslide damage, which is related to the growth in value of assets available to be damaged (Petley et al., 2007), and with the overall cost of landslides which is related to the increasing rate at which business is conducted round the planet.

1.3 UNDERSTANDING LANDSLIDE RISKS

Risk is defined as the product of (hazard) probability and (event) consequence, and so introduces as a variable the societal values potentially affected by a landslide (hazard). The consequence itself comprises products of exposure and vulnerability (fragility), so is conceptually more complex than the hazard. Nevertheless, some of its components are more readily quantifiable than many landslide characteristics—for example, the value of assets in a given area and the number of people in the area, at a given time—but these are of course also time-variable at a wide range of scales into the future. In addition, the cost of a landslide disaster includes a wide range of consequential costs, such as the effect on commerce of the cessation of some activities (including transport and power supply), perhaps for a long time; the effect on commerce nationally, if lifelines and infrastructure have been badly affected; and the disruption of community life and social activity, which can affect commerce even if other inputs are not affected. Landslide disasters also have intangible costs which are evidently difficult to quantify as risk, but which are nevertheless significant to society because they affect the way society functions—deaths, emotional damage, loss of quality of life, and environmental damage are examples.

Probabilistic analysis of landslide risk is the commonly accepted basis for making rational decisions about choosing what measures to take to reduce disaster risks. A fundamental difficulty with this procedure is that in any reasonable future planning time frame (i.e., a time frame that present inhabitants perceive as relevant to their community), the number of landslide events that will occur in a given place will be very small. Thus it is extremely unlikely that the events that actually occur in this time frame will follow the probability distribution used, and this would be the case even if this distribution were perfectly known. Thus the apparent precision deriving from probabilistic risk analysis is to a large extent spurious. As an example, consider a simple and perfect statistic: the average of all the numbers 0−100 is 50.000. If we take, say, 100 numbers at random between 0 and 100, what is their average? Well, it is almost never 50.000. We carried out this exercise, repeating the random number exercise 10 times and then choosing the highest value of the 10 averages of the 100 numbers. We then repeated this exercise for smaller and smaller samples of the numbers 0−100. The results are shown in Figure 1.3.

Here we plot the % error of the highest mean of the 10 repetitions against the size of the sample. With a sample size of 100 the error is about ±10%, with a sample size of 10 it is ±40%, and with a sample size of 2 it is ±95%. So any event that occurs only twice during the planning period is likely to depart from a perfectly known statistical distribution by ±95%. If our planning period is say 100 years, then we can predict 1-year events to ±10%, but 50-year events involve an error of up to ±95%. It is apparent that probabilistic risk analysis involves large and intrinsic imprecision for events that occur less than about

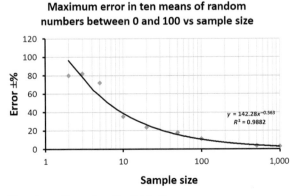

Maximum error in ten means of random numbers between 0 and 100 vs sample size

$y = 142.28x^{-0.563}$
$R^2 = 0.9882$

FIGURE 1.3   Error in small samples of a distribution.

5 times during a planning period—but events of this scale happen so frequently that they can hardly be called disasters, because society will be expecting them and will have taken measures to mitigate them already.

Other difficulties also lie in wait. Suppose we have somehow quantified a realistic landslide risk for a given site, what do we do with it? Two things, generally: if loss of life is a possibility, then there exist levels of acceptable risk that are used worldwide—for example, for an individual a risk of death due to a landslide is acceptable at a level of about 10^{-6} per year (Finlay and Fell, 1997). In this case landslide risk levels need to be managed so that they are below this figure. Alternatively, if the risks are to assets and values, a risk-based cost—benefit or utility—maximization exercise can be undertaken to find the optimal level of expenditure. Both of these, however, have problems.

1. The acceptable risk of loss of life varies dramatically with the number of lives involved. For example, a risk analysis of coseismic landslide-induced tsunami at Milford Sound, New Zealand, showed that the risk of death to an individual tourist is about 10^{-6} per year, so is more or less acceptable; however, such an event will on average kill about 400 people *in one location all at once* every 1000 years, and this is unacceptable by about six orders of magnitude (Figure 1.4). The calculated annual risk of 400 deaths at Milford is $400/1000 = 4 \times 10^{-1}$; the acceptable risk level for this number of deaths is about 2×10^{-7}. So while the individual risk is acceptable, the "societal" risk is grossly unacceptable. What implications this has for risk management requirements—for example, who carries responsibility for "societal"[1] risk—is as yet unclear. The arguments

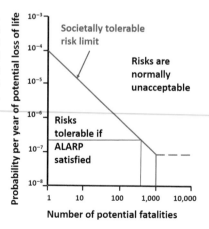

FIGURE 1.4   Acceptable risk for dam failure. Because landslides at Milford Sound are known to occur at a given frequency at a given location (from dated seabed deposits), they are equivalent to known risks of dam failure. The acceptable risk of 400 deaths is 2×10^{-7}, compared with the actual risk of 4×10^{-1}. *After Munger et al. (2009).*

---

1. Given that society is governed by politicians, "political" risk might be a better term: T. Taig, TTAC Ltd, 10 The Avenue, Marston, Cheshire CW9 6EU, UK, personal communication, 2014.

regarding the imprecision of outcomes associated with probabilistically described risks are of little import in this case because of the gross differences between actual and acceptable risks; if these were smaller the implications would be much less clear.

2. Utility optimization involves comparing the net benefits of different levels of mitigation. However, the net benefit is the difference between benefits and costs associated with a given mitigation; these are both large and inevitably imprecise numbers, so subtracting them to derive the net benefit results in a very much smaller number—with a correspondingly very much larger imprecision. This can soon reach the stage where the end result is virtually meaningless—especially considering the imprecision intrinsic to probabilistic planning.

The real value of probabilistic risk analysis appears to lie in insurance, because an insurance (or reinsurance) company spreads local risk across a very large number of locations so that the number of hazard events addressed by the company is very large and widespread; and across all locations, event occurrences will match probabilities fairly well if the probabilities themselves are reliable. This is equivalent to using probabilities to plan mitigation for a given locality over an extremely long time period. However, insurance is a relatively ineffective mitigation for major disasters, because it addresses only with great difficulty the large volumes of claims that result; this is demonstrated by the fact that, several years after the 2011 Christchurch earthquake sequence in New Zealand, only a small proportion of insurance claims have been settled. In addition, insurance has no mitigatory value for many of the types of loss that occur, such as death and distress, and plays no role at all in reducing the physical impacts of disasters.

The only landslide risk management strategy with any realistic chance of being effective against major events is that of avoidance. Substantial landslides—especially earthquake-generated ones—often occur without warning, and are uncontrollable, so the prevention and evacuation procedures used for other hazards such as rainstorm-induced floods are not effective. If a potential landslide has been identified, monitoring of precursory motion may make evacuation a feasible option, but this may be inhibited by political and other considerations, as in the case of the Vaiont tragedy (e.g., Davies, 2013). For small and frequent events, by contrast, prevention is an option (Bowman, 2014)—and, if such events occur frequently in the same locality, a probabilistic risk analysis may be a realistic way of designing a mitigation strategy.

1.4 UNDERSTANDING FUTURE LANDSLIDE DISASTERS

Most of the major landslide disasters that affect society are *unexpected*. This statement is in fact not confined to landslides, but surprise is a particularly consistent feature of landslide disasters. It follows that landslide disaster impacts could probably be reduced substantially if they could be better expected—but how can that be achieved? Again, perhaps relying less on probabilistic information would reduce the natural tendency to underprioritize the inevitable occurrence of major but low-probability events at some time in the future. Identification of precursory behavior to a major landslide would be of great value, and in some places this is being done very effectively (e.g., the Åknes landslide, Norway: Kveldsvik et al., 2009. Here a major landslide seems possible, and would generate an extremely damaging tsunami in a highly populated fiord system). Obviously not all landslide-prone slopes can be identified and monitored, however; an alternative strategy available with modern remote sensing technologies is to search for geomorphic evidence of precursory deformation, such as slumping, developing cracks, and small-scale rockfalls (e.g., Wasowski and Bovenga, 2014). There is evidence that geomorphic detection of precursory motion is also feasible for earthquake-generated landsides, based on the recent realization that many earthquake-affected slopes may deform episodically in a sequence of earthquakes (perhaps hundreds of years apart over thousands of years) until eventually an earthquake (or, indeed, a rainstorm; Chigira et al., 2013) causes further deformation that completes a failure surface and causes catastrophic collapse. An example of this is shown in Figure 1.5. This is a slump at Roche Pass, South Island, New Zealand, that is known to have been mobilized by the 1929 M 7.1 Arthur's Pass earthquake, but has evidently not (yet) failed completely. In hundreds of years' time this feature will be much less conspicuous because the bare rock and debris will have vegetated (unless further slumping occurs), and the slump will only be detectable morphologically. Another example is shown in Figure 1.6, where the near part of the slope has failed in a 0.7-km^3 rock avalanche giving the hummocky deposit at right

FIGURE 1.5   Coseismic slump at Roche Pass, Southern Alps, New Zealand. *Photo by Trevor Chinn.*

FIGURE 1.6   Google Earth image of the 0.7-km^3 Cascade rock avalanche deposit (D), South Westland, New Zealand. The sackung (black dashed line) and the headscarp above it, immediately adjacent to the failed source area (S), indicate a partly deformed slope that could also fail catastrophically in a future earthquake. The Alpine fault is indicated by the white dashed line.

center; however the remainder of the slope, although it has clearly slumped, has not yet failed completely. The 400-km long plate-boundary Alpine fault runs at the base of the slope, and generates M8 earthquakes several times per millennium (Berryman et al., 2012a,b), which probably caused the slumping and the rock avalanche. Identification of slopes with this characteristic morphology thus appears to be promising way to identify future coseismic landslides. Indeed Barth (2013) suggested that a steep slope overlooking Franz Josef Glacier township, Westland, New Zealand might pose a catastrophic collapse threat to the township in a future earthquake; again, the Alpine fault runs at the base of the slope (Figure 1.7). Given that the township hosts thousands of tourists each year, the potential hazard is clear; the volume of the potential landslide appears to be of the order of 10^6 m^3, so a runout of >1 km could occur, obliterating much of the township. As with the Milford Sound example above, the societal risk of this event appears to be unacceptable by several orders of magnitude, and the event itself is unmanageable, so relocation of the township seems to be the only feasible mitigation option. Clearly this is a very difficult sociopolitical situation—but the hazard potential revealed by the science is clear enough to cause the question to be asked.

A peculiarity of some major mountain landslides that is now becoming apparent is that a significant number of them occur with no perceptible trigger. For example, since 1991 seven major (>10^6 m^3) rock avalanches have occurred in the Southern Alps of New Zealand, none of which was associated with either earthquake or rainstorm (e.g., McSaveney, 2002; Hancox and Thompson, 2013). Furthermore, given the relatively frequent occurrence of both earthquakes and severe rainstorms in the Southern Alps, the fact that these landslides did not occur during any previous potential triggering events

FIGURE 1.7   Google Earth image of hillslope overlooking Franz Josef Glacier township, Westland, New Zealand. Sackung indicated by arrows; the Alpine fault is shown by the white dashed line.

suggests that whatever processes were reducing the slope factor of safety (e.g., freeze—thaw, permafrost degradation, stress corrosion), this reduction was occurring quite rapidly. Interestingly, this temporally clustered but spatially dispersed set of landslides could, if not historically recorded, have been interpreted by future scientists as evidence of a widespread trigger earthquake or rainstorm—a potential error that cautions against simplistic paleo-interpretation of landslide deposits (Clague, 2014). In this case a seismic trigger would appear unlikely because all of the landslide source areas are shallow and broad, indicating surficial material failure, whereas major coseismic landslides often have deep-seated source-area scars (Turnbull and Davies, 2006).

Finally, it is important in considering future landslide disasters to realize that the dominant role of landslides in mountain geomorphology means that landslides in mountains are likely to trigger consequential geomorphic events of a variety of types, whose effects can propagate many tens or even hundreds of kilometers across a landscape, and can persist for many decades: Examples include

- A large landslide falling into a mountain valley is likely to affect, or even block, the river in the valley (Korup and Wang, 2014).

- A blockage forms a landslide dam, that can fail when overtopped (immediately or many decades later) causing a short but severe flood to pass along the valley (e.g., Hancox et al., 2005).
- Flooding upstream before the dam fails can also be troublesome, but less catastrophic.
- The large input of sediment to a river from a large landslide (or indeed a large number or small landslides), whether or not the river is blocked, causes the river behavior to alter—typically the river will aggrade to increase its slope and sediment transport capacity, increasing flood risk to the valley floor and to downstream floodplains (e.g., Davies and Korup, 2007; Robinson and Davies, 2013).
- A landslide in a small steep catchment can substantially increase debris flow risk in that catchment and on its fan.
- Large landslides falling into lakes or bays can cause catastrophic tsunami damage to assets close to water level (e.g., Lituya Bay, Alaska, 1954: Weiss et al., 2009).
- A large landslide falling onto a glacier can trigger a far-reaching rock-ice avalanche (e.g., Kolka-Karmodon: Huggel et al., 2005; Sosio, 2014; and Huascaran: Pflaker and Ericksen, 1978).

Thus, assessing the risks consequential on the occurrence of a landslide is no simple task. For example, the 2008 Wenchuan earthquake caused between 56,000 (Parker et al., 2011) and 200,000 landslides (Xu et al., 2013); several hundred of these formed landslide dams that threatened downstream cities with the largest, the Tangjiashan landslide dam, threatening a total of 1.2 million people downstream with a potentially catastrophic outburst flood (Xu et al., 2009; Peng and Zhang, 2012). The largest individual landslide contained $\sim$750 million cubic meters (0.75 km^3) of rock debris and affected a total area of $\sim$7 km^2 (Huang et al., 2012). There is also substantive evidence (Berryman et al., 2001; Davies and Korup, 2007) that past coseismic landslides in the South Island, New Zealand, have caused aggradation of several meters over hundreds of square kilometers of outwash surfaces that are today intensively utilized for farming, towns, and roads, so future earthquake-generated landslides may have severe impacts on society's use these areas for up to some decades after the occurrence of the landslides.

1.5 CONCLUSION

The hazards that landslides pose to society tend to be less well appreciated than the better understood hazards of floods. This is largely because landslides seldom occur in the same places, whereas floods tend to occur adjacent to water bodies where people have lived for a long time, and are thus expected. Communities can nowadays often be provided with hours' or days' warning of major floods, with the opportunity to evacuate people (saving lives) and elevate possessions (reducing damage); by contrast very few landslide locations and times of occurrence are able to be predicted with sufficient reliability to make warning and evacuation a realistic option. Nevertheless, landslides are a major contributor to deaths, damage, and disruption to society (e.g., Sudmeier-Rieux et al., 2013), particularly in association with earthquakes and intense rainstorms. In this context, anticipating the landslide-generated effects of earthquakes and rainstorms is a glaring omission from many if not most hazard management strategies. As a general comment, perhaps the scientific input to developing such strategies has until recently been too discipline-constrained, with seismologists and hydrologists not engaging sufficiently with landslide experts—or indeed with each other—to identify the full suite of geomorphic consequences of a major trigger event.

The currently almost universal risk-based approach to reducing the impacts of specific landslide disasters has serious limitations and is reliable only for minor events. Preparation for, and developing resilience to, larger events is more likely to be successful if based on event and consequence scenarios.

REFERENCES

Barth, N.C., 2014. The Cascade rock avalanche: implications of a very large Alpine Fault-triggered failure, New Zealand. Landslides 11, 327–342. http://dx.doi.org/10.1007/s10346-013-0389-1.

Berryman, K., Alloway, B., Almond, P., et al., 2001. Alpine fault rupture and landscape evolution in Westland, New Zealand. In: Proceedings, 5th International Conference of Geomorphology, Tokyo.

Berryman, K., Cooper, A., Norris, N., Villamor, P., Sutherland, R., Wright, T., Schermer, E., Langridge, R., Baisi, G., 2012a. Late Holocene rupture history of the Alpine fault in South Westland, New Zealand. Seismol. Soc. Am. Bull. 102, 620–638. http://dx.doi.org/10.1785/0120110177.

Berryman, K.R., Cochran, U.A., Clark, K.J., Biasi, G.P., Langridge, R.M., Villamor, P., 2012b. Twenty-four surface-rupturing earthquakes over 8000 years on the Alpine fault, New Zealand. Science 336, 1690–1693. http://dx.doi.org/10.1126/science.1218959.

Bowman, E.T., 2014. Small Landslides: Frequent, Costly, and Manageable. In: Davies, T.R.H. (Ed.), Landslide Hazard and Disasters. Elsevier, pp. 405–439.

Chigira, M., Tsou, C.Y., Matsushi, Y., Hiraishi, N., Matsuzawa, M., 2013. Topographic precursors and geological structures of deep-seated catastrophic landslides caused by Typhoon Talas. Geomorphology 201, 479–493.

Clague, J.J., 2014. Paleo-landslides, Paleolandlsides. In: Davies, T.R.H. (Ed.), Landslide Hazard and Disasters. Elsevier, pp. 321–344.

Davies, T.R.H., 2013. Averting predicted landslide catastrophes: what can be done?. In: International Conference Vajont 1963–2013, Padua, Italy Ital. J. Eng. Geol. Environ. TOPIC2, 229–236.

Davies, T.R.H., Korup, O., 2007. Persistent alluvial fanhead trenching resulting from large, infrequent sediment inputs. Earth Surf. Processes Landforms 32, 725–742. http://dx.doi.org/10.1002/esp.1410.

Finlay, P.J., Fell, R., 1997. Landslides: risk perception and acceptance. Can. Geotech. J. 34, 169–188.

Hancox, G.T., Thompson, R., 2013. The January 2013 Mt Haast rock avalanche and Ball Ridge rockfall in Aoraki/Mt Cook National Park, New Zealand. GNS Sci. Rep. 2013 (33), 26.

Hancox, G.T., McSaveney, M.J., Manville, V.R., Davies, T.R.H., 2005. The October 1999 Mt Adams rock avalanche and subsequent landslide dam-break flood and effects in Poerua River, Westland, New Zealand. N. Z. J. Geol. Geophys. 48, 683–705.

Hewitt, K., 1999. Quaternary moraines vs catastrophic rock avalanches in the Karakoram Himalaya, Northern Pakistan. Quat. Res. 51, 220–237.

Huang, R., et al., 2012. The characteristics and failure mechanism of the largest landslide triggered by the Wenchuan earthquake, May 12, 2008, China. Landslides 9, 131–142.

Huggel, C., Zgraggen-Oswald, S., Haeberli, W., Kääb, A., Polkvoj, A., Galushkin, I., Strom, A., 2005. The 2002 rock/ice avalanche at Kolka/Karmadon, Russian Caucasus: assessment of extraordinary avalanche formation and mobility, and application of QuickBird satellite imagery. Nat. Hazards Earth Sys. Sci. 5 (2).

Korup, O., Clague, J.J., 2009. Natural hazards, extreme events, and mountain topography. Quat. Sci. Rev. 28, 977–990.

Korup, O., Wang, G., 2014. Multiple Landslide-Damming Episodes, Multiple Landslide-Damming Episodes. In: Davies, T.R.H. (Ed.), Landslide Hazard and Disasters. Elsevier, pp. 241–261.

Kritikos, T., Robinson, T.R., Davies, T.R.H. A new GIS-based approach for assessment of regional coseismic landslide susceptibility: the effects of ground shaking and topography on spatial distribution of coseismic slope failures. J. Geophys. Res. (Earth Surf.), in review.

Kveldsvik, V., Einstein, H.H., Nilsen, B., Blikra, L.H., 2009. Numerical analysis of the 650,000 m^2 Åknes rock slope based on measured displacements and geotechnical data. Rock Mech. Rock Eng. 42, 689–728.

Malamud, B.D., et al., 2004. Landslides, earthquakes, and erosion. Earth Planet. Sci. Lett. 229 (1), 45–59.

McColl, S.T., Davies, T.R.H., 2011. Evidence for a rock-avalanche origin for 'The Hillocks' "moraine", Otago, New Zealand. Geomorphology 127, 216–224.

McSaveney, M.J., 2002. Recent rockfalls and rock avalanches in Mount Cook National Park, New Zealand. In: Evans, S.G., DeGraff, J.V. (Eds.), Catastrophic Landslides: Occurrence, Mechanisms and Mobility, vol. 15. Geological Society of America Reviews in Engineering Geology, pp. 35–70.

Munger, D.F., et al., 2009. Interim tolerable risk guidelines for US Army Corps of Engineers dams. In: Managing Our Water Retention Systems: Proceedings of the 29th Annual United States Society on Dams Conference.

Parker, R.N., et al., 2011. Mass wasting triggered by the 2008 Wenchuan earthquake is greater than orogenic growth. Nat. Geosci. 4, 449–452.

Peng, M., Zhang, L.M., 2012. Analysis of human risks due to dam break floods-part 2: application to Tangjiashan landslide dam failure. Nat. Hazards 64, 1899–1923.

Petley, D., 2012. Global patterns of loss of life from landslides. Geology 40, 927–930.

Petley, D., 2010. On the impact of climate change and population growth on the occurrence of fatal landslides in South, East and SE Asia Quarterly J. Eng. Geo. Hydrogeo. 43 (4), 487–496.

Petley, D.N., et al., 2007. Trends in landslide occurrence in Nepal. Nat. Hazards 43, 23–44.

Pflaker, G., Ericksen, G.E., 1978. Nevados huascaran avalanche, Peru. Rockslides Avalanches 1, 277–314.

Reznichenko, N.V., Davies, T.R.H., Shulmeister, J., Larsen, S.H., 2012. A new technique for distinguishing rock-avalanche-sourced sediment in glacial moraines with some paleoclimatic implications. Geology 40 (4), 319–322. http://dx.doi.org/10.1130/G32684.1.

Robinson, T.R., Davies, T.R.H., 2013. Review Article: Potential geomorphic consequences of a future great (Mw = 8.0+) Alpine Fault earthquake, South Island, New Zealand. Natural Hazards and Earth System Science 13, 2279–2299. http://dx.doi.org/10.5194/nhess-13-2279-2013.

Sornette, D., 2009. Dragon-kings, black swans, and the prediction of crises. Int. J. Terraspace Sci. Eng. 2, 1–18.

Sosio, R., 2014. Rock-Snow-Ice Avalanches, Rock-Snow-Ice Avalanches. In: Davies, T.R.H. (Ed.), Landslide Hazard and Disasters. Elsevier, pp. 191–240.

Sudmeier-Rieux, K., Jaquet, S., Baysal, G.K., Derron, M., Devkota, S., Jaboyedoff, M., Shrestha, S., 2013. A neglected disaster: landslides and livelihoods in central-eastern Nepal. In: Margotini, C., et al. (Eds.), Landslide Science and Practice, vol. 4. Springer-Verlag Berlin Heidelberg, pp. 169–176. http://dx.doi.org/10.1007/978-3-642-31337-0_22.

Turnbull, J.M., Davies, T.R.H., 2006. A mass movement origin for cirques. Earth Surf. Processes Landforms 31, 1129–1148. http://dx.doi.org/10.1002/esp.1324.

Wasowski, J., Bovenga,, F., 2014. Remote Sensing of Landslide Motion with Emphasis on Satellite Multi Temporal Interferometry Applications: An Overview, Remote Sensing of Landslide Motion with Emphasis on Satellite Multitemporal Interferometry Applications: An Overview. In: Davies, T.R.H. (Ed.), Landslide Hazard and Disasters. Elsevier, pp. 345–403.

Weiss, R., Hermann, M.F., Wünnemann, K., 2009. Hybrid modeling of the mega-tsunami runup in Lituya Bay after half a century. Geophys. Res. Lett. 36 (9).

Xu, C., et al., 2013. Application of an incomplete landslide inventory, logistic regression model and its validation for landslide susceptibility mapping related to the May 12, 2008. Wenchuan Earthquake China Nat. Hazards 68, 883–900.

Xu, Q., Fan, X.-M., Huang, R.-Q., Westen, C.V., 2009. Landslide dams triggered by the Wenchuan Earthquake, Sichuan Province, south west China. Bull. Eng. Geol. Environ. 68, 373–386.

Chapter 2

Landslide Causes and Triggers

Samuel T. McColl

Physical Geography Group, Institute of Agriculture and Environment, Massey University, New Zealand

ABSTRACT

The stability of a slope is influenced by a variety of time-independent and time-dependent factors, which, when unfavorably aligned, can lead to landsliding. Assessing landslide causes is useful for hazard analysis, mitigation of slope instability, and for considering the role of landslides in landscape systems and evolution. Geological and geomorphological conditions (e.g., material type, strength and structure, and slope angle) predispose slopes to failure; knowledge of these conditions can help to predict the location, types, and volumes of potential failures. Determining when a slope will fail is a considerably more difficult challenge, largely due to the difficulty of observing or predicting the processes of material strength degradation. This chapter describes concepts of stability and explores some of the major causes and triggers of slope failure and opportunities for further research.

2.1 INTRODUCTION

On the 8th of August 1979 in Abbotsford, New Zealand, several weeks of preliminary movements culminated in the rapid failure of an approximately 5 million m^3 block slide in a residential area. The cost of the Abbotsford Landslide amounted to some NZ\$10–13 million and included the loss of 69 houses, but no human lives (Hancox, 2008). When landslides or other hazards result in financial or human losses, a question of blame will often arise, particularly in cases where insurance companies need to make loss adjustments. This, and the desire to prevent similar disasters in the future, necessitates investigation into the causes of a landslide. Furthermore, to gain an appreciation of the role of landslides in landscape systems and in the geomorphic evolution of landscapes, it is necessary to have some knowledge of the factors that determine where and over what timescales landslides are likely to occur or have occurred. As with almost every other landslide, the cause of the Abbotsford Landslide cannot be put down to a single factor; its

Landslide Hazards, Risks, and Disasters. http://dx.doi.org/10.1016/B978-0-12-396452-6.00002-1

17

failure was the result of the convergence of a series of unfavorable stability conditions, both natural and human induced. Hancox (2008) reports on the results of a Government Commission of Inquiry which set out to identify the causes of this landslide. The following factors were considered to be of importance:

1. Unfavorable geology, consisting of weak clay layers in gently tilted strata.
2. Removal of support by quarrying at the toe of the slope 10 years prior to failure.
3. A long-term rise in groundwater resulting from a leaking water main above the slide area and a natural increase in rainfall over the preceding decade.
4. Rainfall during the last few days of movement, which by itself may not have been the difference between failure and stability, but likely influenced the timing and nature of the landslide's final rapid movement.

Several causes of minor importance were also recognized, such as increased loading from urban development and the historical clearance of native vegetation. Despite all of these causes, the disaster, rather than the landslide event, stems from the decision to build (perhaps unknowingly) on terrain susceptible to landsliding without adequate precautionary mitigation measures in place. Although land use decisions were the ultimate cause of the disaster, the purpose of presenting this case study is to highlight that, as with many landslides, none of the recognized factors alone would have been sufficient or necessary to cause the landslide. Assigning responsibility to any person, organization or decision is therefore challenging. Yet, knowledge of the various factors that can cause landslides is invaluable for hazard mitigation. This knowledge helps to identify and avoid human activities that may negatively affect, or be affected by, slope stability, and it improves the capability for predicting natural landslides. For example, providing effective advanced warnings for debris flows requires appropriate triggering thresholds to be set. At which level these thresholds are set, and indeed the identification of the triggers themselves, requires understanding of the various factors causing and triggering debris flows. These factors may include the availability of debris to be mobilized, antecedent soil moisture conditions, rainfall magnitude and intensity, and the condition of vegetative cover. Without sufficient knowledge of contributing instability factors, and the appropriate application of that knowledge, hazard mitigation efforts may be ineffective.

A great deal of scientific observation and enquiry over the past few decades has contributed to more in-depth classification, rationalization, and understanding of the natural and human causes and triggers of landsliding. This has stemmed largely from the disciplines of engineering geology, soil and rock mechanics, and geomorphology. This article does not attempt to list or comment on the history of significant developments in landslide research (which has been done by others; e.g., Crozier, 1986; Petley, 2011), nor does it attempt to describe every known landslide cause. Instead, the work focuses on causes that are of high importance and remain fruitful objectives of scientific research. To begin, slope stability is introduced to provide a conceptual basis for understanding the effects of various slope conditions and destabilizing processes. This is followed by a brief overview of the main types of landslide causes and triggers and our current understanding of these. Finally, a brief synopsis of landslide causes and opportunities for further research is presented.

2.2 CONCEPT OF INSTABILITY

Destabilizing stresses are present within all slopes. Whether or not these stresses (driving stresses) are capable of triggering failure of a given slope at a given moment in time will depend on the relative magnitude of the stresses that resist the tendency for failure; these opposing stresses can be referred to as resisting stresses. Both driving and resisting stresses can change over time. Thus, "stability" is both a relative term and one that refers to a specified time period. The vast majority of existing slopes are stable at this moment in time (otherwise they would be in the process of failing), but every slope has the potential to fail at some future time. The magnitudes of driving and resisting stresses are the result of stability factors, which can be defined as any phenomena that control or influence the forces that determine the stability (Crozier, 1986). Some stability factors are inherent to the slope and unchanging (e.g., lithology), while others may be transient and their influence may vary in magnitude (e.g., porewater pressures).

Stability can be assessed by considering the balance of driving and resisting stresses. The development of shear stresses drives a tendency for failure for most landslide types (with the exception of toppling). The resisting stresses result from reactionary stresses and can be considered as the mobilized shear strength of the slope with respect to the shear stresses. Mobilized strength refers to the strength that resists movement and is distinguished from the "total" strength of the slope, which is arguably an impossible quantity to assess. For example, the movement of an intact slide-prone block (e.g., a rockslide) is influenced by the frictional resistance generated between the sliding surfaces, not the internal strength of the block itself; therefore, only the strength of the failure surface needs to be known to assess the stability of the slide-prone block. However, the assumption that landslides involving sliding are governed by shear stresses alone is an oversimplification. Other (e.g., compressional and tensional) stresses also play a role, especially at the boundaries of the sliding mass and where the landslide mass moves over irregular surfaces.

The balance of resisting and driving forces is often expressed as a ratio, commonly referred to as the factor of safety (FoS):

$$\text{Factor of safety (FoS)} = \frac{\text{resisting stresses (or mobilised shear strength)}}{\text{driving stresses (or shear stresses)}}$$

The FoS equation can be populated with a wide range of parameters which influence the resisting and/or driving stresses. The choice and determination of these parameters depend on the problem being addressed, the physical conditions and processes expected for the slope, and the degree of simplicity or complexity sought. The method of assessing stability by determining the FoS is referred to as limit equilibrium analysis. Examples and explanations of this process and the selection of relevant parameters and variables are provided in numerous texts on slope stability (e.g., Duncan, 1996; Norrish and Wyllie, 1996; Selby, 1993).

The effect of gravity is of fundamental importance in the stability of a slope. The strengths and stresses operating in a slope depend on the way that gravity causes masses within a slope to interact. The weight of material can be resolved into stresses acting normal and parallel to contact surfaces (Figure 2.1). The greater the normal stresses, the greater the frictional strength of the potential failure surface. The relationship between shear strength and the normal stresses is often considered to be linear and governed by the Mohr–Coulomb failure criterion. The greater the stresses acting parallel to the potential failure surfaces, the larger the shear stress. Many of the stability factors discussed below influence the stability of a slope by directly affecting or altering, acutely or chronically, the shear stresses and/or shear strength. Thus, it is useful to keep the relationship between these two stresses in mind.

The term "slope stability" can have different meanings depending on the context of its use or the objectives of its user. It often refers to the inherent

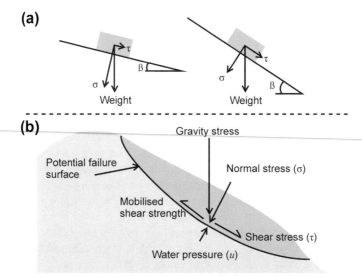

FIGURE 2.1   (a) The influence of slope angle (β) on the relative magnitudes of shear (τ) and normal stress (σ). (b) Stresses acting along a potential failure surface. *Adapted from Selby (1993), by permission of Oxford University Press.*

stability or FoS, as determined by the static physical slope properties; for example, all other stability factors being equal, a high embankment is less stable compared with a low embankment. Landslide susceptibility, which is a measure of the inherent stability of a slope or distribution of slopes, treats the term stability in this way, without any consideration of the likelihood of failure. Alternatively, *slope stability* can, more usefully, be a measure of the probability of a failure of an individual slope, which requires consideration of both the stability factors (i.e., slope stability in the previous sense) and the likelihood of a critical failure threshold being exceeded in a given time span[1]. Assessing the stabilities of the two embankments in this sense requires knowledge of the potential failure triggers and their likelihoods during the time period of interest for each embankment. That knowledge must include the probability of the occurrence of a trigger of sufficient (critical) magnitude to induce failure—the critical threshold for failure itself must also be determined probabilistically because it also fluctuates with time over a frequency/magnitude range. The higher embankment is thus not necessarily more likely to fail than the lower embankment because it may exist in an environment less likely to produce a trigger capable of exceeding the critical threshold for stability for that slope.

In considering the causes of mass movements, a distinction must also be made between the terms "failure" and "movement". Some mass movements only ever exist as a discrete event with the failure resulting in the first and final movement of that material, for example, a rock fall. Other mass movements can exist in a state of transient stability for a long time after failure is first initiated, and be affected by totally different movement triggers or controlling factors to those that initiated the first failure. These slopes may experience discrete periods of instability (movement) and periods of stability (no movement), as is often the case for slopes affected by large, slow-moving rockslides. Other mass movements may evolve from an initially very slowly creeping mass and begin to accelerate as the failure surface develops, culminating in a catastrophic final failure; such mass movements are termed progressive failures (Petley et al., 2005). In such cases, it is not clear to which part of this process the term "failure" applies; the final catastrophic failure only, the initiation of creep, or the entire process (which may take place over an indeterminate amount of time). Perhaps it depends on the scale at which a slope is observed. Petley (2011) explained a distinction between a local and a global FoS for a slope. The global FoS applies to an entire landslide body; although, this in itself may be difficult to define for landslides involving movement of multiple discrete blocks. Failure of the entire mass occurs if the global FoS falls below

---

1. The parameters used in FoS equations can also be treated probabilistically, in recognition of the difficulty and uncertainty of measuring the absolute value of the parameters (e.g., the friction angle value), with the chosen value, at best, representing a range of possible values that may exist for any one slope (e.g., El-Ramly et al., 2002). Additionally, even the type of failure mechanism may be assessed probabilistically (e.g., Hack et al., 2002).

unity. However, even if the landslide as an entire mass may be stable (global FoS > 1), there may be parts of the landslide body that are locally unstable. Petley pointed out that those unstable parts of the landslide mass will locally undergo deformation, and cumulative deformation of these unstable parts may lead to progressively greater reductions in the global FoS (see Section 2.3.2).

Terms such as "stable", "marginally stable", and "actively unstable" have been introduced in recognition of the need to classify the stability state of landslides as a function of the likelihood of failure (Figure 2.2) (Glade and Crozier, 2005). In this context, "stable" refers to a slope that exists in an environment which is unlikely to produce a trigger sufficient to cause failure or movement of that slope (Figure 2.3). A "marginally stable" slope is one for which a trigger of sufficient magnitude to cause failure can be expected to occur in the prevailing environmental conditions (Figure 2.3). Actively unstable slopes are those that are already unstable and actively moving. A slope may shift between these states with a change in any of a multitude of stability factors. Any factor (or collection of factors) that contribute to the reduction in stability to a marginally stable state can be considered to be a landslide cause (or causes). If any of these factors changes the slope to an actively unstable state, it is considered a trigger. While there can be multiple causes, there is usually only one trigger (Varnes, 1978) (Figure 2.3). Others (e.g., Sowers and Sowers, 1970; Zolotarev, 1973; cited in Varnes, 1978) have previously commented that a landslide trigger is often only the final action that initiates failure of a marginally stable slope, and it may be quite trivial in magnitude (Figure 2.3). This can be likened to the idiom "the straw that broke the camel's back." An astonishing example of movement being triggered by a

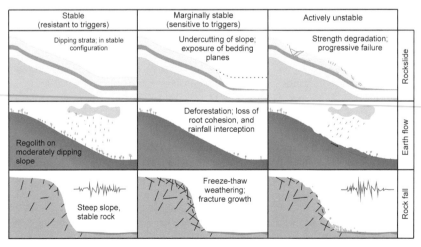

FIGURE 2.2   The relationships between stable, marginally stable, and actively unstable slopes and the interactions of stability factors for a range of landslide types. *Based on Glade and Crozier (2005).*

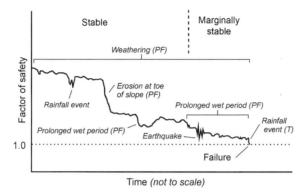

FIGURE 2.3 Schematic representation of the reduction in the FoS of a slope, both through gradual (e.g., weathering) and more discrete (e.g., undercutting) preparatory factors (PF). The resistance of the slope to external processes (potential triggers) lowers through time, eventually bringing the slope to a marginally stable state, whereby it is sensitive to triggering (T) by such events. *Adapted from WG/WLI (1994).*

seemingly trivial force is the Slumgullion Landslide in the United States (Schulz et al., 2009); when the landslide is critically stable due to a combination of other unfavorable stability factors, a tidally-induced state of low atmospheric pressure is sufficient to reduce shear strength and initiate sliding.

The stability factors can be divided into preconditioning, preparatory, and triggering factors (Glade and Crozier, 2005) (Figure 2.2). Preconditioning factors are those that influence the inherent strength of the slope and are generally considered to be temporally unchanging (over human timescales). Factors that reduce the stability over time but do not cause failure or movement are termed preparatory factors, and factors that change a slope to an actively unstable state (i.e., initiate failure or movement) are termed triggers. Descriptions and examples of preconditioning, preparatory, and triggering factors are provided in the following section and examples of each are presented in Table 2.1. Several of these processes can act as either a preparatory factor or a trigger, or both, depending on their degree of activity and the margin of stability of a slope (WG/WLI, 1994). For example, even if an earthquake does not trigger a failure, the shaking may be strong enough to cause deterioration in the strength of material, thus reducing the stability of a slope and potentially allowing a subsequent earthquake of a similar or smaller size to trigger failure. Therefore, earthquake shaking processes can act as both a preparatory factor and a trigger. Some preconditioning and preparatory factors may also be similar. For example, nonzero slope is a precondition for all landslides, but a change in slope angle is considered a preparatory factor. The landslide causes and triggers listed in Table 2.1 include conditions or processes acting both internally and externally to the slope and can reduce the FoS through an increase in the shear stresses, a reduction in the shear strength, or in some cases either of these (Terzaghi, 1950; Varnes, 1978; WG/WLI, 1994).

TABLE 2.1 Examples of Preconditioning, Preparatory, or Triggering Factors and Examples of the Processes Involved

Preconditions

Plastic weak material
Sensitive material
Collapsible material
Weathered material
Sheared material
Jointed or fissured material
Adversely oriented mass discontinuities (including bedding, foliation, cleavage, faults, unconformities, flexural shears, and sedimentary contacts)
Contrast in permeability and its effects on groundwater
Contrast in stiffness (stiff, dense materials overlaying weak plastic materials)

	Processes		
Preparatory Factors	**Geomorphological**	**Physical**	**Human**
Increase in slope height or steepness	Tilting from tectonism, volcanism, or glacial rebound Fluvial, marine or glacial erosion/undercutting		Slope excavation or slope construction
Debuttressing	Glacier retreat		Unloading of toe of slope
Exposure of potential failure surface	Fluvial, marine or glacial erosion/undercutting		Slope excavation

	Geomorphological	Physical	Human
Reduction in inherent strength	Soil piping or solution weathering	Weathering Stress-induced fatigue	Tunneling, underground mining, deforestation
Loading of the slope	Gradual build up of sediments or vegetation		Construction or emplacement of engineering fill or waste deposits
Long-term increase in groundwater levels		Climate change	Infiltration from storm water or broken pipes Irrigation Removal of vegetation
Triggers	**Geomorphological**	**Physical**	**Human**
Rapid increase in porewater pressures	Undrained loading from rapid emplacement of sediment	Rainfall Thaw of snow/ice	Undrained loading from rapid emplacement of fill or heavy loads
Drawdown of groundwater	Natural dam-breach		Lowering of reservoir
Transitory applied stresses		Earthquake Wind	Machinery vibrations
Reduction in strength		Permafrost degradation Weathering Stress-induced fatigue	
Loading of the slope	Other landslides	Precipitation	Building materials

Process examples have been grouped into geomorphological, physical, and human processes following the classification of WG/WLI (1994). More comprehensive lists of landslide causes and triggers can be found in Varnes (1978), Cruden and Varnes (1996), and WG/WLI (1994).

2.3 STABILITY FACTORS

2.3.1 Material Strength and Topography

The inherent strength of material preconditions the stability of all slopes. Material strength determines both the height and angle that a slope can maintain against specific disturbing forces. The influence of material strength and slope geometry on slope stability can be inferred from spatial patterns of landslide distributions (e.g., Dai et al., 2011; Keefer, 2000; Roering et al., 2005) (Figure 2.4). These relationships allow, to a limited degree, determination of landslide susceptibility. Material type (e.g., soils, soft rock, or brittle

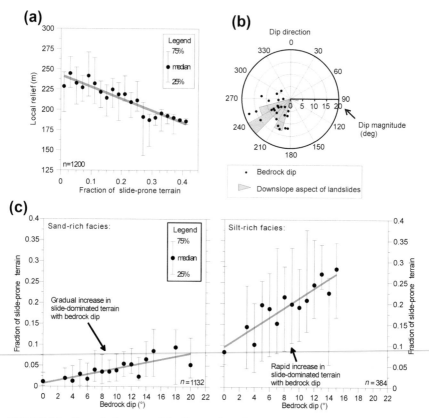

FIGURE 2.4 Example from part of the Oregon Coast Range, U.S.A., showing the relationships between geological and geomorphological conditions and landslide distribution. (a) Relationship between relief and landslide abundance in terrain prone to deep-seated landsliding; (b) Strong correspondence between bedding orientation and the failure of landslides; (c) Moderately strong relationship between landslide abundance and the angle of bedding for sand-rich terrain prone to deep-seated landsliding (left) and a stronger relationship for weaker, silt-rich sedimentary rocks (right). *Reproduced from Roering et al. (2005) with permission by the Geological Society of America.*

rock) and slope geometry also influence the type of landslide that may be expected (e.g., soil flow, slump, rock avalanche), and knowledge of these parameters can also be used to help predict the type of landslides expected (Cruden and Martin, 2013; Gerrard, 1994). Thus, knowledge of material type, slope geometry, and the orientation of structural features relative to slope aspect is essential for assessing slope stability.

Material strength is derived from the strength of the particles or crystals, the interparticle contact forces, the structure produced by fabric or disconti- nuities, and apparent cohesion provided by vegetation. Materials are tradi- tionally divided into (indurated) rock and soils, the latter of which can be further subdivided into cohesive and noncohesive soils. Mohr−Coulomb strength parameters (friction and cohesion) are routinely assessed for each type of soil or rock present at a site; for the latter, it has been recognized that an equivalent Mohr−Coulomb failure criterion that incorporates the effects of discontinuities (e.g., joints) should be used (e.g., Hoek et al., 2002). In rock masses, discontinuity mapping usually involves quantifying the number, orientation, and strength of joints, faults, or bedding layers. In soils, it may be more appropriate to quantify the depth, orientation, strength, and permeability of distinct compositional units.

Discontinuities are perhaps the most important determinant of strength, particularly when dipping out of the slope. They arise from compositional changes (namely bedding or weathering horizons), or structural weaknesses such as cleavage, foliation, fractures, and faults. There are two main effects that discontinuities and compositional changes have on the stability of a slope:

1. They provide structural weaknesses below the (intact) strength of the uniform material, which may become preferential failure surfaces, partic- ularly when planar. For example, the soft rock Tertiary landscapes of New Zealand host many thousands of large landslides; here it is the presence of clay layers, with very low frictional resistance, within the Tertiary sandstones and mudstones that result in landsliding on gently in- clined dip slopes (Thomson, 1982). However, where bedding is horizontal or dipping into the slope, the Tertiary rocks can maintain steep cliffs, which is also due to a distinct lack of tectonic joints.
2. Discontinuities or compositional changes can allow pathways for water seepage or create permeability boundaries, both of which can influence the porewater conditions of the material and modify the slope stresses (Section 2.3.1).

A clear relationship between material strength, slope geometry, and land- slide susceptibility may not always exist. For example, the largest landslide to develop in the greywacke of the Central Southern Alps of New Zealand occurs in a location with some of the highest rock mass strength. Here the 150−200 M m^3 Mueller Rockslide is inferred to have taken advantage of relatively planar bedding surfaces on the limb of a large anticline

(McColl and Davies, 2013). Elsewhere in the greywacke terrain, the folding and jointing are possibly too intense to permit very extensive failure surfaces to develop, and consequently the dominant mode of failure is rock fall. Furthermore, in an analysis of earthquake-induced landslide distributions arising from the 1989 Loma Prieta earthquake in California, Keefer (2000) demonstrated a complex relationship between material type and failure densities. Although the most indurated rocks generally had a lower density of landsliding than less indurated rocks, the least indurated rocks did not have the greatest number of landslides. As well as this, when ranked by estimated or measured shear strength parameters (rather than rock type) there was no statistically significant relationship with landslide density. Although part of the reason for the latter may be problems with the sampling resolution (Keefer, 2000), or variable unaccounted site effects (Section 2.3.4), it may also be partly explained by the concept of strength equilibrium slopes (*sensu* Selby, 1982). These are stable slopes in which the slope angle is in equilibrium with the present stress conditions and strength of that slope. This concept implies that stronger rock masses can attain and maintain steeper angles of slope than weaker rocks, and thus the steepest slopes in a landscape are commonly composed of the strongest rock and vice versa. Therefore, in a landscape dominated by strength equilibrium slopes, it is possible that all slopes have a similar FoS regardless of rock type or slope angle. An increase in slope stresses (or a reduction in material strength may be equally likely to lead to failure of any of these slopes.

One way to increase the slope stresses is through steepening or lengthening the slope. Slopes that have undergone erosional steepening are often said to be "oversteepened" (with respect to their strength equilibrium). A well-recognized process of natural slope steepening is by glacial erosion of valley sides; glacial oversteepening is a commonly recognized preparatory factor for instability of previously glaciated slopes. However, whether or not a slope is oversteepened will depend on its strength equilibrium, and Augustinus (1995) argued that this needs to be quantified to be able to make such a determination. By quantifying the rock mass strength of glaciated slopes in Fiordland, New Zealand, Augustinus (1995) found that most slopes are in strength equilibrium, and despite the glacial steepening, most slopes are not oversteepened. Slopes that were oversteepened were likely to have adjusted quickly back to strength equilibrium. Nevertheless, steepening of the slope can reduce stability. This makes the slope more prone to failure or a given trigger more likely to cause failure, ultimately reducing the life of the slope or the time before the slope will adjust (i.e., undergo failure) to reach a new strength equilibrium profile.

Independently, slope angle and material strength are not reliable predictors of stability; but, if both material strength (including the strength and orientation of discontinuities) and slope angle can be quantified and assessed together, then susceptibility determination can be powerful. These are the two basic parameters of all landslide susceptibility maps. Recent advances in remote sensing (e.g., satellite, airborne, and terrestrial distance ranging) have permitted very high resolution topographic models of slopes at various scales, which allow accurate measurements of slope gradient. Material strength is seldom directly quantified and assessed at a sufficient resolution to detect strength changes within geological units. Instead, most susceptibility maps distinguish between different rock types, rather than distinguishing strength variability within those rock types. This is because the input data usually come from geological maps rather than geotechnical maps and measuring the rock strength at a high spatial resolution is both difficult and expensive.

While the material strength, structure, and topography precondition all landslides and can be used to assess failure susceptibility (i.e., the spatial distribution of potential failures), they do not provide a useful way to predict the timing of those failures. For failure to occur, the FoS needs to drop. This may be caused by an increase in slope angle (considered above), but is most often due to either a decrease in the intrinsic strength or a gradual or sudden increase in an external stability factor. The rest of this chapter focuses on these time-dependent factors.

2.3.2 Intrinsic Strength Degradation

The intrinsic strength of a slope can reduce slowly over time as a result of alteration (weathering) of the slope materials or strain accumulation. The gradual loss of intrinsic strength, or strength degradation, is a phenomenon experienced by all slopes; therefore, there is potentially a lot to be gained from studying the processes involved. Despite its importance, strength degradation is perhaps the least understood of all landslide causes because it operates internally, making it difficult to observe and quantify. Because strength degradation operates in all slopes, it plays a fundamental role in controlling when a slope will fail. This role could be visualized in a model such as that of Figure 2.3, where the gradient of the time-varying stability line would influence the time that FoS = 1 is reached. In the absence of other external factors capable of triggering failure, strength degradation may itself eventually trigger failure. Most of the time, its role is probably to reduce stability, increasing the likelihood for triggering by other instability factors.

Strength degradation can involve any combination of a range of weathering processes, including those of chemical decomposition of materials or stress—strain processes involving the stress-induced growth of fractures or development of a failure surface. While it can operate without external stimuli or fluctuating conditions (e.g., rainfall-induced water pressure changes), the presence of water in a slope or the operation of external stresses may greatly increase the rate of strength degradation.

Strength degradation, or its various processes (e.g., static fatigue, permafrost degradation, weathering), have often been cited as the possible cause of failures that had no apparent trigger. Indeed, there are numerous failures that have occurred without any apparent trigger (Wieczorek and Jäger, 1996), even when the events have been witnessed (e.g., Hauser, 2002; Lipovsky et al., 2008). This is not to say that strength degradation is the only possible trigger for many of these slope failures; changes may occur in the slope by more commonly observed processes that have, for whatever reason, gone undetected or unrecognized, such as unmonitored seepage pressures. For example, recent analysis of temperature conditions by Allen and Huggel (2013) provides support for a plausible alternative hypothesis for the trigger of the 1991 Mt Cook Rock Avalanche in New Zealand, which had previously been cited as an example of a slope failure triggered by strength degradation, or more specifically, a drop in the intrinsic strength of the slope (Glade and Crozier, 2005; McSaveney, 2002; Petley, 2011). Allen and Huggel argued that the trigger was pressurized meltwater. The meltwater was generated during 4 days of extremely warm temperatures, which was then entrapped and pressurized behind seepage outlets that were blocked during two subsequent days of freezing conditions. While careful monitoring and scrutiny of processes may help to reveal the subtle triggers for landslides, strength degradation is no doubt an important process and likely to act as a trigger for some failures.

2.3.2.1 Stress-Induced Fatigue

Stress-induced fatigue encompasses all processes of stress-induced degradation in material strength that operate at stresses below the instantaneous ultimate failure strength of the material (e.g., Attewell and Farmer, 1973; Cruden, 1974; Lajtai and Schmidtke, 1986; Potyondy, 2007). It typically operates at the microscopic scale of interparticle forces and involves the nucleation and growth of fractures within brittle materials. Local failure is thought to be possible at stress magnitudes below the ultimate failure strength because of concentration of stresses around structural imperfections such as fractures; this phenomenon is referred to as subcritical fracture propagation. Within a perfectly homogenous and uniform mass of material, the stresses may be distributed approximately uniformly within the material. However, all materials contain microscopic imperfections and many rock masses are heterogeneous and contain fractures and joints arising from high-magnitude stresses (e.g., tectonic stresses in excess of the ultimate failure strength). In addition, rocks may exhibit unbalanced stresses that develop through rock-forming processes (e.g., cooling or burial) or secondary processes (e.g., tectonic stresses). These imperfections and stress imbalances result in stress concentrations. These concentrations can cause microfractures to nucleate, which further increases the nonhomogeneity of the stress distribution. The stresses become concentrated at fracture tips, causing fracture propagation and further stress concentration. This positive feedback loop can, in theory, involve a nonlinear (rapidly accelerating) increase in stress concentrations as fractures propagate and result in the catastrophic failure of a material.

Stress release fracturing can also occur from unloading of overburden or confining materials. This release of stress can induce stress reorientations, potentially leading to stress concentrations sufficient to cause fracture propagation. It is a process familiar in the mining and tunneling industries where high confining stresses exist and can be altered relatively quickly. In natural slopes the same process occurs, often operating over a longer time period owing to the larger scale of the situation and the longer time frames by which natural processes (erosion) unload the material. The exception to this may be the unloading induced by other mass movements, which can trigger stress release behind the release surface. Stress release fracturing may be important in priming or triggering failures in slopes that have undergone glacial erosion (McColl, 2012).

Static fatigue is the gradual damage to a material as a result of the constant application of stress; for slopes, this stress arises from the self-weight (gravity) stresses, in addition to any constant tectonic or internal/residual stresses. The steeper or higher a slope, the greater the self-weight stresses that are in operation. Thus, the steepening, loading, or heightening of a slope will likely facilitate faster rates of static fatigue. Pore fluids can also help facilitate stress-induced fracturing through chemical reactions at highly stressed fracture tips (this is referred to as stress corrosion). Because the positive feedback mechanism of stress fatigue causes an acceleration of fracture propagation, it may be possible for a slope to fail "out of the blue" with no apparent change in the slope conditions or external factors (e.g., change in slope angle, rainfall, or seismicity). Progressive failure mass movements in brittle materials are thought to involve a stress fatigue process, involving a strain-controlled development of a through-going failure surface from the coalescence of fractures (Petley et al., 2005) (Figure 2.5). This is a process that may operate without external stimuli, involving internal deformation of material, even while the global FoS remains above one. Progressive failures therefore may occur without any other apparent trigger. Petley et al. (2005) discussed how strain monitoring can be used to forecast the timing of progressive failures or failures that are undergoing accelerating creep; but, this relies on the strain being detected. Identifying potential failure sites before they show detectable strain or geomorphic signs of imminent slope failure (e.g., bulging or fractures) presents a greater challenge.

Cyclic fatigue involves the fatigue of materials in a fashion similar to that of static fatigue; but, instead of being static, the stresses acting on a material are temporally varying (cyclic) stresses. In slopes, cyclic stresses may arise from rapid ground motions produced during earthquakes or from vibrations induced by machinery or wind loading. Much lower frequency stress cycles could arise from any changes in the stresses operating in slopes, such as changes associated with water table fluctuations, thermal changes (e.g., Gischig et al., 2011), snow

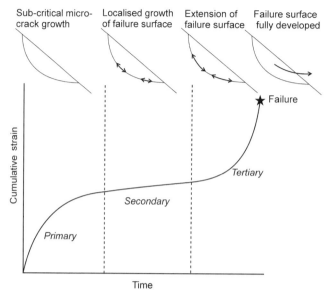

FIGURE 2.5   Schematic representation of the development of a through-going failure surface during the different stages of creep that govern progressive failure. *Based on Petley et al. (2005) and Froude (2011).*

or ice loading and unloading, changes in atmospheric pressure, and tidal forces. Experiments on (mostly nonrock) materials show the following:

1. The number of stress cycles a material can withstand is reduced with increasing levels of stress.
2. There tends to be a stress limit below which no amount of cycles will induce fatigue.

Because of the high-amplitude stress cycles produced by earthquake shaking, most research on cyclic fatigue in slopes has been studied with respect to earthquake ground motions (e.g., Moore et al., 2012). It is thought that repeated earthquake shaking, that does not cause outright failure, can weaken slopes and lead to failure during subsequent (and possibly smaller) earthquakes or as a result of other triggers that would not have otherwise been sufficient to cause failure (Hancox, 2010; Moore et al., 2012).

2.3.2.2 Chemical Weathering

The chemical alteration of rock and sediments can reduce the shear strength and equilibrium angle of slopes (Durgin, 1977). Chemical weathering is most rapid in warm humid (i.e., tropical) environments, but it occurs in all environments. A variety of chemical weathering processes can contribute to the reduction in material strength, for example, dissolution, hydrolysis, hydration, and oxidation. These processes may reduce material strength by weakening the intact particles or the contacts between particles (joints) or change the sensitivity of a material to water (e.g., through the production of some types of clay minerals). The state of weathering of a slope not only controls the stability of the slope but will also influence the type of landslide likely to occur as the structure of the slope material changes. Durgin (1977) related the degree of weathering in granitic rocks to the dominant types of landslides. For example, rock falls are most common on unweathered rocks, whereas debris flows become more prevalent once weathering has begun to disintegrate the rock granules, and completely weathered granites without any structural control host rotational failures (slumps). This slope strength dependence on weathering means that landslide susceptibility based on rock type alone may be misleading, especially because different rock types and rock masses weather at different rates. Indeed, Durgin (1977) described how granite rocks in some regions are relatively unweathered and provide the most stable (i.e., host the least landslides) of all rocks in that region, whereas in regions where chemical weathering is intense, granites can be the most prone to instabilities. The degree of weathering is thus an important factor to quantify. However, the relationship between chemical weathering and strength degradation is not necessarily straightforward. Fan et al. (1996) determined that the strength of fresh rocks with well-developed clay fabrics actually increased during initial weathering because of changes in fabric. Weathering-induced strength degradation did not occur until the sediments themselves were altered. It is also important to bear in mind that many initially cohesionless materials will also increase in strength as weathering progresses because of the additional cohesion provided by weathering products.

2.3.2.3 Cold Environment Processes

Permafrost degradation: In environments cold enough to undergo permanently freezing conditions (permafrost) or ground temperatures fluctuating above and below freezing, another mechanism of strength degradation can exist; one controlled by thawing of frozen groundwater. Frozen water within rock fractures and rock and soil pore spaces can increase shear strength. If this melts, as it may do in a warming climate, the drop in strength may be sufficient to trigger failure if one or both of the following conditions are met:

1. The stability of the slope had been lowered to an otherwise marginally stable state while the water was still frozen.
2. The meltwater within the materials became pressurized and lowered effective normal stresses on potential failure surfaces.

A growing body of empirical evidence suggests a link between warming temperatures, or extremes in warm temperatures, and the occurrence of slope failures (Allen and Huggel, 2013; Gruber and Haeberli, 2007). For rock slopes, Krautblatter et al. (2013) have developed an ice rock mechanical model to explain the loss in strength during thawing of permafrost; and notably, their model explains why permafrost degradation can trigger deep-seated rock slope failures as well as failures in the near surface. Interestingly, their model considers strength degradation arising from a combination of static fatigue, progressive failure, and a drop in the intrinsic (frictional) strength of rock and ice rock contacts.

Freeze–thaw processes: Weathering associated with repeated melting and refreezing of interstitial water has been recognized as another preparatory factor in jointed rock masses (Matsuoka and Murton, 2008). The expansion of water upon freezing (ice wedging), coupled with the movement of water toward the freezing front (ice segregation), can generate sufficient pressures to induce fracture propagation (Matsuoka and Murton, 2008; Selby, 1993). The possible intensification of freeze–thaw conditions during deglaciation may have led to more enhanced rock fall activity and talus development during early deglaciation (e.g., Rapp, 1960). Several researchers have suggested a link between freeze–thaw processes and modern rates of rock fall activity (Gruber et al., 2004; Matsuoka, 2008; Matsuoka and Sakai, 1999; Noetzli et al., 2003; Ravanel et al., 2010) and possibly even a link to large-scale rock slope failures (Davies et al., 2001; Wegmann and Gudmundsson, 1999).

2.3.2.4 Discussion

Every slope must undergo strength degradation during its life span. Strength degradation must therefore play a role in reducing the critical failure threshold required for a host of other more recognizable triggers or may itself be the final failure trigger. To that end, if strength degradation could be quantified accurately, along with the driving and resisting stresses operating for every slope, landslide predictions would become much more accurate. The reality is that we are far from achieving a robust understanding of strength degradation or a means to accurately predict its effect. Two of the limitations to understanding strength degradation are:

1. The difficulty of observing the processes involved and
2. The impracticality of knowing the entire history of the slope; the slope history influences the slope strength and existing stress conditions.

A fuller understanding and predictive capability may best be achieved by integrating various disciplinary approaches. For example, undertaking detailed engineering–geological mapping and strength testing (e.g., Brideau et al., 2009); using powerful numerical techniques to model progressive failure (e.g., modeling progressive failure; Eberhardt et al., 2004; Locat et al., 2013) or stress corrosion/static fatigue processes (e.g., Potyondy, 2007); using geophysical approaches, such as measuring the microseismicity generated by precursory fracture propagation (e.g., Amitrano et al., 2010; Spillmann et al., 2007) or repeat imaging of the material properties to assess reductions in strength through time (e.g., spectral analysis of earthquake ground motion; Moore et al., 2012). Synoptic models that attempt to incorporate a range of processes, such as that of Krautblatter et al. (2013) for permafrost degradation, hold promise for holistic strength degradation models in other types of environments. Because of its pervasive applicability, developing better predictive capability for strength degradation is arguably one of the greatest and potentially most fruitful challenges for slope stability research.

2.3.3 Groundwater Changes

Dynamic triggers are factors that are transient and bring a slope from a marginally stable state to failure or movement when they temporarily exceed the failure threshold. Changes in groundwater (and seismic ground accelerations discussed in the next section) are the most commonly recognized dynamic triggers. A useful point to keep in mind is that most slopes experience numerous dynamic events without undergoing failure, and it is not necessarily the largest of these that causes final failure. The effectiveness of those dynamic events in triggering failure depends not just on event magnitude but on the stability threshold, which can reduce through time as a result of other preparatory factors (e.g., strength degradation).

Groundwater fluctuations arise from either a change in the infiltration of water into a slope (e.g., from rainfall or snowmelt), from a volumetric change in porosity, or from a change in the drainage conditions of the slope (e.g., changes in seepage pressures or flooding at the base). Groundwater fluctuations can alter stability in the following ways:

1. Increased groundwater gradients induce flow or seepage within slope and this adds a driving force (e.g., drawdown following rapid lowering of lakes or reservoirs at the base of slopes).
2. Groundwater fluctuations influence porewater pressures, which change the effective normal stresses, and therefore the shear strength of potential failure surfaces.
2. Saturation of water can change the inherent strength of materials (e.g., saturation of soils and swelling clays).

The relationship between groundwater levels (or more often the magnitude or intensity of rainfall) has been investigated empirically, with the establishment of triggering thresholds, and deterministically using slope stability models.

Particularly for shallow landslides, rainfall is the most common source for groundwater fluctuations that trigger failure or movement (Van Asch et al., 1999). While good empirical relationships can be established between rainfall and triggering thresholds (Glade, 1998)—especially when antecedent soil moisture can be taken into account (Crozier, 1999)—the exact processes controlling these relationships remain uncertain. For example, where effective cohesion, rather than only the frictional properties of a soil, strongly influence stability, it may be peak moisture conditions rather than peak pore-water pressures that trigger failure (Hawke and McConchie, 2011). However, much of the uncertainty with rainfall thresholds arises from uncertainty in the infiltration, flow pathways, and drainage conditions in the slope. Surface topography, the topography of the failure surface (usually the bedrock–soil interface), and spatial variability in the permeability of the slope materials have a strong influence on the spatial relationship between rainfall and groundwater changes. For example, the confluence of water along the bedrock–soil interface of bedrock hollows or fossil gully features (colluvium-filled bedrock depressions) creates additional saturation of the materials, and consequently these are common slope failure sites (Crozier et al., 1990; Wilson and Dietrich, 1987). Although identifying and modeling the surface topography of a slope is relatively easy, the topography of the bedrock–soil interface is more difficult to characterize. Recent work has also shown that the bedrock–soil interface is not as impermeable as often assumed (Brönnimann et al., 2013). The underlying bedrock can store and release water into the soil during rainfall events, and thus may have a considerable influence on both the timing and distribution of soil saturation during a rainfall event. Perhaps the most recognized factor modifying the relationship between soil saturation and rainfall is the interception of rainfall and evapotranspiration provided by vegetation. Vegetation removal is perhaps the most common preparatory factor for the development of shallow landslides, through both loss of rainfall interception and apparent cohesion, and has been recognized as a major contributor to human-induced global soil erosion (Glade, 2003).

Failure or movement of deep-seated landslides is also influenced by rainfall patterns. However, in this case the slopes usually respond over longer timescales than the duration of individual rainfall events, instead responding to seasonal or longer term changes in rainfall or evaporative patterns. For large, slow-moving landslides, porewater pressure and movement are generally positively correlated. However, the advent of high-resolution surface and subsurface monitoring of movement and groundwater fluctuations has shown that this is not always the case. Other internal (e.g., strain hardening during shearing, or development of negative porewater pressures in zones of extension), geometrical, or nonfrictional (e.g., rheological) mechanics can modify or even reverse this relationship (Corominas et al., 2005; Massey et al., 2013; van Asch et al., 2009).

2.3.4 Ground Shaking

Seismic ground shaking or artificial sources of ground vibrations are another type of dynamic failure trigger. Earthquake shaking, as opposed to artificial sources, is the focus of this section because earthquakes are more capable of releasing seismic energy strong enough for slope failure (Figure 2.6). Three main ways exist that a slope failure can arise from earthquake shaking:

1. Sudden ground accelerations that induce instantaneous shear stresses in excess of the mobilized shear strength.
2. Cyclic fatigue-induced propagation of fractures, leading to formation of a failure surface.
3. Seismically induced increases in porewater pressures.

FIGURE 2.6 Example of earthquake-triggered slope failures. These rock avalanches, which failed high on the slopes above, and traveled on to, the Black Rapids Glacier, Alaska, were triggered by the 2002 Denali earthquake. *Photograph: Rod March, US Geological Survey.*

It is possible that more than one of these may operate during any earthquake. If only (2) and/or (3) operate, the failure may occur sometime after shaking has ceased (Jibson et al., 1994), and the effectiveness of both of these will depend on the duration as well as the intensity of shaking.

Despite general uncertainty in the specific processes governing earthquake-triggered failure, good relationships between earthquake magnitude and intensity and the density of landslides have been established for a number of earthquakes (Dai et al., 2011; Hancox et al., 2002; Keefer, 1984; Meunier et al., 2007). In general, the number of landslides reduces with distance from the epicenter, mimicking the attenuation of shaking intensity away from the epicenter; but, the specific spatial distribution depends on a number of other seismic site effects that operate in addition to the static factors influencing failure susceptibility. Ground accelerations also depend on a variety of potential seismic wave characteristics interacting with the unique geologic, structural, and topographic conditions at any given site (Aki, 1988; Alfaro et al., 2012; Buech, 2008; Geli et al., 1988; McColl et al., 2012; Meunier et al., 2008). Interestingly, weakened and fragmented material associated with existing or incipient landslides may even modify and amplify the shaking experienced on those parts of the slope (Burjánek et al., 2012; Moore et al., 2011). The hydrological conditions at the time will also influence the distribution of coseismic failures; a high groundwater table may lower the static FoS but may also mean that a smaller rise in groundwater, through seismic excitation, is required for failure.

2.4 SUMMARY AND CONCLUSION

The geological and geomorphological controls on slope failure (e.g., the type, strength and structure of material, and the slope geometry) play an important role in predisposing slopes to failure and controlling the likely spatial distribution of failures. These factors also largely determine which type of failures will occur and have a strong bearing on the threshold for stability on which other stability factors operate. Spatial analysis using GIS platforms helps to elucidate the strength of the relationships between various stability factors and landslide failure, and is a useful tool for calculating landslide susceptibility. However, even though it is possible to identify the slopes most likely to fail in a regional study, determining the frequency of slope failures on a regional scale is a considerably harder challenge. Even local-scale investigations of single slopes may not be sufficient to achieve this. Precisely predicting the movement patterns for closely-monitored, large, active landslides remains a challenge because of the time-dependent evolution of the stability factors. Many of the slopes investigated in a regional-scale assessment of susceptibility may share apparently similar predisposing factors; but, precisely which slopes will fail during any given period of time, or upon application of an external trigger, may lie beyond our capability of prediction without resorting to methods of statistical probability. Much of this uncertainty probably lies in the operation of the processes of strength degradation. This uncertainty will gradually be overcome through advances in knowledge, such as better recognition and investigation of the multitude of strength degradation processes, more methods becoming available and applied for observing, monitoring, and modeling slopes, and further attempts made to develop long-term time-dependent models of instability. These advances would potentially allow the quantification and useful application of stability models as proposed schematically in Figure 2.3.

REFERENCES

Aki, K., 1988. Local site effects on strong ground motion, earthquake engineering and soil dynamics II – recent advances in ground-motion evaluation. In: Proceedings of the Specialty Conference; 27 June 1988 through 30 June 1988. ASCE, Park City, UT, USA, pp. 103–155.
Alfaro, P., Delgado, J., García-Tortosa, F., Giner, J., Lenti, L., López-Casado, C., Martino, S., Scarascia Mugnozza, G., 2012. The role of near-field interaction between seismic waves and slope on the triggering of a rockslide at Lorca (SE Spain). Nat. Hazards Earth Syst. Sci. 12 (12), 3631–3643.
Allen, S., Huggel, C., 2013. Extremely warm temperatures as a potential cause of recent high mountain rockfall. Global Planet. Change 107, 59–69.
Amitrano, D., Arattano, M., Chiarle, M., Mortara, G., Occhiena, C., Pirulli, M., Scavia, C., 2010. Microseismic activity analysis for the study of the rupture mechanisms in unstable rock masses. Nat. Hazards Earth Syst. Sci. 10 (4), 831–841.
Attewell, P.B., Farmer, I.W., 1973. Fatigue behaviour of rock. Int. J. Rock Mech. Mining Sci. 10, 1–9.

Augustinus, P.C., 1995. Rock mass strength and the stability of some glacial valley slopes. Z. Geomorphol. 39 (1), 55—68.

van Asch, T.W.J., Buma, J., Van Beek, L.P.H., 1999. A view on some hydrological triggering systems in landslides. Geomorphology 30, 25—32.

van Asch, T.W.J., Malet, J.P., Bogaard, T.A., 2009. The effect of groundwater fluctuations on the velocity pattern of slow-moving landslides. Nat. Hazards Earth Syst. Sci. 9 (3), 739—749.

Brideau, M.-A., Yan, M., Stead, D., 2009. The role of tectonic damage and brittle rock fracture in the development of large rock slope failures. Geomorphology 103 (1), 30—49.

Brönnimann, C., Stähli, M., Schneider, P., Seward, L., Springman, S.M., 2013. Bedrock exfiltration as a triggering mechanism for shallow landslides. Water Res. Res. 49 (9), 5155—5167.

Buech, F., 2008. Seismic Response of Little Red Hill — Towards an Understanding of Topographic Effects on Ground Motion and Rock Slope Failure. Ph.D., University of Canterbury, Christchurch, 285 pp.

Burjánek, J., Moore, J.R., Yugsi Molina, F.X., Fäh, D., 2012. Instrumental evidence of normal mode rock slope vibration. Geophys. J. Int. 188 (2), 559—569.

Corominas, J., Moya, J., Ledesma, A., Lloret, A., Gili, J., 2005. Prediction of ground displacements and velocities from groundwater level changes at the Vallcebre landslide (Eastern Pyrenees, Spain). Landslides 2 (2), 83—96.

Crozier, M.J., 1986. Landslides: Causes, Consequences and Environment. Croom Helm Ltd, London.

Crozier, M.J., 1999. Prediction of rainfall-triggered landslides: a test of the antecedent water status model. Earth Surf. Process. Landforms 24 (9), 825—833.

Crozier, M.J., Vaughan, E.E., Tippett, J.M., 1990. Relative instability of colluvium-filled bedrock depressions. Earth Surf. Process. Landforms 15 (4), 329—339.

Cruden, D.M., 1974. The static fatigue of brittle rock under uniaxial compression. Int. J. Rock Mech. Mining Sci. Geomech. Abs. 11 (2), 67—73.

Cruden, D.M., Martin, C.D., 2013. Assessing the stability of a natural slope. In: Wu, F., Qi, S. (Eds.), International Symposium and 9th Asian Regional Conference of the International Association for Engineering Geology and the Environment. Global View of Engineering Geology and the Environment. CRC Press, Beijing, China, pp. 17—26.

Cruden, D.M., Varnes, D.J., 1996. Landslide types and processes: chapter 3. In: Turner, A.K., Schuster, R.L. (Eds.), Landslides: Investigation and Mitigation. Special Report 247. Transportation Research Board, National Research Council, Washington, DC, pp. 36—75.

Dai, F.C., Xu, C., Yao, X., Xu, L., Tu, X.B., Gong, Q.M., 2011. Spatial distribution of landslides triggered by the 2008 Ms 8.0 Wenchuan earthquake, China. J. Asian Earth Sci. 40 (4), 883—895.

Davies, M.C.R., Hamza, O., Harris, C., 2001. The effect of rise in mean annual temperature on the stability of rock slopes containing ice-filled discontinuities. Permafrost Periglacial Process. 12 (1), 137—144.

Duncan, J.M., 1996. Soil slope stability analysis: chapter 13. In: Turner, A.K., Schuster, R.L. (Eds.), Landslides: Investigation and Mitigation. Special Report 247. Transportation Research Board, National Research Council, Washington, DC, pp. 337—371.

Durgin, P.B., 1977. Landslides and the weathering of granitic rocks. Rev. Eng. Geol. 3, 127—131.

Eberhardt, E., Stead, D., Coggan, J.S., 2004. Numerical analysis of initiation and progressive failure in natural rock slopes—the 1991 Randa rockslide. Int. J. Rock Mech. Mining Sci. 41 (1), 69—87.

El-Ramly, H., Morgenstern, N., Cruden, D., 2002. Probabilistic slope stability analysis for practice. Can. Geotech. J. 39 (3), 665—683.

Fan, C.-H., Allison, R.J., Jones, M.E., 1996. Weathering effects on the geotechnical properties of argillaceous sediments in tropical environments and their geomorphological implications. Earth Surf. Process. Landforms 21 (1), 49–66.

Froude, M.J., 2011. Capturing and Characteristing Pre-failure Strain on Failing Slopes. Master of Science, Durham University, 201 pp.

Geli, L., Bard, P.-Y., Jullien, B., 1988. The effect of topography on earthquake ground motion: a review and new results. Bull. Seismol. Soc. Am. 78 (1), 42–63.

Gerrard, J., 1994. The landslide hazard in the Himalayas: geological control and human action. Geomorphology 10 (1–4), 221–230.

Gischig, V.S., Moore, J.R., Evans, K.F., Amann, F., Loew, S., 2011. Thermo-mechanical forcing of deep rock slope deformation – part I: conceptual study of a simplified slope. J. Geophys. Res 116, F04010. http://dx.doi.org/10.1029/2011JF002006.

Glade, T., 1998. Establishing the frequency and magnitude of landslide-triggering rainstorm events in New Zealand. Environ. Geol. 35 (2–3), 160–174.

Glade, T., 2003. Landslide occurrence as a response to land use change: a review of evidence from New Zealand. Catena 51 (3–4), 297–314.

Glade, T., Crozier, M.J., 2005. The nature of landslide hazard impact. In: Glade, T., Anderson, M.G., Crozier, M.J. (Eds.), Landslide Hazard and Risk. John Wiley & Sons Ltd, Chichester, pp. 43–74.

Gruber, S., Haeberli, W., 2007. Permafrost in steep bedrock slopes and its temperature-related destabilization following climate change. J. Geophys. Res. 112, F02S18. http://dx.doi.org/10.1029/2006JF000547.

Gruber, S., Hoelzle, M., Haeberli, W., 2004. Permafrost thaw and destabilization of Alpine rock walls in the hot summer of 2003. Geophys. Res. Lett. 31.

Hack, R., Price, D., Rengers, N., 2002. A new probabilistic approach to rock slope stability – a probability classification (SSPC). Bull. Eng. Geol. Environ. 62 (2), 167–184.

Hancox, G.T., 2008. The 1979 Abbotsford Landslide, Dunedin, New Zealand: a retrospective look at its nature and causes. Landslides 5, 177–188.

Hancox, G.T., 2010. Report on the Landslide that Blocked SH 1 and the Railway Line Near Rosy Morn Stream South of Kaikoura on 10 September 2010. GNS Science Report 2010/59.

Hancox, G.T., Perrin, N.D., Dellow, G.D., 2002. Recent studies of historical earthquake-induced landsliding, ground damage, and MM intensity in New Zealand. Bull. N.Z. Soc. Earthquake Eng. 35 (2), 59–95.

Hauser, A., 2002. Rock avalanche and resulting debris flow in Estero Parraguirre and Rio Colorado, region Metropolitana, Chile. In: Evans, S.G., DeGraff, J.V. (Eds.), Catastrophic Landslides: Effects, Occurence and Mechanisms, pp. 135–148.

Hawke, R., McConchie, J., 2011. In situ measurement of soil moisture and pore-water pressures in an 'incipient' landslide: Lake Tutira, New Zealand. J. Environ. Manage. 92, 266–274.

Hoek, E., Carranza-Torres, C.T., Corkum, B., 2002. Hoek-Brown failure criterion-2002 edition. In: Proceedings of the fifth North American rock mechanics symposium, vol. 1. Toronto, Canada, pp. 267–273.

Jibson, R.W., Prentice, C.S., Borissoff, B.A., Rogozhin, E.A., Langer, C.J., 1994. Some observations of landslides triggered by the 29 April 1991 Racha earthquake, Republic of Georgia. Bull. Seismol. Soc. Am. 84 (4), 963–973.

Keefer, D.K., 1984. Landslides caused by earthquakes. Geol. Soc. Am. Bull. 95 (4), 406–421.

Keefer, D.K., 2000. Statistical analysis of an earthquake-induced landslide distribution – the 1989 Loma Prieta, California event. Eng. Geol. 58 (3–4), 231–249.

Krautblatter, M., Funk, D., Günzel, F.K., 2013. Why permafrost rocks become unstable: a rock–ice-mechanical model in time and space. Earth Surf. Process. Landforms 38 (8), 876–887.

Lajtai, E.Z., Schmidtke, R.H., 1986. Delayed failure in rock loaded in uniaxial compression. Rock Mech. Rock Eng. 19 (1), 11–25.

Lipovsky, P.S., Evans, S.G., Clague, J.J., Hopkinson, C., Couture, R., Bobrowsky, P., Ekström, G., Demuth, M.N., Delaney, K.B., Roberts, N.J., Clarke, G., Schaeffer, A., 2008. The July 2007 rock and ice avalanches at Mount Steele, St. Elias Mountains, Yukon, Canada. Landslides 5 (4), 445–455.

Locat, A., Jostad, H.P., Leroueil, S., 2013. Numerical modeling of progressive failure and its implications for spreads in sensitive clays. Can. Geotech. J. 50 (9), 961–978.

Massey, C., Petley, D., McSaveney, M., 2013. Patterns of movement in reactivated landslides. Eng. Geol.

Massey, C.I., Petley, D.N., McSaveney, M.J., 2013. Patterns of movement in reactivated landslides. Engineering Geology 159, 1–19.

Matsuoka, N., 2008. Frost weathering and rockwall erosion in the southeastern Swiss Alps: long-term (1994–2006) observations. Geomorphology 99 (1–4), 353–368.

Matsuoka, N., Murton, J., 2008. Frost weathering: recent advances and future directions. Permafrost Periglacial Process. 19 (2), 195–210.

Matsuoka, N., Sakai, H., 1999. Rockfall activity from an alpine cliff during thawing periods. Geomorphology 28, 309–328.

McColl, S.T., 2012. Paraglacial rock-slope stability. Geomorphology 153–154, 1–16.

McColl, S.T., Davies, T.R.H., 2013. Large ice-contact slope movements; glacial buttressing, deformation and erosion. Earth Surf. Process. Landforms 38, 1102–1115.

McColl, S.T., Davies, T.R.H., McSaveney, M.J., 2012. The effect of glaciation on the intensity of seismic ground motion. Earth Surf. Process. Landforms 37, 1290–1301.

McSaveney, M.J., 2002. Recent rockfalls and rock avalanches in Mount Cook National Park, New Zealand. Rev. Eng. Geol. 15, 35–70.

Meunier, P., Hovius, N., Haines, A.J., 2007. Regional patterns of earthquake-triggered landslides and their relation to ground motion. Geophys. Res. Lett. 34 (20), L20408.

Meunier, P., Hovius, N., Haines, J.A., 2008. Topographic site effects and the location of earthquake induced landslides. Earth Planet. Sci. Lett. 275, 221–232.

Moore, J.R., Gischig, V., Amann, F., Hunziker, M., Burjanek, J., 2012. Earthquake-triggered rock slope failures: damage and site effects. In: 11th International and Second North American Symposium on Landslides, 3rd–6th June 2012, Banff, Alberta, Canada, p. 6.

Moore, J.R., Gischig, V., Burjanek, J., Loew, S., Fäh, D., 2011. Site effects in unstable rock slopes: dynamic behavior of the Randa instability (Switzerland). Bull. Seismol. Soc. Am. 101 (6), 3110–3116.

Noetzli, J., Hoelzle, M., Haeberli, W., 2003. Mountain permafrost and recent Alpine rock-fall events: a GIS-based approach to determine critical factors. In: Proceedings of the 8th International Conference of Permafrost. International Permafrost Association, Zurich, Switzerland, pp. 827–832.

Norrish, N.I., Wyllie, D.C., 1996. Rock slope stability analysis: chapter 15. In: Turner, A.K., Schuster, R.L. (Eds.), Landslides: Investigation and Mitigation. Special Report 247. Transportation Research Board, National Research Council, Washington, DC, pp. 391–425.

Petley, D., 2011. Hillslopes. In: Gregory, K.J., Goudie, A.S. (Eds.), The SAGE Handbook of Geomorphology. SAGE Publications Ltd, London, pp. 343–359.

Petley, D.N., Higuchi, T., Petley, D.J., Bulmer, M.H., Carey, J., 2005. Development of progressive landslide failure in cohesive materials. Geology 33 (3), 201–204.

Potyondy, D.O., 2007. Simulating stress corrosion with a bonded-particle model for rock. Int. J. Rock Mech. Mining Sci. 44 (5), 677–691.

Rapp, A., 1960. Recent development of mountain slopes in Kärkevagge and surroundings, Northern Scandinavia. Geogr. Ann. 42 (2/3), 65–200.

Ravanel, L., Allignol, F., Deline, P., Gruber, S., Ravello, M., 2010. Rock falls in the Mont Blanc Massif in 2007 and 2008. Landslides 7 (4), 493–501.

Roering, J.J., Kirchner, J.W., Dietrich, W.E., 2005. Characterizing structural and lithologic controls on deep-seated landsliding: implications for topographic relief and landscape evolution in the Oregon Coast Range, USA. Geol. Soc. Am. Bull. 117 (5–6), 654–668.

Schulz, W.H., Kean, J.W., Wang, G., 2009. Landslide movement in southwest Colorado triggered by atmospheric tides. Nat. Geosci. 2 (12), 863–866.

Selby, M.J., 1982. Controls on the stability and inclinations of hillslopes formed on hard rock. Earth Surf. Process. Landforms 7 (5), 449–467.

Selby, M.J., 1993. Hillslope Materials and Processes. Oxford University Press, New York.

Spillmann, T., Maurer, H., Green, A.G., Heincke, B., Willenberg, H., Husen, S., 2007. Microseismic investigation of an unstable mountain slope in the Swiss Alps. J. Geophys. Res. 112 (B7), B07301.

Terzaghi, K., 1950. Mechanism of landslides. In: Paige, S. (Ed.), Application of Geology to Engineering Practice.

Thomson, R.C., 1982. Relationship of Geology to Slope Failures in Soft Rocks of the Taihape-Mangaweka Area, Central North Island, New Zealand. Ph.D., University of Auckland.

Varnes, D.J., 1978. Slope movement types and processes. In: Schuster, R.L., Krizek, R.J. (Eds.), Landslides Analysis and Control. Special Report 176. Transportation Research Board, National Academy of Sciences, Washington, DC, pp. 11–33.

Wegmann, M., Gudmundsson, G., 1999. Thermally Induced Temporal Strain Variations in Rock Walls Observed at Subzero Temperatures, Advances in Cold-region Thermal Engineering and Sciences. Lecture Notes in Physics. Springer Berlin/Heidelberg, 511–518.

WG/WLI, 1994. A suggested method for reporting landslide causes. Bull. Int. Assoc. Eng. Geol. 50 (1), 71–74.

Wieczorek, G.F., Jäger, S., 1996. Triggering mechanisms and depositional rates of postglacial slope-movement processes in the Yosemite Valley, California. Geomorphology 15 (1), 17–31.

Wilson, C.J., Dietrich, W.E., 1987. The contribution of bedrock groundwater flow to storm runoff and high pore pressure development in hollows. In: Beschta, R.L., Blinn, T., Grant, G.E., Ice, G.G., Swanson, F.J. (Eds.), Proceedings of an International Symposium on Erosion and Sedimentation in the Pacific Rim. International Association of Hydrological Sciences, Corvallis, OR, United States, pp. 49–59.

Chapter 3

Mass Movement in Bedrock

Marc-André Brideau and Nicholas J. Roberts

Simon Fraser University, Burnaby, BC, Canada

ABSTRACT

Bedrock mass movements include some of the most common (rockfalls) and most destructive (rock avalanches) slope processes. The characteristic volumes, velocities, runout distances, and frequencies of bedrock mass movement types can vary over many orders of magnitude. An understanding of the multiple attributes, contributing factors, and processes is required to conduct a comprehensive characterization of the hazards associated with bedrock mass movements. The characterization includes the mechanical properties of intact rock, orientation and surface characteristics of discontinuities, geological history of the rock mass, and topography of the slope, including past and currently active tectonic and geomorphic processes. Conditions of rock–slope stability can be classified into three broad categories based on the factor controlling them: rock structures, intact rock strength, and rock mass strength. Applications of the progressive failure concept are presented to explain the behavior of rock masses leading to a slope failure. The range bedrock mass movement types within Varnes-based landslide classification scheme are also reviewed. Three case studies, Seymareh (Iran), La Clapière (France), and Threatening Rock (United States), are presented in detail to illustrate the range of bedrock mass movement types and characteristics presented in this chapter. As with mass movements in other material types (snow, ice, soil, and debris), inventory mapping, monitoring, and modeling (numerical and physical) form the basis of bedrock mass movement recognition and anticipation. Avoidance is the simplest and most effective response to rock–slope instabilities as active mitigation is only effective for small volume rockfalls or is only possible at significant economic cost for large slow-moving features. The chapter closes with a review of risk management strategies as applied to bedrock mass movements. The individual components of the risk equation are presented along with examples of responses that could help reduce the calculated risk value.

3.1 INTRODUCTION

Mass movements in bedrock are an integral component of the landscape's erosional response to tectonic uplift (Densmore et al., 1997; Montgomery and Brandon, 2002; Korup et al., 2007). They involve volumes, velocities, and frequencies that vary over many orders of magnitude (Guzzetti et al., 2012). Bedrock mass movements include very fast mechanisms ranging from small frequent rockfalls to large infrequent rock slides and rock avalanches, but can also occur as slower, continuous, or episodic deep-seated deformation of extensive slopes. Characterizing the hazard associated with bedrock mass movements requires an understanding of multiple attributes, contributing factors, and processes. These include the mechanical properties of intact rock, orientation and surface characteristics of discontinuities, geological history of the rock mass, and topography of the slope, including past and currently active tectonic and geomorphic processes (Griffiths et al., 2012; Frattini and Crosta, 2013). The spatial distribution of bedrock mass movements is strongly influenced by tectonic structures (Agliardi et al., 2001; Badger, 2002; Jackson, 2002; Korup, 2004; Ambrosi and Crosta, 2006; Strom and Korup, 2006; Coe and Harp, 2007; Henderson and Saintot, 2011; Scheingross et al., 2013; Jaboyedoff et al., 2013). The spatiotemporal distribution of bedrock mass movements can also be influenced by external factors such as weather (Wieczorek, 1996; Matsuoka and Sakai, 1999), climate (Stoffel et al., 2014), earthquakes (Keefer, 1984; Tatard and Grasso, 2013; Murphy, 2014), and anthropogenic activities (Heim, 1932; Müller, 1964). Huang and Montgomery (2013) document differing spatial distributions for earthquake-triggered and rainfall-triggered landslides. Their finding supports Densmore and Hovius' (2000) hypothesis that earthquake-triggered landslides tend to occur along ridges, where ground acceleration is amplified by topography, whereas rainfall-triggered landslides occur lower on slopes, corresponding to zones of groundwater discharge. Contributing factors and triggers of mass movements are discussed in more detail by McColl (2014).

3.2 ROCK MATERIALS

In mechanical terms, rock and soil are differentiated based on their intact strength. The boundary between soil and rock is typically considered as 1 MPa intact unconfined compressive strength (UCS), above which geologic materials are classified as rock (ISRM, 1981; British Standard, 1999). The mechanics of rock are distinct from those of soil due to the presence of discontinuities (Jaeger et al., 2007), which are mechanical breaks resulting from geological (e.g., Ramsay, 1967; Matheson and Thomson, 1973; Pollard and Aydin, 1988; Gudmundsson, 2011; Hencher et al., 2011) or anthropogenic (e.g., Nichols, 1980; Adams, 1982; Hoek, 2007) processes. Examples of discontinuities include joints, faults, shear zones, metamorphic foliation, bedding, and intact rock fractures. The strength of a rock mass is controlled by the characteristics of the discontinuities present and the strength of intact rock between them (Hoek and Brown, 1997; Hoek, 1999). Factors such as groundwater, in situ stress, microstructures, weathering, and blasting damage can influence the strength of the rock mass by affecting both intact rock and discontinuities

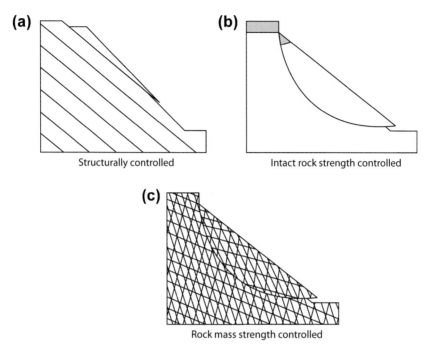

FIGURE 3.1   Main categories of control on slope stability (a) structural, (b) intact rock strength, and (c) rock mass strength. *Modified from Hoek and Bray (1981).*

(e.g., Moon, 1993; Moon and Jayawardane, 2004; Persson and Goransson, 2005; Pola et al., 2014; Pineda et al., 2014). Hoek and Bray (1981) divided the stability of rock masses in three categories based on the factors controlling it (Figure 3.1): rock structure, intact rock strength, or rock mass strength, which includes the effects of both rock structure and intact rock strength.

3.2.1 Structural Control in Strong Rock

Slope stability in strong rock (>25 MPa based on classification from Hungr et al., 2014) is dominantly controlled by the presence, orientation, and geomechanical properties of discontinuities (Terzaghi, 1962). The three simplest structurally controlled failure mechanisms are also the most commonly assessed in rock masses: planar sliding (Figure 3.2(a)), wedge sliding (Figure 3.2(b)), and toppling (Figure 3.3). Planar sliding can occur when a rock mass is present on a single discontinuity that "daylights" (dips less steeply than the slope and, therefore, cross-cuts the slope face) and has a dip greater than the friction angle along its surface. Stereographical and analytical solutions to assess the stability of a rock mass to planar sliding are presented in most hillslope processes, engineering geology, and rock engineering textbooks (e.g., Selby, 1993; Wyllie and Mah, 2004; Pariseau, 2012). Additional

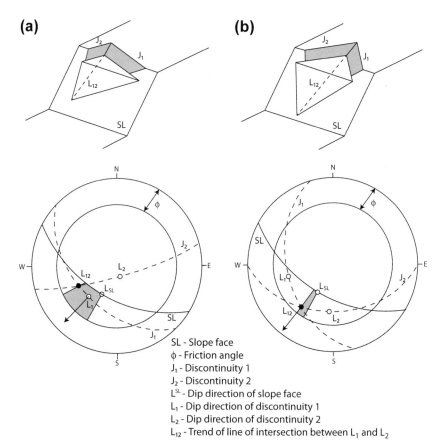

SL - Slope face
ϕ - Friction angle
J_1 - Discontinuity 1
J_2 - Discontinuity 2
L^{SL} - Dip direction of slope face
L_1 - Dip direction of discontinuity 1
L_2 - Dip direction of discontinuity 2
L_{12} - Trend of line of intersection between L_1 and L_2

FIGURE 3.2 Structurally controlled failure mechanism with sliding along (a) one and (b) two planes. *Modified after Yoon et al. (2002).*

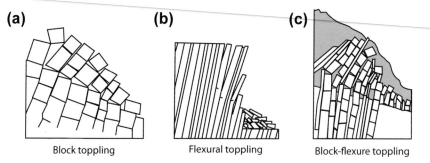

Block toppling   Flexural toppling   Block-flexure toppling

FIGURE 3.3 Primary toppling mechanisms (a) block, (b) flexural and (c) block-flexural. *Modified after Goodman and Bray (1976).*

geometric considerations for evaluating planar sliding include the presence of release surfaces along the sides, release at the rear of the sliding block, and a discontinuity dip direction within ±20° of the slope dip direction (Wyllie and Mah, 2004). Recent numerical modeling and analytical studies by Hungr and Amann (2011) and Brideau and Stead (2012) discuss implications when these geometric criteria are not met. If dip directions of the slope and sliding surface differ by more than 20°, slope stability is increased and sliding may take the form of an asymmetric failures or have a rotational movement component of the unstable rock mass in the plane of the sliding surface (Hungr and Amann, 2011; Brideau and Stead, 2012).

Wedge sliding involves movement along the intersection of two discontinuities (Figure 3.2(b)). In its simple stereographical solution, the intersection must daylight, plunge more steeply than the friction angle, and have a trend within the dip direction of the slope and the dip direction of the individual plane (Wyllie and Mah, 2004). The general analytical solution for the wedge sliding failure mechanism was proposed by Hoek et al. (1973) and is reproduced in Wyllie and Mah (2004).

Toppling (Figure 3.3) occurs when an unstable rock block rotates in a vertical plane about a fixed point (Goodman and Bray, 1976). It is favored in columnar rock blocks on an inclined plane where the center of gravity of the block lies beyond its base (Ashby, 1971). Goodman and Bray (1976) classify three primary types of toppling based on factors controlling their stability: block toppling (stability solely controlled by discontinuity orientation and block shape); flexural toppling (stability predisposed by discontinuity orientation and block shape, but fracturing through intact rock to form a basal release surface is required for failure); and flexural-block toppling (an intermediate case where both bounding discontinuity and intact rock fractures are important to the failure). Goodman and Bray (1976) present a stereographical and analytical solution to assess stability for block toppling.

Additional structurally controlled failure mechanisms—buckling, biplanar sliding, and ploughing—are common in rock masses with a pervasive (sedimentary or metamorphic) fabric (Figure 3.4). Cavers (1981) describes flexural buckling and three-hinge buckling and provides analytical solutions for each type of failure. Examples of buckling in natural slopes include thinly bedded sedimentary rocks in Western Canada (Hu and Cruden, 1993), large-scale toe buckling of schistose rocks on the South Island of New Zealand (Beetham et al., 1992), and of bedded limestone in Italy (Tommasi et al., 2009). A combination of toe-buckling and ploughing has been suggested as the failure mechanism for the gigantic 1.5-Gm^3, prehistoric Sierre Landslide in Switzerland (Pedrazzini et al., 2013). The 1963 Vajont Slide in Italy has a biplanar failure surface (Mencl, 1966). Analytical solutions for biplanar and ploughing failures mechanisms were first presented by Hawley et al. (1986) and updated by Alejano et al. (2011).

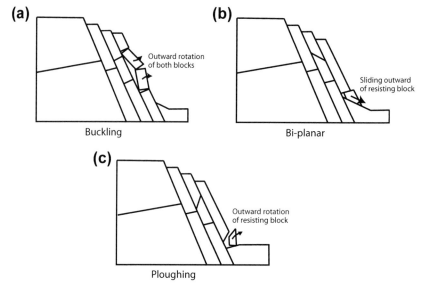

FIGURE 3.4 Structurally controlled failure occurring in thinly bedded/foliated dip slope include the (a) three-hinge buckling, (b) biplanar, and (c) ploughing mechanisms.

Rock slumping, a mechanism proposed by Kieffer (1998, 2003), involves back rotation of hard rock columns akin to backward toppling into a slope. The movement causes original face-to-face contact between blocks to become edge-to-face contacts (Figure 3.5). Like in toppling, three types of rock slumping are recognized based on block shape and strength: block slumping, flexural slumping, and block flexural slumping (Goodman and Kieffer, 2000).

3.2.2 Intact Rock Strength

Bedrock mass movements are not restricted to sliding or rotating failure mechanisms along discontinuities; they can also occur via failure through intact rock (Figure 3.1(b)). Intact rock failures can occur when intact strength is low, when differential stress is high, or when a load is applied dynamically (e.g., earthquake), cyclically (e.g., seasonal variation), or for a long period of time (e.g., infrastructure).

Examples of failure through weak intact rock include the thick chalk deposits on either sides of the English Channel (e.g., Mortimore and Duperret, 2004) and Neogene/Paleogene siltstone in New Zealand (e.g., Read et al., 1981; Thompson, 1982; Prebble, 1995). However, heterogeneity and anisotropy in mechanical and hydrological properties dictated by stratigraphy or discontinuities typically also influence slope failures in weak rocks (e.g., Hutchinson et al., 1980; Duperret et al., 2004; Massey et al., 2013).

In hard rock, failure through intact rock bridges linking rock masses on either side of nonpersistent discontinuities can lead to slope failure. Although

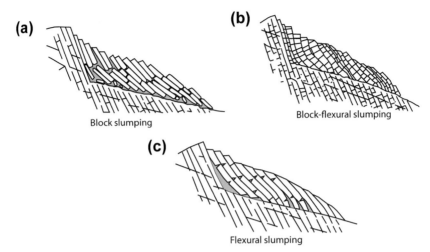

FIGURE 3.5 Rock slumping mechanism: (a) block, (b) block flexural, and (c) flexural. *Modified from Goodman and Kieffer (2000).*

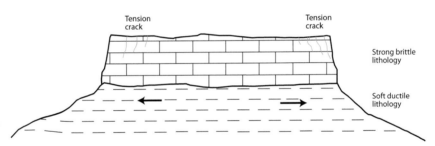

FIGURE 3.6 Rock spread occurs when a strong brittle lithology overlies a soft ductile one. *Modified from Bozzano et al. (2013).*

the failure mode is dictated by the presence of preexisting discontinuities, final loss of stability occurs because of rupture of intact rock where stress is highly concentrated. Rock bridge failure is thought to have played an important role in the development of the 1991 Randa rockslides (Eberhardt et al., 2004). Figure 3.6 demonstrates how a contrast in intact rock strength between geotechnical units can lead to bedrock mass movements such as rock spread (e.g., Evans, 1983; Bozzano et al., 2013).

3.2.3 Rock Mass Strength

Bedrock mass movement along approximately circular failure surfaces (also referred to as slumping) can occur in rock masses with numerous discontinuity sets with close spacing (Figure 3.1(c)). In these cases the rock mass effectively

behaves mechanically as a low-strength homogeneous material. The number
and spacing of the discontinuity sets required in the rock mass for it to be
considered homogeneous depends on the scale of the potentially unstable
slope (Hoek and Brown, 1997). If multiple discontinuity sets are present with
different orientations, the rock mass can also be considered to be isotropic
(Brady and Brown, 2004). Because the large-scale mechanical behavior of
heavily jointed rock masses approaches that of soil, their stability is investigated using continuum slope stability analyses techniques (e.g., limit equilibrium method of slices, finite element, finite difference).

3.3 MASS MOVEMENT CHARACTERISTICS

3.3.1 Volume and Velocity

The volume of material involved in bedrock mass movements varies from
less than 1 m^3 in the case of single rockfall block to more than 10^9 m^3 (Gm^3)
in gigantic rockslides (Hancox and Perrin, 2009; Pedrazzini et al., 2013;
Roberts and Evans, 2013). Large bedrock mass movements can influence the
landscape development for decades to millennia after the slope failure has
occurred (Hewitt et al., 2008). The velocity of bedrock mass movements is
another characteristic that can cover the complete range proposed by Cruden
and Varnes (1996) from extremely slow ($<5 \times 10^{-10}$ m/s) to extremely rapid
(>5 m/s). Historical landslide disasters such as Vajont, Italy (Müller, 1964),
and Khait, Tajikistan landslides (Evans et al., 2009), tragically demonstrate
the risk from large, extremely rapid bedrock slope failures and the damage
they can cause. Large slow-moving bedrocks mass movements can also pose
threats to communities and infrastructure. Glastonbury and Fell (2008)
summarize the geotechnical properties of 45 large slow-moving landslides,
roughly half of which are bedrock mass movements. Through a similar review of over 50 slow-moving landslides (of which roughly half involve
bedrock), Mansour et al. (2011) demonstrate that although slow-moving
landslides do not result in fatalities, they can result in extensive damage to
infrastructure.

3.3.2 Activity

Cruden et al. (1993) first introduced four states to describe the activity of a
landslide: active (currently moving), suspended (has moved over the past year
but not currently moving), dormant (last moved more than 2 years ago), and
reactivated (currently moving after being suspended or dormant). Other classifications of landslide activity are typically based on Cruden et al. (1993),
Cruden and Varnes (1996) and Griffiths and Whitworth (2012) subdivide the
"dormant" to reflect the duration of inactivity. The state of activity of a
landslide is temporally variable (Figure 3.7) and can change in response to
discrete events (e.g., earthquakes (Hadley, 1978), high-intensity/duration

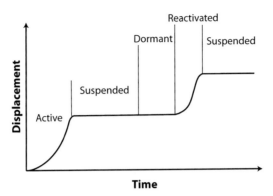

FIGURE 3.7 Displacement as a function of time illustrating the state of activity of a landslide. *Modified from Cruden and Varnes (1996).*

rainfall (Chau et al., 2003)), seasonal variation (Calabro et al., 2010), or distributed events (e.g., years with higher than normal rainfall (Hancox, 2008)). The Portuguese Bend (California; Linden, 1989) and Taihape (New Zealand; Massey, 2010) landslides are examples of large naturally reactivated slow bedrock mass movements that have been identified during and following construction of residential areas, respectively.

3.3.3 Progressive Failure

Progressive failure involves gradual decrease of intact, discontinuity, and rock mass strength (either simultaneously or independently) as rock deforms (i.e., strain-softening mechanical behavior). The concept of strain-softening is based on study of laboratory samples (Brady and Brown, 2004; Knapett and Craig, 2012; Tang and Hudson, 2010) and is applicable to large-scale rock slope failures (Rose and Hungr, 2007; Leroueil et al., 2012). Further enhancing the progressive failure is strain localization that results from damage accumulation as microcracks link up due to local stress conditions to form a continuous failure surface (e.g., Atkinson, 1987; Haied and Kondo, 1997; Tang and Hudson, 2010). Seasonal variation in temperature as well as positive and negative pore pressures can produce cyclic loading of the slope that contributes to strength reduction of rock material (Leroueil et al., 2012). Smithymann et al. (2009) numerically investigate this effect for annual variation of groundwater conditions on a deep-seated, gravitational slope deformation (DSGSD) feature; Watson et al. (2004) and Gischig et al. (2011a) demonstrate the importance of the annual temperature cycle in the deformation of large rock slopes. All of these factors contribute to the reduction of the factor of safety and increase in velocity of the rock mass as the incipient slope failure develops a continuous failure surface (Figure 3.8).

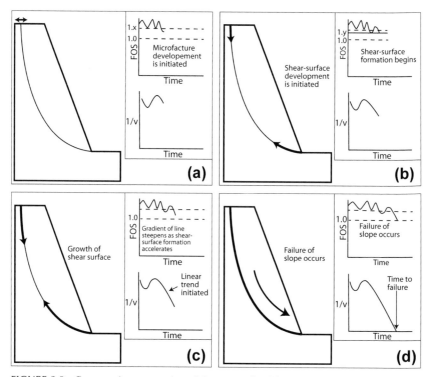

FIGURE 3.8 Conceptual representation of the progressive failure mechanism due to the accumulation of damage leading to the formation of a continuous failure surface. FOS = factor of safety. *Modified from Petley et al. (2005).*

Microseismicity can be used to document the energy release during the formation of fractures in intact rock. The technique has been used to monitor fracture formation in both hard rock (e.g., cliff failure in Normandy, France Amitrano et al., 2005; Senfaute et al., 2009) and soft rock (e.g., clay-shale landslides in France and Italy; Tonnellier et al., 2013). Tonnellier et al. (2013) attribute distinct seismic signals to rockfall in the headscarp of the landslide and to the shearing/deformation of the landslide body. Commonly, the progressive failure of a slope is evidenced in the field by an increase in velocity (Petley et al., 2002, 2005) and/or increase in temporal and spatial distribution of rockfall (Rosser et al., 2007, 2013). Gischig et al. (2009, 2011b) demonstrate the utility of ground-based LiDAR for documenting the temporal and spatial of movement along distinct section of the slope representing the formation of continuous release surfaces. Stead and Eberhardt (2013) regard development of distinct displacement zones, which can be characterized by displacement/velocity plots, as the result of accumulated and localized material damage.

3.3.4 Runout

Where mass movement is suggested by factors favoring slope failure, or past or current deformation, estimation of postfailure runout (travel distance as well as travel path and path-perpendicular spreading) is one of the most important factors in defining the hazard, and risk, it represents. Numerous factors such as failure volume, velocity, rheology, and the characteristics (topography, water content, erodibility) of the terrain over which it travels can all affect the travel distance of a mass movement.

3.4 MASS MOVEMENT TYPES

The most commonly used landslide classification scheme, proposed by Varnes (1978) and updated by Cruden and Varnes (1996), divides mass movements according to three source material types (rock, debris, earth) and five movement types (fall, topple, slide, spread, flow). Bedrock mass movements in this classification include rockfall, rock topple, rock slides, rock spread, and rock avalanches (i.e., rock flows). A suggested update to this classification by Hungr et al. (2014) adds "slope deformation" as a sixth movement type that for rock material includes mountain and rock slope deformation as subtypes of movements based on the scale of the deformation. Both subtypes are commonly referred to as DSGSD.

3.4.1 Rockfalls

Rockfall is a common bedrock mass movement on steep slopes, whether natural or engineered (Figure 3.9(a)). Differential weathering in stratified sedimentary and volcanic units further facilitates the initiation of rockfall. Following detachment, the dynamics of rockfall debris include free falling, rolling, bouncing, and sliding motions (Turner and Duffy, 2012). The initiation of a rockfall failure mechanism can occur via most of the structurally controlled failure modes described above and through the development of intact rock fractures. A rockfall event can include several pieces of rock falling at the same time, but differs from a rock avalanche in that it does not involve significant interactions between individual rock pieces (Hungr and Evans, 1988). Rockfall debris accumulates at the slope base as talus (Figure 3.9(a)). Gravitational sorting during talus deposition results in distal coarsening of rock fragments: larger boulders occur at the distal edge whereas the small fragments are located at the cliff base.

Rockfall, like other bedrock mass movement, can be triggered by earthquakes (Guzzetti et al., 2003; Murphy, 2014), but its temporal frequency is greatest during the rainy season or when a freeze–thaw cycle is pronounced (e.g., Terzaghi, 1962; Piteau, 1977; Wieczorek and Jager, 1996; Stoffel et al., 2005; Dunlop, 2010). Spatiotemporal association between rockfall and alpine

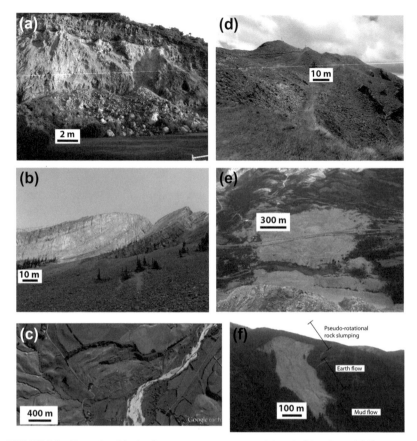

FIGURE 3.9 Example of bedrock mass movement types: (a) rockslide, (b) rockfall, (c) rock spread, (d) deep-seated gravitational slope deformation, (e) rock avalanche, (f) complex.

permafrost degradation has also been noted in Europe (Gruber et al., 2004) and New Zealand (Allen et al., 2009).

3.4.2 Rockslides

Sliding in rock can occur along single planar, circular, and compound surfaces or, as in the case of wedge failure, sliding along the intersection of two planes. Rockslides take place along discrete surfaces or relatively thin failure zones of intense shear strain (Cruden and Varnes, 1996). They can have a wide range of volume from an individual sliding block to some of the largest landslides known on Earth (e.g., Philip and Ritz, 1999; Hancox and Perrin, 2009; Pedrazzini et al., 2013; Roberts and Evans, 2013). These landslides tend to occur at a rapid to extremely rapid velocity. Large rockslides commonly transition in behavior into a rock avalanche (see Section 3.3.4) as the failed mass fragments during the transport phase. Large rockslides that evolve into rock avalanches are the most mobile mass movements on the Earth's surface: Robinson et al. (2014) report a runout of 28 km for a large rock avalanche in Kyrgyzstan; the Avalanche Lake rock avalanche (200 Mm^3) in northern Canada had runup (vertical elevation gain on opposite site of valley measured from valley bottom) of 640 m (Evans et al., 1994) and the Seymareh rock avalanche (44 Gm^3) in Iran ran-out 19.0 km horizontally with a fahrböschung (tangent of ratio between vertical to horizontal travel distance) of only 3.6° (Roberts and Evans, 2013). The failure surfaces of planar slides are generally associated with stratigraphic contacts (e.g., bedding) or metamorphic fabric (e.g., schistosity), which provide mechanical zones of weakness and zones of preferential groundwater flow and/or permeability contrast (Hodge and Freeze, 1977; Loew and Strauhal, 2013).

Circular and compound slides represent a continuum between mass movements controlled by the rock mass strength (circular) to an increasingly important structural control on the failure surface (compound to planar). Compound slides comprise active and passive blocks along a biplanar surface (Hutchinson, 1988) with the boundary between them representing a zone of internal deformation (Mencl, 1966). Compound sliding failures are commonly associated with stratified and/or folded sedimentary rocks (e.g., Gerath and Hungr, 1983). Stead and Eberhardt (2013) suggested that increasing dip angle between the two discontinuity planes forming the biplanar failure surface leads to more intense damage accumulation and internal deformation during a bedrock slope failure. The Vajont landslide is the most famous example of a compound failure surface (Müller, 1964). Recent characterization and numerical modeling of this landslide by Wolter et al. (2013a,b) demonstrate that multiple, compound, three-dimensional surfaces interacted to create zones with complex internal deformation pattern utilizing preexisting discontinuities and deforming through the rock mass.

3.4.3 Rock Spreads

The two types of lateral spreading based on the loading that drives the bedrock mass movement are dynamic and static. Lateral spreading most commonly occurs because of liquefaction of a seismically (dynamically) loaded saturated granular basal layer (or remolded sensitive clay) upon which the overlying material moves laterally (Seed and Wilson, 1967; Updike et al., 1988; Cubrinovski et al., 2012; Hungr et al., 2014). Kinematic freedom of spreading is facilitated by the presence of a free boundary (e.g., riverbank or shoreline) next to a soil and rock mass with a horizontally layered stratigraphy (Kramer, 2013). In surficial material lateral spreading tends to occur rapidly (cm/s), with displacement varying between centimeters and tens of meters. The volume affected can be in the range of millions of cubic meters.

Static lateral spreading has also been described in association with strati-fied rock material. It typically occurs where strong brittle units overlie weaker more ductile units. The spreading mechanism is also more common where the permeability contrast between the overlying units is high (Crozier, 2010). Although it can also incorporate very large volumes (Mm^3) where occurring in rocks, lateral spreading tends to occur slowly (mm/year to cm/year). Large static lateral spreads have been described in Canada (Geertsema et al., 2009), the United States (Reiche, 1937; Watson and Wright, 1963), Europe (Vlcko, 2004; Delgado et al., 2011), and New Zealand (Crozier, 2010). Lateral spreading in bedrock is commonly associated with deep-seated slope defor-mation as large stresses and some degree of material confinement are needed to produce ductile deformation of the weaker underlying material (Dramis and Sorriso-Valvo, 1994; Bozzano et al., 2013).

Static lateral spreading can transport large (hundreds of meters wide, by tens of meters long and thick) intact bedrock blocks of stratified sedimentary rock over large distances. These so-called toreva blocks—named for a type locality in the Colorado Plateau (Reiche, 1937)—can be back-tilted (Reiche, 1937) or remain un-tilted (Watson and Wright, 1963) depending on the degree of strength contrast with weaker underlying units over which they glide. The term has also been applied to large intact remnants of a volcanic edifice preserved in proximal debris field of sector collapses (e.g., Roa, 2003; Ponomareva et al., 2006), but these involve different transport mechanisms (debris avalanche—see Section 3.3.4; van Wyk de Vries and Delcamp, 2014) and are extremely rapid (>5 m/s).

3.4.4 Rock Avalanches

Rock avalanches result from rapid fragmentation of very fast-moving, initially intact rock masses during transport (Hungr et al., 2001). The fragmentation in a rock avalanche means that the term is associated with large-volume ($>1 Mm^3$) bedrock mass movement, in order that overburden stresses are sufficient to cause dynamic fragmentation of intact rock in the moving mass. Some slopes produce repeated, large bedrock failures over time. Rock ava-lanches can occur at the same location after thousands of years since the last discrete event (e.g., Hope Slide, Matthews and McTaggart, 1978; Val Pola, Crosta et al., 2004) or only weeks apart (e.g., Randa, Eberhardt et al., 2004).

It can be difficult to distinguish between failure occurring in an unce-mented granular pyroclastic deposit and in welded pyroclastic rocks (Hungr et al., 2001). Consequently, sector collapses of volcanic edifices have histor-ically been referred to as debris avalanches (Siebert, 1984; Ui et al., 1986; Procter et al., 2009) irrespective of their initial mechanical properties (i.e., whether they are dominated by poorly consolidated pyroclastic material or massive lava flows). See van Wyk de Vries and Delcamp (2014) for more information about landslides associated with volcanic edifices.

The runout of large rock avalanche debris is longer than expected for equivalent volumes of granular material, and mobility of rock avalanche increases with volume. Scheidegger (1973) note that rock avalanche volume is inversely related to the ratio of vertical to horizontal travel distance. Hsu (1975) proposed that the term excess travel distance be used to express the amount by which runout exceeds the expected value for an equivalent volume of granular material. Numerous mechanisms occurring during material transport of the landslide debris have been proposed to explain anomalously long runout (Davies et al., 1999; Legros, 2002; Davies and McSaveney, 2012). Recently broadband seismic signals have been used to locate large rock avalanches occurring in remote area (Ekstrom and Stark, 2013) and to investigate the dynamic processes occurring during transport (Yamada et al., 2013).

3.4.5 Sackungen/DSGSD

Sackungen (Zichinsky, 1966) and gravity faulting (Beck, 1968) were identified as slope "sagging" or gradual adjustment under gravitational forces. The term DSGSD is now commonly used to refer to these slow to very slow slope deformation processes. Geomorphic features that can indicate past or present DSGSD include grabens, trenches, uphill- and/or downhill-facing scarps, split ridges, and toe bulging (Varnes et al., 1989). Observations suggest that DSGSDs typically affect entire mountain slopes from the valley floor to ridge tops or even through the entire rock mass to the slope on the opposite side of the ridge (Agliardi et al., 2012). DSGSD typically occurs in areas with local relief greater than 500 m (Agliardi et al., 2013). The stress distribution derived from the interaction between the rock mass strength and topography plays an important role in the development of DSGSD features (Savage and Varnes, 1987; Varnes et al., 1989; Kinakin and Stead, 2005; Ambrosi and Crosta, 2011). Agliardi et al. (2013) find that DSGSDs in the European Alps were more common in regions with moderate tectonic exhumation rate and lower annual rainfall. A range of failure mechanisms have been proposed to explain the development of DSGSD (Figure 3.10). They form a continuum from dominantly rock mass strength controlled to structurally controlled (Agliardi et al., 2012). The work of Agliardi et al. (2001) and Ambrosi and Crosta (2006) highlights the important role of structural geology and tectonic structures in location and deformation characteristics of DSGSD. Di Maggio et al. (2014) recently documented both types (rock mass strength and structurally controlled) of DSGSD occurring in Sicily, Italy. Glacier retreat appears to influence development of DSGSD due to the oversteepening, debuttressing, unloading, and changes to the groundwater conditions (Ballantyne, 2002; McColl, 2012). Recent work by McLean et al. (2014) in the North Island of New Zealand demonstrates that slope modification by glaciers during the last glaciation is not a prerequisite for the development of DSGSD features.

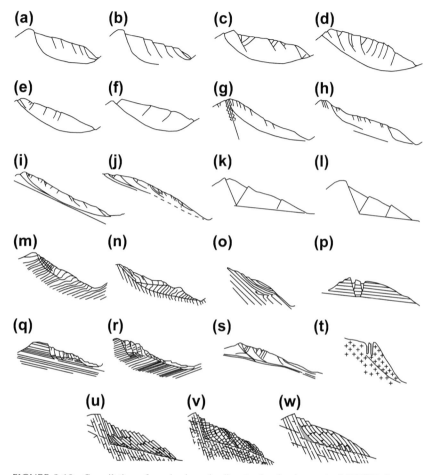

FIGURE 3.10 Compilation of mechanisms leading to the development of DSGSD features. *Modified from Agliardi et al. (2012).*

DSGSDs are regarded as the result of a long-term gravitation deformation of the slope (Soldati, 2013). Detailed stratigraphy and dating of tephra layers occurring in excavation of DSGSD trenches by McCalpin and Irvine (1995) suggest that the temporal displacement pattern associated with these mass movements was episodic. Moro et al. (2011) recently documented the episodic displacement of these features in associated with seismic events. Using pre- and post-event InSAR imagery of the 2009 Aquilla earthquake, they were able to capture movement at two known DSGSDs.

The implications of DSGSDs for hazard characterization are not clear. In some instances DSGSDs have been observed to be precursors to rock avalanches (Japan, Chigira and Kiho, 1994; Chigira et al., 2013; Switzerland, Pedrazzini et al., 2013; Canada, Hewitt et al., 2008), whereas in other scenarios they have been moving at a slow to extremely slow velocity for millennia (Hippolyte et al., 2012). The presence of DSGDSs has also shown to affect the distribution of other types of landslides (Capitani et al., 2013). This is consistent with the results of physical modeling by Bachmann et al. (2004).

3.4.6 Complex Bedrock Mass Movements

Complex landslides incorporate multiple mass movement types, distributed spatially or temporally, at a single site. This change in mass movement type is driven by a change in the kinematic conditions driving the failure mechanism (e.g., removal of material at toe of landslide due to fluvial erosion), material properties (e.g., weathering), or water content (e.g., traveling over a saturated substrate). Geertsema et al. (2006) and Geertsema and Cruden (2008, 2009) document a series of landslides in northern British Columbia where rock slope failure interacted with the surficial material in the runout zone to trigger debris or earth landslides. In particular, the Muskwa landslide of 1979 initiated as a 3 Mm^3 circular rock slide and triggered a $12-15 Mm^3$ earthflow. Geertsema et al. (2006) suggest that the undrained loading from the initial rock slope failure caused mobilization of the surficial material and its subsequent 3.25 km runout. The entrainment (e.g., 1949 Khait landslide, Evans et al., 2009; Mount Meager landslide, Guthrie et al., 2012) or undrained loading (e.g., 2002 McAuley Creek Landslide, Brideau et al., 2012, 2006; Leyte Island landslide, Evans et al., 2007) of saturated sediment in the valley bottom has been observed at numerous sites to influence the rheology of the failed mass and enhanced runout distance.

The East Gate landslide in southern British Columbia (Figure 3.9(f)) is an example of an active complex landslide that initiated as a rock slump in highly fractured rock mass, where the rock material quickly disintegrated and moved as an earthflow (Couture et al., 2004; Couture and Evans, 2006; Brideau et al., 2006). Gullying within the earthflow mass provided water channelization that mobilizes the disintegrated clay-rich rock mass during intense and prolonged precipitation events in the spring and fall (Brideau et al., 2006). Panek et al. (2011) describe a site in the Western Carpathian Mountains where geomorphic evidence suggests that DSGSD, rockslide, debris avalanche, and debris flows have all been active at different times. They show how some of the mass movements can precondition the subsequent slope processes (Panek et al., 2011).

3.4.7 Secondary Hazards Associated with Bedrock Mass Movements

Secondary hazards can arise from the rapid failure of rock slopes, including landslide-dammed lakes and displacement waves. Secondary processes caused by bedrock mass movements can extend the area affected by the initial slope failure, sometimes by an order of magnitude. Landslide dams can impound large volumes of water that can be released suddenly if the dam fails rapidly (e.g., Evans et al., 2011). For example, the damming of the Dadu River in China by an earthquake-triggered landslide and the flood wave associated with the subsequent failure of the landslide dam are estimated to have been responsible for 100, 000 fatalities (Dai et al., 2005). This hazard associated with landslides occurring in confined valleys is discussed in more detail by Korup and Wang (2014).

Displacement waves are produced when a rock mass rapidly enters fjords (e.g., Miller, 1960; Dahl-Jensen et al., 2004; van Zeyl, 2009; Sepulveda et al., 2010), lakes (e.g., Jorstad, 1968; Evans, 1989; Roberts et al., 2013), hydro-electrical reservoirs (e.g., Müller, 1964; Panizzo et al., 2005; Huang et al., 2012), rivers (Huang et al., 2014), or the sea (e.g., Ward and Day, 2003; McMurtry et al., 2004). Many of the over 250 displacement waves generated by subaerial landslides recorded globally (Roberts et al., 2014a) resulted from bedrock mass movement, and many more are likely undocumented. Erismann and Abele (2001) and de Blasio (2011) suggest that the ratio of volume of the unstable rock mass and the volume of water available to be displaced can be used as an indication of the potential height of the displacement wave.

3.5 CASE STUDIES

3.5.1 Seymareh, Iran

The early Holocene Seymareh (Saidmarreh) rock avalanche (Figure 3.11) is the largest known rock avalanche on Earth and one of the largest known subaerial Quaternary landslides of any type. Harrison and Falcon (1937, 1938) identified the landslide during preliminary geologic mapping of the Zagros Simply Folded Belt. Subsequent studies characterize surface features of its debris (Watson and Wright, 1969) and the stratigraphy, structure, and geometry of its source area (Roberts and Evans, 2013). Catastrophic failure of a sequence of Cretaceous-Tertiary limestone and shale initiated as a dip-slope rockslide from the flank of the Kabir Kuh anticline, the largest of the folds in the range. The resulting 38-km^3 scar is 15.5 km wide, extends 6.1 km downslope, and reaches a maximum depth of 680 m in its central region (Roberts and Evans, 2013) (Figure 3.11(a) and (b)). Rapid communion of the rockslide mass produced a highly mobile (3.6° fahrböschung) rock avalanche that covered approximately 220 km^2 and ran out 19.0 km (Figure 3.11(a) and (b)), surmounting the roughly half-kilometer-high nose of the plunging Kuh Dufrush anticline before continuing into the next valley to the northeast (Figure 3.11(a) and (b)). The deposit, which at its thickest is approximately 300 m deep (Figure 3.11(b)), impounded two large lakes whose infill deposits provide some of the most fertile land in the region and have supported settlements for several thousand years. Accelerator mass spectrometer (AMS) radiocarbon ages from

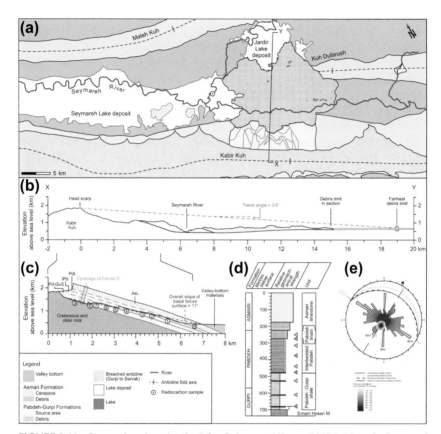

FIGURE 3.11   Seymareh rock avalanche (after Roberts and Evans (2013)), (a) geologic map of the source area, deposit, and associated lacustrine deposits, (b) cross-section showing topography and geology of the source area and deposit, (c) simplified cross-section of the flank of Kabir Kuh showing the broad-scale downward-stepping of the failure surface and the structural features forming it, (d) stratigraphic profile through the failed sequence showing lithologic and geo-mechanical variation *(modified from Alavi, 2004)*, and (e) stereonet summary of structural and topographic features dictating the failed mass.

near the base of one of these lake deposits (Roberts and Evans, 2013: TO-13445) and from sediments of a small lake formed on the debris surface (Griffiths et al., 2001; CAMS-33990, CAMS-33991, CAMS-33992) suggest the rock avalanche occurred between 8710 ± 80 C^{14} and ca. 9800 C^{14} (Roberts and Evans, 2013). Debris morphology and lithologic variation (Roberts and Evans, 2013) as well as continuity of lineaments in the debris surface (Watson and Wright, 1969) suggest that Kabir Kuh failed during a single event.

Roberts and Evans (2013) conclude that the initial rockslide was delineated by the lithology and structure of the northeast flank of Kabir Kuh, consequent on the tectonic and physiographic evolution of the region and of the anticline. Failure surfaces are regularly distributed throughout the sequence and typically correspond to presheared bedding planes between fractured limestone and overlying shale (Figure 3.11(c) and (d)). These positions correspond to expected locations of concentrated strain during fold formation (Skempton, 1985) that would promote interbed shear. Shear stress generated by folding is inversely related to spacing of interlayer slip surfaces (Price and Cosgrove, 1990) and can thus be expected to exceed shear strength at somewhat regular intervals throughout a homogeneous section of rock mass. Additionally, elevated pore-pressure pressure at geologically defined permeability contrasts (Hodge and Freeze, 1977) locally decreases shear strength. Lateral limits of the failed rock mass (Figure 3.11(e)) correspond to bedding-normal joints that were preferentially weakened because of their alignment to stress fields generated by regional tectonic compression and folding of Kabir Kuh. Erosional unroofing of Kabir Kuh's fold axis provided head release (Figure 3.11(a) and (c)).

Failure required breach of the 250-m-thick, thickly bedded Asmari Limestone unit capping the sequence. Buckling of a unit of this thickness requires exceptional heaving loading (Wyllie and Mah, 2004), which may have been partly due to gravitational loading by the incipient rockslide. Harrison and Falcon (1938) suggested prefailure creep of the slope, forming a "knee-fold" (as described elsewhere in the range by Harrison and Falcon, 1936). Geomorphic evidence of such deep-seated deformation is absent in intact parts of the northeast flank of Kabir Kuh (Oberlander, 1965; Roberts and Evans, 2013), although if restricted to the failed rockslope it would have been obliterated. Most likely, breakthrough was aided by erosion by Seymareh River, which Oberlander (1965) suggests was being deflected southwestward by Kashgan River. Erosion may have been by horizontal undercutting or, as Oberlander (1965) proposes, vertical incision resulting from superposition of the channel onto the lower flank of Kabir Kuh following denudation of the overlying marls.

3.5.2 La Clapière, France

The La Clapière landslide is a 60-Mm^3 slow-moving bedrock slope movement that is part of a very large (~ 10 km along strike) DSGSD complex in southeastern France (Figure 3.12). The headscarp is 120 m high and approximately 800 m wide (Lebourg et al., 2011). The depth of the failure surface is interpreted to vary between 80 and 100 m (Jormard et al., 2007). Studies of trench morphology and cosmogenic dating suggest that the bedrock mass movement complex has been active for the last 10,000 years whereas the main part of the La Clapière landslide has been active since 3,600 years BP with notably increased activity since the late 1970s (Bigot-Cormier et al., 2005; El Bedoui et al., 2009, 2011). The entire mass movement complex involves

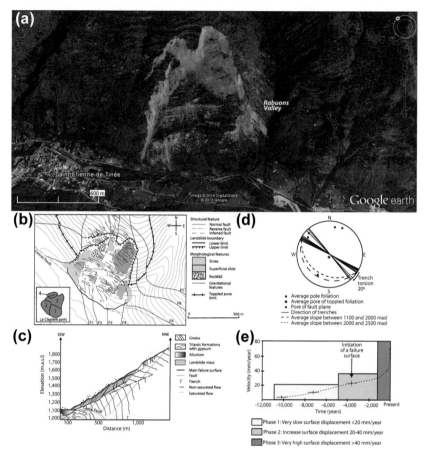

FIGURE 3.12 La Clapière Landslide: (a) overview *(Google Earth 2011 imagery)*, (b) geo-morphology and tectonic map (Jomard et al., 2010), (c) cross-section (Jomard et al., 2010), (d) stereonet of orientation of gneissic fabric *(modified from Vengeon et al. (1999))*, and (e) progressive acceleration of the surface displacement *(modified from El Bedoui et al. (2009))*.

gneissic rocks. The metamorphic foliation dips steeply into the slope in the area next to the landslide (Figure 3.12), but dips more shallowly in the main landslide mass (Follacci, 1987; Vengeon et al., 1999). The initial failure mechanism of the La Clapière landslide is considered to have been caused by toppling along the metamorphic foliation followed by the current sliding motion due to the creation of a continuous failure surface at the base of the toppling column via failure through intact rock (Follacci, 1987; Vengeon et al., 1999). This represents the progressive degradation of the rock mass resulting in the current La Clapière landslide (El Bedoui et al., 2009).

Several tectonic structures play important roles in slope instability at La Clapière. The gneissic foliation is part of a regional-scale overturned synform (Gunzburger and Laumonier, 2002). The influence of the inherited faults at La Clapière is investigated in physical models by Bois et al. (2008) who demonstrates that the deformation pattern and mobilized landslide volume are sensitive to the presence and orientation of tectonic structures. Jomard et al. (2007) consider that the weathering of the rock mass and of the cataclastic material along the gravitationally reactivated faults have played an important role in the development of the La Clapière landslide. Jomard et al. (2007, 2010) use geophysical site investigation techniques to map the extent of these weathered zones. The strength reduction associated with weathering of the rock mass is incorporated in numerical models at both the regional (Guglielmi and Cappa, 2010) and landslide scale to assess its importance to the development of the DSGSD features and the La Clapière landslide (Chemenda et al., 2009; Bouissou et al., 2012). Lebourg et al. (2011) characterize grain size distribution and measure triaxial strength of cataclastic rock samples of the faults inside and outside the main landslide area. They find that the fault materials inside the landslide zone are depleted of fine material and had a larger effective frictional strength (Lebourg et al., 2011).

The influence of groundwater conditions on the stability and failure mechanism of the La Clapière landslide is investigated in numerous studies. Isotopic and hydrochemical studies suggest that a shallow perched aquifer is present near the headscarp of the landslide and that a deep aquifer recharged by a distant source is present near the toe of the landslide (Guglielmi et al., 2002). The presence of the perched aquifer is supported by geophysical investigations (Jormard et al., 2007). These findings, along with estimates of the permeability based on residency time of the water, are used in hydromechanical models to demonstrate the importance of the faults as groundwater conduits and in turn on slope deformation (Cappa et al., 2004; Guglielmi et al., 2005).

Several displacement monitoring techniques have been implemented at the La Clapière landslide. Traditional total station ground surveys and precipitation data have been collected regularly since 1982 (Blanc et al., 1987; Follacci, 1987). On average the landslide moves on the order of 1 cm/day, but several cycles of acceleration and deceleration are documented (Casson et al., 2005). Interpretation of aerial photographs and satellite images provides more complete spatial coverage of the displacement data than obtained from discrete survey monuments (Fruneau et al., 1996; Delacourt et al., 2004; Casson et al., 2005). The use of historical aerial photographs suggests that the landslide has been moving diagonally relative to the main slope direction. A similar result came out of the work by El Bedoui et al. (2009), which investigates the chronology of slope deformation features and found an asymmetric displacement pattern developed at the La Clapière landslide with its southeastern side having moved more. Brideau and Stead (2011) suggest that the asymmetric displacement pattern could be in part influenced by the level of lateral kinematic confinement present at a landslide. At La Clapière the kinematic freedom along its southeastern boundary is facilitated by Rabuons Valley (Figure 3.11(e)).

3.5.3 Threatening Rock, United States of America

On January 22, 1941, following years of slow tilting, a $\sim 12{,}000$-m^3 sandstone monolith aptly named Threatening Rock toppled onto the pre-European housing complex of Pueblo Bonito, destroying a large portion of it (Figure 3.13). The housing complex is one of thirteen "great houses" in Chaco Canyon dating from the eleventh century (Lekson, 1984). These expansive, intricate, stone structures constitute the best preserved record of the ancient inhabitants of the American Southwest and are protected as part of the Chaco

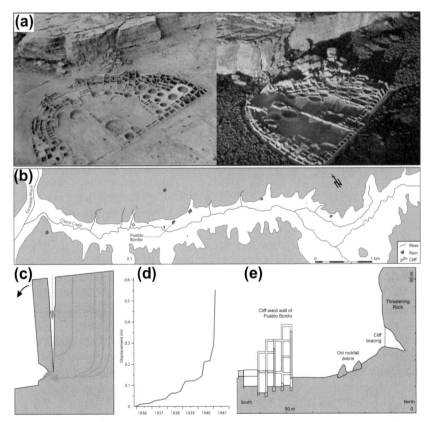

FIGURE 3.13   Multiple panels summarizing various aspects of the Threatening Rock topple. (a) Distribution of cliffs and pre-colonial ruins in central Chaco Canyon before (attributed to Lindbergh, 1929 in Judd, 1964) and after failure. (b) Location/topographic map of Chaco Canyon. (c) Cross-section showing stratigraphic sequence and toe erosion mechanism of the cliff face. (d) Time-to-failure analysis plot *(modified from Voight (1988a))*. (e) Precolonial engineering mitigation works at the toe of the monolith *(modified from Judd, 1964)*.

Culture National Historical Park, designated a UNESCO World Heritage Site in 1987. Prior to the failure, Pueblo Bonito had been only partially excavated during two decades of archeological expeditions (Pepper, 1920; Judd, 1954; Lekson, 1984), and the collapse of Threatening Rock (Figure 3.13(a)) thus also cost part of the record of Ancient American culture.

The fall of Threatening Rock is representative of the ongoing evolution of Chaco Canyon by catastrophic failure of large monoliths (Schumm and Chorley, 1964; Figure 3.13(b)). Prevalence of near-horizontal, clastic sedimentary rock exposed in canyons and mesas in the region (Bryan, 1954), together with cliff-parallel tension cracks, favor failure by toppling. The stratigraphic sequence is dominated by the late Cretaceous Mesaverde Group (Figure 3.13(c)), comprising the massive fine-grained Cliff House Sandstone Formation and the underlying shale Menefee Formation, which includes coal and sandstone members (Bryan, 1928). Failure is further preconditioned by the permeability and geomechanical strength contrast within the cliff-forming sequence (Figure 3.13(c)). Diversion of groundwater flow along the low-permeability Menefee Formation results in sapping near the slope toe (Bryan, 1928). Hard-over-soft geomechanical strength contrast (Bryan, 1928, 1954) likely contributes to failure during late stages of canyon−ward rotation through sliding in the compressed, commonly saturated Menefee shales.

Temporal patterns of deformation during progressive failure provide insight into activity at the basal failure zone and, potentially, timing of failure (Petley et al., 2005). Saito and Uezawa (1961) propose an approach elaborated by Voight (1989) and Fukozono (1990) in which displacement data are plotted in velocity−time space; a linear relationship is typical during acceleration preceding catastrophic failure (Saito and Uezawa, 1961), at least in brittle materials (Voight, 1989). If the reciprocal of velocity is plotted, projection of the trend to the time axis approximates timing of rapid acceleration and consequent catastrophic failure. Using prefailure displacement data collected by the National Parks Service and reported by Schumm and Chorley (1964), Voight et al. (1988a,b) perform time-to-failure prediction of Threatening Rock (Figure 3.13(d)). The linear trend shows that approximate timing of collapse could have been predicted in the years leading to failure (Voight et al., 1988a,b). Prediction accuracy of failure timing increases as velocity increases and catastrophic failure draws nearer; to be used as a reliable prediction tool, this analysis would need to have been repeated with increasing frequency in the year before failure (cf. Crosta and Agliardi, 2003a).

Canyon widening by toppling, and consequent rockfall hazard, occurred throughout the late Quaternary history of Chaco Canyon. Rockfall debris at cliff bases, including some incorporated into architecture of prehistoric housing complexes (Judd, 1964), and currently tilting monoliths similar to Threatening Rock, highlight the landslide risk posed to both the ancient inhabitants of Chaco Canyon and the modern UNSECO Heritage site. However, differing mitigation approaches suggest that the ancient and present in-habitants of Chaco Canyon perceived rockfall risk differently. Precolonial inhabitants of Chaco Canyon recognized the risk posed by Threatening Rock and attempted to stabilize it. Their engineering effort represents one of the oldest known examples of engineering geology in the Americas. They con-structed an embankment capped by a wall, running the length of the monolith and apparently designed to brace it up (Judd, 1959; Keur, 1933; Pepper, 1920; Figure 3.13(e)). Unfortunately, these works were destroyed by the 1941 collapse. By contrast, no major mitigation measures have been implemented at the modern site.

3.6 RECOGNITION AND RESPONSE

3.6.1 Anticipation

Landslides tend to occur at discrete, often nonrecurrent locations, making forecasting their exact spatial location difficult. Many hazard recognition and response measures are similar between bedrock mass movements and land-slides occurring in soil. However, the importance of discontinuities and rock mass characteristics in bedrock landslides, such as discontinuity orientation and strength, are particularly important to consider. These can be considered together with topography and possible triggers to identify areas particularly predisposed to bedrock failure. Alternatively, the distribution of past failures can be used to highlight general areas susceptible to future instability.

Landslide inventory mapping is usually the first step in identifying the spatial distribution of landslide hazard (van Westen et al., 2008). Inventory maps of bedrock landslides record the spatial distribution of bedrock mass movement and can include other characteristics such as date of occurrence, current state of activity, type, and volume (Guzzeti et al., 2012). Inventory maps can be conducted over a wide range of scales from local (tens of square kilometers), to national (up to millions of square kilometers; van Den Eeckhaut and Hervas, 2012), and even continental (Kalsnes and Nadim, 2012).

Detailed rockfall inventories can be used to create rockfall hazard maps at the local scale using various mathematical functions (e.g., Crosta and Agliardi, 2003b; Jaboyedoff et al., 2005; Abbruzzese and Labiouse, 2013). Another form of hazard management that has been successfully applied to rockfalls is the Rockfall Hazard Rating Systems (RHRS). This scheme facilitates sys-tematic and reproducible assessments by attributing scores to various factors such as topography, geology, climate, and historical activity to provide a relative hazard ranking to various sites (Pierson, 2012). Their principal use is prioritization of sites needing attention for mitigation or protection (Pierson, 2012). The RHRSs have been modified and adapted by numerous agencies based on their local knowledge and experience (e.g., Santi et al., 2009; Pierson and Turner, 2012). For slopes above highways and railroads it is possible to incorporate the usage (car or train per day) and driving conditions (e.g., speed, visibility) into the rating system, thereby providing a relative risk ranking between sites (Pierson and Turner, 2012) and at different times. It is more difficult to apply risk zonation based on inventory maps to bedrock mass movement, other than rockfall, as the area affected by the hazard is influenced by the volume mobilized and external factors. For example, water content can influence the landslide runout distance and the temporal aspect of the element at risk (van Westen et al., 2006).

Monitoring of bedrock mass movements can be used to provide early warning of imminent large rock slope failures. This hazard management approach rests on the assumption that the progressive failure behavior—particularly acceleration or other signs of increased instability—of the rock mass can be captured by the monitoring methods. Examples of large natural rock slopes currently being monitored include Åknes in Norway (Blikra, 2012), Gradenbach in Austria (Bruckl et al., 2013), Turtle Mountain in Canada (Froese et al., 2012), and Randa in Switzerland (Loew et al., 2012b). The slope monitoring systems typically consist of combinations of direct displacement measurement (extensometer, tiltmeter, GPS), remote sensing displacement measurement (electronic distance measurements, LiDAR, GB-InSAR), meteorological (temperature and precipitation) and visual observation (photographs or webcam) (Eberhardt, 2012; Petley, 2012; Loew et al., 2012a). The slope monitoring data provide input to an early warning system that incorporates the review of the data against developed thresholds to avoid or minimize the consequence of a slope failure (Intrieri et al., 2012, 2013). Baron and Supper (2013) and Michoud et al. (2013) provided summaries of experiences in developing slope monitoring techniques and early warning systems at sites around the world. As part of the early warning system, an emergency management plan provides the response framework including a communi-cation plan, access restriction, and/or evacuation procedures in case of slope activity above the established background activity level (Froese and Moreno, 2014).

Numerical and physical modeling can be used in conjunction with detailed site investigation and slope monitoring to gain a better understanding of the factors controlling the deformation or failure mechanism at a particular slope or of a specific slope geometry. Recent work by Stead and Coggan (2012) and Stead et al. (2006, 2012) has demonstrated how integrated studies can use state-of-art remote sensing and slope stability modeling to increase knowledge of the processes at work in bedrock mass movements. Brideau et al. (2009) provided a general range of rock mass qualities where the stability is controlled by either the intact rock strength, the presence of structural features, or rock mass strength and the applicability of the various numerical modeling techniques to each scenario. Physical models demonstrate that the three-dimensional topography of a rock slope can greatly influence its stability (Stacey, 1974, 2006). The work of Bachmann et al. (2004, 2006, 2009), and

Bois et al. (2012) using three-dimensional physical models also demonstrated that interaction between the topography and tectonic structures can influence the development of both large-scale and small-scale rock slope instabilities.

A range of numerical and physical modeling techniques is available to assess the postfailure behavior of large rapid bedrock mass movements. The numerical modeling techniques can be based on empirical relationships (e.g., Scheidegger, 1973; Geertsema and Cruden, 2008) continuum mechanics (e.g., Hungr, 1995; McDougall and Hungr, 2004; Pirulli, 2009), or discontinuum mechanics (e.g., Poisel et al., 2009; Thompson et al., 2010). The numerical modeling techniques have been calibrated based on numerous back analyses of various bedrock mass movements and can be used as runout and velocity forecasting tools (e.g., Willenberg et al., 2009). Physical models have been used to investigate the influence of stratigraphy (e.g., Shea and van Wyk de Vries, 2008) and structure (e.g., Manzella and Labiouse, 2009) in the initiation zone on the morphology and stratigraphy of rockslide and rock avalanches deposits. Physical models can also be used to investigate the influence of mechanisms such as entrainment (Dufresne et al., 2010; Dufresne, 2012) and mechanical vibration (Cassie et al., 1988) on the transport behavior of large rapid bedrock mass movements. Empirical relationships and specific software to capture the runout behavior of rockfalls also exist and facilitate site-specific hazard zonation (e.g., Agliardi and Crosta, 2003; Lan et al., 2007; Leine et al., 2013). Calibration of input parameters for these rockfall runout softwares has been provided by physical modeling tests and back calculation of well-documented events (e.g., Chau et al., 2002; Wyllie, 2014). The simplest form of empirical relationship is called the rockfall shadow angle (Evans and Hungr, 1993), which is the angle between the top of the talus (or source area) and the distal edge of the talus/runout zone. Compilations of shadow angle from around the world have found it to range between 17° and 37° depending on how it was defined in the individual study (Frattini et al., 2012). Holm and Jakob (2009) have also reported that the absence of vegetation on large talus slopes can result in a lower than anticipated shadow angle values.

3.6.2 Avoidance

Landslide avoidance consists of locating new infrastructure or moving existing developments away from the hazard area. When possible, avoidance constitutes the simplest and most reliable form of landslide mitigation. Under certain circumstances avoidance can also represent a more economical alternative to active remediation. Avoidance requires predevelopment knowledge of the landslide hazards (location, activity, potential magnitude and frequency, and factors that might change any of these). Landslide susceptibility mapping can be used to guide land-use planning strategies (e.g., Hart and Hearn, 2013) or to provide guidance for the routing at the feasibility stage of linear infrastructure projects (e.g., Couture et al., 2010). In urban and/or residential areas this can be implemented as setback from the edge of a cliff or from the toe of a talus slope as specified by local governments or regional districts (e.g., Santa Barbara County, 2014). It also requires robust urban/infrastructure development plans and compliance with them, which is commonly lacking in many regions with limited socioeconomic development. For example, despite urging by several studies over the past 60 years to curtail development of specific slopes around the city of La Paz, Bolivia, high occupancy use of these slopes has continued (Roberts et al., 2014b).

In the case of linear transportation infrastructure along narrow valleys, avoidance can take the form of realignment or relocation, elevated structures, or tunnels (Badger and Duffy, 2012). Sometimes avoidance results in a project having to be completely relocated or canceled due to the high probability of bedrock mass movement (e.g., Williams and Prebble, 2010). In other situations the spatial extent of the hazard only becomes better understood after a renewed period of bedrock mass movement activity that requires modification to the existing land-use guidelines. An example of this situation is the rockfall activity associated with the 2010 and 2011 Canterbury earthquakes in New Zealand (Dellow et al., 2011; Hancox and Perrin, 2011). After the events, an inventory of the runout distances (Massey et al., 2012a,b) and geomorphic mapping (Townsend and Rosser, 2012) were used to conduct numerical modeling and defined rockfall risk zones for different return intervals (Massey et al., 2012a–d). The local government body used these risk maps (Massey et al., 2012a–d), proposed guidelines (Taig et al., 2012), and recommendations from an independent review committee to define exclusion zones where further construction or repair to existing infrastructure would not be allowed (CERA, 2013, 2014). This example demonstrates the value of conducting landslide hazard zoning prior to residential development and the need to consider potential trigger mechanisms with long return periods.

3.6.3 Prevention

Due to their potentially small volume, spatial reoccurrence over a general area and the difficulty to temporally predict their occurrence via monitoring techniques, the construction of protection structures is a popular mitigation option for rockfall. Numerous structural mitigation types (e.g., embankment, barriers, drapery nets, rock sheds) have been successfully used around the world (see Volkwein et al., 2011; Badger et al., 2012; Lambert and Bourrier, 2013, for recent reviews). Forest management can be a very effective form of prevention (Dorren and Berger, 2012). In rock slope excavation it is possible to use the slope geometry (bench and ditch width) as debris catchment structures (e.g., Ryan and Prior, 2000; Alejano et al., 2007).

Large hydroelectric schemes are a good example of engineering projects that affect large areas where it is often impossible to avoid all unstable rock slopes. The Revelstoke Dam in Canada (Piteau et al., 1978) and the Clyde

Dam in New Zealand (MacFarlane, 2009) have made use of toe buttress and drainage infrastructures to stabilize unstable rock mass. Toe buttresses increase the resisting force of the rock mass, increasing its overall stability. The strategic placement of a toe berm means that a relatively small volume of buttress can drastically improve the stability of a large volume of rock (Corkum and Martin, 2004). Drainage infrastructure dewaters the slope thereby reducing the pore-water pressure and increasing the effective resistive strength of rock mass. The drainage adits at Downie Slide (upstream from Revelstoke Dam, Canada) and at the Nine Mile Creek Landslide (upstream from the Clyde Dam in New Zealand) consist of tunnel networks hundreds to thousands of meters long (Kalenchuk et al., 2012, 2013; MacFarlane, 2009).

3.7 RISK MANAGEMENT IN ROCK SLOPES

Figure 3.14 presents the risk equation used to quantify most natural hazards and provides examples of its application specifically to bedrock mass movements. The components of the risk equation will be introduced here. Further

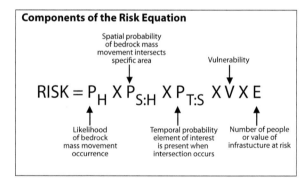

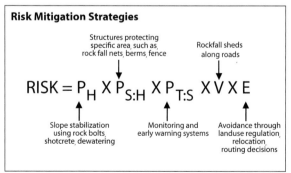

FIGURE 3.14   Components of the risk equation and risk mitigation strategies. *Modified from District of North Vancouver, www.dvn.org.*

details on the definition and quantification of individual terms can be found in
Wise et al. (2004), van Westen et al. (2006) and Corominas et al. (2014). Each
of the mitigation strategies discussed above can be regarded as reducing the
probability of one component of the risk equation. The first term, P_H, represents the likelihood of a bedrock mass movement occurring, and this value is
typically expressed in annual probability. The probability of occurrence can be
reduced by actively reducing the driving forces or increasing the resisting
forces that are acting on the rock slope. Such measures include rock bolting,
dewatering, addition of a berm at the toe of the unstable slope, head unloading,
or any other method employed for slope stabilization (e.g., Andrew and
Pierson, 2012). The second term, $P_{S:H}$, is the conditional probability that if a
bedrock movement initiates it will reach the area of interest. Rockfall fences,
netting, ditches, and berms all represent structural elements that can be added
below an unstable slope to prevent debris from reaching the area of interest
and thereby reduce the $P_{S:H}$ term. The third term, $P_{T:S}$, represents the conditional probability that if the bedrock mass movement reaches the area of interest the element affected by the hazard will be present. The probability
associated with this temporal term can be reduced by the implementation of a
monitoring program and associated early warning system. The fourth term, V,
of the risk equation is the vulnerability of the element affected by the bedrock
mass movement hazard. Vulnerability represents the susceptibility to damage
of the element relative to intersection with a specific bedrock mass movement.
Roberts et al. (2009) discussed the multidimensional aspect of vulnerability to
natural hazards. Structural reinforcement such as a rock shed along road or a
railway for rockfall or strengthening of building foundations for very slow
moving DSGSD are strategies to reduce the vulnerability of infrastructure
(Mavrouli et al., 2014). Public education and awareness of the hazardous
bedrock mass movement can reduce behavior that exposes individuals and
groups to a vulnerable situation. Davies (2013) suggested that the low frequency (long return interval) of the large magnitude events can lead to a
perceived low hazard and additional efforts may be needed to raise awareness
of these type of slope failures. The last term, E, of the risk equation represents
the number of people or the value of the infrastructure at risk. This term can be
reduced by implementing land-use restrictions where hazards associated with
bedrock mass movement have been identified as high. Local, regional, and
national land-use guidelines are developed to reduce the value of this
component of the risk equation (e.g., Australian Geomechanics Society, 2007;
Porter and Morgenstern, 2013). Hermanns and Longva (2012) suggested that
for rapid bedrock mass movement with large volume, slope reinforcement,
capturing or redirecting the debris after initiation, or physical reinforcement of
infrastructure, are not often practical or economical solutions and that monitoring along with an early warning system and public awareness are often
better strategies. Einstein et al. (2010) presented a risk management methodology specific to rock slopes and emphasized the importance of full field,

laboratory, and numerical characterization of the failure mechanism on which the risk assessment will be based. Hermanns et al. (2013) discussed the importance of secondary hazards in the risk assessment of large rock slope failures using examples of displacement wave in Norwegian fjords.

REFERENCES

Abbruzzese, J.M., Labiouse, V., 2013. New Cadanav methodology for quantitative rock fall hazard assessment and zoning at the local scale. Landslides 11, 551–564.

Adams, J., 1982. Stress-relief buckles in the McFarland Quarry, Ottawa. Can. J. Earth Sci. 19, 1883–1887.

Agliardi, F., Crosta, G.B., 2003. High resolution three-dimensional numerical modelling of rockfalls. Int. J. Rock Mech. Mining Sci. 40, 455–471.

Agliardi, F., Crosta, G.B., Frattini, P., Malusa, M.G., 2013. Giant non-catastrophic landslides and the long-term exhumation of the European Alps. Earth Planet. Sci. Lett. 365, 263–274.

Agliardi, F., Crosta, G.B., Frattini, P., 2012. Slow rock-slope deformation. In: Clague, J.J., Stead, D. (Eds.), Landslides: Types, Mechanisms and Modeling. Cambridge University Press, pp. 207–221.

Agliardi, F., Crosta, G.B., Zanchi, A., 2001. Structural constraints on deep-seated slope deformation kinematics. Eng. Geol. 59, 83–102.

Alavi, M., 2004. Regional stratigraphy of the Zagros fold–thrust belt of Iran and its proforeland evolution. Am. J. Sci. 304, 1–20.

Alejano, L.R., Ferrero, A.M., Ramirez-Oyanguren, P., Alvarez Fernandez, M.I., 2011. Comparison of limit-equilibrium, numerical and physical models of wall slope stability. Int. J. Rock Mech. Mining Sci. 48, 16–26.

Alejano, L.R., Pons, B., Bastante, F.G., Alonson, E., Stockhausen, H.W., 2007. Slope geometry design as a means for controlling rockfalls in quarries. Int. J. Rock Mech. Mining Sci. 44, 903–921.

Allen, S.K., Gruber, S., Owens, I.F., 2009. Exploring steep bedrock permafrost and its relationship with recent slope failures in Southern Alps of New Zealand. Permafrost Periglacial Process. 20, 345–356.

Ambrosi, C., Crosta, G.B., 2011. Valley shape influence on deformation mechanism of rock slopes. In: Slope Tectonics. Geological Society, London Special Publication 351, pp. 215–233.

Ambrosi, C., Crosta, G.B., 2006. Large sackung along major tectonic features in the central Italian Alps. Eng. Geol. 83, 183–200.

Amitrano, D., Grasso, J.R., Senfaute, G., 2005. Seismic precursory patterns before a cliff collapse and critical point phenomena. Geophys. Res. Lett. 32, L08314.

Andrew, R.D., Pierson, L.A., 2012. Stabilization of rockfall. In: Turner, A.K., Shuster, R.L. (Eds.), Rockfall: Characterization and Control, pp. 468–493.

Ashby, J., 1971. Sliding and Toppling Modes of Failure in Models and Jointed Rock Slopes (M.Sc. thesis). Imperial College.

Atkinson, B.K., 1987. Fracture Mechanics of Rock. Academic Press, London, 534 pp.

Australian Geomechanics Society, 2007. Landslide risk management: guideline for landslide susceptibility, hazard and risk zoning for land use planning. Austr. Geomech. 42, 13–36.

Bachmann, D., Bouissou, S., Chemenda, A., 2004. Influence of weathering and pre-existing large scale fractures on gravitational slope failure: insights from 3-D physical modelling. Nat. Hazards Earth Syst. Sci. 4, 711–717.

Bachmann, D., Bouissou, S., Chemenda, A., 2006. Influence of large scale topography on gravitational rock mass movements: new insights from physical modeling. Geophys. Res. Lett. 33, L21406.

Bachmann, D., Bouissou, S., Chemenda, A., 2009. Analysis of massif fracturing during deep-seated gravitational slope deformation by physical and numerical modeling. Geomorphology 103, 130–135.

Badger, T.C., 2002. Fracturing within anticlines and its kinematic control on slope stability. Environ. Eng. Geosci. VIII, 19–33.

Badger, T.C., Duffy, J.D., 2012. Avoidance of rockfall areas. In: Turner, A.K., Shuster, R.L. (Eds.), Rockfall: Characterization and Control. Transportation Research Board, pp. 457–467.

Badger, T.C., Duffy, J.D., Schellengerg, K., 2012. Protection. In: Turner, A.K., Shuster, R.L. (Eds.), Rockfall: Characterization and Control. Transportation Research Board, pp. 494–524.

Ballantyne, C.K., 2002. Paraglacial geomorphology. Quat. Sci. Rev. 21, 1935–2017.

Baron, I., Supper, R., 2013. Application and reliability of techniques for landslide site investigation, monitoring and early warning – outcomes from a questionnaire study. Nat. Hazards Earth Syst. Sci. 13, 3157–3168.

Beck, A.C., 1968. Gravity faulting as a mechanism of topographic adjustment. N.Z. J. Geol. Geophys. 11, 191–199.

Beetham, R.D., Moody, K.E., Fergusson, D.A., Jennings, D.N., Waugh, P.J., 1992. Landslide development in schist by toe buckling. In: Bell, D.H. (Ed.), 6th International Symposium on Landslides. Christchurch, New Zealand, pp. 17–24.

Bigot-Cormier, F., Braucher, R., Bourles, D., Guglielmi, Y., Dubar, M., Stéphan, J.-F., 2005. Chronological constraints on processes leading to large active landslides. Earth Planet. Sci. Lett. 235, 141–150.

Blanc, A., Durville, J.-L., Follacci, J.-P., Gaudin, B., Pincent, B., 1987. Méthode de surveillance d'un glissement de terrain de très grande ampleur: La Clapière, Alpes Maritime, France. Bull. Int. Assoc. Eng. Geol. 35, 37–46.

Blikra, L.H., 2012. The Aknes rockslide, Norway. In: Clague, J.J., Stead, D. (Eds.), Landslides: Types, Mechanisms and Modeling. Cambridge University Press, pp. 323–335.

Bois, T., Bouissou, S., Jaboyedoff, M., 2012. Influence of structural heterogeneities and of large scale topography on imbricate gravitational rock slope failures: new insights from 3-D modelling and geomorphological analysis. Tectonophysics 525–529, 147–156.

Bois, T., Bouissou, S., Guglielmi, T., 2008. Influence of major inherited faults zones on gravitational slope deformation: a two-dimensional physical modelling of the La Clapière area (Southern French Alps). Earth Planet. Sci. Lett. 272, 709–719.

Bouissou, S., Darnault, R., Chemenda, A., Rolland, Y., 2012. Evolution of gravity-driven rock slope failure and associated fracturing: geological analysis and numerical modelling. Tectonophysics 526/529, 157–166.

Bozzano, F., Bretschneider, A., Esposito, C., Martino, S., Prestinizi, A., Scarascia-Mugnozza, G., 2013. Lateral spreading processes in mountain ranges: insights from analogue modelling experiment. Tectonophysics 605, 88–95.

Brady, B.H.G., Brown, E.T., 2004. Rock Mechanics: For Underground Mining. Springer.

Brideau, M.-A., Stead, D., 2012. Evaluating kinematic controls on planar translational slope failure mechanisms. Geotech. Geol. Eng. 30, 991–1011.

Brideau, M.-A., Stead, D., 2011. The influence of three-dimensional kinematic controls on rock slope stability. In: Sainsbury, D., Hart, R., Detournay, C., Nelson, M. (Eds.), Continuum and Distinct Element Modelling in Geomechanics. Melbourne, Australia.

Brideau, M.-A., McDougall, S., Stead, D., Evans, S.G., Couture, R., Turner, K., 2012. Three-dimensional distinct element modelling and dynamic runout analysis of a landslide in gneissic rock, British Columbia, Canada. Bull. Eng. Geol. Environ. 71, 467–486.

Brideau, M.-A., Yan, M., Stead, D., 2009. The role of tectonic damage and brittle rock fracture in the development of large rock slope failures. Geomorphology 103 (1), 30–49.

Brideau, M.-A., Stead, D., Couture, R., 2006. Structural and engineering geology of the east gate landslide, Purcell mountains, British Columbia, Canada. Eng. Geol. 84, 183–206.

British Standard, 1999. Code of Practice for Site Investigations, BS5930:1999.

Bruckl, E., Brunner, F.K., Lang, E., Mertl, S., Muller, M., Stary, U., 2013. The Gradenbach observatory – monitoring deep-seated gravitational slope deformation by geodetic, hydrological, and seismological methods. Landslides 10, 815–829.

Bryan, K., 1928. Niches and other cavities in sandstone at Chaco Canyon, N.M. Ann. Geomorphol. 3, 125–140.

Bryan, K., 1954. The geology of Chaco Canyon, New Mexico in relation to the life and remains of the prehistoric peoples of Pueblo Bonito. Smithsonian Miscellaneous Collections, 122, no. 7. The Smithsonian Institution, 65 pp.

Calabro, M.D., Schmidt, D.A., Roering, J.J., 2010. An examination of seasonal deformation at the Portuguese Bend Landslide, southern California, using radar interferometry. J. Geophys. Res. Earth Surf. 115, F02020.

Capitani, M., Ribolini, A., Federii, 2013. Influence of deep-seated gravitational slope deformations on landslide distributions: a statistical approach. Geomorphology 201, 127–134.

Cappa, F., Guglielmi, Y., Merrien-Soukatchoff, V., Mudry, J., Bertrand, C., Charmoille, A., 2004. Hydromechanical modeling of a large moving rock slope inferred from slope levelling coupled to spring long-term hydrochemical monitoring: example for the La Clapière landslide (Southern Alps, France). J. Hydrol. 291, 67–90.

Cassie, J.W., van Gassen, W., Cruden, D.M., 1988. Laboratory analogue of the formation of molards, cones of rock-avalanche debris. Geology 16, 735–738.

Casson, B., Delacourt, C., Allemand, P., 2005. Contribution of multi-temporal remote sensing images to characterize landslide slip surface – application to the La Clapière landslide (France). Nat. Hazards Earth Syst. Sci. 5, 425–437.

Cavers, D.S., 1981. Simple methods to analyze buckling of rock slopes. Rock Mech. 14, 87–104.

CERA, 2013. The Land Use Recovery Plan. Canterbury Earthquake Recovery Authority. New Zealand Government.

CERA, 2014. The CERA Map. Canterbury Earthquake Recovery Authority. New Zealand Government. http://maps.cera.govt.nz/ (accessed 20.01.14.).

Chau, K.T., Wong, R.H.C., Liu, J., Lee, C.F., 2003. Rockfall hazard analysis for Hong Kong based on rockfall inventory. Rock Mech. Rock Engineering 36 (5), 383–408.

Chau, K.T., Wong, R.H.C., Wu, J.J., 2002. Coefficient of restitution and rotation motions of rockfall impacts. Int. J. Rock Mech. Mining Sci. 39, 69–77.

Chemenda, A.L., Bois, T., Bouissou, S., Tric, E., 2009. Numerical modelling of the gravity-induced destabilization of a slope: the example of the La Clapière landslide southern France. Geomorphology 109, 86–93.

Chigira, M., Kiho, K., 1994. Deep-seated rockslide-avalanches preceded by mass rock creep of sedimentary rocks in the Akaishi Mountains, central Japan. Eng. Geol. 38, 221–230.

Chigira, M., Tsou, C.-Y., Matsushi, Y., Hiraishi, N., Matsuzawa, M., 2013. Topographic precursors and geological structure of deep-seated catastrophic landslides caused by Typhoon Talas. Geomorphology 201, 479–493.

Coe, J.A., Harp, E.L., 2007. Influence of tectonic folding on rockfall susceptibility, American Fork Canyon, Utah, USA. Nat. Hazards Earth Syst. Sci. 7, 1–14.

Corkum, A.G., Martin, C.D., 2004. Analysis of a rock slide stabilized with a toe-berm: a case study in British Columbia, Canada. Int. J. Rock Mech. Mining Sci. 41, 1109–1121.

Corominas, J., van Westen, C., Frattini, P., Cascini, L., Malet, J.-P., Fotopoulou, S., Catani, F., van den Eeckhaut, M., Mavrouli, O., Agliardi, F., Pitilaskis, K., Winter, M.G., Pastor, M., Ferlisi, S., Tobani, V., Hervas, J., Smith, J.T., 2014. Recommendation for the quantitative analysis of landslide risk. Bull. Eng. Geol. Environ. 73, 209–263.

Couture, R., Evans, S.G., 2006. Slow-moving disintegrating rockslide on mountain slopes. In: Evans, S.G., Scarascia Mugnozza, G., Strom, A., Hermanns, R. (Eds.), Landslide from Massive Rock Slope Failure, pp. 377–393.

Couture, R., Blais-Stevens, A., Page, A., Kock, J., Clague, J.J., Lipovsky, P.S., 2010. Landslide susceptibility, hazard and risk assessment along pipeline corridors in Canada. In: Proceedings of the 11th IAEG Congress, Auckland, New Zealand, pp. 1023–1031.

Couture, R., Evans, S.G., Polster, A., 2004. Movement and mechanisms of a complex landslide: the case of East Gate Landslide, Glacier National Park, Canada. In: Lacerda, W.A., Ehrlich, M., Fontoura, S.A.B., Sayao, A.S.F. (Eds.), Proceedings of the 9th International Landslide Symposium, pp. 1271–1278.

Crosta, G.B., Chen, H., Lee, C.F., 2004. Replay of the 1987 Val Pola landslide, Italian Alps. Geomorphology 60, 127–146.

Crosta, G.B., Agliardi, F., 2003a. Failure forecast for large rock slides by surface displacement measurements. Can. Geotech. J. 40 (1), 176–191.

Crosta, G.B., Agliardi, F., 2003b. A methodology for physically based rockfall hazard assessment. Nat. Hazards Earth Syst. Sci. 3, 407–422.

Crozier, M.J., 2010. Landslide geomorphology: an argument for recognition, with examples from New Zealand. Geomorphology 120, 3–15.

Cruden, D.M., Varnes, D.J., 1996. Landslide types and processes. In: Turner, A.K., Shuster, R.L. (Eds.), Landslides, Investigation and Mitigation. Transportation Research Board, Special Report 247, pp. 36–75.

Cruden, D.M., Farkas, J., Hutchinson, J.N., Novosad, S., Ting, W.H., Varnes, D.J., Wang, G.X., 1993. A suggested method for describing the activity of a landslide. Bull. Int. Assoc. Eng. Geol. 47, 53–57.

Cubrinovski, M., Robinson, K., Taylor, M., Hughes, M., Orense, R., 2012. Lateral spreading and its impacts in urban areas in the 2010-2011 Christchurch earthquakes. N.Z. J. Geol. Geophys. 55, 255–269.

Dahl-Jensen, T., Larsen, L.M., Pedersen, S.A.C., Perdersen, J., Jepsen, H.F., Pedersen, G.K., Nielsen, T., Pedersen, A.K., von Platen-Hallermund, F., Weng, W., 2004. Landslide and tsunami 21 November 2000 in Paatuut, West Greenland. Nat. Hazards 31, 277–287.

Dai, F.C., Lee, C.F., Deng, J.H., Tham, L.G., 2005. The 1786 earthquake-triggered landslide dam and subsequent dam-break flood on the Dadu River, southwestern China. Geomorphology 65, 205–221.

Davies, T., 2013. Averting predicted landslide catastrophes: what can be done? In: International Conference Vajont 1963–2013, Padua, Italy. Ital. J. Eng. Geol. Environ., 229–236.

Davies, T., McSaveney, M., 2012. Mobility of long-runout rock avalanches. In: Clague, J.J., Stead, D. (Eds.), Landslides: Types, Mechanisms and Modeling. Cambridge University Press, pp. 50–58.

Davies, T., McSaveney, M., Hodgson, K.A., 1999. A fragmentation-spreading model for long-runout rock avalanches. Can. Geotech. J. 36, 1096–1110.

De Blasio, F.V., 2011. Introduction to the Physics of Landslides: Lecture Notes on the Dynamics of Mass Wasting. Springer, 408 pp.

Delacourt, C., Allemand, P., Casson, B., Vadon, H., 2004. Velocity field of the "La Clapière" landslide measured by the correlation of aerial and QuickBird satellite images. Geophys. Res. Lett. 31, L15619.

Delgado, J., Vicente, F., Garcia-Tortosa, F., Alfaro, P., Estevez, A., Lopez-Sanchez, J.M., Tomas, R., Mallorqui, J.J., 2011. A deep seated compound rotational rock slide and rock spread in SE Spain: structural control and DInSAR monitoring. Geomorphology 129, 252−262.

Dellow, G., Yetton, M., Massey, C., Archibald, G., Barrell, D.J.A., Bell, D., Bruce, Z., Campbell, A., Davies, T., De Pascale, G., Easton, M., Forsyth, P.J., Gibbons, C., Glassey, P., Grant, H., Green, R., Hancox, G., Jongens, R., Kingsbury, P., Kupec, J., Macfarlane, D., McDowell, B., McKelvey, B., McCahon, I., McPherson, I., Molloy, J., Muirson, J., O'Halloran, M., Perrin, N., Price, C., Read, S., Traylen, N., Van Dissen, R., Villeneuve, M., Walsh, I., 2011. Landslides caused by the 22 February 2011 Christchurch Earthquake and management of landslide risk in the immediate aftermath. Bull. N.Z. Soc. Earthquake Eng. 44, 227−238.

Densmore, A.L., Hovius, N., 2000. Topographic fingerprint of bedrock landslides. Geology 28, 371−374.

Densmore, A.L., Anderson, R.S., McAdoo, B.G., Ellis, M.A., 1997. Hillslope evolution by bedrock landslides. Science 275, 369−372.

van Den Eeckhaut, M., Hervas, J., 2012. State of the art of national landslide databases in Europe and their potential for assessing landslide susceptibility, hazard and risk. Geomorphology 139−140, 545−558.

Di Maggio, C., Madonia, G., Vattano, M., 2014. Deep-seated gravitational slope deformation in western Sicily: controlling factors, triggering mechanisms and morphoevolutionary models. Geomorphology 208, 173−189.

Dorren, L.K.A., Berger, F., 2012. Integrating forest in the analysis and management of rockfall risks: experiences from research and practice in the Alps. In: Eberhardt, E., Froese, C., Turner, A.K., Leroueil, S. (Eds.), Landslides and Engineered Slopes: Protecting Society through Improved Understanding. CRC Press, pp. 117−128.

Dramis, F., Sorriso-Valvo, M., 1994. Deep-seated gravitational slope deformations, related land-slides and tectonics. Eng. Geol. 38, 231−243.

Dufresne, A., Davies, T.R., McSaveney, M.J., 2010. Influence of runout-path material on emplacement of the Round Top rock avalanche, New Zealand. Earth Surf. Process. Landforms 35, 190−201.

Dufresne, A., 2012. Granular flow experiments on the interaction with stationary runout path materials and comparison to rock avalanche events. Earth Surf. Process. Landforms 37, 1527−1541.

Dunlop, S.W., 2010. Rockslides in a Changing Climate: Establishing Relationship between Meteorological Conditions and Rockslides in Southwestern Norway for the Purposes of Developing a Hazard Forecast System (M.Sc. thesis). Queen's University, Kingston, Ontario.

Duperret, A., Genter, A., Martinez, A., Mortimore, R.N., 2004. Coastal chalk cliff instability in NW France: role of lithology, fracture pattern and rainfall. In: Mortimore, R.N., Duperret, A. (Eds.), Coastal Chalk Cliff Instability Geological Society, Engineering Geology Special Publications 20, pp. 33−55.

Eberhardt, E., Stead, D., Coggan, J.S., 2004. Numerical analysis of initiation and progressive failure in natural rock slopes − the 1991 Randa rockslide. Int. J. Rock Mech. Mining Sci. 41, 69−87.

Eberhardt, E., 2012. Landslide monitoring: the role of investigative monitoring to improve understanding and early warning of failure. In: Clague, J.J., Stead, D. (Eds.), Types, Mechanisms and Modeling. Cambridge University Press, pp. 222–234.

Einstein, H.H., Sousa, R.L., Karam, K., Manzella, I., Kveldsvik, V., 2010. In: Zhao, J., Labiouse, V., Dudt, J.-P., Mathier, J.-F. (Eds.), Rock Slope from Mechanics to Decision Making. Rock Mechanics in Civil and Environmental Engineering. CRC Press, pp. 2–13.

Ekstrom, G., Stark, C.P., 2013. Simple scaling of catastrophic landslide dynamic. Science 339, 1416–1419.

El Bedoui, S., Bois, T., Jomard, H., Sanchez, G., Lebourg, T., Trics, E., Guglielmi, Y., Bouissou, S., Chemenda, A., Rolland, Y., Corsini, M., Perez, J.L., 2011. Paraglacial gravitational deformation in the SW Alps: a review of field investigations, 10Be cosmogenic dating and physical modelling. In: Jaboyedoff, M. (Ed.), Slope Tectonics Geological Society of London Special Publication 351, pp. 11–25.

El Bedoui, S., Guglielmi, Y., Lebourg, T., Perez, J.-L., 2009. Deep-seated failure propagation in a fractured rock slope over 10,000 years: the La Clapière slope, the south-eastern French Alps. Geomorphology 105 (3–4), 232–238.

Erismann, T.H., Abele, G., 2001. Dynamics of Rockslides and Rockfalls. Springer, 332 pp.

Evans, S.G., 1983. Landslides in Layered Volcanic Successions with Particular Reference to the Tertiary Rocks of South Central British Columbia (Ph.D. thesis). University of Alberta, Edmonton, Alberta.

Evans, S.G., 1989. The 1946 Mount Colonel Foster rock avalanche and associated displacement wave, Vancouver Island, British Columbia. Can. Geotech. J. 26, 447–452.

Evans, S.G., Hungr, O., 1993. The assessment of rockfall hazard at the base of talus slopes. Can. Geotech. J. 30, 620–636.

Evans, S.G., Hermanns, R.L., Strom, A., Scarascia-Mugnozza, G. (Eds.), 2011. Natural and Artificial Rockslide Dams. Lectures in Earth Sciences, vol. 133, p. 662.

Evans, S.G., Roberts, N.J., Ischuk, A., Delaney, K.B., Morozova, G.S., Tutubalina, O., 2009. Landslides triggered by the 1949 Khait earthquake, Tajikistan, and associated loss of life. Eng. Geol. 108, 96–118.

Evans, S.G., Guthrie, R.H., Roberts, N.J., Bishop, N.F., 2007. The disastrous 17 February 2006 rockslide-debris avalanche on Leyte Island, Philippines: a catastrophic landslide in tropical mountain terrain. Nat. Hazards Earth Syst. Sci. 7, 89–101.

Evans, S.G., Hungr, O., Enegren, E.G., 1994. The avalanche Lake rock avalanche, Mackenzie mountains, Northwest Territories, Canada: description, dating, and dynamics. Can. Geotech. J. 31, 749–768.

Follacci, J.-P., 1987. Les mouvements du versant de la Clapière à Saint-Etienne-de-Tinée (Alpes Maritimes). Bull. Liaison Laboratoires Ponts Chaussées 150/151, 39–54.

Frattini, P., Crosta, G.B., 2013. The role of material properties and landscape morphology on landslide size distributions. Earth Planet. Sci. Lett. 361, 310–319.

Frattini, P., Crosta, G.B., Agliardi, F., 2012. Rockfall characterization and modelling. In: Clague, J.J., Stead, D. (Eds.), Landslides: Types, Mechanisms and Modeling. Cambridge University Press, pp. 267–281.

Froese, C.R., Moreno, F., 2014. Structure and components for the emergency response and warning system on Turtle Mountain, Alberta, Canada. Nat. Hazards 70, 1689–1712.

Froese, C.R., Charriere, M., Humair, F., Jaboyedoff, M., Pedrazzini, A., 2012. Characterization and management of rockslide hazard at Turtle Mountain, Alberta, Canada. In: Clague, J.J., Stead, D. (Eds.), Landslides: Types, Mechanisms and Modeling. Cambridge University Press, pp. 310–322.

Fruneau, B., Achache, J., Delacourt, C., 1996. Observation and modelling of the Saint-Étienne-de-Tinée landslide using SAR interferometry. Tectonophysics 265, 181–190.

Fukozono, T., 1990. Recent studies on time prediction of slope failure. Landslide News 4, 9–12.

Geertsema, M., Cruden, D.M., 2009. Rock movements in northeastern British Columbia. In: Landslide Processes: From Geomorphologic Mapping to Landslide Modelling. Strasbourg, France, pp. 31–36.

Geertsema, M., Cruden, D.M., 2008. Travels in the Canadian Cordillera. In: Fourth Canadian Conference on Geohazards, Quebec, Canada, pp. 383–390.

Geertsema, M., Schwab, J.W., Blais-Stevens, A., Sakals, M.E., 2009. Landslides impacting linear infrastructure in west central British Columbia. Nat. Hazards 48, 59–72.

Geertsema, M., Clague, J.J., Schwab, J.W., Evans, S.G., 2006. An overview of recent large catastrophic landslides in northern British Columbia, Canada. Eng. Geol. 83, 120–143.

Gerath, R.F., Hungr, O., 1983. Landslide terrain, Scatter river Valley, northeastern British Columbia. Geosci. Can. 10, 30–32.

Gischig, V.S., Moore, J.R., Evans, K.F., Amann, F., 2011a. Thermomechanical forcing of deep rock slope deformation: 1. Conceptual study of a simplified slope. J. Geophys. Res. 116, F04010.

Gischig, V., Amann, F., Moore, J.R., Loew, S., Eisenbreiss, H., Stempfhuber, W., 2011b. Composite rock slope kinematics at the current Randa instability, Switzerland, based on remote sensing and numerical modeling. Eng. Geol. 118, 37–53.

Gischig, V., Loew, S., Kos, A., Moore, J.R., Raetzo, H., Lemy, F., 2009. Identification of active release planes using ground based differential InSAR at the Randa rock slope instability, Switzerland. Nat. Hazards Earth Syst. Sci. 9, 2027–2038.

Glastonbury, J., Fell, R., 2008. Geotechnical characteristics of large slow, very slow, and extremely slow landslides. Can. Geotech. J. 45, 984–1005.

Goodman, R.E., Kieffer, D.S., 2000. Behaviour of rock in slopes. J. Geotech. Geoenviron. Eng. 126 (8), 675–684.

Goodman, R.E., Bray, J.W., 1976. Toppling of rock slopes. In: Specialty Conference on Rock Engineering for Foundations and Slopes. American Society of Civil Engineering, Boulder, United States, pp. 201–234.

Griffiths, H.I., Schwalb, A., Stevens, L.R., 2001. Environmental change in southwestern Iran: the Holocene ostracod fauna of Lake Mirabad. Holocene 11, 757–764.

Griffiths, J.S., Stokes, M., Stead, D., Giles, D., 2012. Landscape evolution and engineering geology: results from IAEG Commission 22. Bull. Eng. Geol. Environ. 71, 605–636.

Griffiths, J.S., Whitworth, M., 2012. Engineering geomorphology of landslides. In: Clague, J.J., Stead, D. (Eds.), Landslides: Types, Mechanisms and Modeling. Cambridge University Press, pp. 172–186.

Gruber, S., Hoelzle, M., Haeberli, W., 2004. Permafrost thaw and destabilization of Alpine rock wall in the hot summer of 2003. Geophys. Res. Lett. 31, L13504.

Gudmundsson, A., 2011. Rock Fractures in Geological Processes. Cambridge University Press, 578 pp.

Guglielmi, Y., Cappa, F., 2010. Regional-scale relief evaluation and large landslides: insights from geomechanical analyses in the Tinée Valley (southern French Alps). Geomorphology 117, 121–129.

Guglielmi, Y., Cappa, F., Binet, S., 2005. Coupling between hydrogeology and deformation of mountainous rock slopes: insights from La Clapière area (southern Alps, France). Comp. Rend. Geosci. 337, 1154–1163.

Guglielmi, Y., Vengeon, J.M., Bertrand, C., Mudry, J., Follacci, J.P., Giraud, A., 2002. Hydro-chemistry: an investigation tool to evaluate infiltration into large moving rock masses (case study of La Clapière and Séchilienne alpine landslides). Bull. Eng. Geol. Environ. 61, 311–324.

Gunzburger, Y., Laumonier, B., 2002. Origine tectonique du pli supportant le glissement de terrain de la Clapière (North-Ouest du massif de l'Argentera-Mercantour, Alpes du Sud, France) d'après l'analyse de la fracturation. Comp. Rend. Géosci. 334, 415–422.

Guthrie, R.H., Friele, P., Allstadt, K., Roberts, N., Evans, S.G., Delaney, K.B., Roche, D., Clague, J.J., Jakob, M., 2012. The 6 August 2010 Mount Meager rock slide-debris flow, Coast Mountains, British Columbia: characteristics, dynamics, and implications for hazard and risk assessment. Nat. Hazards Earth Syst. Sci. 12, 1–18.

Guzzetti, F., Mondini, A.C., Cardinali, M., Fiorucci, F., Santangelo, M., Chang, J.-T., 2012. Landslide inventory maps: new tools for an old problem. Earth-Sci. Rev. 112, 42–66.

Guzzetti, F., Reichenbach, P., Wieczorek, G.F., 2003. Rock fall hazard and risk assessment in Yosemite Valley, California, USA. Nat. Hazards Earth Syst. Sci. 3, 491–503.

Hadley, J.B., 1978. Madison Canyon rockslide, USA. In: Voight, B. (Ed.), Rockslides and Avalanches, pp. 167–180.

Haied, A., Kondo, D., 1997. Strain localization in Fontainebleau Sandstone: macroscopic and microscopic investigations. Int. J. Rock Mech. Mining Sci. 34. Paper 161.

Hancox, G.T., 2008. The 1979 Abbotsford landslide, Dunedin, New Zealand: a retrospective look at its nature and causes. Landslide 5, 177–188.

Hancox, G.T., Perrin, N.D., 2011. Report on Landslide Reconnaissance Flight on 24 February 2011 Following the Mw 6.3 Christchurch Earthquake of 22 February 2011. GNS Science Immediate Report LD8, 44 pp.

Hancox, G.T., Perrin, N.D., 2009. Green Lake landslide and other giant and very large postglacial landslides in Fiordland, New Zealand. Quat. Sci. Rev. 28, 1020–1036.

Harrison, J.V., Falcon, N.L., 1936. Gravity collapse structures and mountain ranges as exemplified in southwestern Iran. Quart. J. Geol. Soc. Lond. 92, 91–102.

Harrison, J.V., Falcon, N.L., 1937. The Saidmarreh landslip, southwest Iran. Geogr. J. 89, 42–47.

Harrison, J.V., Falcon, N.L., 1938. An ancient landslip at Saidmarreh in southwestern Iran. J. Geol. 46, 296–309.

Hart, A.B., Hearn, G.J., 2013. Landslide assessment for land use planning and infrastructure management in the Paphos District of Cyprus. Bull. Eng. Geol. Environ. 72, 173–188.

Hawley, P.M., Martin, D.C., Acott, C.P., 1986. Failure mechanics and design considerations for footwall slope. CIM Bull. 79, 47–53.

Heim, A., 1932. Landslides and Human Lives (N. Skermer, Trans.). BiTech Publisher.

Hencher, S.R., Lee, S.G., Carter, T.G., Richards, L.R., 2011. Sheeting joints: characterisation, shear strength and engineering. Rock Mech. Rock Eng. 44, 1–22.

Henderson, I.H.C., Saintot, A., 2011. Regional Spatial Variations in Rockslide Distribution from Structural Geology Ranking: An Example from Storfjorden, Western Norway. Slope Tectonics Geological Society, London, Special Publication 351, 79–95.

Hermanns, R.L., Oppikofer, T., Anda, E., Blikra, L.H., Bohme, M., Bunkholt, H., Crosta, G.B., Dahle, H., Devoli, G., Fischer, L., Jaboyedoff, M., Loew, S., Saetre, S., Yugsi Molina, F.X., 2013. Hazard and risk classification for large unstable rock slopes in Norway. In: International Conference Vajont 1963–2013, Padua, Italy Ital. J. Eng. Geol. Environ., 245–254.

Hermanns, R.L., Longva, O., 2012. Rapid rock-slope failures. In: Clague, J.J., Stead, D. (Eds.), Landslides: Types, Mechanisms and Modeling. Cambridge University Press, pp. 59–70.

Hewitt, K., Clague, J.J., Orwin, J.F., 2008. Legacies of catastrophic rock slope failures in mountain landscapes. Earth Sci. Rev. 87, 1−38.

Hippolyte, J.-C., Bourles, D., Leanni, L., Braucher, R., 2012. ^{10}Be ages reveal >12 ka of gravitational movement in a major sackung of the Western Alps (France). Geomorphology 171−172, 139−153.

Hodge, R.A.L., Freeze, R.A., 1977. Groundwater flow systems and slope stability. Can. Geotech. J. 14, 466−476.

Hoek, E., 2007. Chapter 17: blasting damage in rock. In: Practical Rock Engineering. www.rocscience.com (accessed 29.05.13.).

Hoek, E., 1999. Putting numbers to geology − an engineer's viewpoint. Quart. J. Eng. Geol. 32, 1−19.

Hoek, E., Bray, E.T., 1981. Rock Slope Engineering, third ed. Routledge.

Hoek, E., Brown, E.T., 1997. Practical estimates of rock mass strength. Int. J. Rock Mech. Mining Sci. 34, 1164−1186.

Hoek, E., Bray, J.W., Boyd, J.M., 1973. The stability of a rock slope containing a wedge resting on two intersecting discontinuities. Quart. J. Eng. Geol. 6, 1−55.

Holm, K., Jakob, M., 2009. Long rockfall runout, Pascua Lama, Chile. Can. Geotech. J. 46, 225−230.

Hsu, K.J., 1975. Catastrophic debris streams (sturzstroms) generated by rockfalls. GSA Bull. 86, 129−140.

Hu, X.-Q., Cruden, D.M., 1993. Buckling deformation in the Highwood Pass, Alberta, Canada. Can. Geotech. J. 30, 276−286.

Huang, A.Y.-L., Montgomery, D.R., 2013. Topographic location and size of earthquake- and typhoon-generated landslides, Tachia River, Taiwan. Earth Surf. Process. Landforms 39, 414−418.

Huang, B., Yin, Y., Wang, S., Liu, G., 2014. Analysis of waves generated by the Zhaojun bridge rockfall in Xingshan county, Three Gorges reservoir, on December 28, 2012. In: Sassa, K., Canuti, P., Yin, Y. (Eds.), Landslide Science for a Safer Geoenvironment, pp. 609−614.

Huang, B., Yin, Y., Liu, G., Wang, S., Chen, X., Zhitao, H., 2012. Analysis of waves generated by Gongjiafang, landslide in Wu Gorge, Three Gorges reservoir, on November 23, 2008. Landslides 9, 395−405.

Hungr, O., 1995. A model for the runout analysis of rapid flow slides, debris flows and avalanches. Can. Geotech. J. 32, 610−623.

Hungr, O., Amann, F., 2011. Limit equilibrium of asymmetric laterally constrained rockslides. Int. J. Rock Mech. Mining Sci. 48, 748−758.

Hungr, O., Evans, S.G., 1988. Engineering evaluation of fragmental rockfall hazard. In: Bonnard, C. (Ed.), Proceedings of the 5th International Symposium on Landslides. Balkema, Lausanne, Switzerland, pp. 685−690.

Hungr, O., Evans, S.G., Bovis, M.J., Hutchinson, J.N., 2001. A review of the classification of landslides of the flow type. Environ. Eng. Geosci. VII, 221−238.

Hungr, O., Leroueil, S., Picarelli, L., 2014. Varnes classification of landslide types an update. Landslides 11, 167−194.

Hutchinson, J.N., 1988. Morphological and geotechnical parameters of landslides in relation to geology and hydrogeology. In: Bonnard, C. (Ed.), Proceedings of the 5th International Symposium on Landslides. Balkema, Lausanne, Switzerland, pp. 3−35.

Hutchinson, J.N., Bromhead, E.N., Lupini, J.F., 1980. Additional observations on the Folkestone Warren landslides. Quart. Eng. Geol. 13, 1−31.

Intrieri, E., Gigli, G., Casagli, N., Nadim, F., 2013. Landslide early warning system: toolbox and general concepts. Nat. Hazards Earth Syst. Sci. 13, 85−90.

Intrieri, E., Gigli, G., Mugnai, F., Fanti, R., Casagli, N., 2012. Design and implementation of a landslide early warning system. Eng. Geol. 2012, 124−136.

ISRM, 1981. Basic geotechnical description of rock masses. Int. J. Rock Mech. Mining Sci. Geomech. Abstr. 18, 87−110.

Jaboyedoff, M., Penna, I., Pedrazzini, A., Baron, I., Crosta, G.B., 2013. An introductory review on gravitational-deformation induced structures, fabrics and modelling. Tectonophysics 605, 1−12.

Jaboyedoff, M., Dudt, J.P., Labiouse, V., 2005. An attempt to refine rockfall hazard zoning based on the kinematic energy, frequency and fragmentation degree. Nat. Hazards Earth Syst. Sci. 5, 621−632.

Jackson, L.E., 2002. Landslides and landscape evolution in the Rocky Mountains and adjacent Foothills area, southwestern Alberta, Canada. Rev. Eng. Geol. 15, 325−344.

Jaeger, J.C., Cook, N.G.W., Zimmerman, R.W., 2007. Fundamentals of Rock Mechanics, fourth ed. Blackwell Publishing. 475 pp.

Jomard, H., Lebourg, T., Guglielmi, T., Tric, E., 2010. Electrical imaging of sliding geometry and fluids associated with a deep seated landslide (La Clapière, France). Earth Surf. Process. Landforms 35, 588−599.

Jomard, H., Lebourg, T., Tric, E., 2007. Identification of the gravitational boundary in weathered gneiss by geophysical survey: La Clapière landslide (France). J. Appl. Geophys. 62, 47−57.

Jørstad, F.A., 1968. Waves generated by landslides in Norwegian fjords and lakes. Norweg. Geotech. Inst. Publ. 79, 13−32.

Judd, N.M., 1954. The Material Culture of Pueblo Bonito. Smithsonian Miscellaneous Collections, 124. Smithsonian Institution, Washington, 398 pp.

Judd, N.M., 1959. The Braced-up Cliff at Pueblo Bonito. Annual Report of the Smithsonian Institution.

Judd, N.M., 1964. The Architecture of Pueblo Bonito. Smithsonian Miscellaneous Collections, 147 no. 1. Smithsonian Institution, Washington, 349 pp.

Kalenchuk, K., Hutchinson, D., Diederichs, M., 2013. Geomechanical interpretation of the Downie Slide considering field data and three-dimensional numerical modelling. Landslides 10, 737−756.

Kalenchuk, K., Hutchinson, D.J., Diederichs, Moore, D., 2012. Downie slide, British Columbia, Canada. In: Clague, J.J., Stead, D. (Eds.), Landslides: Types, Mechanisms and Modeling. Cambridge University Press, pp. 345−358.

Kalsnes, B., Nadim, F., 2012. Safeland: changing pattern of landslides risk and strategies for its management. In: Sassa, K., Rouhban, B., Briceno, S., McSaveney, M., He, B. (Eds.), Landslides: Global Risk Preparedness. Springer-Verlag, pp. 95−114.

Keefer, D.K., 1984. Landslides caused by earthquakes. Geol. Soc. Am. Bull. 95, 406−421.

Keur, J.Y., 1933. A Study of Primitive Indian Engineering Methods Pertaining to Threatening Rock.

Kieffer, D.S., 1998. Rock Slumping: A Compound Failure Mode of Jointed Hard Rock Slopes (Ph.D. thesis). University of California at Berkeley.

Kieffer, D.S., 2003. Rotational instability of hard rock slopes. Felsbau 21 (2), 31−38.

Kinakin, D., Stead, D., 2005. Analysis of the distributions of stress in natural ridge forms: implications for the deformation mechanisms of rock slope and the formation of a sacking. Geomorphology 65, 85−100.

Knappett, J., Craig, R.F., 2012. Craig's Soil Mechanics. Spon Press.

Korup, O., Wang, G., 2014. Landslide dams. In: Davies, T.R.H. (Ed.), Landslide Hazard and Disasters. Elsevier, pp. 241–261.

Korup, O., 2004. Geomorphic implication of fault zone weakening: slope instability along the Alpine Fault, South Westland to Fiordland. N.Z. J. Geol. Geophys. 47, 257–267.

Korup, O., Clague, J.J., Hermanns, R.L., Hewitt, K., Strom, A.L., Weidinger, J.T., 2007. Giant landslides, topography, and erosion. Earth Planet. Sci. Lett. 261, 578–589.

Kramer, S.L., 2013. Lateral spreading. In: Bobrowsky, P.T. (Ed.), Encyclopedia of Natural Hazards, p. 623.

Lambert, S., Bourrier, F., 2013. Design of rockfall protection embankments: a review. Eng. Geol. 154, 77–88.

Lan, H., Martin, D.C., Lim, H.C., 2007. Rockfall analyst: a GIS extension for three-dimensional and spatially distributed rockfall hazard modelling. Computer Geosci. 33, 262–279.

Lebourg, T., Hernandez, M., Jomard, H., El Bedoui, S., Bois, T., Zerathe, S., Tric, E., Vidal, M., 2011. Temporal evolution of weathered cataclastic material in gravitational faults of the La Clapière deep-seated landslide by mechanical approach. Landslides 8, 241–252.

Leine, R.I., Schweizer, A., Christen, M., Glover, J., Bartelt, Gerber, W., 2013. Simulation of rockfall trajectories with consideration of rock shape. Multibody Syst. Dyn. 32, 241–271.

Legros, F., 2002. The mobility of long-runout landslides. Eng. Geol. 63, 301–331.

Lekson, S.H., 1984. Great Pueblo Architecture of Chaco Canyon, New Mexico. National Park Service, U.S. Department of the Interior, 299 pp.

Leroueil, S., Locat, A., Eberhardt, E., Kovacevic, N., 2012. Progressive failure in natural and engineered slopes. In: Eberhardt, E., Froese, C., Turner, A.K., Leroueil, S. (Eds.), Landslides and Engineered Slopes: Protecting Society through Improved Understanding. CRC Press, pp. 31–46.

Linden, K.V., 1989. The Portuguese bend landslide. Eng. Geol. 27, 301–373.

Loew, S., Strauhal, T., 2013. Pore pressure distributions in brittle translational rockslides. In: International Conference Vajont 1963–2013, Padua, Italy Ital. J. Eng. Geol. Environ., 181–191.

Loew, S., Gishig, V., Moore, J.R., Keller-Signer, A., 2012a. Monitoring of potentially catastrophic rockslides. In: Eberhardt, E., Froese, C., Turner, A.K., Leroueil, S. (Eds.), Landslides and Engineered Slopes: Protecting Society through Improved Understanding. CRC Press, pp. 101–116.

Loew, S., Gishig, V., Willenberg, H., Alpiger, A., Moore, J.R., 2012b. Randa: kinematics and driving mechanisms of a large complex rockslide. In: Clague, J.J., Stead, D. (Eds.), Landslides: Types, Mechanisms and Modeling. Cambridge University Press, pp. 297–309.

MacFarlane, D.F., 2009. Observations and predictions of the behaviour of large, slow-moving landslides in schist, Clyde Dam reservoir, New Zealand. Eng. Geol. 109, 5–15.

Mansour, M.F., Morgenstern, N.R., Martin, C.D., 2011. Expected damage from displacement of slow-moving slides. Landslides 8, 117–131.

Manzella, I., Labiouse, V., 2009. Flow experiments with gravel and blocks at small scale to investigate parameters and mechanisms involved in rock avalanches. Eng. Geol. 109, 146–158.

Massey, C.I., 2010. The Dynamics of Reactivated Landslides: Utiku and Taihape, North Island, New Zealand (Ph.D. thesis). Durham University.

Massey, C.I., Petley, D.N., McSaveney, M.J., 2013. Pattern of Movement in Reactivated Landslides, vol. 159, 1–19.

Massey, C.I., McSaveney, M.J., Herron, D., Lukovic, B., 2012a. Canterbury Earthquakes 2010/11 Port Hills Slope Stability: Pilot Study for Assessing Life-safety Risk from Rockfalls (Boulder Rolls). GNS Science Consultancy. Report 2011/311.

Massey, C.I., McSaveney, M.J., Yetton, M.D., Heron, D., Lukovic, B., Bruce, Z.R.V., 2012b. Canterbury Earthquakes 2010/11 Port Hills Slope Stability: Pilot Study for Assessing Life-safety Risk from Cliff Collapse. GNS Science Consultancy. Report 2012/57.

Massey, C.I., McSaveney, M.J., Lukovic, B., Heron, D., Ries, W., Moore, A., Carey, J., 2012c. Canterbury Earthquakes 2010/11 Port Hills Slope Stability: Life-safety Risk from Rockfalls (Boulder Rolls) in the Port Hills. GNS Science Consultancy. Report 2012/123.

Massey, C.I., McSaveney, M.J., Heron, D., 2012d. Canterbury Earthquakes 2010/11 Port Hills Slope Stability: Life-safety Risk from Cliff Collapse in the Port Hills. GNS Science Consultancy. Report 2012/124.

Matheson, D.S., Thomson, S., 1973. Geological implications of valley rebound. Can. J. Earth Sci. 10, 961−978.

Matthews, W.H., McTaggart, K.C., 1978. Hope rockslides, British Columbia, Canada. In: Voight, B. (Ed.), Rockslides and Avalanches. Elsevier, pp. 259−275.

Matsuoka, N., Sakai, H., 1999. Rockfall activity from an alpine cliff during thawing periods. Geomorphology 28, 309−328.

Mavrouli, O., Fotopoulou, S., Pitilakis, K., Zuccaro, G., Corominas, J., Santo, Cacace, F., De Gregoria, D., Di Crescezo, G., Foerster, E., Ulrich, T., 2014. Vulnerability assessment for reinforced concrete buildings exposed to landslides. Bull. Eng. Geol. Environ. 73, 265−289.

McCalpin, J.P., Irvine, J.R., 1995. Sackungen at the Aspen Highlands Ski area, Pitkin County, Colorado. Environ. Eng. Geosci. 1, 277−290.

McColl, S.T., 2014. Landslide causes and triggers. In: Davies, T.R.H. (Ed.), Landslide Hazard and Disasters. Elsevier, pp. 17−42.

McColl, S.T., 2012. Paraglacial rock-slope stability. Geomorphology 153−154, 1−16.

McDougall, S., Hungr, O., 2004. A model for the analysis of rapid landslide motion across three-dimensional terrain. Can. Geotech. J. 41, 1084−1097.

McLean, M.C., Brideau, M.-A., Augustinus, P.C., 2014. Deep-seated gravitational slope deformation in greywacke rocks of the Tararua Range, North Island, New Zealand. In: Proceedings of the IAEG Congress, Torino, Italy.

McMurtry, G.M., Fryer, G.J., Tappin, D.R., Wilkinson, I.P., Williams, M., Fietzke, J., Garbe-Schoenberg, D., Watts, P., 2004. Megatsunami deposits on Kohala volcano, Hawaii, from flank collapse of Mauna Loa. Geology 32, 741−744.

Mencl, V., 1966. Mechanics of landslides with non-circular sliding surfaces with special reference to the Vaiont Slide. Géotechnique 16, 329−337.

Michoud, C., Bazin, S., Blikra, L.H., Derron, M.-H., Jaboyedoff, M., 2013. Experiences from site-specific landslide early warning systems. Nat. Hazards Earth Syst. Sci. 13, 2659−2673.

Miller, D.J., 1960. Giant Wave in Lituya Bay, Professional Paper 345. U.S. Geological Survey, Washington, 86 pp.

Montgomery, D.R., Brandon, M.T., 2002. Topographic controls on erosion rates in tectonically active mountain ranges. Earth Planet. Lett. 201, 481−489.

Moon, V.G., 1993. Microstructural controls on the geomechanical behaviour of ignimbrite. Eng. Geol. 35, 19−31.

Moon, V., Jayawardane, J., 2004. Geomechanical and geochemical changes during early stages of weathering of Karamu Basalt, New Zealand. Eng. Geol. 74, 57−72.

Moro, M., Chini, M., Saroli, M., Atzori, S., Stramondo, S., Salvi, S., 2011. Analysis of large, seismically induced, gravitational deformations imaged by high-resolution COSMO-SkyMed synthetic aperture radar. Geology 39, 527−530.

Mortimore, R.N., Duperret, A., 2004. Coastal Chalk Cliff Instability. Geological Society Engineering Geology Special Publication No. 20.

Müller, 1964. The rock slide in the Vaiont valley. Rock Mech. Eng. Geol. 2, 148—212.

Murphy, W., 2014. Coseismic landslides. In: Davies, T.R.H. (Ed.), Landslide Hazard and Disasters. Elsevier, pp. 91—129.

Nichols, T.C., 1980. Rebound, its nature and effect on engineering works. Quart. J. Eng. Geol. Hydrogeol. 13, 133—152.

Oberlander, T.M., 1965. The Zagros Streams; a New Interpretation of Transverse Drainage in an Orogenic Zone. Syracuse Geographical Series. Syracuse University Press.

Panek, T., Silhan, K., Taborik, P., Hradecky, J., Smolkova, V., Lenart, J., Brazdil, R., Kasickova, L., Pazdur, A., 2011. Catastrophic slope failure and its origins: case of the May 2010 Girova Mountain long-runout rockslide (Czech Republic). Geomorphology 130, 352—364.

Panizzo, A., de Girolamo, P., di Ridio, M., Maistri, A., Petaccia, A., 2005. Great landslide events in Italian artificial reservoirs. Nat. Hazards Earth Syst. Sci. 5, 733—740.

Pariseau, W.G., 2012. Design Analysis in Rock Mechanics, second ed. Taylor and Francis. 682 pp.

Pedrazzini, A., Jaboyedoff, M., Loye, A., Derron, M.-H., 2013. From deep seated slope deformation to rock avalanche: destabilization and transportation models of the Sierre landslide (Switzerland). Tectonophysics 605, 149—198.

Pepper, G.H., 1920. Pueblo Bonito. Anthropological Papers of the American Museum of Natural History, vol. 27. The American Museum of Natural History, New York, 398 pp.

Petley, D.N., 2012. Remote sensing techniques and landslides. In: Clague, J.J., Stead, D. (Eds.), Types, Mechanisms and Modeling. Cambridge University Press, pp. 159—171.

Petley, D.N., Higuchi, T., Petley, D.J., Bulmer, M.H., Carey, J., 2005. Development of progressive landslide failure in cohesive materials. Geology 33, 201—204.

Petley, D.N., Bulmer, M.H., Murphy, W., 2002. Patterns of movement in rotational and translational landslides. Geology 30, 719—722.

Persson, L., Goransson, M., 2005. Mechanical quality of bedrock with increasing ductile deformation. Eng. Geol. 81, 42—53.

Philip, H., Ritz, J.-F., 1999. Gigantic paleolandslide associated with active faulting along the Bogd fault (Gobi-Altay, Mongolia). Geology 27, 211—214.

Pierson, L.A., 2012. Rockfall hazard rating systems. In: Turner, A.K., Shuster, R.L. (Eds.), Rockfall: Characterization and Control. Transportation Research Board, pp. 56—71.

Pierson, L.A., Turner, A.K., 2012. Implementation or rock slope management systems. In: Turner, A.K., Shuster, R.L. (Eds.), Rockfall: Characterization and Control. Transportation Research Board, pp. 72—112.

Pineda, J.A., Alonso, E.E., Romero, E., 2014. Environmental degradation of claystones. Géotechniques 64, 64—82.

Pirulli, M., 2009. The Thurwieser rock avalanche (Italian Alps): description and dynamic analysis. Eng. Geol. 109, 80—92.

Piteau, D.R., 1977. Regional slope-stability controls and engineering geology of the Fraser Canyon British Columbia. Rev. Eng. Geol. 3, 85—112.

Piteau, D.R., Mylrea, F.H., Blown, I.G., 1978. Downie slide, Columbia river, British Columbia, Canada. In: Voight, B. (Ed.), Rockslides and Avalanches. Elsevier, pp. 365—392.

Poisel, R., Angerer, H., Pollinger, M., Kalcher, T., Kittl, H., 2009. Mechanics and velocity of the Larcheberg-Galgenwald landslide (Austria). Eng. Geol. 109, 57—66.

Pola, A., Crosta, G.B., Fusi, N., Castellanza, R., 2014. General characterization of the mechanical behaviour of different volcanic rocks with respect to alteration. Eng. Geol. 169, 1—13.

Pollard, D.D., Aydin, A., 1988. Progress in understanding jointing over the past century. GSA Bull. 100, 1181—1204.

Ponomareva, V.V., Melekestsev, I.V., Dirksen, O.V., 2006. Sector collapses and large landslides on Late Pleistocene-Holocene volcanoes in Kamchatka, Russia. J. Volcanol. Geotherm. Res. 158, 117−138.

Porter, M., Morgenstern, N., 2013. Landslide risk evaluation. Canadian technical guidelines and best practices related to landslides: a national initiative for loss reduction. Geol. Surv. Can. Open File 7312, 19 p.

Prebble, W.M., 1995. Landslides in New Zealand. In: Bell, D.H. (Ed.), Sixth International Symposium on Landslides, Christchurch, New Zealand, pp. 2101−2123.

Price, N.J., Cosgrove, J.W., 1990. Analysis of Geological Structures. Cambridge University Press.

Procter, J.N., Cronin, S.J., Zernack, A.V., 2009. Landscape and sedimentary response to catastrophic debris avalanches, western Taranaki, New Zealand. Sediment. Geol. 220, 271−287.

Ramsay, J.G., 1967. Folding and Fracturing of Rocks. McGraw-Hill, 568 pp.

Read, S.A.L., Millar, P.J., White, T., Riddolls, B.W., 1981. Geomechanical properties of New Zealand soft sedimentary rocks. In: International Symposium on Weak Rock, Tokyo, Japan, pp. 33−38.

Reiche, P., 1937. Toreva-block − a distinctive landslide type. J. Geol. 45, 538−548.

Roa, K., 2003. Nature and origin of toreva remnants and volcaniclastics from La Palma, Canary Islands. J. Volcanol. Geotherm. Res. 125, 191−214.

Roberts, N.J., McKillop, R., Hermanns, R.L., Clague, J.J., Oppikofer, T., 2014a. Preliminary global catalogue of displacement waves from subaerial landslides. In: Proceedings of the Third World Landslide Forum, vol. 3, p. 6.

Roberts, N.J., Rabus, B., Hermanns, R.L., Guzm'an, M.-A., Clague, J.J., Minaya, E., 2014b. Recent landslide activity in La Paz, Bolivia. In: Proceedings of the Third World Landslide Forum, vol. 3, p. 7.

Roberts, N.J., Evans, S.G., 2013. The gigantic Seymareh (Saidmarreh) rock avalanche, Zagros fold − thrust Belt, Iran. J. Geol. Soc. Lond. 170, 685−700.

Roberts, N.J., McKillop, R.J., Lawrence, M.S., Psutka, J.F., Clague, J.J., Brideau, M.-A., Ward, B.C., 2013. Impacts of the 2007 landslide-generated wave in Chehalis Lake, Canada. In: Margottini, C., Canuti, P., Sassa, K. (Eds.), Proceedings of the Second World Landslide Forum, vol. 6, pp. 133−140.

Roberts, N.J., Nadim, F., Kalnes, B., 2009. Quantification of vulnerability to natural hazards Georisk. Assess. Manage. Risk Eng. Syst. Geohazards 3, 164−173.

Robinson, T.R., Davies, T.R.H., Reznichenko, N.Y., De Pascale, G.P., 2014. The extremely long-runout Komansu rock avalanche in the Trans Alai range, Pamir Mountains, southern Kyrgyzstan. Landslides. http://dx.doi.org/10.1007/s10346-014-0492-y.

Rose, N.D., Hungr, O., 2007. Forecasting potential rock slope failure in open pit mines using the inverse-velocity method. Int. J. Rock Mech. Mining Sci. 44, 308−320.

Rosser, N.J., Brain, M.J., Petley, D.N., Lim, M., Norman, E.C., 2013. Coastline retreat via progressive failure of rocky coastal cliffs. Geology 41, 939−942.

Rosser, N., Lim, M., Petley, D., Dunning, S., Allison, R., 2007. Patterns of precursory rockfall prior to slope failure. J. Geophys. Res. 112, F04014.

Ryan, T.M., Pryor, P.R., 2000. Designing catch benches and interramp slopes. In: Hustrulid, W.A., McCarter, M.K., Van Zyl, D.J.A. (Eds.), Slope Stability in Surface Mining. Society for Mining, Metallurgy, and Exploration, Inc., pp. 27−38.

Saito, M., Uezawa, H., 1961. Failure of soil due to creep. In: 5th International Conference on Soil Mechanics and Foundation Engineering.

Santa Barbara County, 2014. Article II Coastal Zoning Ordinance. County of Santa Barbara Planning and Development, 362 pp.

Santi, P.M., Russell, C.P., Higgins, J.D., Spriet, J.I., 2009. Modification and statistical analysis of the Colorado Rockfall Hazard rating system. Eng. Geol. 104, 55–65.

Savage, W.Z., Varnes, D.J., 1987. Mechanics of gravitational spreading of steep-sided ridges ("sackung"). Bull. Int. Assoc. Eng. Geol. 35, 31–36.

Scheidegger, A.E., 1973. On the prediction of the reach and velocity of catastrophic landslides. Rock Mech. 5, 231–236.

Scheingross, J.S., Minchew, B.M., Mackey, B.H., Simons, M., Lamb, M.P., Hensley, S., 2013. Fault-zone controls on the spatial distribution of slow-moving landslides. GSA Bull. 125, 473–489.

Schumm, S.A., Chorley, R.J., 1964. The fall of threatening rock. Am. J. Sci. 262 (9), 1041–1054.

Seed, H.B., Wilson, S.D., 1967. The Turnagain heights landslide, Anchorage Alaska. ASCE J. Soil Mech. Found. Div. 93, 325–353.

Selby, M.J., 1993. Hillslope Materials and Processes, second ed. Oxford University Press. 451 pp.

Senfaute, G., Duperret, A., Lawrence, J.A., 2009. Micro-seismic precursory cracks prior to rock-fall on coastal chalk cliffs: a case study at Mesnil-Val, Normandie, NW France. Nat. Hazards Earth Syst. Sci. 9, 1625–1641.

Sepulveda, S.A., Serey, A., Lara, M., Pavez, A., Rebolledo, S., 2010. Landslides induced by the April 2007 Aysen Fjord earthquake, Chilean Patagonia. Landslides 7, 483–492.

Shea, T., van Wyk de Vries, B., 2008. Structural analysis and analogue modeling of the kinematics and dynamics of rockslide avalanches. Geosphere 4, 657–686.

Siebert, L., 1984. Large volcanic debris avalanches: characteristics of source areas, deposits, and associated eruptions. J. Volcanol. Geotherm. Res. 22, 163–197.

Skempton, A.W., 1985. Residual strength of clays in landslides, folded strata and the laboratory. Géotechnique 35, 3–18.

Smithyman, M., Eberhardt, E., Hungr, O., 2009. Characterization and numerical modelling of intermittent slope displacement and fatigue in deep-seated fractured crystalline rock slopes. In: 62nd Canadian Geotechnical Conference, Halifax, Canada, pp. 504–511.

Soldati, M., 2013. Deep-seated gravitation slope deformation. In: Bobrowsky, P.T. (Ed.), Encyclopedia of Natural Hazards, pp. 151–155.

Stacey, T.R., 2006. Consideration of failure mechanism associated with rock slope stability and consequences for stability analysis. J. Southern Afr. Instit. Mining Metall. 106, 485–493.

Stacey, R.T., 1974. The behaviour of two and three dimensional model rock slopes. Quart. J. Eng. Geol. 8, 67–72.

Stead, D., Coggan, J.S., 2012. Numerical modeling of rock-slope instability. In: Clague, J.J., Stead, D. (Eds.), Landslides: Types, Mechanisms and Modeling. Cambridge University Press, pp. 144–158.

Stead, D., Eberhardt, E., 2013. Understanding the mechanics of large landslides. In: International Conference on Vajont, Padova, Italy Ital. J. Eng. Geol. Environ., 85–112.

Stead, D., Jaboyedoff, M., Coggan, J.S., 2012. Rock slope characterization and geomechanical modelling. In: Eberhardt, E., Froese, C., Turner, A.K., Leroueil, S. (Eds.), Landslides and Engineered Slopes: Protecting Society through Improved Understanding. CRC Press, pp. 83–100.

Stead, D., Eberhardt, E., Coggan, J.S., 2006. Developments in the characterization of complex rock slope deformation and failure using numerical modelling techniques. Eng. Geol. 83, 217–235.

Stoffel, M., Tiranti, D., Huggel, C., 2014. Climate change impacts of mass movements − case studies from the European Alps. Sci. Total Environ. 493, 1255–1266.

Stoffel, M., Lievre, I., Monbaron, M., Perret, S., 2005. Seasonal timing of rockfall activity on a forested slope at Taschgufer (Swiss Alps) − a dendrochronological approach. Z. Geomorphol. 49, 89–106.

Strom, A.L., Korup, O., 2006. Extremely large rock slides and rock avalanches in the Tien Shan Mountains, Kyrgyzstan. Landslides 3, 125–136.

Taig, T., Massey, C., Webb, T., 2012. Canterbury Earthquakes Port Hills slope stability: principles and criteria for the assessment of risk from slope instability in the Port Hills. GNS Science Consultancy Report 2011/319, Christchurch.

Tang, C., Hudson, J.A., 2010. Rock Failure Mechanism, Explained and Illustrated. CRC Press, 322 pp.

Tatard, L., Grasso, J.R., 2013. Controls of earthquake faulting style on near field landslide triggering: the role of coseismic slip. J. Geophys. Res. Solid Earth 118 (6), 2953–2964.

Terzaghi, K., 1962. Stability of steep slopes on hard unweathered rock. Géotechnique 12, 251–270.

Thompson, R.C., 1982. Relationship of Geology to Slope Failures in Soft Rock of the Taihape-Mangaweka Area, Central North Island, New Zealand (Ph.D., thesis). The University of Auckland, Auckland, New Zealand.

Thompson, N., Bennett, M.R., Petford, N., 2010. Development of characteristic volcanic debris avalanche deposit structures: new insight from distinct element simulations. J. Volcanol. Geotherm. Res. 192, 191–200.

Tommasi, P., Verrucci, L., Campedel, P., Veronese, L., Pettinelli, E., Ribacchi, R., 2009. Buckling of high natural slopes: the case of Lavini di Marco (Trento-Italy). Eng. Geol. 109, 93–108.

Tonnelier, A., Helmstetter, A., Malet, J.-P., Schmittbuhl, J., Corsini, A., Joswig, M., 2013. Seismic monitoring of soft-rock landslides: the Super-Sauze and Valoria case studies. Geophys. J. Int. 193, 1515–1536.

Townsend, D.B., Rosser, B., 2012. Canterbury Earthquakes 2010/2011 Port Hills Slope Stability: Geomorphology Mapping for Rockfall Risk Assessment. GNS Science Consultancy Report 2012/15, 112 pp.

Turner, A.K., Duffy, J.D., 2012. Evaluation of rockfall mechanics. In: Turner, A.K., Shuster, R.L. (Eds.), Rockfall: Characterization and Control. Transportation Research Board, pp. 285–333.

Ui, T., Yamamoto, H., Suzuki-Kamata, H., 1986. Characterization of debris avalanche deposits in Japan. J. Volcanol. Geotherm. Res. 29, 231–243.

Updike, R.G., Olsen, H.W., Schmoll, H.R., Kharaka, Y.K., Stoke, K.H., 1988. Geologic and geotechnical conditions adjacent to the Turnagain Heights Landslide, Anchorage, Alaska. U.S. Geol. Surv. Bull. 1817, 40.

Varnes, D.J., 1978. Slope movement types and processes. In: Schuster, R.L., Krizek, R.J. (Eds.), Landslides, Analysis and Control. Special Report 176: Transportation Research Board, pp. 11–33.

Varnes, D.J., Radbruch-Hall, D., Savage, W.Z., 1989. Topographic and structural conditions in areas of gravitational spreading of ridges in the western United States. U.S. Geol. Surv. Profession. Paper 1496, 28.

Vengeon, J.-M., Giraud, A., Antoine, P., Rochet, L., 1999. Contribution à l'analyse de la déformation et de la rupture des grands versants rocheux en terrain cristallophyllien. Can. Geotech. J. 36, 1123–1136.

Vlcko, J., 2004. Extremely slow slope movements influencing the stability of Spis Castle, UNESCO site. Landslides 1, 67–71.

Voight, B., 1989. A relation to describe rate-dependent material failure. Science 243 (4888), 200–203.

Voight, B., Voight, B.A., Voight, M.A., Voight, L., 1988a. Failure Prediction for Soil and Rock Slopes in Protection of Architectural and Archaeological Monuments and Historical Sites. A. A. Balkema, Rotterdam, 253–259.

Voight, B.A., Voight, M.A., Voight, L., Voight, B., 1988b. Architecture and Engineering at Pueblo
    Bonito, New Mexico; 11th−12th and 20th Century Hazard Forecasts and Mitigation Practices
    Compared. A. A. Balkema, Rotterdam, 65−73.

Volkwein, A., Schellenberg, K., Labiouse, V., Agliardi, F., Berger, F., Bourrier, F., Dorren, L.K.A.,
    Gerber, Jaboyedoff, M., 2011. Rockfall characterisation and structural protection − a review.
    Nat. Hazards Earth Syst. Sci. 11, 2617−2651.

Ward, S.N., Day, S., 2003. Ritter Island Volcano − lateral collapse and the tsunami of 1888.
    Geophys. J. Int. 154, 891−902.

Watson, A.D., Moore, D.P., Stewart, T.W., 2004. Temperature influence on rock slope movements
    at Checkerboard Creek. In: Lacerda, W., Ehrlich, M., Fontoura, S.A.B., Sayao, A.S.F. (Eds.),
    Ninth International Symposium on Landslides. Rio de Janeiro, Brazil, pp. 1293−1298.

Watson, R.A., Wright Jr., H.E., 1969. The Saidmarreh landslide, Iran. In: Schumm, S.A.,
    Bradley, W.C. (Eds.), United States Contributions to Quaternary Research. Geological Society
    of America, Special Papers 123, pp. 115−139.

Watson, R.A., Wright Jr., H.E., 1963. Landslides on the east flank of the Chuska Mountains,
    Northeastern New Mexico. Am. J. Sci. 261, 525−548.

van Westen, C.J., Castellanos, E., Kuriakose, S.L., 2008. Spatial data for landslide susceptibility,
    hazard, and vulnerability assessment: an overview. Eng. Geol. 102, 112−131.

van Western, C.J., van Asch, T.W.J., Soeters, R., 2006. Landslide hazard and risk zonation − why
    is it still so difficult? Bull. Eng. Geol. Environ. 65, 167−184.

Wieczorek, G.F., 1996. Landslide triggering mechanisms. In: Turner, A.K., Schuster, R.L. (Eds.),
    Landslides Investigation and Mitigation, pp. 76−90.

Wieczorek, G.F., Jager, S., 1996. Triggering mechanisms and depositional rates of
    postglacial slope-movement processes in the Yosemite Valley, California. Geomorphology 15,
    17−31.

Willenberg, H., Eberhardt, E., Loew, S., McDougall, S., Hungr, O., 2009. Hazard assessment and
    runout analysis for an unstable rock slope above an industrial site in the Riviera valley,
    Switzerland. Landslides 6, 111−116.

Williams, A.L., Prebble, W.M., 2010. Subsurface investigation of defects; proposed sanitary
    landfill, Whitford, New Zealand. In: Williams, A.L., Pinches, G.M., Chin, C.Y.,
    McMorran, T.J., Massey, C.I. (Eds.), Proceedings of the 11th IAEG Congress. Auckland,
    New Zealand, CD-ROM.

Wise, M., Moore, G., vanDine, D., 2004. Landslide Risk Case Studies in Forest Development
    Planning and Operations. British Columbia Ministry of Forests, Resource Branch, Victoria,
    BC. Land Management Handbook, No. 56.

Wolter, A., Stead, D., Clague, J.J., 2013a. A morphologic characterisation of the 1963 Vajont Slide,
    Italy, using long-range terrestrial photogrammetry. Geomorphology 206, 147−164.

Wolter, A., Havaej, M., Zorzi, L., Stead, D., Clague, J.J., Ghirotti, M., Genevois, R., 2013b.
    Exploration of the kinematics of the 1963 Vajont Slide, Italy, using a numerical modelling
    toolbox. In: International Conference Vajont 1963−2013, Padua, Italy Ital. J. Eng. Geol.
    Environ., 599−612.

van Wyk de Vries, B., Delcamp,, A., 2014. Volcanic landslides and sector collapse. In:
    Davies, T.R.H. (Ed.), Landslide Hazard and Disasters. Elsevier, pp. 131−157.

Wyllie, D.C., 2014. Calibration of rock fall modeling parameters. Int. J. Rock Mech. Mining Sci.
    67, 170−180.

Wyllie, D.C., Mah, C.W., 2004. Rock Slope Engineering, Civil and Mining, fourth ed. Spon Press.
    431 pp.

Yamada, M., Kumagai, H., Matsushi, Y., Matsuzawa, T., 2013. Dynamic landslide processes revealed by broadband seismic records. Geophys. Res. Lett. 40. http://dx.doi.org/10.1002/grl. 50437.

Yoon, W.S., Jeong, U.J., Kim, J.H., 2002. Kinematic analysis for sliding failure of multi-faced rock slopes. Eng. Geol. 67, 51—61.

van Zeyl, D., 2009. Evaluation of Subaerial Landslide Hazards in Knight Inlet and Howe Sound, British Columbia (M.Sc., thesis). Simon Fraser University, 184 pp.

Zischinsky, U., 1966. On the deformation of high mountain slopes. In: Proceedings of the International Society of Rock Mechanics Symposium. Lisbon, Portugal, pp. 179—185.

Chapter 4

Coseismic Landslides

Bill Murphy

Leeds University, UK

ABSTRACT

Coseismic landslides are important effects of strong earthquakes and one of the most poorly understood. Such landslides are driven by inertial forces acting on slopes and can vary in size from individual rock blocks of a few m^3 in volume to catastrophic rock avalanches of many Mm^3, which can claim many thousands of lives. The capability of an earthquake to trigger a landslide is a function of energy arriving at the site, which is itself related to earthquake magnitude and epicentral distance. However, as much as these two parameters can be shown to exert a limit on the ability of the earthquake to cause landslides, it is clear from the spatial distribution of landslides caused by earthquakes that these are not the controlling factors. A detailed consideration of ground accelerations recorded during earthquakes indicates that threshold or yield accelerations are required to initiate movement on slopes. These yield accelerations are strongly dependent on the degree of stability of the slope under ambient stress conditions, development of pore-water pressures during shaking, and degradation of strength and stiffness under cyclic loading. Traditional engineering analyses have tended to focus on horizontal accelerations; however, a growing body of evidence points to the importance of vertical accelerations in seismic slope stability. In addition to the polarity of ground motions, there is an important influence from topographic amplification of shaking. This can give rise to ground motions that are 5−10 times higher than free-field accelerations, meaning that ground shaking at slope crests and ridges can be considerably higher than the vibration at the foot, or behind the crest, of the slope. This variation in ground motions means that shaking recorded on free-field instruments is unlikely to be representative of ground motions driving slope instability. Therefore, several significant unknowns severely limit understandings of this group of dynamic slope processes.

4.1 SEISMICALLY TRIGGERED LANDSLIDES

4.1.1 Introduction

Earthquake-induced landslides (EILs) are those that are triggered by the transient stresses associated with strong seismic shaking. The processes that cause earthquakes can also result in changes to ground conditions, such as

Landslide Hazards, Risks, and Disasters. http://dx.doi.org/10.1016/B978-0-12-396452-6.00004-5

groundwater level changes, which may also result in landslides. The latter are often difficult to ascribe with confidence to single events, and these will be excluded from the current discussion. In addition to the transient loads imposed on slopes by strong shaking, the generation of excess pore fluid pressures by shaking may result in the liquefaction of either surface material or a stratum within slopes. These may develop into landslides depending on the gradient of the slope in question. In fact, the first question to be asked by the investigator considering the seismic stability of slopes is "will this slope-forming material lose a significant component of its shear strength when shaken?" In rock slopes, the answer to this question is generally "no." However, in loose soils and some clay deposits, answering that question can be difficult. The range of behaviors of slopes formed in liquefiable soils, collapsible materials, or sensitive clays is large and will receive only cursory consideration in this chapter.

The study of EILs is not new. After the "Great Earthquake of Calabria" Giovanni Vivenzio (1788) described the effects of the earthquake, as did Sir William Hamilton who was at the time British Envoy to the Court of Naples. Hamilton's (1783) letter to the Royal Society (based largely on a letter from Count Francesco Ippolito) and his later descriptions of the landslides and what were probably liquefaction effects of the earthquake were in effect the first published systematic study of landslides caused by earthquakes. This was largely reprinted in Lyell's classic text "Principles of Geology."

While it is difficult to interpret the detail of the mechanics of these landslides, what is clear is that there were far-reaching effects. Descriptions exist of houses being moved largely intact on large sliding blocks, as well as diagrams of what we now recognize as sand volcanoes testifying to liquefaction, and the formation of some 200 or so new lakes due to valley-blocking landslides. One of the subsequent consequences of this was the reintroduction of malaria into Calabria. Many of the impacts described by Vivenzio, Hamilton, and Count Ippolito in 1783 would be instantly recognizable to scientists and engineers working in Kashmir, Pakistan, in 2005 and Wenchuan, China, in 2008.

4.1.2 A Note on Terminology

The study of EILs crosses geology, geotechnical engineering, and seismology. Therefore, terms in common usage in one field may not be readily known to other readers. Table 4.1 provides a glossary of terms that includes the abbreviation, full name, and a brief description (along with a citation as appropriate). Where possible, the same units will be used throughout this chapter. For example, peak ground acceleration (PGA) will always be given as a fraction of the acceleration due to gravity (i.e., a PGA of 0.5 g is ~ 4.9 m/s^2).

4.1.3 Landslides Caused by Earthquakes

In his classic 1984 paper, David K. Keefer of the US Geological Survey carried out a systematic and painstaking review of landslides caused by

TABLE 4.1 Glossary and Abbreviations

Abbreviation	Term	Brief Description
α	Slip surface angle	The angle of the sliding surface in a planar stability model, or an angle of thrust.
A_{crit}	Critical acceleration	The threshold acceleration for a slope to accumulate displacement. Sometimes given the notation K_c.
D_N	Newmark displacement	The calculated displacement, measured in centimeters, of a sliding block subjected to a forced vibration.
D	Duration	The time over which shaking occurs. This is often expressed as a "bracketed" duration that is the duration between the first and last occurrences of a given acceleration (normally 0.01 or 0.05 g).
d	Epicentral distance	The distance between the earthquake epicenter and the landslide (or any given site of interest). Given in kilometers. Some authors will give this the notation of "r". "d" is also sometimes used to describe "hypocentral" distance, that is, the distance in kilometers between the hypocenter and the site of interest.
	Epicenter	The geographic location of an earthquake. Normally given in latitude and longitude.
F	Factor of safety	The ratio of shear strength to shear stress in slope-stability calculations for the ambient (nonseismic) state of stress.
F_{dyn}	Dynamic factor of safety	The ratio of shear strength to shear stress for seismic loading conditions.
f	Frequency	The number of cycles of loading per second of an earthquake ground motion. Frequency is expressed in Hertz (Hz). The same data are often presented as "period" ($1/f$), which is measured in seconds.
	Focal depth	The depth from the ground surface to the point of nucleation of an earthquake.
G	Shear modulus	Elastic property. G_0 is the shear modulus at the start of shaking. G_{max} is the maximum observed value of shear modulus.
g	Acceleration due to gravity	Normally given as a decimal fraction, for example, 0.5 g is a ground motion of 4.9 m/s^2.

Continued

TABLE 4.1 Glossary and Abbreviations—cont'd

Abbreviation	Term	Brief Description
	Hypocenter	The location of an earthquake in three dimensions—normally described in terms of latitude, longitude, and focal depth. This is normally the point of rupture initiation.
I_a	Arias intensity	A description of earthquake ground motions, which is a function of the middle 90 percent of the energy recorded arriving at a site. Expressed as meters per second.
k	Seismic coefficient	Additional stress component applied to a pseudostatic slope-stability analysis to represent additional loading from earthquake shaking. k_h refers to the seismic coefficient in a horizontal direction and k_v refers the loads applied in the vertical direction.
M	Earthquake magnitude	This is a description of the amount of energy released by an earthquake. Numerous different magnitude scales are used. The most common are M_M, M_S, and M_L. These are moment magnitude, surface wave magnitude, and local magnitude
PGA	Peak ground acceleration	The maximum ground motion recorded on an acceleration–time (accelerogram) history.
PGD	Peak ground displacement	The maximum ground motion recorded on a displacement–time (seismogram) history.
PGV	Peak ground velocity	The maximum ground motion recorded on a velocity–time history.
SH	Horizontally polarized S wave	An S wave that is traveling in a vertical direction with the particle motion polarized into the horizontal plane.
V_s	Shear wave velocity	The speed at which shear waves propagate through the ground. It is often used as an indicator of ground conditions on site. A shear wave velocity of 760 m/s is the boundary between engineering soils and engineering rocks.

earthquakes from a database of 40 historical earthquakes. These included a wide range of different seismic events from small shocks to great earthquakes. Keefer was able to establish that there were limiting distances, areas, and numbers of landslides that could be triggered by earthquakes of any given magnitude. Casual consideration of the physics of the process indicates that this should be relatively unsurprising. To cause a landslide, sufficient energy must arrive at the slope to be capable of overcoming the resisting forces acting on the slope. The more the energy arriving at the site, the greater the work that can be done. Correspondingly, the greater the instability under ambient stress conditions of the slope, the less the energy required to induce movement. This is of course a very simplistic view as the attenuation of energy with distance is a three-dimensional effect; it is also a function of simple geometric spreading and the ability of the crust, and especially the shallow subsurface, to dampen seismic waves, and depends on amplification and deamplification by topography and structure.

To consider landslides in a broadly similar mechanical grouping, Keefer (1984) divided them into three main categories: These are falls (Figures 4.1 and 4.2) and topples; disrupted landslides; and coherent landslides (e.g., Figure 4.3), flow slides, and lateral spreads (see Table 4.2 for illustration).

Figure 4.4 shows the curves produced by Rodriguez et al. (1999) by adding additional data to the work of Keefer (1984). These relationships show a remarkable degree of agreement even with an expanded database of earthquake events, many of which contain data of much higher precision. Some revision is made in the limiting curve for the area affected by landslides as a function of earthquake magnitude (given in terms of moment magnitude here) (Figure 4.4(a)), which may be a result of better observations of landslides triggered at large distances than of any change in the physical process. It is worth noting that for more recent (post 1970s) earthquakes, epicentral data are much more tightly constrained. This means that the geography of EILs, which is largely governed by the geography of earthquake sources and the size of the earthquakes those sources are capable of producing, is more precisely known. However, in the absence of sloping ground and liquefiable materials, even the

FIGURE 4.1   Single block rockfall near La Libertad triggered by the January 13, El Salvador earthquake. *Photograph courtesy of Professor Julian Bommer, Imperial College London.*

FIGURE 4.2   Multiple block rockfall in Taroko Gorge, Taiwan, triggered by the $M = 6.8$ Hualien earthquake in 2002. Note the landslide scar on the slope to the right (south) of the dust cloud. *Photograph courtesy of Dr Mark Bulmer.*

FIGURE 4.3   Landslide damage in the town of Comosagua triggered by the 2001 $M = 7.6$ El Salvador earthquake.

biggest of seismic events will not produce landslides. This alludes to the principal limitation on studies that investigate the geography of landslides caused by earthquakes. That is, that no real account is taken of the in situ stability conditions of the slopes subject to shaking. Keefer (1984) and, subsequently, Rodriguez et al. (1999) also established limiting slope angles below which landslides could not happen. For example, rockfalls require a rock slope of $>42°$, but lateral spreads can be triggered at slope angles as low as $2°$. This gives the geoscientist or engineer a first-order approach to considering whether

TABLE 4.2 Landslide Classifications

Landslide Types (Cruden and Varnes, 1996)	Keefer's (1984) Grouping of Landslides	Examples
Rockfalls (single and multiple blocks) Rock topples (single and multiple blocks) Debris falls and topples Earth falls and topples.	Disrupted slides and falls.	 Debris topple in pyroclastic ashfall deposits, in San Salvador (El Salvador earthquake $M = 7.6$).
Earth and mud slumps Debris slumps Rock slumps; (rotational. i.e., circular or mostly circular slip surfaces).	Coherent slides Slumps that fragment can fall into the "disrupted slides and falls" category.	 Earth slump on the Ruamahanga River near Masterton, New Zealand (1855 Wairarapa earthquake $M = 8.2$).

Continued

TABLE 4.2 Landslide Classifications—cont'd

Landslide Types (Cruden and Varnes, 1996)	Keefer's (1984) Grouping of Landslides	Examples
Earth and mud slides Debris slides Rock slides (translational, i.e., planar failure surface).	Coherent slides Slides that fragment can fall into the "disrupted slides and falls" category.	 Debris slide near the town of Comasagua, El Salvador (El Salvador earthquake $M = 7.6$).
"Complex" landslides (e.g., rock, debris, and earth (not volcanic) avalanches).	Coherent slides.	 Rock avalanche triggered by the 1928 Murchison ($M = 7.9$) earthquake, New Zealand.
"Complex" landslides—flow slides and lateral spreads.	Lateral spreads and flow slides or occasionally coherent slides.	 Turnagain heights landslide, Anchorage, USA—an "earth" flow in the Bootlegger cove clay due to liquefaction of sand lenses in the succession (1964 Alaska earthquake, $M = 9.3$) (photograph courtesy National Geophyisical Data Center, NOAA.).

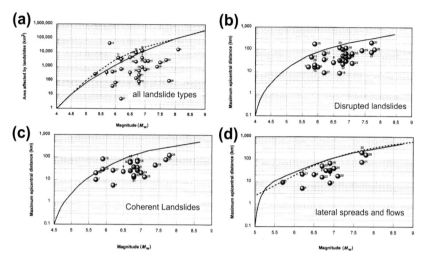

FIGURE 4.4   Updated bounding curves for different types of landslides. (a) For all landslide types; (b) For disrupted landslides; (c) For coherent landslides and (d) For lateral flows and spread i.e. liquefaction related, landslides. *From Rodriguez et al. (1999).*

a slope is susceptible to landsliding during earthquakes. Therefore, the investigator can add a second question to that outlined in opening paragraph, that is: "Is there a slope of sufficient gradient for failure to occur?"

One of the challenges encountered when investigating seismically triggered landslides is establishing with certainty that a landslide has been triggered by an earthquake. In many areas, landslide inventories are only made after a catastrophic event (e.g., earthquake or storm); therefore, identifying what was caused by an earthquake and what existed previously presents a challenge for the investigator. Geomorphologically, little discernible difference exists between a landslide caused by an earthquake and one that is caused by rainfall or some other triggering phenomenon. A useful illustration of this is the Ruamahanga Slump shown in Table 4.2. This landslide shows all the characteristics of a rotational landslide (a slump according to Cruden and Varnes, 1996) and nothing to indicate that any external triggering mechanism is involved. In fact, the presence of the Ruamahanga River at the toe of the landslide would indicate erosion and toe unloading as the principal driving mechanism in the absence of historical documentary evidence for the seismic trigger (Grapes, 1988). Back-analysis has been used to infer the stability of slopes prior to a known earthquake to establish a causative link (e.g., Jibson and Keefer, 1993). A number of challenges need to be overcome, which include reconstructing the prior slope geometry. Given the role of water pressures in the stability of slopes, trying to extend such analyses to landslides formed under a different morphoclimatic regime is challenging. Landslides that have occurred on very low slope angles have been used as evidence of a seismic trigger for paleoseismology (e.g., Lagerbäck and Sundh, 2008).

4.1.4 Geological Materials and EILs

No geological unit occurs that is not prone to seismically induced landslides, depending on the size of the ground motions. However, the type of material and mass (rock masses are a combination of rock material and the fracture systems) will constrain the type of landslide that can occur. This is reflected in Table 4.2, which describes the different types of landslides and links those to the characterization of landslides by Keefer (1984). Again it can be seen that landslides caused by earthquakes cover the full spectrum of landslide phenomena, although rockfalls tend to be the most common landslide type during earthquakes. Combinations of material types produce some interesting challenges in dealing with seismically induced landslides. The presence of liquefiable materials in the stratigraphic succession can lead to significant differences of slope failure type (e.g., Turnagain Heights, Anchorage—see Table 4.2). Assessing the stability conditions of slopes composed of more than one material, especially liquefiable soils, can be difficult. If a material is considered as liquefaction prone, a range of tools is available to assist with the assessment of this phenomenon (Youd et al., 2001).

Keefer's (1984) work identified that threshold levels exist for different groups of landslides. Figures 4.4(b—d) are the updated versions of the original bounding curves from Rodriguez et al. (1999). Figure 4.4(b) shows the maximum distance between the epicenter and disrupted landslides. Such landslides can be triggered by events $M > 4$ (although there is some anecdotal evidence that the $M = 3.8$ Lleyn Peninsula earthquake in May 2012 may have triggered single rock block falls on the North Wales, UK coast; Boon personal communication, 2013). At $M > 4.5$ (Figure 4.4(c)), coherent slides will occur, and at $M > 5.0$, lateral spreads and flows (Figure 4.4(d)) can be induced. It is worth noting that soil and debris avalanches will generally not occur until $M > 6.0$ and rock avalanches will generally only happen at $M > 6.5$. This can be thought of as a function of energy available to do work. The energy released during an earthquake E is

$$\log E = 1.5 M_W + 11.8, \qquad (4.1)$$

where E is the energy released by the earthquake and M_W is the moment magnitude (Kanamori, 1977). This means that rockfalls and topples can be triggered on susceptible slopes with small amounts of energy but to mobilize the large volume of material required to cause rock and debris avalanches requires significantly greater amounts of energy. An important caveat can be put on this, which is that energy can be filtered and focused, resulting in higher ground motions than those that would have been expected based on attenuation models. Therefore, the proximity of the earthquake to the slope both spatially and in depth (Figure 4.5) is a major influence on the distribution of landslides caused by earthquakes. It is also worth noting that in earthquakes of $M > c. 5$ thinking of the seismic source as a "point source" is likely to be incorrect, and

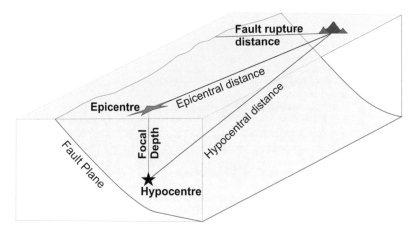

FIGURE 4.5   Basic spatial relationships between source and site.

any analysis should be carried out with respect to the proximity of the fault rupture, and not to the epicenter.

4.2 MECHANICS OF EARTHQUAKE-INDUCED LANDSLIDES

4.2.1 Earthquake Energy, Magnitude, and Attenuation

Given that energy release and the attenuation of energy between the source and landslides are significant in the occurrence of landslides triggered by earthquakes, it will be worth considering these concepts. The relationship shown in Eqn (4.1) highlights that the energy released by an earthquake is large, but finite. This energy is transmitted through the crustal rock by seismic (elastic) waves. Those waves can be described like any waveform in that they have a velocity, frequency, and a wavelength. A number of different types of waves are generated by earthquakes. The main ones are P (compressional) waves; S (shear) waves; and surface waves (Love and Raleigh waves mainly). The partitioning of energy among these different wave types will largely be a function of the characteristics of the seismic source. The characteristics of these waves can be found in any basic textbook on seismology (e.g., Aki and Richards, 2002). The speed at which waves travel through the ground is a function of the density and elastic properties of the material through which they are passing; the frequency content is commonly a function of the seismic source and the epicentral distance, although it can be significantly modified by the local geological conditions. Note that wave velocity should not be confused with particle velocity recorded in the ground, which is one measure of earthquake shaking.

Engineering applications generally do not distinguish between different wave forms when considering ground motions. Figure 4.6 shows an acceleration−time history recorded at the Heathcote Valley School during the

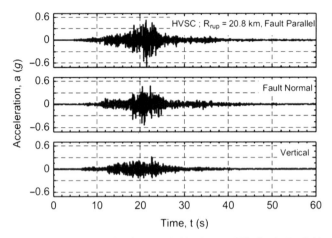

FIGURE 4.6   Acceleration–time histories recorded in the Port Hills for the Darfield earthquake. Resolved into vibrations recorded in the horizontal plane orientations parallel and normal to the orientation of the Greendale Fault. The bottom trace shows the shaking recorded in the vertical direction. "HVSC" is the instrument reference - Heathcote Valley Primary School. *From Bradley (2012).*

September 4, $M = 7.0$ Darfield earthquake, New Zealand. The epicentral distance between the source and site was 20.8 km. The accelerogram in Figure 4.6 shows the first wave arrival at about 5 s on the trace. The peak horizontal ground acceleration (PGA_H) is recorded at 0.61 g (~ 21 s on the trace) and the peak vertical ground acceleration (PGA_V) is recorded at 0.31 g (~ 22 s on the trace shown in Figure 4.6). The significant duration (the duration over which the middle 90 percent of the energy arrived at the station; Dobry et al., 1978) was 14.3 s. The difficult parameter to evaluate from these data is the dominant frequency of shaking. It is therefore common in earthquake engineering and seismology to present data not with respect to time but with respect to frequency to understand the pattern of applied loading during shaking. These are known as response spectra, or sometimes Fourier spectra (depending on how they are computed). It is less common to use data from such analyses to understand the forces acting on landslide systems. Bradley (2012) showed that the ground motion record shown in Figure 4.6 has modes of vibration at approximately 8, 5, and 3 Hz, with some differences between the fault-normal and fault-parallel directions. For safety factor analysis (i.e., assessing the stability of the slope), a single parameter, such as PGA, seismic coefficient, or Arias Intensity, is commonly used as a description of seismic forces.

Earthquake shaking is commonly measured as ground acceleration (e.g., Figure 4.6), ground velocity, and ground displacement. These are normally measured in three orientations: two perpendicular horizontal motions and a vertical motion. Probably the most commonly used indicator of shaking is the peak horizontal ground acceleration. Although this is a crude measure of ground motions, it is commonly used in engineering applications. The decay of PGA with distance is normally described by an attenuation model. In practical terms, such attenuation models can be used to predict PGA or peak ground velocity (PGV). The terms that are most commonly used in attenuation models are magnitude and epicentral distance. These models generally do not take account of different wave forms and are normally derived from ground motion observations made on flat ground, so they may not be representative of ground motions occurring on slopes. Attenuation models normally assume elastic behavior and may only apply to a specific range of vibration frequencies. It is also obvious that such models are region specific, and care should be exercised when applying such models to other regions. Other issues that are not normally considered are fault directivity and nonlinear soil response. Figure 4.7 shows attenuation models for two earthquakes in Taiwan. Both attenuation models show that PGV and PGA decay with distance. It is also evident that a reasonably large scatter occurs within the observations.

Earthquake ground motions can be divided into near-field and far-field motions. Far-field ground motions are normally used for the derivation of attenuation models. It is recognized that the prediction of ground motions in close proximity to seismic sources is difficult. Given the significant distance between source and site, the different velocities of different seismic waves means that P waves arrive first, then S waves, and subsequently surface waves. However, in the near field, which is often taken as being <25 km, all the different seismic waves arrive largely at the same time, and the patterns of

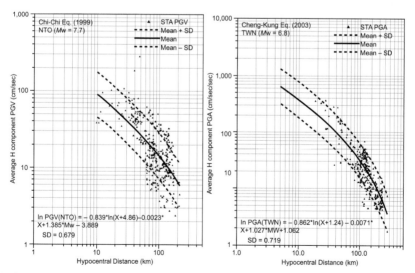

FIGURE 4.7   Attenuation models showing the decay of PGV (left) and PGA (right) for two Taiwanese earthquakes. The dashed lines show the mean relationship ±1 standard deviation. The scatter of data show the significant variations of such relationships, which highlights the complexity of factors influencing site response. PGV, peak ground velocity; PGA, peak ground acceleration. *From Liu and Tsai (2005). Image courtesy Seismological Society of America.*

constructive and destructive interference make prediction of ground motions difficult. Additionally, near-field ground motions are normally dominated by high-frequency (short period) vibrations.

Generally, high-frequency modes of vibration are attenuated with distance, and near-field records are dominated by high-frequency, large amplitude, short duration, shaking, whereas far-field ground motions are characterized by low frequency, low amplitude, and long durations of shaking. Long-duration earthquakes often have many cycles of applied loading and will often give rise to liquefaction, even if the accelerations are relatively small.

In addition to epicentral distance, site geology and the geometry of the soil−rock interface have a significant impact on the ground motions recorded at a site. Different geological materials have different resonant frequencies (or fundamental periods of vibration), which means that energy arriving at the site will focus energy into distinct frequency bands. This is known as soil amplification. Additionally, the soil−rock interface can result in the focusing of waves onto specific areas. These issues were illustrated by Hadley et al. (1991) who investigated the spatial distribution of ground motions during the 1985 $M_W = 8.2$ Michoacan earthquake, which caused a significant loss of life in Mexico City. This identified the importance of both resonance of the soil column and focusing of waves being refracted through the stratigraphy, as well as those being reflected off rockhead. It is also worth noting that soil properties are not steady throughout an earthquake. Sugito et al. (2000) observed a time-dependent change in contractive soils during the 1995 Hyogo Ken Nambu earthquake. As shaking developed, a systematic increase in pore pressure led to liquefaction and the onset of strain. The resulting drop in shear modulus (G) caused a change in the ability of these materials to transmit seismic waves. The combination of these factors means that ground shaking during earthquakes varies spatially and temporally. This is further complicated by the simple observed fact that recorded ground motions at the ground surface are higher than those recorded at depths. Therefore, the stress−free (sometimes referred to as the free-field) ground motion may not in fact be the vibration that actually drives a landslide.

Although these issues are of limited importance for rock slope failures, the challenges associated with understanding ground motions in soil slopes makes a detailed investigation of EILs difficult. This is further complicated by limited information on which to base a geological model for many landslides triggered by earthquakes. Bazzurro and Cornell (2004) noted that the uncertainty in ground conditions can have significant impacts on predicted site response with significant increases in amplification factor due to unknown factors in the geological column.

One way in which some of these problems have been addressed is to use Arias Intensity (I_a), which represents the middle 90 percent of the acceleration−time history. I_a describes ground motions at the site as a single value (measured in m/s) and acknowledges the fact that the damaging part of the earthquake is not right at the beginning or at the end of the earthquake. Arias Intensity can be described as

$$I_a = \frac{\pi}{2g} \int\limits_0^\infty a_x(t)^2 dt, \qquad (4.2)$$

where g is the acceleration due to gravity and $a_x(t)$ is the ground motion at any given point in the acceleration time history. The fact that this reduces the whole ground motion record to a single value means that an attenuation model can be used to describe the ground motions by the use of the magnitude and epicentral distance. Arias Intensity has been used extensively in the simplified Newmark Sliding block model (see Jibson, 1987, 1993) (Section 4.3).

4.2.2 Topographic Amplification and Landslides

It was recognized as early as the late nineteenth century that there was a zone of enhanced damage around ridges and hillcrests during strong earthquakes. However, the systematic study of what has become known as topographic amplification did not occur until the work of Boore (1972, 1973) and Bouchon (1973). These authors demonstrated that horizontally polarized S waves could be significantly amplified by topography, and the review of the subject by Géli et al. (1988) demonstrated that these effects could involve focusing of waves not only on individual hill slopes but also on ridges within chains. These observations were critical to the consideration of how landslides are triggered by earthquakes. The work of Ashford and Sitar (1997) and Ashford et al. (1997) quantified these observations in terms of the geometry of the slopes in question and the wavelength of the transmitted wave and presented data in terms of a normalized wavelength (h/λ where h is the height of the slope and λ is the wavelength of the incident seismic wave). Poppeliers and Pavlis (2002), in a study of SH waves (S waves that have particle motions polarized in the horizontal direction), demonstrated that slopes also have a resonant frequency (in this case ~9 Hz in sandstones; most rock masses have resonant frequencies between 8 and 20 Hz) and that this also contributes to the pattern of ground motions during earthquakes. Buech et al. (2010) also observed amplification of weak motions by topography (in this case in sandstones and mudstones) over a range of frequencies at different points on the hill slope. Gao et al. (2012) demonstrated a three-dimensional effect and attributed this to patterns of constructive and destructive interference rather than resonance. However, the latter model did not recreate the kind of deamplification effects observed at the foot of the slope indicated by field observations at the time of the landslide shown in Figure 4.2; neither the author nor his colleagues were aware of the earthquake, and there was no indication that anyone else had felt significant shaking, but clearly the ground motions at the ridge crest were strong enough to trigger this rockfall.

The significance of topographic amplification was identified in studies of landslides produced by the 1994 Northridge ($M_S = 6.7$) California, earthquake and the 1999 Chi Chi ($M_W = 7.6$), Taiwan earthquake, where different geomorphological units responded differently to forced vibration (Sepulveda et al., 2005a,b); similar geomorphological controls were observed after mapping landslides triggered by the Straits of Messina earthquake ($M_S = 7.1$) in 1908 (Murphy, 1995). Bouckovalas and Papadimitriou (2005) noted that topographic effects are likely to be significant at slope inclinations >17°. The magnitude of topographic amplification is such that ground accelerations could be between three and five times those of free-field ground motions (Faccioli et al 2002) or up to 10 times the free-field accelerations (Buech et al., 2010). This makes for a very nonuniform distribution of earthquake loading along slopes.

Consider the situation shown in Figure 4.8. The diagram shows calculated displacement time histories for different points on a hill slope subjected to a strong earthquake with intensities ranging from MCS IX–X. The calculated displacements in the north–south (NS) direction for the different elements of the terrain show that the particle displacement at the slope foot is clearly different from the amount of motion at the slope crest, and at a point behind the crest of the slope. These differential motions are a significant issue in slope stability.

Although topographic amplification has now been identified and investigated, the link to seismically triggered landslides remains complicated. What should be clear is that a single ground motion parameter, such as PGA, is a poor description of the shaking along a long slope profile and with regards to effects that extend some, normally unspecified, distance beneath the surface of the ground. The majority of numerical parametric studies of topographic amplification use relatively simple slope geometries and simple stratigraphic relationships. Figure 4.9 shows the basic slope used by Tripe et al. (2013) for a parametric study of ground motions. Contrast this with the natural slopes shown in Figure 4.10, which exhibit significant landsliding during the Chi Chi earthquake, and it is apparent that there are significant variations. It can be seen that slopes are not uniform along their length and often have a three-dimensional character that is likely to impact on the pattern of near-surface shaking. Although it is tempting to ascribe earthquake-triggered landslides to topographic amplification, the very locations where landslides tend to occur under ambient states of stress, namely, steep slopes, are those that are most likely to experience topographic amplification.

One of the most infamous seismically triggered landslides was the Huascaran rock avalanche that was triggered by the May 31, 1970 ($M = 7.75$), Peru earthquake (Figure 4.11). The landslide traveled approximately 16 km to devastate the towns of Yungay and Ranrahirca (Cluff, 1971). The landslide originated as a rockfall/topple and became a fluidized sliding mass with an estimated volume of $3-5 \times 10^7$ m³ traveling at speeds

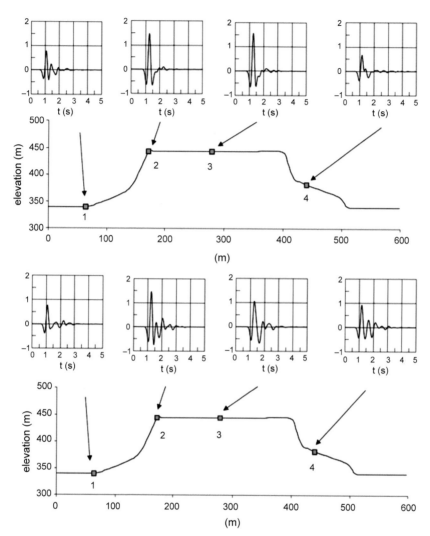

FIGURE 4.8 Displacement–time histories at selected receivers on an EW cross-section of Civita di Bagnoregio hill. Top: EW polarization of input motion and EW displacement components. Bottom: NS polarization of input motion and NS displacement components. *From Paolucci (2002).*

of up to 475 kmph and an average velocity of approximately 280 kmph. Geomorphological evidence suggests that the slide mass was airborne for parts of the track. Although the mechanics of the landslide movement are intriguing, the question of whether the shaking felt in Yungay was topographically amplified at the summit will remain unanswered. Cluff (1971) noted that a similar rock avalanche had happened in 1962 without an earthquake trigger, so clearly stability issues existed.

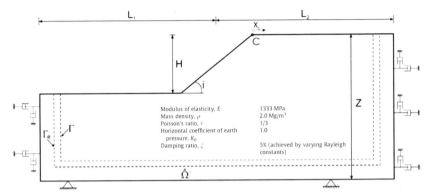

FIGURE 4.9   Simple slope used for parametric study of topographic amplification by Tripe et al. (2013).

FIGURE 4.10   Rock slides and rockfalls on the Central Cross Island Highway triggered by the 1999 Chi-Chi earthquake. Note the landslides in question initiate at significant breaks of slope.

4.2.3 Shaking and Pore Water Pressures

Topographic amplification is a significant factor on landslide activity in slopes formed in rocks. However, most soil slopes are not steep enough to cause significant topographic amplification. In soil slopes, water is the single most important factor that affects landslides. The response of water in the ground to strong shaking is twofold. First, there can be groundwater movements associated with elastic rebound. Second, there are dynamic effects associated with the transfer of elastic waves through the ground.

The issue of elastic rebound and postseismic groundwater level changes has not received a lot of attention in the context of slope stability. The

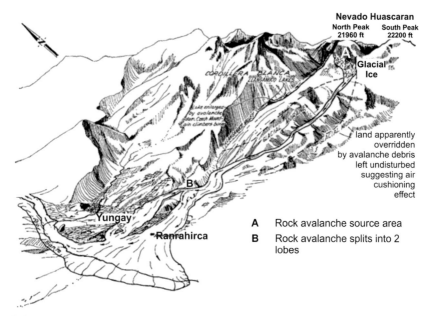

Nevado Huascaran
North Peak   South Peak
21960 ft   22200 ft

Glacial
Ice

land apparently
overridden
by avalanche debris
left undisturbed
suggesting air
cushioning
effect

Yungay

Ranrahirca

A   Rock avalanche source area
B   Rock avalanche splits into 2
     lobes

FIGURE 4.11   Sketch of the Huascaran rock avalanche in 1970. This landslide was triggered by a magnitude 7.75 Peru earthquake (Clough, 1971). *Image courtesy of Lloyd Cluff.*

principles involved are outlined by Sibson et al. (1975) and Muir-Wood and King (1993), and the effects can be substantial. Schuster and Murphy (1996) noted significant changes in groundwater after the relatively modest $M_W = 5.9$ Draney Peak earthquake in 1994. Descriptions of water changes by Vivenzio (1788) and Hamilton (1783) are sufficiently vague to be attributed to this mechanism and could in fact be related to liquefaction and dynamic compaction. The role of such changes in slope-stability studies remains difficult to assess. Any permanent changes to groundwater level will contribute toward changes in the stability conditions in slopes. However, transient effects are even more difficult to assess. Hutchinson and Del Prete (1985) reported the reactivation of the landslide at Calitri during the 1980 ($M_S = 6.7$) Irpinia earthquake in southern Italy. Contemporary accounts however remain unclear as to when the landslide started moving, in that the slip on the landslide was only observed on the morning after the earthquake, so it could in fact not have been coseismic. The analysis by Martino and Scarascia Mugnozza (2005), however, presents a convincing argument for a coseismic trigger.

Although postseismic groundwater changes may influence stability conditions for some time after an earthquake, dynamic pore-water pressures contribute significantly to landslide movement during shaking. The majority of research in this field has been carried out to investigate liquefaction, and therefore, many of the following comments apply to relatively loose sands and silty sands. What is apparent however is that with increasing loading cycles a systematic increase occurs in pore fluid pressures. The fact that P and S wave phases could be seen in dynamic pore-pressure response (Harp et al., 1984; Mavko and Harp, 1984) clearly indicates that this is a function of deformation of the soil skeleton, not an effect of the water. The details of the seismic response of soils are demonstrated by Holzer and Youd (2007). Figure 4.12 shows the ground motions displayed as a shearing stress (Figure 4.12(a)) and the strain response in the soil element (Figure 4.12(b)). The pore pressures

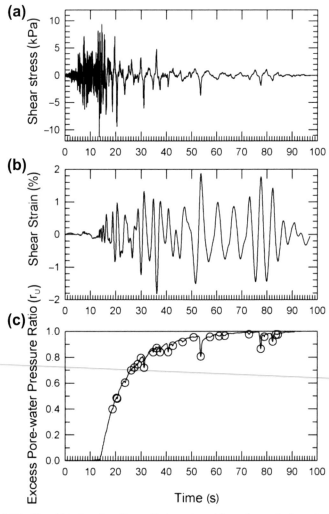

FIGURE 4.12 Time histories of north–south shear stress (a), north–south shear strain (b), and excess pore-water pressure ratio (c), pore-water pressure drops are circled. *From Holzer and Youd (2007). Image courtesy of the Seismological Society of America.*

(Figure 4.12(c)) expressed as a pore-pressure ratio ($r_u = u/\sigma'_v$, where u is the pore-water pressure and σ'_v is the effective vertical stress) increase systematically as shaking continues. Some small drops in water pressure can be seen (e.g., ~ 55 s on the record), but these appear to be associated with unusually large strains that probably allow for some partial drainage or dilation of the soil skeleton. Ultimately, the systematic increase in pore-water pressure leads to a reduction in the effective normal stress and decrease in the frictional strength mobilized in the soil.

Liquefaction-related landslides such as the failures at Turnagain Heights (the image for the flow slides part of Table 4.2, see also Figure 4.13) during the 1964 Alaska earthquake are the result of the complete loss of shear strength of loose, saturated, normally geologically young materials subjected to shaking. The landslide was triggered by an earthquake of $M_w = 9.2$. The earthquake shaking in Anchorage was up to 4 min in duration. Although referred to as a single event, this was in fact two smaller fault rupture episodes separated by about 44 s. This complexity at least partly explains the long duration of shaking. The landslide occurred in the Bootlegger Cove Clay, which is a rather

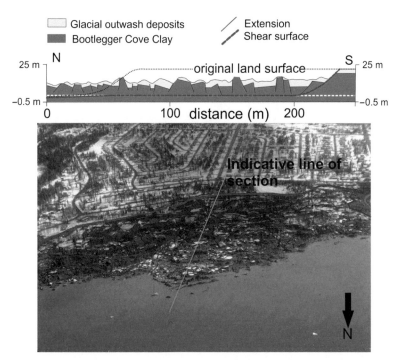

FIGURE 4.13 The Turnagain Heights landslide triggered by the 1964 Alaska Earthquake. *Photograph Karl V. Steinbrugge Collection, Earthquake Engineering Research Center, University of California, Berkeley. Section modified from Seed and Wilson (1967).*

heterogeneous geological unit (Figure 4.13) containing sand lenses that liquefied and caused the sensitive marine clays to deform (Figure 4.14).

Liquefaction has been extensively studied, and a large number of methods are available for assessing liquefaction potential, which range from geotechnical methods (e.g., the use of Standard Penetration Testing or Cone Penetration Testing) to more descriptive geological methods (e.g., studying particle size distributions, age of deposition, and saturation state). The interested reader should refer to Seed et al. (1983) and Youd et al. (2001) for more information.

What is clear however is that pore-water pressures rise during earthquakes and take some time to dissipate after shaking has ceased. The magnitudes of such rises depend on the soil skeleton, amplitude of shaking, and the number of effective loading cycles during shaking (see Hancock and Bommer, 2005 for more information). Effective loading cycles within a landslide context of slope behavior are those where the shear resistance of the slope has been overcome. The significance of this will be discussed in Section 5.3.

4.2.4 Summary

It is worth while pulling together the various strands of information presented in this section. In Section 4.2, the seismological, geotechnical, and geomorphological information has been reviewed, so it will be valuable to put those different disciplines in context.

EILs are caused by earthquake shaking. The level of that shaking is governed by the magnitude of the causative earthquake and the distance

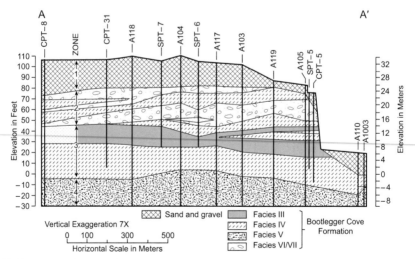

FIGURE 4.14 Cross-section through the Bootlegger Cove Clay. *(From Updike et al., 1988.)* Facies III is sensitive clay silt and silty clay. Facies IV is silty clay and clayey silt, with silty fine sand lenses; Facies V is silty clay and clayey silt, with random stones; Facies VI is silty fine sand with silt and clay and Facies VII is fine to medium sand with occasional gravel. *Image courtesy of the United States Geological Survey.*

between source and site. The ground motions recorded on site can be strongly modified by site geology and local topography, and this is a function of the frequency and wavelength of the incoming waves, the geometry of the site, and the dynamic properties of the materials. Material properties are not constant throughout the earthquake as the development of strain and pore-water pressures will result in changes in the seismic response of geological materials due to decreases in the elastic properties of soils, specifically shear modulus. This observation is particularly true of landslides associated with liquefaction. In slopes formed in rock, these problems generally do not exist. However, velocity effects also occur on rock discontinuities (Crawford and Curran, 1981), and the opening of discontinuities affects the wave propagation across joints (Pyrak-Nolte et al., 1990; Li and Ma, 2009) and through rock masses (Burjánek et al., 2010), which would modify the dynamic properties of the rock mass. It is recognized that engineering soils show hysteresis during earthquakes, but it is less clear whether rock masses show similar responses under dynamic conditions. The biggest single issue in considering the stability of rock slopes is topographic amplification. Given that natural slopes are not uniform, this is likely to result in differential shaking in three dimensions.

4.3 STABILITY ANALYSIS AND HAZARD ASSESSMENT

Given the complexities described in Section 4.2, any form of stability analysis for seismic conditions has to balance the available science against the practicalities of the available data and available methods. The key requirements for slope-stability analysis are

1. an adequate model of the ground and therefore the resisting forces—for natural slopes this equates to a good geological model and a reasonable constraint on groundwater conditions. It is seen in Section 4.2.3 that this is likely to change during earthquakes in response to pore-pressure effects or strain.
2. an adequate model of driving forces—again it can be demonstrated that this is not only not constant during shaking as the vibrations change with time, but also the dynamic response of slopes may change due to changing material properties and the geographic position along the slope.

   In reality, the calculation of actual ground motion records at a given site for stability analysis is challenging. So many factors ranging from the seismic source, through ray path effects to site-specific conditions, will influence particle amplitude that these could almost be considered as a stochastic process, albeit one bounded by empirical observation. A variety of ways exist in which the stability of slopes can be assessed, ranging from simple pseudostatic methods of analysis through to more complex finite elements and finite difference models.

4.3.1 Pseudostatic and Limit State Models

The common conceptual model for assessing the stability of a slope during an earthquake is that of a sliding block. The "block" that is the landslide mass is stable until a yield acceleration (or critical acceleration) is exceeded, at which point it starts to move. As a broad conceptual model, this is a useful way to think about EILs. However, it is one that has some significant limitations, as will be seen.

One way in which the stability of slopes can be assessed is by using a modification of limit-equilibrium modeling called pseudostatic analysis. The basics of slope-stability analysis can be found on any basic text on soil or rock engineering (Wyllie and Mah (2004) provide an excellent overview for rock slopes and outline the key principles). Pseudostatic analysis makes use of a seismic coefficient to represent the additional loading applied by earthquake shaking. The seismic coefficient adds an additional component to the weight of the sliding block, which resolves into an additional component of effective normal stress and also additional shear loading. As this is still a classical slope-stability analysis, a factor of safety of <1 indicates instability. As can be seen in Table 4.3, a wide range of seismic coefficients are available

TABLE 4.3 Recommended Seismic Coefficients for Various Conditions. Estimates of Earthquake Magnitudes Given the Descriptions are Approximate

Horizontal Seismic Coefficient (k_h)	Description	Approximate Magnitude	References
0.05–0.15	USA	NA	Melo and Sharma (2004)
0.12–0.25	Japan	NA	Melo and Sharma (2004)
0.1	"Severe" earthquakes	$M_w > 6$	Terzaghi (1950)
0.2	"Violent, destructive" earthquakes	$M_w > 7$	
0.5	"Catastrophic" earthquakes	$M_w > 7.5$	
0.1–0.2	For $F \geq 1.15$		Seed (1979)
0.1	"Major" earthquake	$M_w > 6$	Corps of engineers (1982)
0.15	"Great" earthquake	$M_w > 8$	
50 percent of PGA			Hynes-Griffin and Franklin (1984)

Source: After Melo and Sharma (2004).

for use, with limited guidance on how they should be selected. The challenge in using such methods is to select a seismic coefficient that is representative of the whole earthquake record. The use of PGA would be exceptionally conservative as the peak value occurs, by definition, only once in an acceleration–time history. Further, such values tend to be large by comparison to the seismic coefficients shown in Table 4.3, and this would lend itself to still more conservatism in slope design. It is interesting that Hynes-Griffin and Franklin (1984) attempted to describe a seismic coefficient in terms of a proportion of PGA. More rigorous methods will allow discrimination between horizontal and vertical coefficients. This is normally approximated to a 3:2 ratio between horizontal and vertical motions. However, this is a very broad approximation as was seen in the February 22, Christchurch earthquake where vertical accelerations were significantly higher than the horizontal motions.

4.3.2 The Newmark Sliding Block Model

Newmark (1965) set out a procedure whereby the coseismic stability of an embankment could be calculated. This involved a number of stages that sought to replicate the process by which slopes fail during earthquakes. The first stage was the calculation of preearthquake stability conditions. This is often done using the infinite slope model (Haefeli, 1948) as it makes for simplicity in subsequent calculations. More recently, a variety of studies has used circular or noncircular slip models to calculate the initial factor of safety (F) and assume the forces act through an angle of thrust (e.g., Jibson and Keefer, 1993). F is given by

$$F = \frac{c + (\gamma - m\gamma_w)z \cos^2 \alpha \tan (\varphi)}{\gamma z \cos \alpha \sin \alpha}, \tag{4.3}$$

where c is the cohesion, γ is the unit weight of the material, γ_w is the unit weight of water, m is the pore-pressure ratio, α is the inclination of the slide mass, z is the thickness of the slide mass, and φ is the coefficient of friction. This broad model can be used for either soil or rock slopes. For soil slopes, the cohesion and coefficient of friction of the soil are used in the calculation. In rock slopes, the cohesion and coefficient of friction of the rock "discontinuities" are used.

Regardless of the method used to calculate the static factor of safety, the critical acceleration for sliding to occur can be defined as shown in Eqn (4.3):

$$A_{crit} = (F - 1)g \sin (\alpha), \tag{4.4}$$

where A_{crit} is the threshold acceleration for movement, g is the acceleration due to gravity (in units appropriate to the answer, so if the critical acceleration is needed as a fraction of the acceleration due to gravity, $g = 1$). Whenever the acceleration acting on a slope exceeds A_{crit}, the slope will move and accumulate displacement. For a given acceleration—time history, therefore, a double integration between A_{crit} and the ground acceleration will yield a displacement of the sliding mass. This is shown graphically in Figure 4.15 (after Romeo, 2000). The first exceedence of the threshold acceleration occurs at about 4 s on the accelerogram. This results in the first movement of the slide mass seen in Figure 4.15(b) (velocity) and Figure 4.15(c) (displacement). The initial velocity pulses are small as the recorded acceleration is only slightly greater than A_{crit}. However, after approximately 6 s of shaking, the prolonged strong shaking results in a significant accumulation of displacement between 6 and 15 s. Continued vibration and numerous occasions where the acceleration exceeds the threshold value results in further displacement. This coseismic displacement is known as the Newmark Displacement. If the displacement is sufficiently large to reduce the strength on the slip surface to its residual or fully softened state, then the slope may no longer have sufficient strength to remain stable at the existing angle and the slide mass may continue to move. For safety case analysis (i.e., for the design of slopes), a new calculation of F is required using residual (or fully softened), rather than peak, strength parameters.

This method is in effect a bilinear form of analysis. That is, the mobilized strength is static over the duration of the earthquake. This is a useful approximation, but most materials do not show this kind of behavior. Figure 4.16 shows the results of a cyclic direct shear test on a granite discontinuity (modified from Lee et al., 2001). Under the first cycle of loading, a clear peak shear strength can be seen. The displacement—stress curve is rough and undulating, which is a function of the asperity geometry on the discontinuity surface. However, as the sample is loaded and reloaded, the asperities are sheared, so by the 15th cycle of loading no clear peak is seen, the curve is smooth and a general reduction in strength occurs. Similar patterns occur in soils during shear. However, in the Newmark sliding block model, the reduced strength criteria only come into effect after the earthquake is over. Peak strength conditions are normally mobilized at small displacements and after that there will be a gradual decline in mobilized strength with displacement of the slide mass. As the strength declines, A_{crit} will also decrease, and therefore the number of exceedances of the threshold value will increase, as will the resultant displacement. Additionally, any strain within the slide mass will change the wave transmission properties and the seismic response of the ground (Moore et al., 2011). Although the assumption of more or less constant material strength properties makes for ease of calculation, it is unlikely to reflect reality.

This technique clearly requires an acceleration—time history to carry out the analysis, and this can be selected either on the basis of empirical evidence or as a synthetic record. The difficulties in ground motion selection can be avoided by the use of a simplified analysis using Arias Intensity (Arias, 1970) as a descriptor of ground motion that can be obtained from attenuation models.

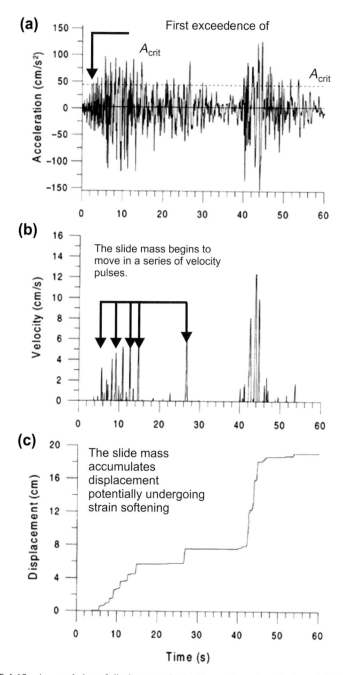

FIGURE 4.15 Accumulation of displacement in the Newmark Sliding block model. (a) Acceleration time history showing yield acceleration. (b) Block velocity as yield acceleration is exceeded. (c) Accumulated displacement of the sliding block. *After Romeo (2000).*

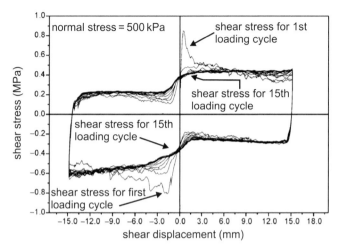

FIGURE 4.16 Degradation of shear strength with repeated loading cycles of a rock joint in granite. *Modified from Lee et al. (2001).*

4.3.3 Coupled Analyses

Coupled analyses (see Bray and Travasarou, 2007) are basically those that contain an element of seismic-response analysis for the landslide mass. This is a concept that works well for soil slopes where a thick soil column may require 1D or 2D site-response analyses (in most cases, 1D site response is assumed to be sufficient). The fundamental period of vibration of a soil slope as shown in Figure 4.17 can be estimated as

$$T_o = \frac{4H}{V_S}. \tag{4.5}$$

For slopes broadly similar to that shown in Figure 4.17(a), and

$$T_o = \frac{2.6H}{V_S}. \tag{4.6}$$

for slopes broadly similar to that shown in Figure 4.17(b).

Here T_o is the fundamental period of the soil column (seconds), H is the thickness of the soil column (meters), and V_s is shear wave velocity of the soil

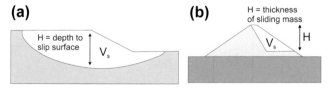

V_s= the velocity of shear wave propagation in the slide mass

FIGURE 4.17 Generic landslide shapes for different types of coupled analysis for equations 4.17A and 4.17B.

(m/s). Logically, this allows a broad differentiation of slopes into those that are likely to be dominated by thin soil columns that will normally not require their own site response, and those with a thick soil column that "may" require a site-response analysis. However, it is worth noting that such site responses relate to the slide mass, and not to the thickness of the soil column. The geometry of the critical slip circle (or plane) therefore is largely a function of the geological characteristics of the site, and the method of analysis used for the calculation of the critical slip circle used for the ambient state of stress. Wartman et al. (2003) noted in a comparison of coupled analyses with rigid sliding block models that the latter tend to underestimate displacement of the slide mass.

This raises significant concerns about the precision of this analysis and the way in which shaking is distributed throughout the earthquake; for example, two strong motion records could contain very similar energies but have very different patterns of shaking. Given that any amplification of the motions by topography is a function of the wavelength of the seismic wave, the geometry of the slope and the speed at which the wave travels through the ground, several variables are at play here, which makes a "precise" assessment of the stability of slopes difficult. In addition to this, it is normal to assume that ground accelerations only act in the horizontal direction. This ignores the importance of vertical stresses in resolving the effective normal stresses acting on a slide mass. Huang et al. (2001) noted the significance of vertical stresses in the analysis of the Jih-Feng-Erh-Shan landslide. Equally, Ling et al. (2014) noted that replicating the stability conditions of the Nigawa landslide was only possible by incorporating the vertical accelerations.

It should be clear, therefore, that the stresses imposed on slope by seismic shaking are complicated, nonuniform throughout the potential slide mass, and will show considerable variations in three dimensions. Therefore, pseudostatic analyses, coupled analyses, or the Newmark sliding block models, although useful indicators of slope performance, contain significant assumptions, many of which are difficult to justify.

Thus one may ask, if such significant challenges occur in using these methods, is there any value in them? The advantage these methods have is that there is considerable experience in the "interpretation" of the results and anecdotal evidence tends to suggest a certain broad agreement between analytical results and predicted performance. That is, slopes that are engineered to stand up during a design event continue to stand. It is not clear, however, whether this is because of any rigor in the methodology, or because these methods are largely conservative (see Table 4.4, Jibson, 2011). The lack of vertical motions is a significant omission; however, the complexity of seismic wave forms means that a reasoned description of wave forms would be difficult to reproduce and would still be subject to large errors. Some of these issues may of course cancel each other out. Ultimately, the key issue is the selection of appropriate parameters, and given the large uncertainties in ground motion inputs, a tendency toward conservatism occurs. This may in fact be the key to the apparent success.

TABLE 4.4 Observations on Appropriateness of Different Analytical
Methods (Jibson, 2011)

Slide Type	T_s/T_m	Rigid-Block Analysis	Decoupled Analysis
Stiffer, thinner slides (e.g., rock slides, rock block slides).	0–0.1	Best results	Good results
	0.1–1	Unconservative	Conservative to very conservative
Softer, thicker slides (e.g., deep seated earth slumps).	1.0–2.0	Conservative	Conservative
	>2.0	Very conservative	Conservative to unconservative

T_s is the site period. T_m is the mean period of earthquake record.

Given that most seismic hazard assessment involves a calculation of probability, it appears logical to consider slope-stability analysis in a probabilistic framework. Examples of such methods include Christian and Urzua (1998) and Murphy and Mankelow (2004). The former uses a statistical method to assist in ground motion selection, whereas the latter uses Monte Carlo simulation to account for the challenges associated with parameter selection. Probabilistic methods are usefully employed in conjunction with geographic information systems (GISs). Additionally, statistical methods have been used to deal with the "randomness" of ground motions and the variance of material properties in three-dimensional slope-stability analysis giving useful results (Al-Hamoud and Tahtamoni, 2000). Given the uncertainties associated with the spatial variations in ground motions on slopes, it is debatable whether the additional complexity associated with 3D stability analysis provides a significantly better assessment of potential problems.

4.3.4 Statistical Models, Hazard Mapping, and GIS

Given the large number of uncertainties in the assessment of slope stability, a reasonable case can be made for using statistical observations rather than making a large number of assumptions about the mechanical factors that result in a landslide happening during an earthquake. This becomes even more rational when the fact that most design earthquakes are statistical constructs based on the regional seismicity and the design life of the structure under consideration is considered. Therefore, using either a statistical description of input parameters rather than discrete values (e.g., Mankelow and Murphy, 1998; Rafice and Capolongo, 2002), or statistical regression combined with sliding block models, appears logical. Data normally used in a predictive

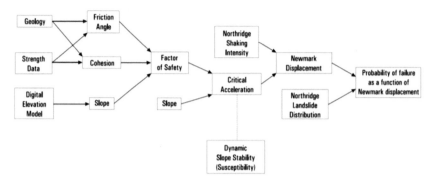

FIGURE 4.18   Outline methodology used to create seismically triggered landslide hazard maps. *From Jibson et al. (1998).*

capacity include inter alia elevation, slope angle, slope morphology, topographic position, some measure of distance from the seismic source, magnitude, PGA (or Arias Intensity), and lithology. Commonly, some form of "weights of evidence" is given to more qualitative data. Seismic inputs may be the result of a seismic hazard assessment or some form of scenario model. Jibson et al. (1998) used the methodology highlighted in Figure 4.18 to produce probabilistic hazard maps and tested these against the incidence of landslides triggered by the 1994 ($M_W = 6.7$) earthquake. Figure 4.19(a) shows the expected Newmark Displacements for the scenario used and Figure 4.19(b) shows the distribution of landslides triggered by the earthquake. It can be seen that although landslides occur in areas with the highest calculated Newmark Displacement, large areas in Figure 4.19(a) are indicated as showing some degree of movement; however, the estimated values of Newmark Displacement are below the 5-cm threshold for the assumption of residual-strength conditions.

The most common outcome of the use of GIS in the study of earthquake-triggered landslides is the production of a hazard or susceptibility map. Some attempts to add a "risk" component have been made to such assessment by considering, for example, distances to vulnerable targets (e.g., infrastructure; Xu et al., 2012). The most commonly used data are slope height and slope angle. This is a de facto recognition that higher, steeper slopes have a lower state of stability under ambient stress conditions and that topographic amplification plays a critical role in the assessment of stability of EILs.

4.4 LIMITATIONS OF CURRENT UNDERSTANDING

A number of factors occur that limit our understanding of coseismic landslides. These can be divided into two broad groups: seismological factors (i.e., ground motions and forcing functions) and geotechnical factors (i.e., the dynamic properties of soil and rock masses "in the field").

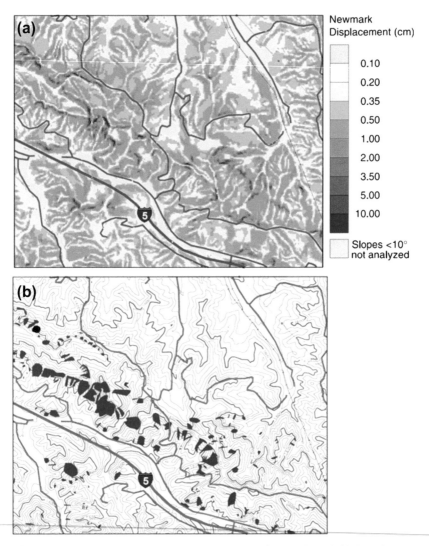

FIGURE 4.19   A map showing a probabilistic assessment of coseismic landslide hazard (a) and mapped landslides triggered by the 1994 Northridge earthquake (b). *From Jibson et al. (1998). Image courtesy of the United States Geological Survey.*

4.4.1 Seismological Unknowns

The majority of seismological research is carried out on flat terrain. Attenuation models are derived from observations made from instruments deployed away from slopes to evaluate site effects and building response. This means that predictions of peak ground motions or Arias Intensity may be not be representative of shaking experienced on slopes. Limited rigorous data sets remain that capture the variation in shaking along slope long profiles, especially in three dimensions. A significant lack of work occurs, however, in which ground motions are linked to site geology on slopes.

Consider the conceptual model shown in Figure 4.20. This is broadly based on slopes that showed landslide activity during the February 22, 2011 ($M_W = 6.3$), Christchurch earthquake. This takes the observations made by Ashford et al. (1997) and applies them to the slope shown. This highlights the increase in ground motions at the hillcrest and the variation in these ground motions behind the crest and across the slope. This applies the model for vertically propagating, horizontally polarized S waves, which is likely to be the case only close to the seismic source. Although this figure deals only with peak accelerations, it is clear that variations and differences occur in acceleration that will result in differential stresses in slopes. What is evident is that for this geological model, the geometry shown and a mode vibration with a high-frequency, free-field ground motion characterized by a PGA of 1 g, then ground motions along the crest vary from 0.99 to 1.3 g. Further back from the slope at distances up to twice the height of the ridge, motions are effectively deamplified and drop to a PGA of 0.8 g. The inset photograph highlights ground cracking behind the cliff crest at Redcliffs in Christchurch. It is therefore an interesting question whether such a

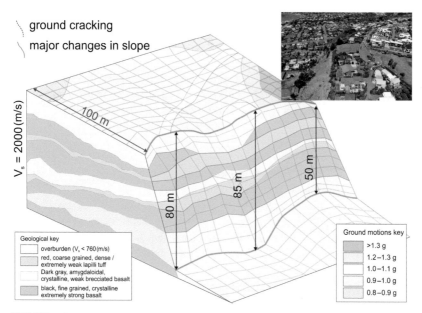

FIGURE 4.20 Hypothetical hill slope illustrating the role of topographic amplification in creating spatial variations in ground shaking along a ridge line and behind the slope. *Inset photograph courtesy of GNS Science, New Zealand.*

cracking is a function of differential ground motions related to topographic amplification.

Given the monolithological nature of parametric studies, the work of Tripe et al. (2013) is interesting. The investigation of "soft" layers in the geological column indicates that a difference occurs in topographic amplification when weaker layers are incorporated in the analysis. The results indicate that an effect only appears when the soft layer is >250 m in thickness. Interestingly, given the material properties used, this is likely to be close to the wavelength of the transmitted shear wave. The consequence of this study is that the pattern of crest amplification changes, and the patterns of shaking away from the crest show significant differences. This observation is potentially significant in terms of stresses imposed on slopes and the geometry of a potential landslide. Additionally, the depth distribution of topographic amplification is very unclear. Is the topographic effect limited to the top 0.1, 1, or 10 m? This is important information for the engineer trying to assess the stability of slopes, but one which cannot at present be answered.

The majority of parametric studies of topographic amplification assume a very simple model of wave propagation. Normally, this approximation is a horizontally polarized S wave. Ashford et al. (1997) also highlight the significance of variation in incidence angle for the effect of seismic shaking on monolithological slopes. The current understanding does not consider the importance of surface waves (e.g., Love or Rayleigh waves) or interface waves (e.g., Stoneley or Scholte) waves. The latter are potentially significant as particle velocities are often higher due to energy attenuation only occurring in two dimensions.

To understand EILs, a better understanding of the interaction of waves and slopes is required. A predictive tool for engineers to make appropriate parameter selection to represent ground motions on slopes is needed.

4.4.2 Geotechnical Considerations

A better understanding of the deterioration of strength of geological materials under dynamic loads is needed. If yield (or critical) acceleration is reached early in the earthquake, a progressive weakening of the slope means that the yield acceleration will get progressively lower, and therefore even if shaking is declining, the yield acceleration may still be exceeded. If the yield acceleration is not reached until later in the earthquake, there may be no or little significant weakening.

The role of water pressures during shaking in liquefiable materials is well understood. However, dynamic pore-pressure responses in cohesive materials or in jointed rock masses remain unclear. The role of water in any landslide is significant, but this is normally a very poorly constrained issue when it comes to seismically triggered landslides.

Finally, current analyses assume that coseismic ground failure occurs on the same failure surface that is critical in static conditions. Given that coseismic stresses within a slope are dramatically difefrent from static stresses, this assumption is questionable.

4.4.3 Concluding Comments

Little doubt exists that seismically triggered landslides involve a large number of variables that can best be described as "uncertain." In addition to the difficulties in characterizing the ground, the ability to predict ground motions with rigor means that, at best, the forcing functions are poorly understood. The current state of understanding of the variables involved in the initiation of landslides during earthquakes varies considerably. It is possible to produce a reasonable characterization of the ground in terms of materials, masses, and water content. It is relatively easy to characterize the geomorphology of the land surface; however, our understanding of the interaction of sloping ground, propagating waves and variable lithology, and moisture content is limited in two dimensions. The complexities associated with shaking in three dimensions make for a greater challenge. In engineered slopes, the current methods work appear to give a reasonable indication of performance. Where the material can be considered uniform and the slope heights are relatively small, the forces can be reasonably described. However in high, complex (e.g., natural) slopes with varying seismic properties, the current lack of understanding makes for a difficult problem.

It is tempting therefore to say that the problem is intractable. However, at some point, it is necessary to evaluate the stability of natural (i.e. not engineered) slopes in the field. Practitioners do not have the luxury of saying "it is too complex." It is clear that the development of better tools to represent the forces acting on slopes is required.

REFERENCES

Aki, K., Richards, P.G., 2002. Quantitative Seismology: Theory and Methods. University Science Books, ISBN 0-935702962.

Al-Hamoud, A.S., Tahtamoni, W.W., 2000. Reliability analysis of three-dimensional dynamic slope stability and earthquake-induced permanent displacement. Soil Dyn. Earthquake Eng. 19, 91−114.

Ashford, S.A., Sitar, N., 1997. Analysis of topographic amplification of inclined shear waves in a steep coastal bluff. Bull. Seismol. Soc. Am. 87, 692−700.

Ashford, S.A., Sitar, N., Lysmer, J., Deng, N., 1997. Topographic effects on the seismic response of steep slopes. Bull. Seismol. Soc. Am. 87, 701−709.

Arias, A., 1970. A measure of earthquake intensity. In: Hansen, R. (Ed.), Seismic Design for Nuclear Power Plants. MIT Press, Cambridge, MA, pp. 439−483.

Bazzurro, P., Cornell, C.A., 2004. Ground-motion amplification in nonlinear soil sites with uncertain properties. Bull. Seismol. Soc. Am. 94 (6), 2090−2109.

Buech, F., Davies, T.R., Pettinga, J.R., 2010. The Little Red Hill seismic experimental study: topographic effects on ground motion at a bedrock-dominated mountain edifice. Bull. Seismol. Soc. Am. 100 (5A), 2219−2229. http://dx.doi.org/10.1785/0120090345.

Boore, D.M., 1972. A note on the effect of simple topography on seismic SH waves. Bull. Seismol. Soc. Am. 62, 275−284.

Boore, D.M., 1973. The effect of simple topography on seismic waves: implications for the accelerations recorded at Pacoima Dam, San Fernando Valley, California. Bull. Seismol. Soc. Am. 63, 1603−1609.

Bouchon, M., 1973. Effect of topography on surface motion. Bull. Seismol. Soc. Am. 63, 615−632.

Bouckovalas, G.D., Papadimitriou, A.G., 2005. Numerical evaluation of slope topography effects on seismic ground motion. Soil Dyn. Earthquake Eng. 25, 547−558.

Bradley, B., 2012. Strong ground motion characteristics observed in the 4 September 2010 Darfield, New Zealand earthquake. Soil Dyn. Earthquake Eng. 42, 32−46.

Bray, J.D., 2007. Simplified seismic slope displacement procedures. In: Pitilakis, K.D. (Ed.), 4th International Conference on Geotechnical Earthquake Engineering − Invited Lectures, vol. 488. Springer-Verlag, pp. 327−353. ISBN-13: 9781402058929.

Bray, J.D., Travasarou, T., 2007. Simplified procedure for estimating earthquake-induced deviatoric slope displacements. J. Geotech. Geoenviron. Eng. ASCE 33 (4), 381−392.

Burjánek, J., Gassner-Stamm, G., Poggi, V., Moore, J.R., Fäh, D., 2010. Ambient vibration analysis of an unstable mountain slope. Geophys. J. Int. 180, 820−828. http://dx.doi.org/10.1111/j.1365-246X.2009.04451.x.

Christian, J.T., Urzua, A., 1998. Probabilistic evaluation of earthquake-induced slope failure. J. Geotech. Geoenviron. Eng. ASCE 124 (11), 1140−1143.

Cluff, L.S., 1971. Peru earthquake of 31 May 1970; engineering geological observations. Bull. Seismol. Soc. Am. 61 (3), 511−533.

Crawford, A.M., Curran, J.H., 1981. The influence of shear velocity on the frictional resistance of rock discontinuities. Int. J. Rock Mech. Min. Sci. Geomech. Abstr. 18 (6), 505−515. http://dx.doi.org/10.1016/0148-9062(81)90514-3.

Cruden, D.M., Varnes, D.J., 1996. Landslide types and processes. In: Special Report 247: Landslides: Investigation and Mitigation. Transportation Research Board, Washington D.C.

Dobry, R., Idriss, I.M., Ng, E., 1978. Duration characteristics of horizontal components of strong motion earthquake records. Bull. Seismol. Soc. Am. 68 (5), 1487−1520.

Faccioli, E., Vanini, M., Frassine, L., 2002. "Complex" site effects in earthquake ground motion, including topography. 12th European Conference on Earthquake Engineering. Paper number 844.

Gao, Y., Zhang, N., Li, D., Liu, H., Cai, Y., Wu, Y., 2012. Effects of topographic amplification induced by a U-shaped canyon on seismic waves. Bull. Seismol. Soc. Am. 102 (4), 1748−1763. http://dx.doi.org/10.1785/0120110306.

Géli, L., Bard, P.Y., Jullien, B., 1988. The effect of topography on earthquake ground motion: a review and new results. Bull. Seismol. Soc. Am. 78, 42−63.

Grapes, R.H., 1988. Geology and revegetation of an 1855 landslide, Ruamahanga River, Kopuaranga, Wairarapa. Tuatara 30, 77−83.

Hadley, P.K., Askar, A., Cakmak, A.S., 1991. Subsoil geology and soil amplification in Mexico Valley. Soil Dyn. Earthquake Eng. 10 (2), 101−109.

Haefeli, R., 1948. The stability of slopes acted upon by parallel seepage. In: Proceedings of the 2nd International Conference on Soil Mechanics, Rotterdam, vol. 1, pp. 57−62.

Hancock, J., Bommer, J.J., 2005. The effective number of cycles of earthquake ground motion. Earthquake Eng. Struct. Dyn. 34, 637–664. http://dx.doi.org/10.1002/eqe.437.

Harp, E.L., Sarmiento, J., Cranswick, E., 1984. Seismic-induced pore-water pressure records from the Mammoth lakes, California, earthquake sequence of 25 to 27 May 1980. Bull. Seismol. Soc. Am. 74 (4), 1381–1393.

Hamilton, W., 1783. "An account of the earthquakes which happened in Italy, from February to May 1783". By Sir William Hamilton, Knight of the Bath, F. R. S.; in a letter to Sir Joseph Banks, Bart. P. R. S. Philos. Trans. R. Soc. London 73, 169–208.

Holzer, T.L., Youd, T.L., 2007. Liquefaction, ground oscillation, and soil deformation at the wildlife array, California. Bull. Seismol. Soc. Am. 97 (3), 961–976. http://dx.doi.org/10.1785/0120060156.

Huang, C.C., Lee, Y.H., Liu, H.P., Keefer, D.K., Jibson, R.W., 2001. Influence of surface-normal ground acceleration on the initiation of the Jih-Feng-Erh-Shan landslide during the 1999 Chi-Chi, Taiwan, earthquake. Bull. Seismol. Soc. Am. 91, 953–958.

Hutchinson, J.N., Del Prete, M., 1985. Landslide at Calitri, southern Apennines, reactivated by the earthquake of 23rd November 1980. Geol. Appl. Idrogeol. XX (1), 9–38.

Hynes-Griffin, M.E., Franklin, A.G., 1984. Rationalizing the Seismic Coefficient Method. U.S. Army Corps of Engineers Waterways Experiment Station, Vicksburg, Mississippi. Miscellaneous Paper GL-84–13, 21 pp.

Jibson, R.W., 1987. Summary of Research on the Effects of Topographic Amplification of Earthquake Shaking on Slope Stability. U.S. Geological Survey. Open-File Report 87-268, 166 pp.

Jibson, R.W., 1993. Predicting Earthquake Induced Landslide Displacements Using Newmark's Sliding Block Analysis. Trans. Res. Rec, Nat. Acad. Press, 1411, pp. 9–17.

Jibson, R.W., 2011. Methods for assessing the stability of slopes during earthquakes—a retrospective. Engi. Geol. 122 (1–2), 43–50.

Jibson, R.W., Keefer, D.K., 1993. Analysis of the seismic origin of landslides: examples from the New Madrid seismic zone. Geol. Soc. Am. Bull. 105 (4), 521–536.

Jibson, R.W., Harp, E.L., Michael, J.A., 1998. A Method for Producing Digital Probabilistic Seismic Landslide Hazard Maps: An Example from the Los Angeles, California, Area. USGS. Open-File Report, pp. 98–113.

Kanamori, H., 1977. The Energy release in great earthquakes. J. Geophys. Res. 82, 2981–2987.

Keefer, D.K., 1984. Landslides caused by earthquakes. Geol. Soc. Am. Bull. 95 (4), 406–421.

Lagerbäck, R., Sundh, M., 2008. Early Holocene Faulting and Paleoseismicity in Northern Sweden. Research Paper C 836 Sveriges Geologiska Undersökning. 80 pp. ISBN: 978-91-7158-859-3.

Lee, H.J., Park, Y.J., Cho, T.F., You, K.H., 2001. Influence of asperity degradation on the mechanical behaviour of rough rock joints under cyclic shear loading. Int. J. Rock Mech. Min. Sci. 38, 967–980.

Ling, H., Ling, H.I., Kawabata, T., 2014. Revisiting Nigawa landslide of the 1995 Kobe earthquake. Géotechnique 64 (5), 400–404.

Li, J.C., Ma, G.W., 2009. Experimental study of stress wave propagation across a filled rock joint. International Journal of Rock Mechanics and Mining Sciences 46 (3), 471–478. http://dx.doi.org/10.1016/j.ijrmms.2008.11.006.

Liu, K.S., Tsai, Y.B., 2005. Attenuation relationships of peak ground acceleration and velocity for crustal earthquakes in Taiwan. Bull. Seismol. Soc. Am. 95 (3), 1045–1058. http://dx.doi.org/10.1785/0120040162.

Mankelow, J.M., Murphy, W., 1998. Using GIS in the probabilistic assessment of earthquake triggered landslides hazards. J. Earthquake Eng. 2 (4), 593−623.

Mavko, G., Harp, E.L., 1984. Analysis of wave-induced pore pressure changes recorded during the 1980 Mammoth Lakes, California, earthquake sequence. Bull. Seismol. Soc. Am. 74 (4), 1395−1407.

Martino, S., Scarascia Mugnozza, G., 2005. The role of the seismic trigger in the Calitri landslide (Italy): historical reconstruction and dynamic analysis. Soil Dyn. Earthquake Eng. 25, 933−950. http://dx.doi.org/10.1016/j.soildyn.2005.04.005.

Melo, C., Sharma, S., 2004. Seismic coefficients for pseudostatic slope analysis. In: 13th World Conference on Earthquake Engineering, August 1−6, 2004 Vancouver, B.C., Canada. Paper No. 369.

Moore, J.R., Gischig, V., Burjanek, J., Loew, S., Fäh, D., 2011. Site effects in unstable rock slopes: dynamic behavior of the Randa instability (Switzerland). Bull. Seismol. Soc. Am. 101 (6), 3110−3116. http://dx.doi.org/10.1785/0120110127.

Muir-Wood, R., King, G.C.P., 1993. Hydrological signatures of earthquake strain. J. Geophys. Res. 98, 22,035−22,068.

Murphy, W., 1995. The geomorphological controls in seismically triggered landslides during the 1908 Straits of Messina earthquake, Southern Italy. Q. J. Eng. Geol. 28, 61−74.

Murphy, W., Mankelow, J.M., 2004. Obtaining probabilistic estimates of displacement on a landslide during future earthquakes. J. Earthquake Eng. 8 (1), 133−157.

Newmark, N.M., 1965. Effects of earthquakes on dams and embankments. Geotechnique 15 (2), 139−159.

Paolucci, R., 2002. Amplification of earthquake ground motion by steep topographic irregularities. Earthquake Eng. Struct. Dyn. 31, 1831−1853. http://dx.doi.org/10.1002/eqe.192.

Poppeliers, C., Pavlis, G.L., 2002. The seismic response of a steep slope: high-resolution observations with a dense, three-component seismic array. Bull. Seismol. Soc. Am. 92, 3102−3115. http://dx.doi.org/10.1785/0120010186.

Pyrak-Nolte, L.J., Myer, L.R., Cook, N.G.W., 1990. Transmission of seismic waves across single natural fractures. J. Geophys. Res. 95 (B6), 8617−8638.

Rafice, A., Capolongo, D., 2002. Probabilistic modelling of uncertainties in earthquake-induced landslide hazard assessment. Computers and Geosciences 28, 735−749.

Rodriguez, C.E., Bommer, J.J., Chandler, R.J., 1999. Earthquake-induced landslides: 1980−1997. Soil Dyn. Earthquake Eng. 18, 325−346.

Romeo, R., 2000. Seismically induced landslide displacements: a predictive model. Eng. Geol. 58, 337−351.

Sepulveda, S.A., Murphy, W., Petley, D.N., 2005a. Topographic controls on coseismic rock slides during the 1999 Chi-Chi earthquake, Taiwan. Q. J. Eng. Geol. Hydrogeol. 38, 189−196.

Sepulveda, S.A., Murphy, W., Jibson, R.W., Petley, D.N., 2005b. Seismically induced rock slope failures resulting from topographic amplification of strong ground motions: the case of Pacoima Canyon, California. Eng. Geol. 80, 336−348.

Schuster, R.L., Murphy, W., 1996. Structural damage, ground failure and hydrologic effects of the magnitude (M_W) 5.9 Draney Peak, Idaho, earthquake of February 3, 1994. Seismol. Res. Lett. 67 (3), 20−29.

Seed, H.B., 1979. Considerations in the earthquake-resistant design of earth and rockfill dams. Geotechnique 29, 215−263.

Seed, H., Idriss, I., Arango, I., 1983. Evaluation of liquefaction potential using field performance data. J. Geotech. Geoenviron. Eng. ASCE 109 (3), 458−482.

Seed, H.B., Wilson, S.D., 1967. The Turnagain Heights Landslide in Anchorage, Alaska. Journal of the Soil Mechanics and Foundation Division (American Society of Civil Engineers) 93, 325—354.

Sibson, R.H., Moore, J.Mc.M., Rankin, A.H., 1975. Seismic pumping—a hydrothermal fluid transport mechanism. J. Geol. Soc. 131 (6), 653—659. http://dx.doi.org/10.1144/gsjgs.131.6.0653.

Sugito, M., Oka, F., Yashima, A., Furumoto, Y., Yamada, K., 2000. Time-dependent ground motion amplification characteristics at reclaimed land after the 1995 Hyogoken Nambu Earthquake. Eng. Geol. 56, 137—150.

Terzaghi, K., 1950. Mechanisms of Landslides. Engineering Geology (Berkeley) Volume. Geological Society of America.

Tripe, R., Kontoe, S., Wong, T.K.C., 2013. Slope topography effects on ground motion in the presence of deep soil layers. Soil Dyn. Earthquake Eng. 50, 72—84.

Updike, R.G., Olsen, H.W., Schmoll, H.R., Kharaka, Y.F., Stokoe, K.H., 11,1988. Geologic and geotechnical conditions adjacent to the Turnagain Heights landslide. Anchorage, Alaska. Reston, VA: U.S Geological Survey Bulletin 1817.

Vivenzio, G., 1788. Istoria de' tremuoti avvenuti nella provincia di Calabria ulteriore e nella cittá di Messina nell'anno 1783. Napoli.

Wartman, J., Bray, J.D., Seed, R.B., 2003. Inclined plane studies of the Newmark sliding block procedure. J. Geotech. Geoenviron. Eng. ASCE 129 (8), 673—684.

Wyllie, D.C., Mah, C., 2004. Rock Slope Engineering, fourth ed. Taylor and Francis. 431 pp. ISBN-13: 978-0415280013.

Xu, C., Xu, X., Lee, Y.H., Tan, X., Yu, G., Dai, F., 2012. The 2010 Yushu earthquake triggered landslide hazard mapping using GIS and weight of evidence modelling. Environ. Earth Sci. 66 (6), 1603—1616.

Youd, T., Idriss, I., Andrus, R., Arango, I., Castro, G., Christian, J., Dobry, R., Finn, W., Harder Jr., L., Hynes, M., Ishihara, K., Koester, J., Liao, S., Marcuson III, W., Martin, G., Mitchell, J., Moriwaki, Y., Power, M., Robertson, P., Seed, R., Stokoe II, K., 2001. Liquefaction resistance of soils: summary report from the 1996 NCEER and 1998 NCEER/NSF workshops on evaluation of liquefaction resistance of soils. J. Geotech. Geoenviron. Eng. ASCE 127 (10), 817—833.

Chapter 5

Volcanic Debris Avalanches

Benjamin van Wyk de Vries [1] and Audray Delcamp [2]

[1] *Laboratoire Magmas et Volcans, Univeristé Blaise Pascal, Clermont-Ferrand, France,*
[2] *Department of Geography, Faculty of Sciences, Vrije Universiteit Brussel, Brussel, Belgium*

ABSTRACT

Volcanoes are growing mountains with hydrothermal and magmatic systems, which have strong controls on volcanic landslides and debris avalanches. Such landslides are conditioned by the nature of volcanic rock, which is highly fractured, usually in granular form, often clay-rich and water-saturated. In consequence, volcanic landslides are generally more fractured, have more fine material, are more variably saturated than non-volcanic landslides, and they have a tendency to transform into large debris flows. Volcanic landslides vary in size from small failures of a valley side ($<$million m^3) to a large portion of the edifice (tens of km^3). The larger landslides are generally more deep-seated, because weak hydrothermal and magmatic systems in the volcano core are involved.

Volcanoes undergo significant gravitational and tectonic deformation, creating faulting and fracturing in the edifice. The structures form the framework for landslides, and the resulting debris avalanches tend to form hummocky horst and graben topography that reflects the initial structure. As many volcanic landslides descend onto flat plains, this type of topography is often well preserved.

Volcanoes of all types in all geological settings suffer landslides, and even extinct volcanoes are landslide prone. About four volcanic landslides occur world wide per century, meaning they are a significant hazard, specially as they are associated with secondary tsunami, volcanic eruptions, debris flows and lahars.

5.1 INTRODUCTION

Volcanic landslides and their associated debris avalanches differ from other landslides on mountains and hills created by uplift and erosion, in that they occur on mountains constructed by eruptions that have active hydrothermal and magmatic systems (Siebert, 1984). This particular set of circumstances means that the rocks are generally highly fractured or in granular form at the start, are often clay-rich, and may contain large amounts of water.

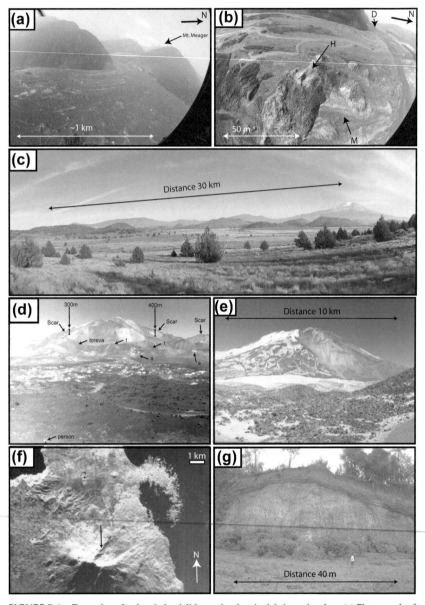

FIGURE 5.1 Examples of volcanic landslides and volcanic debris avalanches. (a) Photograph of the Mt Meager landslide and avalanche of 2010 (Photograph DB Steers), (b) Close-up image of the Mt Meager debris avalanche surface (Photograph DB Steers). "H" indicates a hummock with block facies and "M" a highly stretched interhummock area with matrix or mixed facies. Note in the middle ground, thin lahars are forming through the dewatering of the deposit. The Meager Creek dam is indicated as "D." (c) Photograph of Mt Shasta (with snow) and the Shasta debris avalanche, California. The foreground is a smooth interhummock area, and middle ground shows a

Volcanic landslides have been also called "sector collapse," "flank collapse," "flank failure," or rockslide–debris avalanche, in the volcanological literature (e.g., Siebert, 1984; Siebert et al., 1987; Glicken, 1996; van Wyk de Vries and Francis, 1997). A volcanic landslide can involve a small proportion of a volcano's flank that fails on the side or headwall of a preexisting valley, as at Mt Meager, Canada, in 2009 (Guthrie et al., 2012), can involve part of an active dome and valley as at Montserrat in 2000 (Sparks et al., 2002), or can involve a large portion of the edifice as at Socompa (Wadge et al., 1995), and Mt Shasta (Crandell, 1989) (Figure 5.1).

Volcanic landslides have formed some of the largest catastrophic landslide events on the Earth, especially from the large oceanic islands such as Hawai'i, the Canaries, and in island arcs, such as Japan (Sekiya and Kikuchi, 1890; Miyachi, 1992) and the Philippines (Paguican et al., 2014). A major hazard associated with such sea-bound volcanoes is the generation of tsunamis, for example, that formed by a landslide at Ritter Island volcano, Papua New Guinea, in 1888 (Ward and Day, 2003). The secondary hazards from volcanic landslides also comprise volcanic eruptions, when sudden decompression created directed blasts and pyroclastic flows, as at Bezymyanny and Mt St Helens. Debris flows and lahars can also be generated as water in the volcanic landslide is liberated during or after emplacement.

Volcanic landslides occur at the rate of about four per century (van Wyk de Vries and Francis, 1997) and historic examples for the twentieth and twenty-first centuries are Mt Meager in 2009 (Guthrie et al., 2012), Stromboli in 2002 (Tinti et al., 2005), Montserrat in 2000 (Sparks et al., 2002), Casita in 1998 (Vallance et al., 2004; van Wyk de Vries et al., 2001), Mt St Helens in 1980 (Voight et al., 1981, 1983; Glicken, 1996), and Bezymianny in 1957 (Gorshkov, 1959). Before this, many volcanic landslides may have gone unreported, or were interpreted as volcanic eruptions. Examples that were reported are Bandai-San in 1888 (Sekiya and Kikuchi, 1890), the Shimabara collapse at Mayu-yama in 1794 (Miyachi, 1992), and Ritter in 1888 (Ward and Day, 2003). Most volcanoes experience a large landslide at least once during their lifetime, some several times. Examples of multiple collapses are Mt Augustine, Alaska (Swanson and Kienle, 1988), Mombacho, Nicaragua (Shea et al., 2008),

---

10-km-wide view of hummocks. (d) Photograph of the Socompa debris avalanche, N Chile, taken from the Median Scarp, a large backthrusted area in the middle of the debris avalanche 20 km from the summit. The Scar (12 km wide) is indicated; "t" stands for toreva block, and "s" for white outcrops of the substrate that were involved in the original failure volume. (e) View of the Socompa Volcano showing the scar, now partially refilled with lava. The white material is pumice deposited during an eruption that was triggered by the landslide. The foreground is hummocky terrain of the avalanche deposit. Arrows indicate the 15 km diameter of cone. (f) Radar image of the Las Isletas Landslide deposit at Mombacho, Nicaragua (Shea et al., 2008). The black arrow points to the scar on the volcano, and the deposit forms a fan of several hundreds of islands. (g) Hummock cut in half by quarrying on the Iriga Debris avalanche, Philippines. This shows normal faulting cutting conglomerate bands. The original landslide involved the volcano substrata.

Mt Meager, Canada (Friele et al., 2008; Hickson et al., 1999), and large oceanic islands, such as Hawai'i (Moore et al., 1994; Morgan et al., 2003), the Canaries (Paris et al., 2013), and the lesser Antilles volcanoes (Lebas et al., 2011).

Volcanic landslide events occur on all types of volcanoes, be they oceanic, continental, monogenetic, or polygenetic (Ui, 1983; Siebert, 1984; Voight and Elseworth, 1997). Landslides occur at volcanoes in all geodynamic contexts, such as mid-ocean rift, hot spot, arc, or intraplate, and evidence also exists that they have occurred on other planets. Collapse may occur long after the volcano has become extinct, as with the Shimabara landslide at Mayu-yama in 1792 (Miyachi, 1992), or after a long repose period as a volcano is reactivated, as at Mt St Helens (Voight et al., 1983). Even small monogenetic volcanoes are prone to landslides (Delcamp et al., 2014, Figure 5.2).

5.2 VOLCANIC DEBRIS AVALANCHES

A volcanic debris avalanche forms when a volcanic landslide disaggregates following its initial motion. It accelerates as a fast-moving, deforming mass over the area below the volcano. The resulting debris avalanche deposit (DAD) can extend many kilometers from the foot of the edifice (Figures 5.1 and 5.2). Such events are generally somewhat longer in runout compared to their nonvolcanic equivalents of a similar volume. Both volcanic and nonvolcanic events are able to travel a long way (L) with respect to the total height loss (H). The H/L ratio is often around 0.6 for small events, but can be very low (~ 0.1) for the largest events with volumes of several cubic kilometers (Heim, 1932; Erismann, 1979). This volumetric influence over distance traveled is shown in Figure 5.3(a) (Shea et al., 2008). The area covered by the debris avalanche (A) is related to its volume (V) by the relationship $V = C \cdot A^{2/3}$, where C varies according to the geometry and topography of the deposition area (Figure 5.3; Dade and Huppert, 1998; Kilburn and Sorensen, 1998). Figure 5.3(b) demonstrates the scale-independent relationship between area and volume, where both are cast into dimensionless numbers using H as the denominator (Shea et al., 2008).

5.3 TYPES OF VOLCANIC LANDSLIDES

5.3.1 Large-Scale Volcano and Substrata Landslides

Volcanic landslides can occur on many different scales, from small valley-side failures to those that include many cubic kilometers of material. At the smallest scale, monogenetic volcanoes may suffer landsliding, but being much smaller in volume than their polygenetic counterparts, their avalanches do not travel very far (Figure 5.2). Landslides may occur in association with cryptodome formation, and the resulting lava eruption may raft the landslide debris away from the cone to a far greater distance than a landslide would have carried it. Examples of this were observed at Paricutin (with accent on last "i")

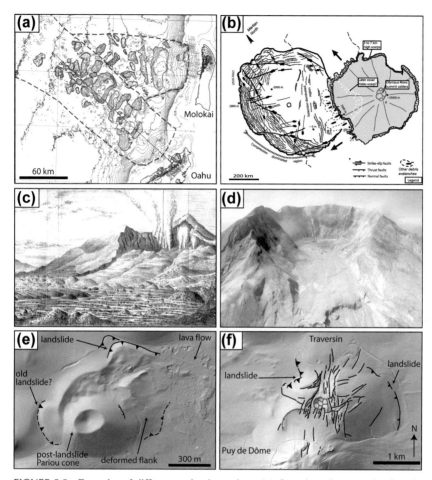

FIGURE 5.2   Examples of different avalanche settings: (a) Oceanic volcano avalanche: the Oahu and Molokai landslides in the Hawaiian Islands (Moore et al., 1994), (b) Large extraterrestrial landslide at Olympus Mons (Shea et al., 2008). (c) The Bandai-San landslide and avalanche associated with hydrothermal system disruption (Sekiya and Kikuchi, 1890). (d) The Mt St Helens scar from the 1980 landslide and the subsequent dome growth. *(Image from USGS.)* (e) Monogenetic volcano collapse at the Pariou Volcano, Chaîne des Puys, France. In this case, the northeast flank has slid away and a lava flow has broken out of the scar. (f) Uplifted and deformed flank of the Petit Puy de Dôme, Chaîne des Puys, France, showing an intrusion-generated volcanic landslide that stabilized before developing into a debris avalanche.

(Mexico: Luhr and Simkin, 1993), Red Mountain, Arizona (Riggs and Duffield, 2008), and are well preserved in the Lemptégy Scoria Cone in the Chaîne des Puys, France (Delcamp et al., 2014).

On large volcanoes, landslides may develop in near-surface strata, without involving the volcano core. These landslides are smaller in volume (up to

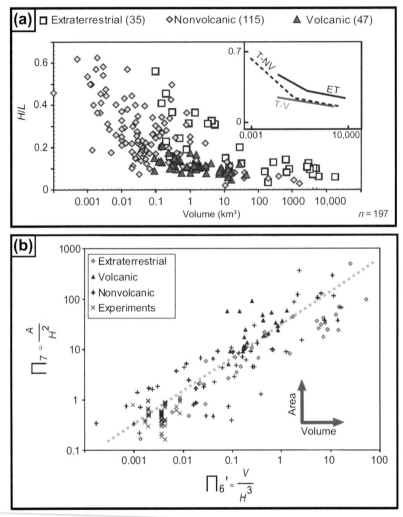

FIGURE 5.3 Plots of the geometric relationships of volcanic and nonvolcanic landslides.
(a) Height/length (H/L) versus volume. (b) Area (A) versus volume (V), relationship cast in
dimensionless ratios of A/H^2 versus V/H^3. *From Shea et al. (2008).*

several million cubic meters). Good examples are at Lastarria (Chile, Naranjo
and Francis, 1987) and at Mt Meager (Canada, Friele et al., 2008). In the case
of Lastarria, slope-parallel scoria beds failed on the flank, leaving a shallow
scar, and a thin DAD was emplaced. At Mt Meager, deep glacial incision,
coupled with glacial retreat, has eroded the edifice, leaving large cliffs that are
prone to frequent landslides.

At a larger scale, the core of a volcano may be involved: landslides generated
from hydrothermally altered cores, and from magma intrusion, will generally

involve just the volcano. This limits their volume to the available portions of the edifice affected by this localized deformation (e.g., Mombacho and Mt St Helens, ∼1 km³). Some larger landslides, such as those of Parinacota (Clavero et al., 2002) and Chimborazo (Bernard et al., 2008), attain volumes of >10 km³, by incorporating large amounts of their basal volcaniclastic aprons.

At the very largest scale, the landslide also involves the strata beneath the volcano. If the substrata of the volcano are involved in the initial landslide, the debris avalanche can contain not only the volcano but also a considerable part of the substrate. Such collapses are often the largest (e.g., Socompa), although domes growing on hydrothermally altered rock can form a similar smaller-volume versions, such as at Montserrat (Sparks et al., 2002). Oceanic volcanoes that grow up on pelagic sediments or on their own fine detrital material can also have a similar type of landslides. The huge slides of Hawai'i and La Réunion are examples of this type.

The largest volcanic landslides occur in the huge oceanic volcanoes. These can involve many tens of cubic kilometers of material, because these volcanoes are generally the largest, such as Hawai'i, the Canaries, and La Réunion Island. The Hilina Slump, Kilauea, the Enclos Fouqué of Piton de la Forunaise, and the western side of El Hierro, Canaries, are examples of ongoing landslides that may eventually cause catastrophic debris avalanches. The Las Canadas, Orotava, and Guimar scars on Tenerife are good examples of collapsed sectors of oceanic volcanoes.

Arc volcanoes in oceanic settings may also suffer large collapses, such as the 1888 collapse of Mt Ritter (Papua New Guinea). Stromboli and the Antilles volcanoes also provide examples of such events (Tibaldi, 2001; Lebas et al., 2011).

Probably the largest, and some of the best-exposed, DADs are preserved in the aureole around Olympus Mons, Mars (Figure 5.2(b)). As most other extraterrestrial volcanoes are flat shields, landslides and debris avalanches are restricted to the inner edges of their craters.

5.4 DEEP-SEATED VOLCANIC LANDSLIDE DEFORMATION: PRIMING AND TRIGGERS

A volcano is composed of primary eruption products, secondary epiclastic sediments, intrusions, and hydrothermally altered material. Water, magma, and gas occupy pores, fractures, and other spaces. Within the volcano, these elements combine to create zones of weakness that contribute to slow deep-seated deformation.

Further, during volcano growth, the load on the underlying rocks is increased, and deforms them. This gravitational deformation causes the volcano to sink and/or spread outward in the substrata (Nakamura, 1980; Borgia et al., 1992, 2000; van Wyk de Vries and Borgia, 1996). For example, the Big Island of Hawai'i has caused about 5 km of flexure in the oceanic lithosphere, and accent on last "c" of Concepcion (?), Nicaragua, has created a range of folds around its base (Borgia and van Wyk de Vries, 2003, Figure 5.4). If outward movement is oriented in one main direction, the volcano may develop landslides. This inherent propensity to collapse means that volcanic landslides are not necessarily related to eruptions, in fact little evidence occurs that eruptions themselves generate failures: it is the internal activities of intrusion, hydrothermal system, and seismicity that are the more likely triggers, whereas loading and internal long-term and short term weakening prime the volcano for failure.

5.5 DEEP-SEATED VOLCANO GRAVITATIONAL DEFORMATION

Loading of substrata by a growing volcano may cause material to flow outward from underneath the edifice. This flow will drag the edifice flanks outward, creating a summit graben area, linked to outer strike-slip faults and possible thrusts and folds at the base (Figures 5.4 and 5.5).

Alternatively, if the amount of sinking is greater than the outward flow (mainly due to thicker ductile material in the strata below the volcano), the central part is constricted, causing thrusting, whereas at the base, an annular extensional zone is formed. The former is termed volcano spreading and the

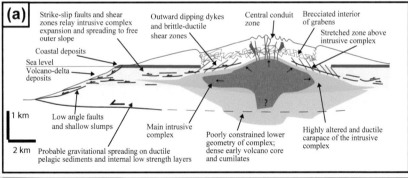

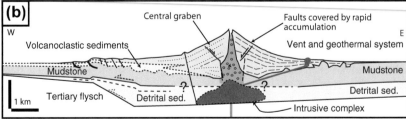

FIGURE 5.4   Examples of gravitational deformation of volcanoes. (a) Cross-section of Piton des Neiges volcano, La Reunion, showing internal structure, deformation and its relationship to structures. *(From Delcamp et al. (2012).)* (b) Concepcion, Nicaragua, also seen in the cross-section, showing large-scale gravitational spreading and sagging on weak lake sediments (Borgia and van Wyk de Vries, 2003).

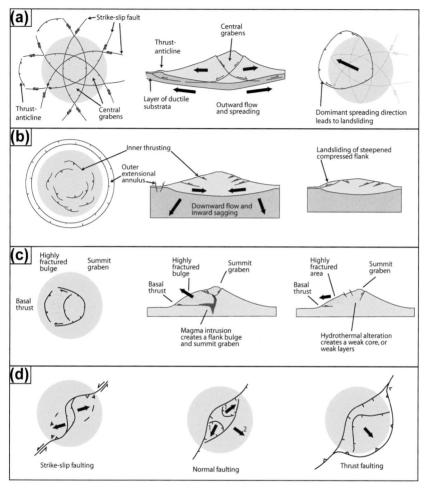

FIGURE 5.5   Styles of deep-seated volcanic gravitational deformation and landsliding shown in simple sketches. (a) Spreading, with the plan view and cross-section of ideal fault geometry around a stratocone with central grabens and basal strike-slip or thrust faults. Spreading in a predominant direction can favor landsliding, whereas radial spreading does not. (b) Sagging, with the plan view and cross-section showing the compression created by downward flow in the substrata, and thrusting in the edifice. Landslides may be generated on the constrained flanks. (c) Plan view and cross-sections of structural configurations of flank spreading triggered by magma intrusion or hydrothermal alteration (Cecchi et al., 2005; Donnadieu and Merle, 1998). (d) Deformation in volcanoes influenced by regional tectonics, with the consequent landslide directions. Note for normal faulting, near parallel small-scale landslides are created first. For vertical faulting, the situation is similar to thrusting.

latter volcano sagging (Borgia et al., 2000; van Wyk de Vries and Borgia, 1996; Byrne et al., 2013).

The deformation and stress states imposed by the gravitational deformation can have major impacts on magmatic and eruptive activity. Sagging may tend to impede magma rise by constricting the upper edifice. Spreading, in contrast, allows rifting and possibly more rapid or earlier eruption of magma. Changes in composition of magmas and eruptive style have been attributed to changes in gravitational stress fields, such as at Concepcion (Borgia and van Wyk de Vries, 2003) and Etna (Borgia et al., 1992; Longpré et al., 2009). Clear links between flank movement and eruption have also been observed at Mt Etna (e.g., Lundgren et al., 2004).

The volcanic edifice is built by the superposition of deposits and intrusion of magma according to depositional and intrusive processes. Primary volcanic deposits are formed by lava flows, including their autobreccias, airfall (tephra), and pyroclastic density currents. Erosion and modification of these occur by aeolian, marine, fluvial, and glacial activity, and by mass movements, these form epiclastic sediments. Internal changes, such as hydrothermal alteration, can alter rock properties and weaken the edifice. The load of the volcano may thus become larger than the rock's resistance to deformation, and flanks can deform. Such deformation has been termed flank spreading, and may cause a flattening of the summit area, and bulging of the flanks (Cecchi et al., 2005).

5.6 REGIONAL TECTONIC INFLUENCES

Regional tectonic movement around a volcano will combine with gravitational deformation. The lifetime of a volcanic edifice is long enough for several hundreds of meters of fault movement to occur. In extensional settings, the gravitational load may focus rifting and faults, and these may lead to landslide generation. In areas of strike-slip movement, volcanoes may initiate pull-apart grabens, and volcano-tectonic depressions and calderas and landslides may result (Lagmay et al., 2000; van Wyk de Vries and Merle, 1996; Merle et al., 2001). In compressional settings, thrusts may be deflected around volcanoes (Galland et al., 2007).

5.7 PRIMING OF VOLCANIC LANDSLIDES

A volcanic landslide is largely caused by the presence of weak zones in a volcano. These can be original weaknesses in the volcanic materials, such as weakly consolidated pyroclastics, or lavas that fractured during cooling. Weak material can also be formed by weathering and alteration. Soils can be incorporated by burial from eruptive products, and rock can be altered by hydrothermal activity. Rock may also be fractured due to deformation, by intrusion, by tectonic movements, or due to gravitationally induced deformation.

Water is an important cause of landslides in granular materials because its pore pressure can reduce the strength of a rock mass. For this reason, the distribution of water in a volcano is an important factor in instability. Volcanoes that hold large volumes of water, for example, in wet climates or with glaciers, and so may have increased landslide susceptibility; for example,

Glicken (1996) estimated that the precollapse edifice of Mt St Helens had an initial porosity of about 14% and was about 92% saturated. Changes in climate may alter this susceptibility, so increased aridity may lead to more stable volcanoes, for example. Sea-level changes may also affect the stability of oceanic volcanoes, in either altering the gravitational forces on the edifice or changing the hydrological conditions. Erosion of a volcano (e.g., by minor landsliding) can steepen slopes and remove retaining material, thus making a larger landslide possible.

Eruptions will progressively load weak zones with new material, and thus can contribute to instability. Ultimately, eruptions add the material that will eventually collapse, and by adding material rapidly, the likelihood of landsliding may be increased.

5.8 TRIGGERING VOLCANIC LANDSLIDES

A volcano can evolve to an unstable state over long time scales (tens to tens of thousands of years) with slow deformation characterizing the progressive weakening (van Wyk de Vries and Francis, 1997). This slow creep within the whole edifice could in itself lead to eventual collapse, but various triggers can cause a landslide and produce a debris avalanche.

Intrusion of new magma, such as occurred at Mt St Helens, can lead to large amounts of internal and surface deformation. This phenomenon steepens flanks, and reduces rock strength by creating shear zones, together with pervasive brecciation of the deforming mass.

Changes in the hydrothermal system may trigger collapse by increasing pore pressures and causing fluid migration. The hydrothermal system is ultimately driven by magma intrusion, and new intrusions can thus disturb, and trigger a landslide. Internal changes in the hydrothermal system, such as self-sealing, where pores are blocked by precipitation of minerals, can also alter pore fluid pressures and lead to landsliding.

Large rainfall events, or ice melting, could contribute to triggering at least smaller collapses, such as at Casita (rain in 1988) and Mt Meager (ice melt in 2009). Eruptions load the edifice and rapid pressure increase, such as occurs during dome growth, may lead to landsliding. This has been observed at Montserrat in 2009 and during the many growth phases of St Augustine, Aleutians (Swanson and Kienle, 1988), and Shiveluch, Kamchatka (Belousov et al., 1999).

Earthquakes may cause failure by providing an extra force to trigger the landslide, by increasing pore pressures, or both. Regional faults cutting the volcano may also trigger failure by directly causing deformation as a result of fault movement. During caldera formation, stress changes and seismic release may also generate failures, as suggested for the island of Tenerife in the Canaries (Hurlimann et al., 1999). Eruptions themselves, however, may not have much effect on edifice stability, because their direct energy release does not directly affect most of the edifice. Conversely, there are plenty of cases where landslides have set off eruptions, such as at Mt St Helens in 1980.

5.9 THE STRUCTURE OF VOLCANIC LANDSLIDES

Volcanic landslides share general similarities with nonvolcanic landslides in having a head region, where there is normal faulting, and often a graben, side flanks with strike-slip movement and a thrusting toe (Figures 5.1 and 5.6).

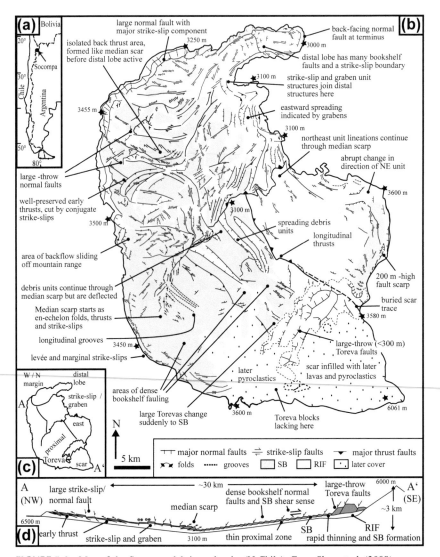

FIGURE 5.6   Map of the Socompa debris avalanche (N Chile). *From Shea et al. (2008).*

The landslide can affect the whole flank of a volcano, such as the Hilina Slump on Kilauea (Moore et al., 1994; Morgan et al., 2003), the Eastern Flank of Etna (Borgia et al., 1992), and the east—southeast side of Piton de la Fournaise (Upton and Wandsworth, 1965), or it may be a much more restricted area, such as the southeast flank of Casita (Scott et al., 2004), the 2009 Mt Meager landslide, and those at St Augustine (Swanson and Kienle, 1988).

The clearly intrusion-related landslides at Mt St Helens and Bezymianny have the same general geometry, with the difference that the central part of the landslide is also uplifted by the intruding cryptodome. The crucial difference between volcanic and nonvolcanic landslides is that the former often involve clastic rocks with a high water content, whereas a landslide in nonvolcanic rock often involves massive impermeable rock containing relatively little water—especially with earthquake triggering.

5.10 VOLCANIC LANDSLIDE DEPOSITS

Volcanic debris avalanches develop directly from landslides, and the structure of the deposit represents a link with the initial landslide configuration. This has been well studied with analog models of volcanic landslides (Andrade and van Wyk de Vries, 2010; Paguican et al., 2014). On the largest landslides, the head is often preserved as Toreva blocks (large rotational blocks), as at Socompa (Wadge et al., 1995; van Wyk de Vries et al., 2001, Figure 5.6). The head of the landslide, the landslide graben, and the frontal area are often preserved, but are highly stretched in the deposit (Figure 5.7). The graben area is often preserved as an area of elongate, transverse-to-motion parallel hummocks, and the frontal area as a wide zone of variously aligned hummocks, ridges, and/or fault structures.

A DAD can be divided into proximal, medial, distal, and marginal regions that roughly correspond to the head of the landslide, the graben, and the frontal part (Figure 5.8). Most volcanic DADs have hummocky and ridged surfaces. Hummock size can vary from several kilometers wide and hundreds of meters high to meter-sized mounds.

5.10.1 Scar

The scar can be horseshoe shaped or triangular and may cut only the edifice, or extend out into the piedmont and reach into the substrata.

5.10.2 Toreva Blocks

These are the backward-rotated slide blocks (Reiche, 1937) coming from the scar head, often preserved in the largest-volume avalanches that cut deep into the volcano and substrata. For avalanches that drop over the flank of the volcano, or into deeply eroded terrain, these are not preserved.

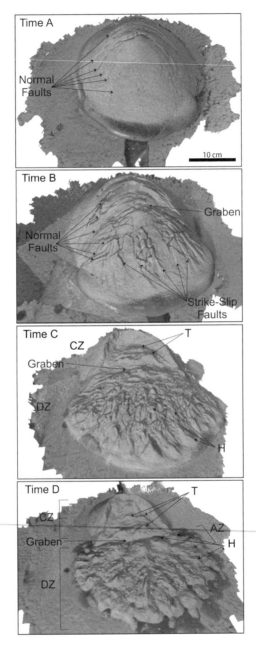

FIGURE 5.7 Example of landslide to debris avalanche evolution provided by an analogue model (Paguican et al., 2014). (A)−(D) Progressive development of structures, from the initial landslide phase (A), the formation of a back zone of arcuate normal faults; a central graben area, and a distal zone cut by numerous transtensional faults, with a thrusting front. The individualization of hummocks is seen in the later two models (C and D). CZ = Collapse zone; DZ = Deposit zone, H = Hummock; T = Toreva, AZ = Accumulation zone.

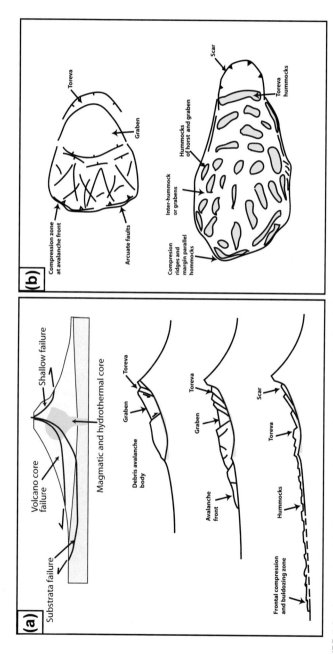

FIGURE 5.8 A general model of a volcanic debris avalanche that has developed from a landslide shown as a sketch. (a) Cross-sections, showing the early conditions in the landslide mass, and the development of structures during avalanche spreading. (b) Plan view of the development of hummocks and surface structures as the avalanche spreads.

5.10.3 Hummocks

These are characteristic features of debris avalanches and landslides. Hummocks are formed during the spreading of an avalanche mass, and are extensional features (Davies and McSaveney, 2009). They are commonly arranged linearly in the direction of motion, and reduce in size with increasing distance from the source (McColl and Davies, 2011). Two types of hummock have been identified: those formed as horst and graben features, and those by boudinage, or isolation of individual pieces of an edifice. Horst and graben hummocks include Torevas, and can range from several kilometer-wide to tens of meters. Figure 5.1(g) shows the internal structure of a hummock exposed in a quarry at Iriga, Philippines, and images of hummocks at Mt Shasta (California) and Mt Meager (Canada). For a review of hummock formation, see Paguican et al. (2014).

5.10.4 Interhummock Areas

The zones between hummocks in a debris avalanche generally form a greater proportion of the avalanche surface area. The proportion generally increases away from the source, as hummocks become sparser. These areas form originally from grabens, that separate the hummock horsts, and it is in these grabens that most of the extension is accommodated. Thus, the interhummock areas are zones of intense extension. They are generally made of matrix and mixed facies (see below) with significant concentrations of incorporated substrate. These areas are generally where water is concentrated, and secondary lahars may be generated from them (as at Mt St Helens). If the water-saturated volume is a sufficiently large proportion of the debris avalanche, it may transform from a sliding mass into a debris flow.

5.10.5 Ridges

Longitudinal ridges may form on avalanches, due to the production of long hummocks bordered by strike-slip faults in lateral areas. A family also occurs of highly elongate ridges of some meters high, tens of meters wide and up to a kilometer long that may be related to differential granular sorting during high-velocity motion (Dufresne and Davies, 2009). Transport-normal (lateral) ridges are formed by avalanche compression (commonly in distal regions), so that folds and thrusts are produced.

5.10.6 Marginal Zones

Debris avalanche sides and fronts can be steep sided; large lobes are in some places present, and commonly the front is ridged with compressional structures. In exposed fronts, the ridges are commonly seen to contain, or be entirely composed of, substrate bulldozed at the front of the mass (Belousov et al., 1999).

5.10.7 Deposit Facies

Within a volcanic DAD, distinct facies occur (Figure 5.9). They are generally separated into toreva-and-block facies, matrix facies, and mixed facies. Although this is a simple classification (Glicken, 1996; Siebert, 1984), it serves well as a first separation.

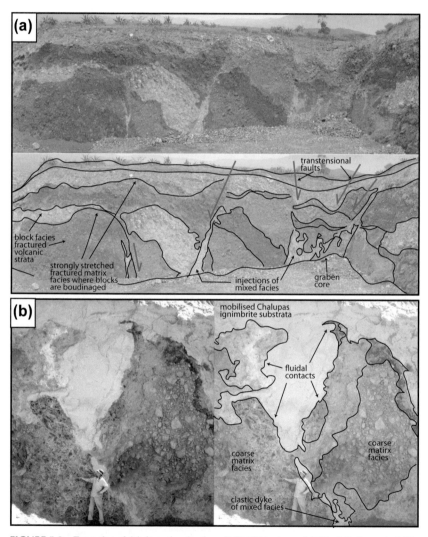

FIGURE 5.9   Examples of debris avalanche deposit (DAD) textures. (a) Block facies material in the Imbabura debris avalanche, Ecuador. This photograph shows the well-preserved bedding, of lavas and pyroclastics. Some large faults displace these, along which there are injections of light-colored mixed facies. The lithologies are of very different thicknesses across the faults, due to large amounts of strike-slip displacement. (b) Matrix and facies and the incorporated Chalupas ignimbrite in the Chimborazo DAD, Ecuador.

5.10.8 Block Facies

Block facies are composed of intact to highly fractured parts of the original slide mass. The classic texture is that of jig-saw crack, or jigsaw fit, where the original position of the elements is maintained, but the mass is shattered. A complete grading occurs between block facies and matrix facies.

5.10.9 Matrix Facies

Matrix facies are composed of highly brecciated block facies, and generally surround or occur within block facies of the same origin.

5.10.10 Mixed Facies

Mixed facies are composed of sand to silt-size fragments and isolated blocks (of up to a few meters in diameter) of a highly polylithologic character. The mixing of the elements may partly be related to the incorporation of originally mixed volcaniclastic and fluviatile sediments, but can be also due to strong mixing in the moving mass. Fluidial textures are common features at the boundary of mixed facies and other facies, and migration of individual grains or volumes of one facies are observed. This indicates that the mixed and matrix facies behave as a granular fluidized mass, at least part of the time during emplacement. Faults also occur, and clastic dikes may cross the matrix, indicating brittle behavior at some stages during emplacement.

Clastic dikes and injected material are commonly seen as injections of mixed facies into the matrix and block facies, although larger areas of injection can form significant areas within the deposit. Injections again point to a granular fluid phase existing as part of the mass at some periods during emplacement.

5.10.11 Basal Facies

A basal layer commonly occurs beneath a volcanic debris avalanche that has a finer matrix than the overlying mass (Figure 5.10). Frictionite and abrasion textures may occur in this, and commonly evidence exists of substrata incorporation. This occurs either by intershearing of material, diapiric rise of material, or by abrasion. This fine layer has been interpreted to indicate that much of the shear of the moving mass is concentrated basally (Davies et al., 2010) as required for formation of hummocks.

5.11 DEBRIS AVALANCHE TEXTURES AND STRUCTURES

All debris avalanche facies have a characteristic brecciated character. Jigsaw crack and jigsaw-fit textures are common. Clasts tend to be angular, but shape is dependent on the lithological type, so clays and other ductile materials can have

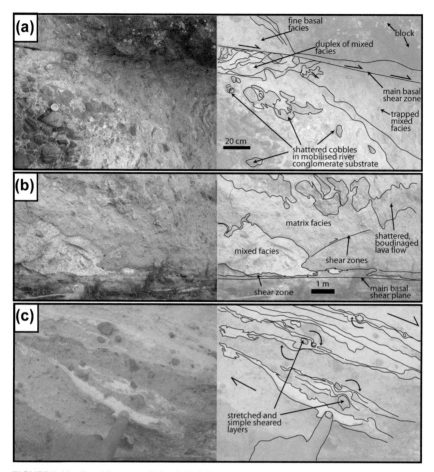

FIGURE 5.10   Basal Layers at Volcanic Debris Avalanches. (a) Deformed and mobilized cobbles at the base of the Perrier debris avalanche, France. (b) Highly deformed and planar basal zone at the PichuPichu debris avalanche, Peru. (c) Closeup of stretched and rotated elements in the PichuPichu debris avalanche.

rounded shapes. Sorting is poor, but size differences exist, generally relating to variations in brecciation (finer material in crush zones) or to original lithologies. As all types of volcanic material are incorporated into the debris avalanche, all types of textural variation are possible within the block facies, and partly within the mixed facies. Thus, volcanic debris avalanches in the field are commonly difficult to tell apart from other volcanic in-place deposits, unless the brecciated zones are apparent or other contextual information is available.

Faults and shear zones are common in deposits. The nature of these struc-tures depends on the rheology of the materials. Contacts between rigid blocks and matrix can be smoothed and contain gouge zones and slickensides, for example. In the finer, granular deposits, faults tend to occur as broad shear zones with no discrete surface of failure. Folds are commonly present in basal parts where sediments are incorporated. Faults are commonly related to clastic dikes.

5.12 SECONDARY HAZARDS OF VOLCANIC LANDSLIDES

Volcanic debris avalanches may create hazardous secondary effects. The most immediate may be the destabilization of hydrothermal systems and magmatic systems. Collapses accompanied by eruptions such at Mt St Helens in 1980 and Bandai-San in 1888 are examples. In both cases, it is likely that the magmatic and hydrothermal activity played a part in preparing and triggering the original landslide, and that once failure occurred, decompression caused large blasts. In the case of Mt St Helens, and Bezymyanny in 1956, large plinian eruptions were also triggered by the landslides.

Debris avalanches may convert to debris flows, and they may travel even further than the initial mass (Caballero and Capra, 2011; Capra and Macías, 2000, 2002; Capra et al., 2002; Vallance and Scott, 1997; Stoopes and Sheridan, 1992; Iverson et al., 1997). Debris avalanches entering water bodies may create tsunamis, such as at Mayu-Yama, Japan, in 1792 (Miyachi, 1992); a small tsunami was created by a minor landslide on Stromboli in 2002 (Tinti et al., 2005), and larger ones occur in the prehistorical records around large oceanic volcanoes (Paris et al., 2013). As the tsunamis are created by an effective point source, rather than on a very long fault line as in earthquake-generated tsunamis, the energy dissipates more rapidly and large-scale destruction is generally restricted to areas close to the volcano.

Once a landslide and avalanche have occurred, the pressure drop in the hydrothermal system can generate precipitation of epithermal mineral deposits (Silitoe, 1994), preceding phreatic eruptions, such as at Bandai-San in 1888. Decompression may occur even deeper and result in the mobilization of magma even down to the mantle, as suggested for anakramite eruptions in the Canary islands (Longpré et al., 2009; Manconi et al., 2009). Renewed volcanic activity leads to a rapid masking of the landslide scar, and many volcanoes with known debris avalanches do not show this in their morphometry (Grosse et al., 2009). This shows that landsliding is commonly a response to, or cause of, increased activity at a volcano.

5.13 VOLCANIC LANDSLIDE TRANSPORT MECHANISMS

Explaining the long runout of volcanic (and nonvolcanic) avalanches has posed a major and still unresolved problem. The debris avalanches clearly flow much further than expected if the motion is constrained by simple granular friction. However, there is no general agreement on what process or processes cause this. This is partly because the evidence of transport processes is hidden in the hard-to-study (especially basal) deposits, where the appropriate analytical tools have yet to be developed or properly applied.

Data used to describe runout involve many geometric and geological factors, leading to a wide scatter on any plot of geometric variables (Figure 5.3). The role of constraining topography and the profile of the runout path, and the rheology of the mass, are the major factors controlling the deposit geometry (Figure 5.4).

Numerical, analytical, and analog models (Kilburn and Sorensen, 1998; Dade and Huppert, 1998; Kelfoun and Druitt, 2006; Iverson and Denlinger, 2001; Denlinger and Iverson, 2001; Davies and McSaveney, 2009; Legros, 2002; Thompson et al., 2010; Shea et al., 2008) thus do not yet have the necessary foundation to discriminate between the possible processes. Proposed mechanisms for long runout distances are transport on an air cushion (Shreve, 1966); travel on a watery base (Shaller, 1991), or incorporation of water rich clays (Vallance and Scott, 1997; Stoopes and Sheridan, 1992; Iverson et al., 1997; Capra and Macías, 2000); fluidization by vibration (Campbell, 1989; Campbell et al., 1995); and lubrication by release of elastic fracture energy during intense basal fragmentation (Davies et al., 1999; Davies et al., 2010, 2012). Numerical models for debris avalanches generally require some retarding stress and a reduced basal friction, but, with the exception of Davies et al. (2010), do not specify the exact physical process that causes this (e.g., Kelfoun and Druitt, 2006; Dondin et al., 2012).

5.14 HAZARDS FROM VOLCANIC LANDSLIDES

The above summary gives an indication of the types and magnitudes of volcanic landslides, which can provide an idea of the hazards and risks that they pose. A possibility exists for detecting precursory signs through monitoring of increased or intense slope movement, surface fracturing, changes to the hydrothermal and hydrogeological system, or increased seismicity on volcanoes. If such symptoms are detected, evacuation is probably the most effective mitigation, but would depend on assessment of the likelihood of an event, something that is still poorly constrained.

Our knowledge of landslide generation in volcanoes is still rudimentary and some triggering mechanisms may still not be known. At present, it is thus possible that volcanic landslides could occur with no or little warning. Most volcanoes are not monitored for slope stability, so the likelihood of undetected landslide and debris avalanche events is high. However, if adequate monitoring systems are in place, then the acceleration of ground deformation that often occurs before a landslide should be detected. Nevertheless, sudden failure could be triggered by sudden changes in a volcano, such as hydrothermal system disruption, and sudden magmatic intrusion, or as a result of seismic shaking (Reid, 2004; Reid et al., 2010). Such landslides could be small, but the risk depends on how many people are in the potential impacted area, how suddenly the event occurs, and what prior mitigation measures have been taken.

Response to detected prelandslide deformation requires an assessment of the possible runout and area to be covered by the impending debris avalanche. Numerical models (Iverson and Denlinger, 2001; Kelfoun and Druitt, 2006; Davies et al., 2010) require a knowledge of the volume and location of the landslide, and the detailed topography of the runout area. These estimations are still based on untested physical assumptions, and poorly constrained material properties, so caution should be exercised in hazard predictions.

Anticipating secondary hazards (e.g., the Mt St Helens lateral blast) requires integrated landslide and volcanological expertise. Similarly, volcanic landslide-generated tsunamis require both subaerial and submarine modeling. For this, the location, the volume of landslide and the emplacement rate are required. To create large tsunamis, rapid emplacement into water of a large mass is required, whereas if the landslide is slow, and/or piecemeal, then the tsunami hazard is low (Kelfoun et al., 2010).

Some volcanoes have large cities at their feet (e.g., Mexico City, Tokyo, Arequipa), or a park frequented by many tourists (e.g., Fuji, Yellowstone, Tongariro national parks Kilimanjaro), and thus, even small events may have serious risks.

5.15 SUMMARY

Due to their internal dynamics, growth, and style of evolution, volcanoes are particularly prone to landslides. Slow loading due to edifice growth causes both substrata and edifice to deform, an effect amplified by hydrothermal alteration, elevated fluid pressures, and magma intrusion. Slow gravitational deformation is common on volcanoes and can generate landslides. These can transform into rapid, long runout debris avalanches.

The structure of these debris avalanches is inherited from the edifice, via the initial landslide conditions, and consists of horsts (hummocks) and grabens (interhummock) areas. The lithology of DADs is categorized into a block facies of deformed edifice rock, grading into matrix facies of highly fractured rock, and a mixed facies where all types of lithologies are combined. The base of the mass characteristically has a finer-grained layer that is highly fragmented and can show extreme abrasion, melting, and stretching.

All types of volcanoes undergo landslides and debris avalanches, the largest examples of which originate from the oceanic hot spot volcanoes, and the giant Olympus Mons Volcano on Mars. Debris avalanches can cause severe consequential effects, such as eruption and tsunami.

The physical processes in volcanic landslides and the mobility of volcanic debris avalanches are still not well constrained and will be a major research challenge in the next few years. Equally, the prediction of the occurrence and runout of the volcanic landslides and debris avalanches is a challenge for the coming years.

REFERENCES

Andrade, D., van Wyk de Vries, B., 2010. Structural analysis of the early stages of catastrophic stratovolcano flank-collapse using analogue models. Bull. Volcanol. 72, 771–789.

Belousov, A., Belousova, M., Voight, B., 1999. Multiple edifice failures, debris avalanches and associated eruptions in the Holocene history of Shiveluch volcano, Kamchatka, Russia. Bull. Volcanol. 61 (5), 324–342.

Bernard, B., van Wyk de Vries, B., Barba, D., Leyrit, H., Robin, C., Alcaraz, S., Samaniego, P., 2008. The Chimborazo sector collapse and debris avalanche: deposit characteristics as evidence of emplacement mechanisms. J. Volcanol. Geotherm. Res. 176, 36–43.

Borgia, A., Delaney, P.T., Denlinger, R.P., 2000. Spreading volcanoes. Annu. Rev. Earth Planet. Sci. 28, 3409–3412.

Borgia, A., Ferrari, L., Pasquarè, G., 1992. Importance of gravitational spreading in the tectonic and volcanic evolution of Mount Etna. Nature 357, 231–235.

Borgia, A., van Wyk de Vries, B., 2003. The volcano-tectonic evolution of Concepción, Nicaragua. Bull Volcanol 65, 248–266.

Byrne, P.K., Holohan, E.P., Kervyn, M., van Wyk de Vries, B., Troll, V.R., Murray, J.B., 2013. A sagging-spreading continuum of large volcano structures. Geology 41, 339–342.

Caballero, L., Capra, L., 2011. Textural analysis of particles from El Zaguán debris avalanche deposit Nevado de Toluca volcano, Mexico: evidence of flow behavior during emplacement. J. Volcanol. Geotherm. Res. 200, 75–82.

Campbell, C.S., 1989. Self-lubrication for long runout landslides. J. Geol. 97 (6), 653–665.

Campbell, C.S., Cleary, P.W., Hopkins, M.J., 1995. Large-scale landslide simulations: global deformation, velocities and basal friction. J. Geophys. Res. 100, 8267–8283.

Capra, L., Macías, J.L., 2000. Pleistocene cohesive debris flows at Nevado de Toluca Volcano, central Mexico. J. Volcanol. Geotherm. Res. 102, 149–167.

Capra, L., Macías, J.L., Scott, K.M., Abrams, M., Garduño-Monroy, V.H., 2002. Debris avalanches and debris flows transformed from collapses in the Trans-Mexican Volcanic Belt, México—behavior, and implication for hazard assessment. J. Volcanol. Geotherm. Res. 113 (1–2), 81–110. Elsevier.

Capra, L., Macías, J.L., 2002. The cohesive Naranjo debris-flow deposit (10 km^3): a dam breakout flow derived from the pleistocene debris-avalanche deposit of Nevado de Colima Volcano (Mexico). J. Volcanol. Geotherm. Res. 117, 213–235. Elsevier.

Cecchi, E., van Wyk de Vries, B., Lavest, J.M., 2005. Flank spreading and collapse of weak-cored volcanoes. Bull. Volcanol. 67, 72–91.

Clavero, J.E., Sparks, R.S.J., Huppert, H.E., 2002. Geological constraints on the emplacement mechanism of the Parinacota avalanche, northern Chile. Bull. Volcanol. 64 (1), 40–54.

Crandell, D.R., 1989. Gigantic debris avalanche of Pleistocene age from ancestral Mount Shasta volcano, California, and debris-avalanche hazard zonation. U.S. Geol. Surv. Bull. 1861, 32.

Dade, W.B., Huppert, H.E., 1998. Long-runout rockfalls. Geology 26 (9), 803–806.

Davies, T.R.H., McSaveney, M.J., Hodgson, K.A., 1999. A fragmentation-spreading model for long runout rock avalanches. Can. Geotech. J. 36 (6), 1096–1110.

Davies, T.R.H., McSaveney, M.J., 2009. The role of dynamic rock fragmentation in reducing frictional resistance to large landslides. Eng. Geol. 109, 67–79. http://dx.doi.org/10.1016/j.enggeo.2008.11.004.

Davies, T.R.H., McSaveney, M.J., Kelfoun, K., 2010. Runout of the Socompa volcanic debris avalanche, Chile: a mechanical explanation for low basal shear resistance. Bull. Volcanol. 72, 933–944. http://dx.doi.org/10.1007/s00445-010-0372-9.

Davies, T.R.H., McSaveney, M.J., Boulton, C.J., 2012. Elastic strain energy release from fragmenting grains: effects on fault rupture. J. Struct. Geol. 38, 265−277.

Denlinger, R.P., Iverson, R.M., 2001. Flow of variably fluidized granular masses across three-dimensional terrain 2. Numerical predictions and experimental tests. J. Geophys. Res. 106 (B1), 553−566.

Delcamp, A., van Wyk de Vries, B., James, M., Gailler, L., Lebas, E., 2012. Relationships between volcano gravitational spreading and magma intrusion. Bull. Volcanol. 74, 743−765.

Delcamp, A., Van Wyk de Vries, B., Kervyn, M., 2014. The Lemptégy Scoria Cone, Chaîne des Puys France. Geosphere; October 2014, 10 (5), 1−22. http://dx.doi.org/10.1130/GES01007.1 (Geological Society of America, Geosphere accepted March 2014).

Donnadieu, F., Merle, O., 1998. Experiments on the indentation process during cryptodome intrusions: new insights into Mount St. Helens deformation. Geology 26 (1), 79−82.

Dondin, F., Lebrun, J.-F., Kelfoun, K., Forunier, N., Randriansolo, A., 2012. Sector collapse at Kick 'em Jenny submarine volcano (Lesser Antilles): numerical simulation and landslide behaviour. Bull. Volcanol. 74, 595−607.

Dufresne, A., Davies, T.R., 2009. Longitudinal ridges in mass movement deposits. Geomorphology 105 (3−4), 171−181.

Erismann, T.H., 1979. Mechanisms of large landslides. Rock Mech. Rock Eng. 12 (1), 5−46.

Friele, P., Jakob, M., Clague, J., 2008. Hazard and risk from large landslides from Mount Meager volcano, British Columbia, Canada. Georisk: Assess. Manage. Risk Eng. Sys. Geohazards 2, 48−61.

Galland, O., Hallot, E., Cobbold, P.R., Ruffet, G., de Bremond d'Ars, J., 2007. Volcanism in a compressional Andean setting: a structural and geochronological study of Tromen volcano (Neuquén province, Argentina). Tectonics 26. http://dx.doi.org/10.1029/2006TC002011.

Glicken, H., 1996. Rockslide-debris avalanche of May 18, 1980, Mt. St. Helens volcano, Washington. Open-file Report 96−677. US Geological Survey. 90 pp.

Gorshkov, G.S., 1959. Gigantic eruption of the volcano Bezymianny. Bull. Volcanol. 20, 77−109.

Grosse, P., van Wyk de Vries, B., Petrinovic, I.A., Euillades, P.A., Alvarado, G.E., 2009. The morphometry and evolution of arc volcanoes. Geology 37 (7), 651−654.

Guthrie, R.H., Friele, P., Allstadt, K., Roberts, N., Evans, S.G., Delaney, K.B., Roche, D., Clague, J.J., Jakob, M., 2012. The 6 August 2010 Mount Meager rock slide-debris flow, Coast Mountains, British Columbia: characteristics, dynamics, and implications for hazard and risk assessment. Nat. Hazards Earth Syst. Sci. 12, 1277−1294.

Guthrie, R.H., Friele, P., Allstadt, K., Roberts, N., Evans, S.G., Delaney, K.B., Roche, D., Clague, J.J., Jakob, M., 2012. The 6 August 2010 Mount Meager rock slide-debris flow, Coast Mountains, British Columbia: characteristics, dynamics, and implications for hazard and risk assessment. Natural Hazards and Earth System Science 12 (5), 1277−1294.

Heim, A., 1932. Bergsturz und Menchenleben, Zurich. Vierteljahrsschrift 77 (20), 218.

Hickson, C.J., Russell, J.K., Stasiuk, M.V., 1999. Volcanology of the 2350 B.P. eruption of Mount Meager volcanic complex, British Columbia, Canada: implications for hazards form eruptions in topographically complex terrain. Bull. Volcanol. 60, 489−507.

Hürlimann, M., Turon, E., Marti, J., 1999. Large landslides triggered by caldera collapse events in Tenerife, Canary Islands. Phys. Chem. Earth Part A Solid Earth Geodesy 24, 921−924.

Iverson, R.M., Denlinger, R.P., 2001. Flow of variably fluidized granular masses across three-dimensional terrain 1. Coulomb mixture theory. J. Geophys. Res. 106 (B1), 537−552.

Iverson, R.M., Reid, M.E., La Husen, R.G., 1997. Debris-flow mobilization from landslides. Annu. Rev. Earth Planet. Sci. 25, 85−138.

Kelfoun, K., Druitt, T.H., 2006. Numerical modelling of the emplacement of Socompa rock avalanche, Chile. J. Geophys. Res. 110 (B12), 12202.

Kelfoun, K., Giachetti, T., Labazuy, P., 2010. Landslide-generated tsunamis at Réunion Island. J. Geophys. Res. 155, F04012. http://dx.doi.org/10.1029/2009JF001381.

Kilburn, C., Sorensen, S.-A., 1998. Runout lengths of sturzstroms: the control of initial conditions and of fragment dynamics. J. Geophys. Res. 103 (B8), 17877.

Lagmay, A.M.F., van Wyk de Vries, B., Kerle, N., Pyle, D.M., 2000. Volcano instability induced by strike-slip faulting. Bull. Volcanol. 62 (4−5), 331−346.

Legros, F., 2002. The mobility of long-runout landslides. Eng. Geol. 63, 301−331.

Lebas, E., Le Friant, A., Boudon, G., Watt, S.F.L., Talling, P.J., Feuillet, N., Deplus, C., Berndt, C., Vardy, M.E., 2011. Multiple widespread landslides during the long-term evolution of a volcanic island: Insights from high-resolution seismic data, Montserrat, Lesser Antilles. Geochem. Geophys. Geosyst. 12 (5), 1−20, 05/2011.

Longpré, M.A., Troll, V.R., Walter, T.R., Hansteen, T.H., 2009. Volcanic and geochemical evolution of the Teno massif, Tenerife, Canary Islands: some repercussions of giant landslides on ocean island magmatism. Geochem. Geophys. Geosyst. 10 (n/a-n/a). http://dx.doi.org/10.1029/2009GC002892 (Impact Factor: 2.94). 12/2009.

Luhr, J.F., Simkin, T., 1993. Paracutin: the Volcano Born in a Mexican Cornfield. Geoscience Press, 427 p.

Lundgren, P., Casu, F., Manzo, M., Pepe, A., Bernadino, P., Sonsosti, E., Lanari, R., 2004. Gravity and magma induced spreading of Mount Etna volcano revealed by satellite radar interferometry. Geophys. Res. Lett. 31, L04602. http://dx.doi.org/10.1029/2003GL018736.

Manconi, A., Longpre, M.A., Walter, T.R., Troll, V.R., Hansteen, T.H., 2009. The effects of flank collapses on volcano plumbing systems. Geology 37, 1099−1102.

McColl, S.T., Davies, T.R.H., 2011. Evidence for a rock-avalanche origin for 'The Hillocks' "moraine," Otago, New Zealand. Geomorphology 127, 216−224.

Merle, O., Vidal, N., van Wyk de Vries, B., 2001. Experiments on vertical basement fault reactivation below volcanoes. J. Geophys. Res. 106, 2153−2162.

Moore, J.G., Nomak, W.R., Holcomb, R.T., 1994. Giant Hawaiian landslides. Science 264, 46−47.

Morgan, J.K., Moore, G.F., Clague, D.A., 2003. Slope failure and volcanic spreading along the submarine south flank of Kilauea volcano, Hawaii. J. Geophys. Res. 108, 2415.

Miyachi, M., 1992. Geological examination of the two old maps from the Tokugawa Era concerning the Shimabara Catastrophe. In: Yanagi, T., Okada, H., Ohta, K. (Eds.), Unzen Volcano, the 1990−1992 Eruption. Nishinippon & Kyushu University Press, pp. 99−102.

Nakamura, K., 1980. Why do long rift zones develop in Hawaiian volcanoes—a possible role of thick oceanic sediments. Bull. Volcanol. Soc. Jpn. 25, 255−269.

Naranjo, J.A., Francis, P.W., 1987. High velocity débris avalanche deposit at Lastarria volcano in the North Chilean Andes. Bull. Volcanol. 49, 509−514.

Paguican, E.M.R., Van Wyk de Vries, B., Lagmay, A.M.F., 2014. Hummocks: how they from in and how they evolve in rockslide-debris avalanches. Landslides 11 (1), 67−80.

Paris, R., Kelfoun, K., Giachetti, T., 2013. Marine conglomerate and reef megaclasts at Mauritius Island (Indian Ocean): evidences of a tsunami generated by a flank collapse of Piton de la Fournaise volcano, Réunion Island? Sci. Tsunami Hazards 32 (4), 281−291.

Reid, M.E., 2004. Massive collapse of volcano edifices triggered by hydrothermal pressurization. Geology 32, 373−376.

Reid, M.E., Keith, T.E.C., Kayen, R.E., Iverson, N.R., Iverson, R.M., Brien, D.L., 2010. Volcano collapse promoted by progressive strength reduction: new data from Mount St. Helens. Bull. Volcanol. 72, 761−766.

Reiche, P., 1937. The Toreva-block, a distinctive landslide type. J. Geol. 45 (5), 538–548.

Riggs, N.R., Duffield, W.A., 2008. Record of complex scoria cone eruptive activity at Red Mountain, Arizona, USA, and implications for monogenetic mafic volcanoes. J. Volcanol. Geotherm. Res. 178, 763–776.

Scott, K., Vallance, K.M., Kerle, J.W., Macias, J.L., Strauch, W., Devoli, G., 2004. Catastrophic precipitation-triggered lahar at Casita volcano, Nicaragua—flow bulking and transformation. Earth Surf. Processes Landforms 30, 59–79.

Shaller, P.J., 1991. Analysis and implications of large Martian and terrestrial landslides (Ph.D. thesis), California Institute of Technology, 586 pp.

Shea, T., van Wyk de Vries, B., Pilato, M., 2008. Emplacement dynamics of contrasting debris avalanches at Volcán Mombacho (Nicaragua), provided by structural and facies analysis. Bull. Volcanol. 70, 899–921.

Shreve, R.L., 1966. Sherman landslide, Alaska. Science 154, 1639–1643.

Siebert, L., 1984. Large volcanic debris avalanches: characteristics of source areas, deposits, and associated eruptions. J. Volcanol. Geotherm. Res. 22 (3–4), 163–197.

Sekiya, S., Kikuchi, Y., 1890. The eruption of Bandai-san. Trans. Seismol. Soc. Jpn. 13 (2), 139–222.

Siebert, L., Glicken, H., Ui, T., 1987. Volcanic hazards from Bezymianny- and Bandaï-type eruptions. Bull. Volcanol. 1, 435–459.

Silitoe, R.H., 1994. Erosion and collapse of volcanoes: causes of telescoping in intrusion-centered ore deposits. Geology 10, 945–948.

Stoopes, G.R., Sheridan, M.F., 1992. Giant debris avalanches from the Colima volcanic complex, Mexico: implication for long-runout landslides (100 km) and hazard assessment. Geology 20, 299–302.

Swanson, S.E., Kienle, J., 1988. The 1986 eruption of Mount St. Augustine: field test of a hazard evaluation. J. Geophys. Res. 93, 4500–4520.

Sparks, R.S.J., Barclay, J., Calder, E.S., Herd, R.A., Luckett, R., Norton, G.E., Ritchie, L.J., Voight, B., Woods, A.W., 2002. Generation of a debris avalanche and violent pyroclastic density current on 26 December (Boxing Day) 1997 at Soufrière Hills Volcano, Montserrat. In: Druitt, T.H., Kokelaar, B.P. (Eds.), The Eruption of Soufrière Hills Volcano, Montserrat, from 1995 to 1999 (GSL Memoir 21). Geological Society, pp. 409–434.

Thompson, N., Bennet, M.R., Petford, N., 2010. Development of characteristic volcanic debris avalanche deposit structures: new insights from distinct element simulations. J. Volcanol. Geotherm. Res. 192, 191–200.

Tibaldi, A., 2001. Multiple sector collapses at Stromboli volcano, Italy: how they work. Bull. Volcanol. 63, 121–125.

Tinti, S., Manucci, A., Pagnoni, G., Armigliato, A., Zaniboni, F., 2005. The 30th December 2002 tsunami in Stromboli: sequence of the events reconstructed from the eyewitness accounts. Nat. Hazards Earth Sys. Sci. 5, 763–775.

Upton, B.G.J., Wadsworth, W.J., 1965. Geology of Réunion Island, Indian ocean. Nature 207, 151–154.

Ui, T., 1983. Volcanic dry avalanche deposits. Identification and comparison with non-volcanic debris stream deposits. J. Volcanol. Geotherm. Res. 18 (1–4), 135–150.

Vallance, J.W., Scott, K., 1997. The Osceola Mudflow from Mount Rainier: Sedimentology and hazard implications of a huge clay-rich debris flow. Geol. Soc. Am. Bull. 109, 143–163.

Vallance, J.W., Schilling, S.P., Devoli, G., Reid, M.E., Howell, M.M., Brien, D.L., 2004. Lahar hazards at Casita and San Cristobal volcanoes, Nicaragua, U. S. Geol. Surv. Open File, 01-468.

van Wyk de Vries, B., Borgia, A., 1996. The role of basement in volcano deformation. In: McGuire, W.J., Jones, A.P., Neuberg, J. (Eds.), Volcano Instability on the Earth and other Planets, Geological Society of London Special Publication, 110, pp. 95–110.

van Wyk de Vries, B., Merle, O., 1996. The effect of volcanic constructs on rift fault patterns. Geology 24 (7), 643–646.

van Wyk de Vries, B., Francis, P.W., 1997. Catastrophic collapse at stratovolcanoes induced by gradual volcano spreading. Nature 387, 387–390.

van Wyk de Vries, B., Self, S., Francis, P.W., Keszthelyi, L., 2001. A gravitational spreading origin for the Socompa debris avalanche. J. Volcanol. Geotherm. Res. 105, 225–247.

Voight, B., Glicken, H., Janda, R.J., Douglass, P.M., 1981. Catastrophic rockslide avalanche of May 18. In: Lipman, P.W., Mullineaux, D.R. (Eds.), The 1980 Eruptions of Mount St. Helens. Geological Survey Professional, Washington, U.S, pp. 347–377. Paper no.1250.

Voight, B., Janda, R.J., Glicken, H., Douglass, P.M., 1983. Nature and mechanics of the Mount St Helens rockslide avalanche of 18 May 1980. Géotechnique 33, 243–273.

Voight, B., Elseworth, D., 1997. Failure of volcanic slopes. Géothechnique 47, 1–31.

Wadge, G., Francis, P.W., Ramirez, C.F., 1995. The Socompa collapse and avalanche event. J. Volcanol. Geotherm. Res. 66, 309–336.

Ward, S.N., Day, S., 2003. Ritter Island Volcano—lateral collapse and the tsunami of 1888. Geophys. J. Int. 154 (3), 891–902.

Chapter 6

Peat Landslides

Jeff Warburton

Durham University, UK

ABSTRACT

Peat landslides form a distinct suite of slope failures that are characteristic of land-scapes where organic soils dominate. Six main types of peat mass movement are recognized: bog burst, bog flow, bog slide, peat slide, peaty-debris slide, and peat flow. Such failures have been prevalent in the British Isles, but their occurrence globally is far more widespread than hitherto reported. Peat has distinct geotechnical properties that influence its stability and govern the range of impacts of landslide events. Geo-technically, peat is a low density, organic-rich, nonmineral soil, which has a high water content, significant fiber content, high voids ratio, high compressibility, and low shear strength. Peat landslides cause significant environmental impacts at-a-site, and their runout is far traveled causing considerable downstream devastation to infrastructure and stream ecology. Peat landslides triggered by construction in upland areas demonstrate the importance of surface/subsurface drainage and surface loading in contributing to failure. Although the general mechanisms of peat failure are now well understood, considerable uncertainties associated with determining geotechnical properties of peat, and adequately assessing the hydrological conditions relating to peat instability remain.

6.1 INTRODUCTION AND BACKGROUND

Peat landslides are a characteristic rapid mass movement in areas dominated by organic soils (histosols). Commonly reported as peat slides and bog bursts, the mass movement of peat on slopes has been well documented for several centuries (Molyneux, 1697; Ouseley, 1788; Griffiths, 1821; Bailey, 1879; Crofton, 1902; Feehan and O'Donovan, 1996), yet the fundamental controls of this form of shallow instability have only relatively recently been examined in a systematic way (Warburton et al., 2004; Dykes and Kirk, 2006; Dykes and Warburton, 2007a; Boylan et al., 2008; Dykes, 2008a, 2009). Peat landslides are most common in the British Isles, and have been reported from many locations in northern parts of the United Kingdom, and throughout Ireland, from the sixteenth century to the present (Bowes, 1960; Crisp et al., 1964; Tomlinson and Gardiner, 1982; Alexander et al., 1986; Kirk, 2001; Mills,

2002; Warburton et al., 2004; Dykes and Kirk, 2006; Boylan et al., 2008; Dykes and Jennings, 2011). Although most frequently reported from the British Isles, numerous examples also occur in other peatlands throughout the world including Germany (Vidal, 1966); Switzerland (Feldmeyer-Christe and Mulhauser, 1994); Canada (Hungr and Evans, 1985); Argentina (Gallart et al., 1994); Nyika Plateau, Malawi (Shroder, 1976); Wingecarribee Swamp, Australia (Tranter, 1999); and the sub-Antarctic Islands (Selkirk, 1996).

Peat landslides tend to occur in association with heavy and/or prolonged rainfall. Some peat mass movements are very small and geomorphologically relatively insignificant involving displacement of perhaps only a few hundreds of cubic meters of peat over a short distance, but their cumulative impact on the ever-decreasing area of intact peatland cannot be ignored (Evans and Warburton, 2007). Conversely, many peat failures are very large, sometimes involving tens or hundreds of thousands of cubic meters of peat, and the short-term impacts of individual events can be extremely severe. Such events have a major impact on organic material transfer in upland drainage basins, present a hazard locally, and can have a devastating effect on stream ecology, for example, the Baldersdale slide of 1689 in Northern England caused considerable flood damage and poisoned fish for several kilometers downstream.

Figure 6.1 shows a range of examples of peat landslides. Figure 6.1(a) shows a typical peat slide on Dooncarton Mountain, Co. Mayo, Ireland, September 19, 2003; the slide was triggered by an extreme rainfall event (~90 mm in 90 min, Dykes and Warburton (2007b)). The second example, Figure 6.1(b), in contrast with Figure 6.1(a), is an example of a peat landslide triggered during the construction of a moorland road at Burnhope in the North Pennines, United Kingdom. Eye-witness accounts indicate the failure occurred during construction immediately after the passage of a large truck carrying road ballast. Figure 6.1(c) shows the aftermath of the peat slope failures that occurred in Channerwick, South Shetland, on September 19, 2003 (Dykes and Warburton, 2008b). The river system became completely choked with peaty debris and slurry due to multiple peat landslides entering the upstream water courses. This caused inundation of properties, blocked culverts and bridges, and contaminated the stream ecosystem. The final example, Figure 6.1(d) shows the main failure track of the Derrybrien peat landslide, Cashlaun-drumlahan Mountain, Co. Galway, Ireland, October 16, 2003. This peat flow appears to have been triggered by shearing within the basal peat associated with the construction of the foundations for a wind turbine at the head of the instability (Boylan et al., 2008; Dykes, 2008a).

These examples clearly demonstrate the catastrophic nature of peat land-slides both at the site of failure (Figure 6.1(a)) and their impact downstream (Figure 6.1(c)) and also the sensitive nature of upland peat soils to anthro-pogenic disturbance associated with road construction and wind farm devel-opment. Three of the peat landslides illustrated in Figure 6.1(a,c,d) were part

FIGURE 6.1   Examples of peat landslides. (a) Peat slide on Dooncarton Mountain, Co. Mayo, Ireland, September 19, 2003. (b) Peat landslide triggered during the construction of a moorland road, Burnhope, North Pennines, United Kingdom. (c) River system choked with peaty debris and slurry following peat slope failures South Shetland, Scotland, September 19, 2003. (d) Runout track from the Derrybrien peat landslide, Cashlaundrumlahan Mountain, Co. Galway, Ireland, October 16, 2003. *Photograph courtesy of Olivia Bragg.*

of a series of damaging mass movements at Dooncarton Mountain, Co. Mayo, Ireland, and South Shetland, Scotland, United Kingdom, in September 2003 and at Derrybrien, Co. Galway, Ireland, in October 2003, which brought the issue of peat instability into sharp public focus and partly stimulated renewed research into this phenomenon and a more systematic approach to the analysis of this type of failure (Evans and Warburton, 2007).

The main objectives of this chapter are to describe the key factors that control the stability of peat covered hill slopes; discuss the main morphological features of peat mass movements and their classification; and outline the main impact of peat landslides on the environment.

6.2 THE NATURE OF PEAT, ITS STRUCTURE, AND MATERIAL PROPERTIES

In many respects, peat hill slopes can be viewed like any hill slope, and in this sense, established concepts of hydrology, soil mechanics, and slope stability still broadly apply (Selby, 1993). However, the special characteristics of peat, particularly its hydraulic behavior and unusual geotechnical properties (Boelter, 1968; Huat et al., 2014), mean that the processes operating on peat hill slopes can be significantly different (Warburton et al., 2004).

6.2.1 Peat Properties

Peat is a low-density, organic, nonmineral soil that is highly compressible. Peat results from the decay of organic matter and accumulates wherever suitable conditions occur, such as in areas of high (excess) rainfall and where ground drainage is poor leading to high water tables. In these waterlogged areas, peat accumulates where the rate of dry vegetative matter deposited exceeds the rate of decay. Physiochemical and biochemical processes associated with wetland conditions ensure that the accumulating organic matter decays very slowly, meaning that plant structures remain partially intact for long periods of time (Bell, 2000). For example, in the United Kingdom, temperate peat accumulates slowly, typically 0.2−1 mm/yr with local rates varying depending on the topography and hydrology of the peat mire (Charman, 2002).

A number of characteristics distinguish peat as an engineering material. These include a high natural water content ($\sim$ 1,500 percent), very high organic content (loss on ignition 25−100 percent), significant fiber content (Fc), low specific gravity (bulk density), high voids ratio (5−15), high initial permeability, high compressibility, and low strength (Edil, 2001). The ash content (mineral content) provides one of the most common criteria for defining peat. For example, Carlsten (1993) presented a geotechnical classification, originally proposed by Landva et al. (1983), in which "peat" is defined by an ash content of <20 percent, whereas "peaty organic soils" have ash contents >20 percent but contain up to 50 percent fibers. The advantage of this definition is that it can be determined on the basis of a simple "loss on ignition" test (550 °C for 3 h: Skempton and Petley, 1970) and an estimate of the Fc. Another important property of peat is the degree of humification (the extent of biochemical decomposition of plant remains), which is a key factor that determines the overall behavior of peat. The 10-point "H" (humification) categorization used in the von Post scheme (von Post, 1924) for characterizing peat deposits is widely used in peat landslide studies (Evans and Warburton, 2007). The end members of this scale range from low values (1−2) for highly fibrous deposits with insignificant decomposition to amorphous peat (high values, 9−10) with few or no discernible plant remains. A similar distinction has been used by Macfarlane and Radford (1965) and Macfarlane (1969) to broadly categorize the engineering behavior of peat into fibrous and amorphous granular deposits, and recognizes three main categories that characterize broad divisions of peat based on Fc and the von Post scale (Edil, 2001). These are Fibric (>67 percent Fc, von Post H1—H3); Hemic (33—67 percent Fc, von Post H4—H6); and Sapric (<33 percent Fc, von Post H7—H10). Numerous other classifications of organic soils exist, but there is currently no overall standardized scheme for peat soils (Myślińska, 2003).

6.2.2 Peat Deposits and Peat Depths

Peat deposits are normally defined in terms of the depth of peat present at a location. However, the depth of these deposits is highly variable, and peat accumulates according to the local topographic context and related vegetation assemblages (Hobbs, 1986). Most surviving peat in the United Kingdom and Ireland comprises ombrotrophic blanket bog that only occasionally exceeds 2—3 m in thickness and which typically grades into thin peaty soils at the margins. In a global sense, the majority of peat occurs outside the British Isles in the histosols and "muck" deposits of the great Northern peatlands; however, >80 percent of all known peat failures have occurred in the British Isles (Mills, 2002; Dykes and Kirk, 2006). Therefore, Dykes and Warburton (2007a) have argued it is appropriate to consider British and Irish criteria for separating peats from mineral soils. Even in the restricted geographical area of the British Isles, peat deposit definitions vary. The Soil Survey of England and Wales uses 40 cm as the minimum depth for a peat deposit (Cruikshank and Tomlinson, 1990; Burton, 1996), whereas 50 cm is used in Scotland (Burton, 1996) and 45 cm (undrained) in Ireland (JNCC, 2011).

Although peat deposits are usually classified in accordance with depth, Dykes and Warburton (2007a) advocate a mass-based criterion for defining what constitutes a "peat failure," on the basis that this is closely related to standard limit-equilibrium failure models that rely on the unit weights (derived from bulk densities) of the slope materials. Measured saturated bulk densities of mineral substrates typically vary between around 1.6 and 1.8 Mg/m^3. Assuming a representative bulk density of 1.0 Mg/m^3 for peat, Dykes and Warburton (2007a) define a peat failure in terms of the mass of failed peat exceeding the mass of failed substrate. Using equivalent depths for ease of application in the field, Dykes and Warburton (2007a), defined a peat failure as a mass movement involving

1. Peat (<20 percent ash content) ≥0.4 m deep, with failure occurring entirely within the peat or at a depth in the mineral substrate below the peat <0.6 × peat depth (i.e., total failure depth <1.6 × peat depth);

   and/or

2. Peat (<20 percent ash content) 0.2—0.4 m deep with failure occurring entirely within the peat or at the peat substrate interface.

Therefore, Dykes and Warburton (2007a) recommended that the lower limit of 0.4 m for defining a peat deposit should be used for the definition of a peat failure so as to reflect the environmental importance of peat loss and thus to minimize the possible exclusion of failures involving very thin peat/organic soil.

6.2.3 "Peat" or "Bog" Mass Movements?

The terms "peat" and "bog" are frequently used and are included as part of Hutchinson's (1988) landslide classification as "peat slides" and "bog slides." However, in formal terms "peat" refers to any partly decomposed or undecomposed plant material, whereas a "bog" is an accumulation of peat in a particular hydromorphological setting that receives its water from rain or snowmelt (Evans and Warburton, 2007). This is akin to an "ombrotrophic peat bog" more commonly described as a "raised bog" or "blanket bog." Most of the recorded peat failures have occurred in raised bog or blanket bog (Mills, 2002; Dykes and Kirk, 2006; Boylan et al., 2008), and in many of these cases, the bog has failed, that is, failure has occurred "within" the mass of peat that comprises the peatland. This distinction provides a basis for distinguishing between "bog" failures and "peat slides" (Dykes and Kirk, 2001; Evans and Warburton, 2007). However, although this scheme provides a simple division of peat failures, it does not adequately align the range of peat failure types with existing landslide classification schemes familiar to most geomorphologists, geologists, and engineers (Dykes and Warburton, 2007a).

6.3 MORPHOLOGY AND CLASSIFICATION OF PEAT LANDSLIDES

Although differences exist between different types of peat mass movement (Dykes and Warburton, 2007a), four key morphological elements occur that can be observed in most peat landslide settings (Figure 6.2). These consist of a source zone where the initial failure began; a zone of chaotic peat debris comprising a mixture of large peat rafts, smaller blocks, and debris; a downslope runout debris track of finer peat material commonly forming a peaty organic slurry; and a series of secondary features in proximity to the main landslide that show evidence of tension (cracks and tears) and compression (ridges and bulges) in the adjacent peat mass. These are well illustrated in Figure 6.2, which is a map of the Harthope peat slide in the North Pennines, United Kingdom (Warburton et al., 2003). The landslide was triggered by heavy rain and rapid snowmelt in February 1995 along the line of an active hill slope drainage pathway (flush). Detailed observations of the slide area and downstream deposits show that the landslide was initiated as a blocky mass that disintegrated into a downstream debris flow (Figure 6.2). The landslide pattern was complex, with areas of extending and compressive movement. Bare mineral substrate extends throughout much of the scar area

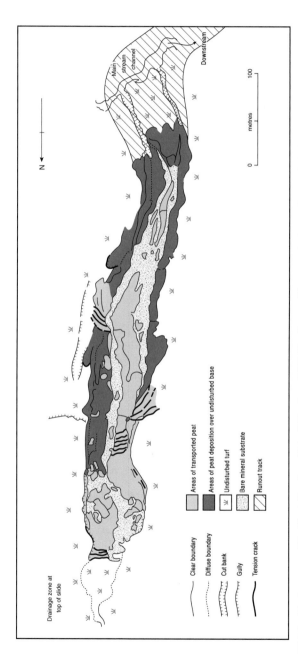

FIGURE 6.2   Harthope peat slide, North Pennines, United Kingdom, 1995. The diagram illustrates the four main morphological elements that are common in peat landslides: a source zone; a zone of chaotic peat debris comprising large peat rafts and smaller blocks; a downslope runout track of finer peat debris and slurry; and secondary features in proximity to the main landslide, which show evidence of tension and compression in the adjacent peat mass.

where the surficial peat (up to 2 m thick) slid from the hill slope. Within the area of instability, large rafts of peat, some up to 50 m in lateral extent, have been displaced. In places, these have been overridden by smaller rafts and blocks creating a complex topography of broken blocks. Toward the base of the scar, peat is deposited on the undisturbed blanket peat particularly to the margins of the main failure track. At the proximal end of the scar, the deposited peat ends fairly abruptly at a break of the slope, but finer debris continues into the stream greatly increasing the capacity of the flow and leaving superelevated peat debris on the valley sides (Figure 6.2). In this example and many other peat failures, these key morphological elements form overlapping zones that differ in detail leading to a complex variety of landslide forms, which relate partly to differences in the material properties of the peat and different mechanisms of failure.

6.3.1 A Confused Terminology

Previous reviews of peat landslide events highlighted the fragmented and disconnected literature on peat failure mechanisms, and a general lack of coherence with the wider landslides literature (Dykes and Warburton, 2007a; Evans and Warburton, 2007). This is particularly evident in the lack of consistent terminology to describe peat failures. In the past, peat landslides have been variously described as "debacles" (Kinahan, 1897) and "cloudbursts" (Hudleston, 1930), and use of some terms such as "peat flow" by Hungr and Evans (1985) has been inconsistent with previously published accounts. The peat failures of June 1986 in the valley of the Yellow River, County Leitrim, Ireland, highlight the problem. Coxon et al. (1989) described two "peat slides," whereas Large (1991) refers to "a bog-burst" comprising "three separate slides." This confused terminology has partly arisen due to the difficulties in determining the particular type of failure, especially some time after the event, but equally no formal classification existed until the work of Dykes and Warburton (2007a). Prior to 2007, the only landslide classification scheme that explicitly identified peat as a failure material was that of Hutchinson (1988). The scheme is based primarily on the morphology of slope movements, although the materials involved and the mechanism and rate of movement are also considered. Although Hutchinson's (1988) landslide classification scheme was the first to place peat failures into the wider framework of all mass movements the "peat slides" and "bog slides" categories of the scheme were not rigorously defined.

6.3.2 A Formal Classification of Peat Landslides (Dykes and Warburton, 2007a)

Peat failures comprise a variety of forms and mechanisms, from small shallow failures on steep slopes involving thin blanket peat being displaced by failure within the mineral substrate, to the creation of shallow valleys through the massive failure of a raised bog involving millions of cubic meters of peat escaping and flooding the land downstream. Evans and Warburton (2007) suggested that the mechanisms behind these extremes, that is, basal shearing and loss of peat matrix strength, respectively, provide a simple framework for classifying peat failures whereby intermediate forms and processes tend to converge toward flow-type movements of the failed mass irrespective of the initial failure condition. Within this framework, however, distinct types of peat failure can now be more reliably distinguished according to the type of peat deposit, failure morphology, the material that failed (i.e., peat or mineral substrate) and the dominant failure mechanism (Dykes and Warburton, 2007a).

Therefore, to develop an unambiguous terminology for the classification of peat failures, it is important to distinguish between different types of peat deposit in which failures occur.

Dykes and Warburton (2007a) developed a classification scheme for peat failures (Table 6.1, Figure 6.3) that enabled for the first time a systematic approach to the identification and recording of peat landslides. The definitions developed allowed peat failures to be systematically classified in a manner consistent with existing landslide classification frameworks. Figure 6.3 shows in schematic form the classification of peat landslides proposed by Dykes and Warburton (2007a). Definitions are provided for six main peat landslide types: bog bursts (flow failure of raised bogs), bog flows (flow failure of blanket bogs), bog slides (shear failure and sliding of blanket bogs), peat slides (shear failure at peat—mineral interface in blanket bogs), peaty-debris slides (shear failure within mineral substrate beneath blanket bogs), and peat flows (natural failures of other types of peat deposits including flow failure caused by head loading). Table 6.1 provides a fuller definition in a simple look-up table but for full definitions, explanations, and examples, readers should consult the original paper (Dykes and Warburton, 2007a). The aim of this classification was to provide a reference tool for more reliable identification and analysis of site conditions associated with different types of failure. The scheme can also be regarded as a formal modification of Hutchinson's (1988) classification of mass movements, thus placing peat instability within the broader framework of landslides research. The classification was designed to guide future research into failure mechanisms, encourage better linking of failure types and associated peat properties, and thus improve hazard assessment of peat instability.

6.4 RELATIONSHIP BETWEEN LANDSLIDE TYPE AND PEAT STRATIGRAPHY

Figure 6.3 is a modification of the original classification diagram from Dykes and Warburton (2007a) in that it distinguishes between: (1) bog flows and bog slides occurring in a bog with a peat stratigraphy comprising a two-layer system of fibrous peat overlying a more amorphous lower peat and (2) peat

TABLE 6.1 Look-up Table of the Definitions of Peat Landslides

Peat Landslide Type	Definition
1. Bog burst	Failure of a raised bog (i.e., bog peat[1]) involving the breakout and evacuation of (semi) liquid basal peat
2. Bog flow (or "Bog flow")	Failure of a blanket bog (i.e., bog peat) involving the breakout and evacuation of semiliquid highly humified basal peat from a clearly defined source area
3. Bog slide (or peat slide)	Failure of a blanket bog (i.e., bog peat) involving sliding of intact peat on a shearing surface within, or at the upper or lower surface of, the catotelm[2]
4. Peat slide	Failure of a blanket bog involving sliding of intact peat on a shearing surface at the interface between the peat and the mineral substrate material or immediately adjacent to the underlying substrate
5. Peaty-debris slide	Shallow translational failure of a hill slope with a mantle of blanket peat in which failure occurs by wholly shearing within the mineral substrate (including palaeosols or other materials of exogenous origin) and at a depth below the interface with the base of the peat such that the peat is only a secondary influence on the failure
6. Peat flow	1. Flow failure in any type of peat caused by head loading 2. Failure of any type of peat deposit not covered by the preceding terms, by any mechanism

[1]Bog peat—peat formed from the accumulation of organic matter above the level of the natural groundwater table due to saturated conditions maintained from the excess of precipitation over evapotranspiration.
[2]Catotelm—is the deeper peat that is generally below the level of seasonal water table fluctuations/ active root growth and remains permanently saturated.
After Dykes and Warburton (2007a).

slides and peaty-debris slides that occur in a dominantly one-layer peat profile. Such generalization are of course an approximation, but this is supported by documented examples from the literature and field (Mills, 2002; Evans and Warburton, 2007, Figure 6.4). This distinction is important as bog bursts, slides, and flows involve failure within the basal peat, whereas peat slides and peaty-debris flows involve failure below the peat (Table 6.1). This classification of peat failures (Figure 6.3) provides a systematic approach to identifying peat landslides and offers a tool for more reliable identification and analysis of site conditions associated with the different types of failure. With regard to Hutchinson's widely accepted 1988 classification of mass movements (Hutchinson, 1988), these definitions can easily be accommodated within that

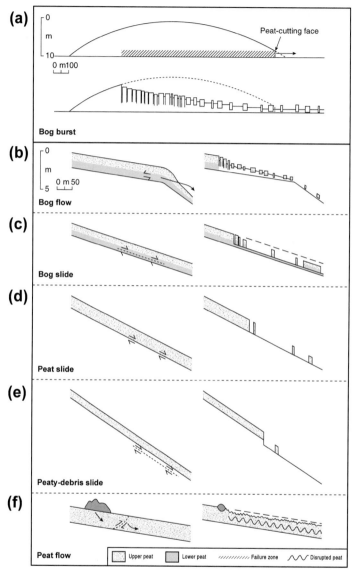

FIGURE 6.3   Classification of mass movements in peat based on Dykes and Warburton (2007). (a) Bog burst, (b) bog flow, (c) bog slide, (d) peat slide, (e) peaty-debris slide, and (f) peat flow (head-loaded). Scales are approximate.

framework, thus providing a coherent overall landslide classification (Dykes and Warburton, 2007a).

Figure 6.4 shows examples of the general peat stratigraphy at a selection of British peat landslide sites. Five examples are shown from two main

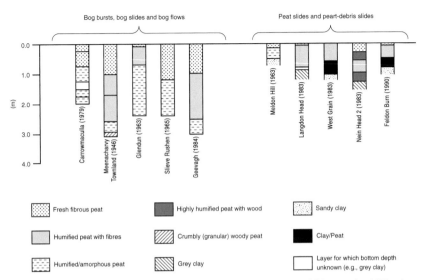

FIGURE 6.4 Examples of the general stratigraphy occurring at British peat landslide sites. Divided between bog burst and bog flows (failures within the basal peat) and peat slides and peaty-debris slides (failures below the peat). *Modified from Mills (2002) and Evans and Warburton (2007).*

groupings, which are divided between bog bursts, bog slides and bog flows (failures within the basal peat) and peat slides, and peaty-debris slides (failures below the peat). Examples are defined in the literature or logged in the field (Mills, 2002; Evans and Warburton, 2007). The importance of Figure 6.4 is that it makes the important link between the material properties/stratigraphy of the peat mass and the form and mode of failure. Although obviously some significant variability occurs in the stratigraphy of these broad groupings, in general terms "bog"-type failures that involve failure in a basal peat layer are broadly characterized by a "two-layer" peat stratigraphy, which generally consists of a less humified fibrous upper peat (often up to 1 m in thickness) overlying a similar depth or deeper layer of amorphous well-humified peat. However, in the "peat" failure stratigraphies fibrous less humified peat dominates throughout the peat profile.

6.5 IMPACTS OF PEAT LANDSLIDES

In general, the impact of peat landslides on environmental processes and human activities is similar to that of many shallow landslides. However, because of the distinct physical and biogeochemical characteristics of peat the "effect" and consequent range of hazards differs from those of mineral dominated/soil landslides (Evans and Warburton, 2007). Table 6.2 provides a concise summary of the environmental impacts of peat landslides, which can be categorized into short/long-term impacts and at-a-site and off-site effects.

TABLE 6.2 Environmental Impacts of Peat Landslides

	Short-term Impacts	Long-term Impacts
At the landslide	• Loss of farm stock and danger to wildlife	• Danger to juvenile livestock and wildlife
	• Loss of agricultural standing crop	• Loss of agricultural land
	• Burial of agricultural land in local deposition zone	• Loss of land of ecological and paleoecological importance
	• Disruption of surface and subsurface drainage	• Alteration of natural drainage patterns
	• Large peat/sediment loss	• Site of continued erosion—engineering intervention may be required
	• Loss of terrestrial carbon	• Reduced upland carbon store (nonrenewable)
Offsite effects	• Damage to infrastructure and housing	• Slope modification above important infrastructure
	• Damming of culverts and bridge crossings	• Redesign/maintenance of stream crossings
	• Damage to stream ecosystems—fish kill	
	• Alteration of stream morphodynamics	• Changing patterns of local stream sedimentation
	• Deterioration in stream water quality	
	• Contamination of upland reservoirs and lakes	• Increased lake and reservoir sedimentation rates—peat/carbon capture

Modified from Evans and Warburton (2007).

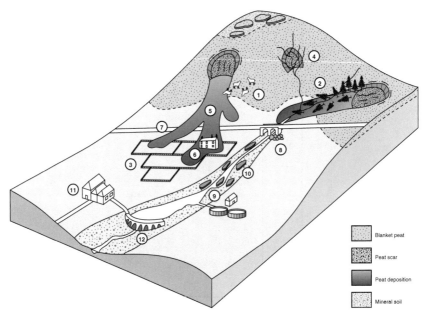

FIGURE 6.5 Schematic representation showing the range of peat landslide impacts (1−12) in a blanket peat environment (Table 6.2). (1) Loss of livestock; (2) loss of standing crop; (3) burial of productive agricultural land; (4) modified drainage patters; (5) peat (carbon)/sediment loss; (6) damage to housing and buildings; (7) damage and blocking of roads; (8) damage/blockage to stream/river crossings (culverts and bridges); (9) ecosystem/fisheries damage (fish kills); (10) alteration of stream courses (morphodynamics); (11) reduced water quality and water supply contamination; and (12) deposition in lakes and reservoirs. Fatalities and serious injury to the human population have also been reported following peat landslides, but these are not shown in the figure.

These are also shown schematically in Figure 6.5, which provides a spatial representation of the range of peat landslide impacts (1−12) in a blanket peat environment. Broadly speaking, peat failures affect the ecology of the immediate failure site, but if peat debris reaches water courses, substantial fish kills and loss of other aquatic life can result (McCahon et al., 1987; Wilson et al., 1996). Damage can also occur to property and infrastructure (Colhoun et al., 1965), as well as significant alterations to natural drainage channels (Alexander et al., 1986; Coxon et al., 1989), and other geomorphic impacts.

The immediate short-term effects at the landslide site involve a range of direct impacts resulting in loss of livestock; loss of standing crop; burial of productive agricultural land in the runout zone adjacent to the source of failure; modification of local drainage patterns often involving the exhumation of sub-surface drainage patterns in the landslide scar; and large losses of surface materials in the form of both peat, which is rich in organic carbon, and mineral sediment loss. Away from the initial failure and runout zone, landslide debris has a number of significant off-site impacts. In the distal reaches of the runout track, houses and buildings are frequently damaged by direct impact of peat blocks, peaty debris, or ingress of peat slurry inside properties. Mobile peat debris also causes direct damage to the wider infrastructure as well as burying and blocking roads, culverts, and bridges. Stream and river systems are particularly at risk as excess loading by peaty debris causes ecosystem and fisheries damage often resulting in extensive fish kills due to the high concentration of suspended solids (McCahon et al., 1987). Peat landslides not only contribute to extremely high suspended loads in streams but they also introduce low-density peat debris and large peat blocks that are transported by the river flow in very large quantities, and which once deposited may alter stream courses and redirect flow. Because of the low density of organic peat debris, its rapid abrasion within the fluvial system and its in-stream chemical reactions, water quality is often degraded and water supplies are either contaminated or disrupted as intake valves to water treatment plants are shut down. Finally, although peat is easily transported in fluvial systems due to its low density, it is nevertheless eventually deposited either on the margins of floodplain/river channels or in lakes and reservoirs where it can increase sedimentation. Fatalities and serious injury to the human population have also been reported following peat landslides, but these are mostly restricted to historical events (Figure 6.5).

Over the longer term, peat landslide scars will persist in the landscape for decades resulting in the loss of agricultural land, reduction in the upland carbon store, and a loss of valuable ecological resource. Landslide scars may also continue to erode as surface drainages incise into bare soils leading to gullying. These persistent erosion forms present a hazard to juvenile livestock and wildlife that may be trapped in the ground cracks or drowned in local pools. Downstream of landslide source zones, remedial works may also be needed to engineer unstable slopes; rebuild and redesign stream crossings, bridge structures, and culverts; and clean up contaminated floodplains (Table 6.1).

From Table 6.2, Figure 6.5 and the above discussion, it is clear that although peat landslides are generally similar in their impacts to many shallow landslides, they have a distinct character that is due to the physical and biogeochemical characteristics of peat. In particular,

1. They are a low-density material that affects both the mechanism(s) of failure and landslide runout dynamics, but more significantly increases the efficiency of downstream transport if a landslide connects (couples) with a stream or river. This results in extremely high suspended solid loads and large volumes of peat material that can easily block culverts and jam in bridge arches.
2. Peat failures also represent a significant polluter of river systems due to high suspended sediment concentration, associated with contaminant loads and the efficient biogeochemical reactions that occur rapidly in the water column due to in-stream processing of the organic debris.

3. Peat landslides transport significant amounts of organic carbon and can be effective in tipping the balance of upland carbon budgets, at least in the short term and at the local catchment scale. Further, once peat is lost from a landslide site, it is unlikely to be replaced by natural processes.

Although contemporary peat landslides rarely cause fatalities, their capacity to inflict damage remains a significant threat to many peatland communities and economies. Recent estimates of the economic disruption caused by peat failures run into millions of dollars. In September 2003, a cluster of 20 peat slides in Channerwick, South Shetland (Scotland), caused an estimated $5M of damage, and on the same day at Dooncarton Mountain in Co. Mayo, Ireland, a group of 35 landslides caused damage estimated at $11M. Then, approximately a month later in October 2003 during the construction of a wind farm development close to Derrybrien in Co. Galway, a major peat landslide was triggered in forested blanket peat on Cashlaundrumlahan Mountain. The landslide was estimated to add $7M to the construction costs of the works and delayed completion of the project by a year (Irish Independent, 2006).

6.5.1 Example: Cashlaundrumlahan Peat Flow, Derrybrien, Ireland (October 2003)

The catastrophic peat landslide that occurred at the Derrybrien wind farm development (Co. Galway, Republic of Ireland) on October 16, 2003, captured national attention and put on hold a wind farm industry that was rapidly expanding across the British uplands (Lindsay and Bragg, 2005). Figure 6.6 shows the Derrybrien wind farm development site and peat landslide. Part (a) shows the geographical location of the site and illustrates the connectivity between the landslide on Cashlaundrumlahan Mountain and the Owenda-lullegh River system. Part (b) shows the track of the landslide that originated at the excavation of Turbine base T68 before flowing downslope through the forest and connecting with the local stream drainage system. The main peat slide, estimated volume 450,000 m^3, failed on October 16 and traveled 2.5 km downslope from the south-facing slope of Cashlaundrumlahan Mountain (358 m) having originated in the vicinity of two partially constructed turbine bases (Figure 6.6). The main runout track of the landslide extended in a south/south-easterly direction approximately 1.2 km downslope before entering a stream gorge at an elevation approximately 270 m (Figure 6.6(b)). The peat came to rest on October 19 at 195-m altitude but was reactivated by heavy rain on October 28. It then moved another 1.5 km, blocking two roads, before eventually reaching the Owendalulleegh River where it became highly fluidized and continued to flow downstream 30 km into Lough Cutra, which was the source of a domestic water supply (Figure 6.6(a)). Earlier on October 2, a small ($\sim 2,000$ m^3) peat slide had occurred at another turbine excavation site, 230 m west of the main T68 failure site.

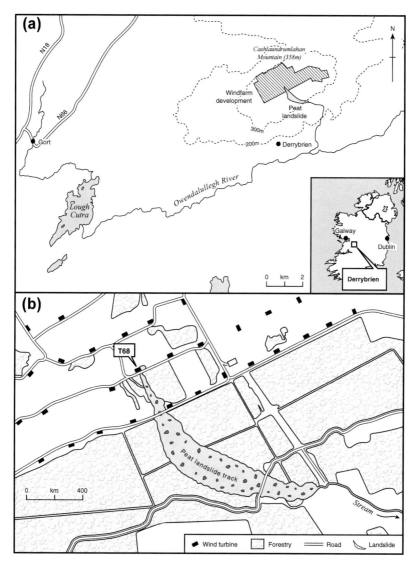

FIGURE 6.6   The Derrybrien wind farm development site and peat landslide. (a) Geographical location and connectivity to the Owendalullegh River system. (b) Track of the peat landslide, originating at Turbine base T68 and flowing downslope through the forestry.

Although engineering investigations were undertaken to evaluate the cause of the failure, there was no clear conclusion reached among the consulting engineers about the exact cause of the failure, but it was clear that loading of the peat surface played a significant role in the instability. Investigations by peatland specialists however (Lindsay and Bragg, 2005) suggest that the failure may have resulted from sudden loading of the peat surface by either excavation machinery or soil excavated from the wind turbine foundation site. Locally, the peat was heavily fissured in the failed zone. Under such conditions, a bearing type failure may be initiated in the peat below the area of loading. Shear planes develop in the peat and the peat loses strength that increases the active pressure downslope. This can lead to progressive downslope failure of the peat that has the potential to result in a runaway failure (Boylan et al., 2008). Drainage was also thought to be a significant factor. At the head of the failure in the vicinity of T68, the wind turbine excavation was located in a drainage depression where water seeping through the peat began to converge into a main drainage line. This coincided with locally deeper peat (e.g., peat depths vary across the site (0.4−5.5 m), but in the vicinity of the failure, depths were relatively deep 3.3−3.4 m; Bragg, 2007), which together with the local surface loading, will have all contributed to the failure.

The initial impacts included loss of land, obstruction of roads, pollution of the domestic water supply, and the death of an estimated 100,000 fish (Lindsay and Bragg, 2005; Bragg, 2007). An assessment by the Shannon Regional Fisheries Board in November 2003 estimated that 50,000 fish had died as a result of the landslide and the spawning gravel in the Owendalulleegh River system had become heavily silted with peaty material. Water quality was seriously degraded, and it was anticipated that during heavy rain, large volumes of fine sediment would continue to be evacuated from the affected tributary.

Following the incident at Derrybrien, far greater awareness emerged of the construction problems that are inherent when building on peat. Although these were well known elsewhere (MacFarlane, 1969), a greater consciousness was triggered in the United Kingdom resulting in much better planning and implementation of improved guidelines, for example, Guidelines were developed for the risk management of peat slips on the construction of low-volume/low-cost roads over peat (Munro and McCulloch, 2006) and for greater research into the mechanism that can trigger peat failures (Long, 2005; Dykes and Warburton, 2007a,b; Long and Boylan, 2012).

6.5.2 Example: Failure during Road Construction, North Pennines, United Kingdom (August 2006)

Figure 6.7 shows a peat landslide triggered by moorland road construction in an area of upland blanket peat in the North Pennines, United Kingdom. The failure occurred in August 2006 during the construction phase of the road. Road foundations were excavated directly into the peat, which in the vicinity of the failure, was approximately 1.1−1.3 m deep. The road was approximately 3.8 m wide and back filled with a coarse aggregate rubble and gravel to a depth of about 0.5 m over a wire mesh that was laid in the construction

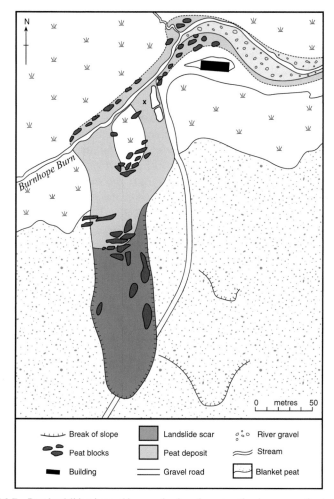

FIGURE 6.7   Peat landslide triggered by moorland road construction in an area of upland blanket peat in the North Pennines, United Kingdom. Point x marks a 25-m section of intact road that was transported downslope during the failure.

trench. The stratigraphy of the peat consisted of an upper fibrous peat over-lying a more humified basal peat unit.

Based on eye-witness testimony, the failure occurred just after a fully laden aggregate truck carrying road fill had passed over the section of the peat where the landslide was initiated. The slope failed rapidly, and the driver was fortunate not to get caught up in the landslide. The failed peat mass slid downslope and entered the local stream course (Burnhope Burn) at the base of the slope. Peat debris blocked the channel and spilled out on to the adjacent floodplain. This debris continued downstream 2 km into a drinking water reservoir that was forced to shut down its water intake due to high suspended solid concentrations. Marked in Figure 6.7 ("x") is a 25-m section of road that was transported down the slope during the failure but remained largely intact in transit, supporting the assumption that failure was by a shallow translational slide mechanism.

In this particular example, the moorland road was of a simple construction designed to be of a low cost for low volumes of traffic (mainly for shooting parties) but illustrates some of the potential challenges facing road building on peat. At this site, high water content, high compressibility, and low strength of the peat all appear to be significant factors in the failure. First, at the point of failure, the road traversed a natural moorland "flush" where the peat was slightly deeper and was very wet, even during the summer, due to preferred drainage of groundwater along the flush. Second, the construction method used conventional coarse aggregate as road ballast that was laid on the peat. This did not take advantage of using lightweight fill materials, and it is estimated that the haulage trucks were running 20-t loads over the newly constructed road with no period of consolidation (Munro, 2004). Third, the peat in this locality consisted of approximately 0.7 m of fibrous peat over about 0.6 m of amorphous peat that overlaid a coarse stone clayey substrate. Approximately 0.3 m above the base of the amorphous peat, a pronounced water seepage zone occurred, which appeared to be the zone of the failure. Under these conditions, the landslide would be appropriately described as a bog slide (Dykes and Warburton, 2007a). Given these characteristics, it is clear that the road was potentially very unstable at this point along the road corridor and without additional engineering would be susceptible to failure.

6.6 THE RUNOUT OF PEAT LANDSLIDES

Continued movement of a disaggregated peat mass downslope usually results in the breakdown of the peat by mechanical abrasion, splitting, and rolling (Colhoun et al., 1965). The debris consists of abundant abraded and fractured peat blocks with a trail of peat slurry and uprooted vegetation (Latimer, 1897). The resultant peat fragments often form small peat peds with a spindle form. At the same time, much of the abraded material mixes with local water sources (both runoff and stored water) forming a peaty slurry. The composition of the slurry and its rheological properties depend on the nature of the peat source, its initial condition (wet or dry), and the relative proportions of the peat/water mix (Mills, 2002). The runout track is often bounded by distinct levees or lobes of peat blocks. If the flow becomes confined to a channel, evidence of superelevation often occurs at bends in the flow track; under these conditions, peat levees can be used to reconstruct flow velocities (Warburton et al., 2003).

Direct evidence for the speed of peat mass movements is summarized by Evans and Warburton (2007) who concluded that the estimated rates of movement are highly variable. Much of the variability relates to the definition of what is meant by "the period of the failure" or "area of the flow" that is

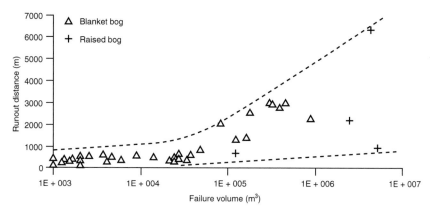

FIGURE 6.8   Runout distance versus failure volume for 44 peat failures (Boylan et al., 2008).

being described. For example, extremely slow rates are often associated with prefailure creep (Sollas et al., 1897) or postfailure readjustment (Praeger, 1897); moderate rates are usually directly observed in the zone of failure; and rapid rates in the runout track (Alexander et al., 1986). Although the data are limited, bursts appear to fail over longer timescales than peat slides.

Boylan et al. (2008) were the first to present a general survey of peat landslide runout distance relationships (Figure 6.8). They considered the runout distance of 44 documented peat failures and plotted this versus failure volume. It is well known that peat landslides can travel large distances and that their impacts can be recorded for many kilometers downstream, for example, peat debris from the Knockmageeha failure of 1896 was documented 15 km from the failure source (Cole, 1897). Figure 6.8 show the plot of runout distance versus failure volume for 44 peat failures reported in Boylan et al. (2008). The authors distinguish between blanket bog and raised bog failures. In general, runout distance increases with failure volume, although large inherent variability occurs in the data. Larger failure volumes and consequently larger runout distances seem to occur more in raised bogs, which supports the hypothesis that raised bogs are deeper and contain larger volumes of peat and when failed generate a large downstream impact (Mills, 2002). The long runout of peat landslides is often associated with the rapid breakdown of peat in transit, which increases the capacity for conveyance. However, equally important is the low density of peat, which means that when it enters a stream course or river and mixes with floodwater, it can be transported large distances.

The data presented in Figure 6.8 must be regarded as preliminary because of several key factors:

1. Patterns are reconstructed from historical data/publications not written to capture information on peat runout extents.
2. Runout distance is influenced by a range of factors, which include topography, slope angle, degree of confinement, material properties of the

peat (especially water content), degree of coupling with stream systems, and the roughness of the runout track (Boylan et al., 2008).

3. Of critical importance is the degree of coupling with stream systems. Once peat enters a stream in fragments it abrades extremely rapidly and by virtue of its low density (~ 1.0–1.1 Mg/m^3) can be transported as a "floating" load. By this mechanism, peat debris may be transported large distances, and much of the debris may pass entirely through the fluvial system (Warburton and Evans, 2011). This essentially means that for many peat landslides no clear downstream limit occurs to debris dispersal.

6.7 SLOPE STABILITY ANALYSIS OF PEAT LANDSLIDES AND GEOTECHNICAL PROPERTIES

Slope stability analysis of peat landslides has been undertaken in relatively few cases. Where this has been done, the peat failure is usually treated as a translational planar slide and a simple infinite slope analysis is used to back calculate strength parameters of the slope at the time of failure (Hendrick, 1990; Carling, 1986; Dykes and Kirk, 2001; Warburton et al., 2003). Boylan et al. (2008) described the factor of safety (FOS) in an effective stress strength analysis for a planar translational slide where steady seepage of groundwater occurs parallel to the ground surface:

$$\text{FOS} = \frac{c\prime}{\gamma z \cos\beta \sin\beta} + \frac{\left[\gamma - \gamma_w \left(\dfrac{z_w}{z}\right)\right]\tan\Phi\prime}{\gamma\tan\beta}, \tag{6.1}$$

where β is the slope angle of the slide surface, γ is the bulk unit weight, z is the depth of the failure surface, $c\prime$ is the apparent drained cohesion of peat, $\Phi\prime$ is the effective angle of shearing resistance, γ_w is the bulk unit weight of water, and z_w is the height of the water table above the failure surface.

Figure 6.9 shows the sensitivity analysis of $c\prime$ and z_w on the calculated FOS for bulk unit weights (γ) between 10 and 15 kN/m^2. For low bulk unit weights, $c\prime$ and z_w play an important role in the calculated FOS with $c\prime$ accounting for approximately 50 percent of the value at z_w/z of 0.5 and over 90 percent of the FOS when the water table is at the surface ($z_w/z = 1.0$). In many peat settings where the unit weight of peat and unit weight of water are similar ($\gamma \approx 10$ kN/m^2) and the water table is near the surface, the cohesion of the peat ($c\prime$) controls the FOS. In this case, equation (7.2) is similar to the FOS equation for undrained peat, and the results are dependent on the shear strength of the peat (S_u) (Boylan et al., 2008):

$$\text{FOS} = \frac{S_u}{\gamma z \sin\beta \cos\beta}, \tag{6.2}$$

where S_u is the undrained shear strength of peat.

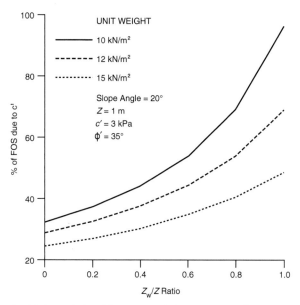

FIGURE 6.9   Sensitivity analysis of the apparent drained cohesion (c') and height of the water table above the failure surface (z_w) on the calculated FOS for three values of unit weight. z is the depth of the failure surface (Boylan et al., 2008).

Estimating the shear strength of peat is a problem that engineers have struggled with for several decades (Landva and Pheeney, 1980; Long and Boylan, 2012). If peat is treated as a cohesive material, the relationship between shear strength and effective stress can be represented by the Mohr−Coulomb strength criterion:

$$S_u = c' + \sigma'\tan\Phi', \tag{6.3}$$

where S_u is the shear strength of peat, c' is the cohesion, σ' is the effective stress on the failure plane, and Φ' is the effective stress angle of internal friction.

First, estimating c' and Φ' for peat soils is notoriously difficult because published values of the shear strength properties of peat are relatively few; and second, testing of peat using standard geotechnical tests is fraught with problems especially when trying to remove intact samples from the field without disturbance or field testing using conventional vane and penetrometer tests in the presence of multiple fibers (Long, 2005; Dykes and Kirk, 2006; Long and Boylan, 2012). Published data (Dykes and Kirk, 2006; Dykes, 2008b) suggest values for cohesion in the range of 0−11 kPa and internal friction angles of 21−58°. These values vary with the vegetation composition of the peat (particularly Fc) and the degree of humification and are also affected by the testing procedures employed (Long, 2005; Boylan et al., 2008).

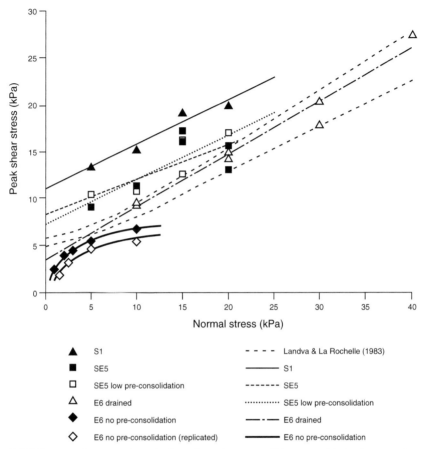

FIGURE 6.10   Shear strength results from direct shear tests of Irish upland blanket bog peat (S1, SE5—Dykes and Warburton (2008a); E6—(Dykes and Kirk, 2006) together with ring shear test results (envelope curves) from Canadian H3-4 Sphagnum peat (Landva and La Rochelle, 1983).

Figure 6.10 shows experimental results from direct shear tests of Irish upland blanket bog peat (S1, SE5—Dykes and Warburton (2008a); E6—Dykes and Kirk, 2006) together with ring shear test results (envelope curves) from Canadian H3-4 Sphagnum peat (Landva and La Rochelle, 1983). In these experiments, Dykes (2008b) attempted to simulate blanket peat failure conditions wherein the effective normal stresses in the basal peat was very low. Dykes compared shear strength results from standard testing procedures where samples were preconsolidated with the same samples that did not undergo preconsolidation (Figure 6.10). Samples were sheared rapidly (0.2—0.5 mm/min) to simulate rapid failure. Preliminary results indicated that in an undisturbed, low effective stress environment, the basal peat had very low in situ strength (~5 kPa) (Figure 6.10).

6.8 HISTORICAL PERSPECTIVE ON THE FREQUENCY OF PEAT LANDSLIDES

Pearsall (1950) was one of the first authors to recognize the general importance of peat mass movements for landscape development. He suggested that peat mass movements were becoming less common than was previously reported because of the widespread drainage of peatlands and changes in land use, which have resulted in vegetation that is more resistant to lateral tearing. This observation was made without the benefit of a database of known peat failures, which is now available from the literature (Figure 6.11).

Figure 6.11 shows the changing cumulative frequency of peat landslides and fatalities in British peatland environments from 1700 to 2003, based on Mills (2002), Evans and Warburton (2007), and Boylan et al. (2008). Records contradict Pearsall's early assertion and show nearly 100 failures in the last approximately 300 years with an increased frequency toward the present day. Pre-1800 events are less well reported. Over the same period, the estimated number of fatalities caused by peat landslides is 36; this is taken from Boylan et al. (2008) who report that the number of deaths has been approximated due to uncertainties in earlier records. Considering peat landslides that caused fatalities, the most notable are the 1708 failure of Castlegarde Bog, Co. Limerick that buried three cottages accounted for approximately 21 fatalities (Praeger, 1897); Owenmore Valley failure, Erris, Co. Mayo in 1819 (six deaths); the 1896 Knockmageeha landslide, Co. Kerry in 1896, which resulted in eight deaths in one family as their house was swept away in the event

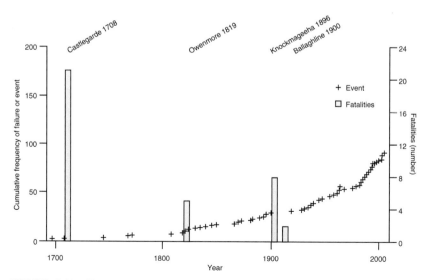

FIGURE 6.11 Changing frequency of peat landslides and fatalities in British peatland environments from 1700 to 2003. Based on Evans and Warburton (2007) and Boylan et al. (2008).

(Cole, 1897; Sollas et al., 1897); and the Ballaghline (Lisdoonvarna) Co. Clare event of 1900, which killed two people (Colhoun et al., 1965).

The graphic account of the 1819 Owenmore failure describes the impact of the event: "A mountain tarn burst its banks, and heaving the bog that confined it, came like a liquid wall a-down, forcing everything along boulders, bog timber, and sludge, until, as it were in an instant, it broke upon the houses [of a small village], carrying all before it, stones, timbers, and bodies; and it was only some days after, that at the estuary of the river in Tullohan Bay, the bodies of the poor people were found." (from: Otway, "Sketches in Erris and Tirawley," p. 14, 1841, in Praeger (1897)). Outside of Ireland, significant historical peat landslides are less well recorded. However, a notable exception is the 1771 "eruption" of Solway Moss on the English/Scottish border, which caused extensive damage to farmland, inundation of settlements, and pollution of nearby rivers (McEwen and Withers, 1989). The event, estimated to involve approximately 3.7 M m^3 of peat, is generally interpreted as a bog burst triggered by heavy rainfall on an already saturated bog. However, details on the morphology and stratigraphy of the bog are not well documented, so the exact mechanism is unlikely to be known.

These examples illustrate that great care should be exercised when interpreting chronologies of landslide events from historical records as the observed details and patterns are likely to be distorted. In the past, the historical record becomes sparser as only significant events tend to be documented. Towards the present, the volume and quality of documentary sources increase, giving an apparent increase in the frequency of landslides, which may not represent the true occurrence of such events. Typically, before the seventeenth/eighteenth centuries, few landslides and floods are recorded. However, paleoecological evidence suggests that variations in peat mass movement activity have operated over longer timescales than indicated in Figure 5.9 (Tallis et al., 1997; Ashmore et al., 2000); hence, recent data need to be placed in the context of a longer time series.

Mills (2002) attempted to calculate the significance of peat slides in the overall sediment budget of the North Pennines peatland over the last century by comparing sediment losses from peat slides with background fluvial activity. Calculations showed that just 3 percent of the total sediment yield could be associated with rapid peat mass movements over 1914—2014 years. Although this in a regional context is relatively minor, locally, peat slide events remain highly significant and may have disastrous short-term impacts on fluvial systems (McMahon et al., 1987).

6.9 THE FUTURE INCIDENCE OF PEAT LANDSLIDES

Several important factors govern peatland slope processes and include slope topography and form; hill slope hydrology; hydraulic properties of the peat; and peat material properties.

In terms of sites of peat mass movement, several characteristics predispose them to failure, but arguably the most important relate to hill slope hydrological processes that either directly or indirectly trigger landslides (Warburton et al., 2004). All peat landslides, with the possible exception of loading or excavation failures, will include one or more of the following preconditions:

1. impeded drainage where a peat layer overlies an impervious clay or mineral base (hydrological discontinuity);
2. a convex slope or a slope with a break of slope at its head (concentration of subsurface flow);
3. proximity to local drainage either from flushes, pipes, or streams; and
4. connectivity between surface drainage and the peat/impervious interface.

Given the importance of hill slope hydrology in peat stability it is hardly surprising that a clear association exists between heavy rainfall and the occurrence of peat landslides (Warburton et al., 2004; Boylan et al., 2008), which often occur after dry periods that stress the peat profile (Dykes and Warburton, 2008a). Such events often result in clusters of landslides that locally have a significant destabilizing effect on the peat blanket (Dykes and Warburton, 2008a,b). Therefore, the future incidence of peat landslides will be dictated by the changing climate of peatland areas (particularly rainfall), the local surface/near surface water hydrology, and changing use of peatland areas.

In the future, these drivers are likely to be affected in three main ways:

1. The changing intensity and distribution of rainfall patterns will affect the distribution and frequency of peat landslides. In the United Kingdom, it is predicted that the pattern of rainfall will change with a shift to wetter winters. Such changes are likely to create a change in the pattern of general landslide activity (Dijkstra and Dixon, 2010). These changes will have an impact on sensitive upland areas, which may act to destabilize the peat blanket in areas hitherto unaffected. Such patterns will be complex, and trends may only be recognized retrospectively.
2. The subsurface hydrology of the peat, and in particular the presence of macropores (peat pipes) of the peat is very important when considering peat failure mechanisms (Dykes and Warburton, 2008a).
3. Thawing ground ice in many Northern peatlands will contribute to the instability of the peat blanket resulting in more slope failures (active layer detachments) (Lewkowicz, 1990; Jorgenson, 2013). Under these conditions, a knowledge of peat failures will be useful in assessing the conditions of failure and assessing the impacts of such events.

In addition, the history of the use of peatlands is one of evolution and changing land management whether this is for peat cutting/extraction, forestry, drainage, or wind farm development. All these changes have an impact of the hydrology and stability of peatlands. For some of these land uses, the impacts are well understood, but a number of significant issues remain that will continue to cause concern in managing the development of peatlands:

1. The impact of roads and foundations on the hydrology/drainage of blanket peat, for example, Derrybrien (Lindsay and Bragg, 2005).
2. Assessing the importance of loading (both on roads and during excavations) on the bearing capacity of different types of peat, for example, Burnhope Burn peat failure (Section 6.4.2).
3. The need for clear guidelines for recognizing peat failure types and the necessary tools to test/assess the stability of the peat and predict areas of risk from future failures.

6.10 CONCLUSION

Peat landslides include a distinctive range of mass movement types that include bog burst, bog flow, bog slide, peat slide, and peaty-debris slide and peat flow failures. Although the general mechanisms of peat failure are now well documented and clearly classified (Warburton et al., 2004; Dykes and Warburton, 2007a; Boylan et al., 2008), the uncertainties associated with determining geotechnical properties of peat, and key hydrological issues relating to peat instability, need further clarification. Peat is a difficult material to characterize using standard geotechnical methods due to the significant microstructural components that make up its complex fabric. The low undrained in situ shear strength of peat makes any attempt to carry out a rigorous geotechnical investigation of peat slope stability fraught with difficulties and subject to considerable uncertainty. Further, little research has been undertaken on the runout dynamics of peat landslides, which arguably cause the largest environmental impact due to their extensive downstream dispersal of peaty debris and severe impacts on stream ecology. This research is needed so that more reliable assessments of the stability of peat deposits can be made (Boylan et al., 2008).

Continued development of peatland areas for access and wind farm construction presents a real need for improved peat landslide hazard assessments. Peat landslide risk assessment currently lacks a sound basis in research and a coherent methodology. Partly driven by the peat landslide activity that occurred across the British Isles in September/October 2003, in 2007, the Scottish Government published "The Peat Hazard and Risk Assessment Guide" to provide best practice methods to identify, mitigate, and manage peat landslide hazards and associated risks in respect to consent applications for electricity projects in Scotland. This made it a legal requirement to carefully consider the potential impacts of development in triggering peat landslides and provided a standardized approach for assessing peat landslide risk. Although this was an important first step, in practice, the scheme has proved to be difficult to apply in all peatland settings and some revision is necessary.

REFERENCES

Alexander, R.W., Coxon, P., Thorn, R.H., 1986. A bog flow at Straduff Townland, county Sligo. Proc. R. Ir. Acad. 86B, 107−119.

Ashmore, P., Brayshaw, B.A., Edwards, K.J., Gilbertson, D.D., Grattan, J.P., Kent, M., Pratt, K.E., Weaver, R.E., 2000. Allochthonous and autochthonous mire deposits, slope instability and palaeoenvironmental investigations in the Borve Valley, Barra, Outer Hebrides, Scotland. The Holocene 10 (1), 97−108.

Bailey, A., 1879. A letter from acting Governor Bailey to Governor Callaghan. Q. J. Geol. Soc. London XXXV, 96−97.

Bell, F.D., 2000. Organic soils: peat (Chapter 7). In: Engineering Properties of Soils and Rocks, fourth ed. Blackwell Science, Oxford, pp. 202−222.

Boelter, D.H., 1968. Important physical properties of peat materials. In: Proceedings of the 3rd International Peat Congress, Quebec 1968, pp. 150−154.

Bowes, D.R., 1960. A bog-burst in the Isle of Lewis. Scott. Geogr. J. 76, 21−23.

Boylan, N., Jennings, R., Long, M., 2008. Peat slope failure in Ireland. Q. J. Eng. Geol. Hydrogeol. 41 (1), 93−108.

Bragg, O., 2007. Derrybrien: where the questions began. No.4 Int. Mires Conserv. Group Newsl., 3−8.

Burton, R., 1996. The peat resources of Great Britain (Scotland, England, Wales and Isle of Man). In: Lappalainen, E. (Ed.), Global Peat Resources. International Peat Society, Jyskä (Finland), pp. 79−86.

Carling, P.A., 1986. Peat slides in Teesdale and Weardale, northern Pennines, July 1983: description and failure mechanisms. Earth Surf. Processes Landforms 11, 193−206.

Carlsten, P., 1993. Peat—Geotechnical Properties and Up-to-date Methods of Design and Construction. State-of-the-art-report, second ed. Swedish Geotechnical Institute, Linköping. Report No. 215.

Charman, D., 2002. Peatlands and Environmental Change. John Wiley & Sons, Chichester.

Cole, G.A.J., 1897. The bog-slide of Knocknageeha, in the county of Kerry. Nature 55 (1420), 254−256.

Colhoun, E.A., Common, R., Cruikshank, M.M., 1965. Recent bog flows and debris slides in the north of Ireland. Sci. Proc. R. Dublin Soc., Ser. A 2, 163−174.

Coxon, P., Coxon, C.E., Thorn, R.H., 1989. The Yellow River (County Leitrim, Ireland) flash flood of June 1986. In: Beven, K., Carling, P. (Eds.), Floods: Hydrological, Sedimentological and Geomorphological Implications. Wiley, Chichester, pp. 199−217.

Crisp, D.T., Rawes, M., Welch, D., 1964. A Pennine peat slide. Geogr. J. 130 (4), 519−524.

Crofton, M.T., 1902. How Chat Moss broke out in 1526. Trans. Lancashire and Cheshire Antiquarian Society 20, 139−144.

Cruickshank, M.M., Tomlinson, R.W., 1990. Peatland in Northern Ireland: inventory and prospect. Ir. Geogr. 23, 17−30.

Dijkstra, T.A., Dixon, N., 2010. Climate change and slope stability in the UK: challenges and approaches. Q. J. Eng. Geol. Hydrogeol. 43 (4), 371−385.

Dykes, A.P., 2008a. Geomorphological maps of Irish peat landslides created using hand-held GPS. J. Maps, 258−279.

Dykes, A.P., 2008b. Properties of peat relating to instability of blanket bogs. In: Chen, Z., Zhang, J.-M., Ken, H., Fa-Quan, W., Zhong-Kui, L. (Eds.), Landslides and Engineered Slopes: From the Past to the Future. CRC Press, Boca Raton (USA), pp. 339−345.

Dykes, A.P., 2009. Geomorphological maps of Irish peat landslides created using hand-held GPS—second edition. J. Maps, 179—185.

Dykes, A.P., Jennings, P., 2011. Peat slope failures and other mass movements in western Ireland, August 2008. Q. J. Eng. Geol. Hydrogeol. 44 (1), 5—16.

Dykes, A.P., Kirk, K.J., 2006. Slope instability and mass movements in peat deposits. In: Martini, I.P., Martínez Cortizas, A., Chesworth, W. (Eds.), Peatlands: Evolution and Records of Environmental and Climatic Changes. Elsevier, Amsterdam, pp. 377—406 (Chapter 16).

Dykes, A.P., Kirk, K.J., 2001. Initiation of a multiple peat slide on Cuilcagh Mountain, Northern Ireland. Earth Surf. Processes Landforms 26, 395—408.

Dykes, A.P., Warburton, J., 2007a. Mass movements in peat: a formal classification. Geomorphology 86 (1—2), 73—93.

Dykes, A.P., Warburton, J., 2007b. Geomorphological controls on failures of peat-covered hillslopes triggered by extreme rainfall. Earth Surf. Process Landforms 32, 1841—1862.

Dykes, A.P., Warburton, J., 2008a. Failure of peat-covered hillslopes at Pollatomish, Co.Mayo, Ireland: analysis of topographic and geotechnical influences. Catena 72, 129—145.

Dykes, A.P., Warburton, J., 2008b. Characteristics of the Shetland Islands (UK) peat slides of September 19, 2003. Landslides 5, 213—226.

Edil, T.B., 2001. Site characterization in peat and organic soils. In: International Conference on Insitu Measurement of Soil Properties and Case Histories. Parhyangan Catholic University, Bandung, pp. 49—60.

Evans, M.G., Warburton, J., 2007. Geomorphology of Upland Peat: Erosion, Form and Landscape Change. Blackwell Publishing, Oxford, 262pp.

Feehan, J., O'Donovan, G., 1996. Bog bursts. In: The Bogs of Ireland: An Introduction to the Natural, Cultural and Industrial Heritage of Irish Peatlands. University College Dublin, Dublin, pp. 399—419.

Feldmeyer-Christe, E., Mulhauser, G., 1994. A moving mire—the bog burst of la Vraconnaz. In: Grünig, A. (Ed.), Mires and Man. Mire Conservation in a Densely Populated Country—the Swiss Experience. Excursion Guide and Symposium Proceedings of the 5th Field Symposium of the International Mire Conservation Group (IMCG) to Switzerland 1992. Swiss Federal Institute for Forest, Snow and Landscape Research, Birmensdorf, pp. 181—186.

Gallart, F., Clotet-Perarnau, N., Bianciotto, O., Puigdefabregas, J., 1994. Peat soil flows in Gahia del Buen Sucesco, Tierra del Fuego (Argentina). Geomorphology 9, 235—241.

Griffith, R., 1821. Report relative to the moving bog of Kilmaleady, in the King's County, made by Order of the Royal Dublin Society. J. R. Dublin Soc. 1, 141—144.

Hendrick, E., 1990. A bog flow at Bellacorrick Forest, Co. Mayo. Ir. For. 47, 32—44.

Hobbs, N.B., 1986. Mire morphology and the properties and behaviour of some British and foreign peats. Q. J. Eng. Geol. 19, 7—80.

Huat, B.K., Prasad, A., Asadi, A., Kazemian, S., 2014. Geotechnics of Organic Soils and Peat. Taylor & Francis Group, London.

Hudleston, F., 1930. The cloudbursts on Stainmore, Westmorland, June 18, 1930. Br. Rainfall, 287—292.

Hungr, O., Evans, S.G., 1985. An example of a peat flow near Prince Rupert, British Columbia. Can. Geotech. J. 22, 246—249.

Hutchinson, J.N., 1988. General Report: morphological and geotechnical parameters of landslides in relation to geology and hydrogeology. In: Bonnard, C. (Ed.), Proceedings, Fifth International Symposium on Landslides, vol. 1. A. A. Balkema, Rotterdam, pp. 3—36.

Irish Independent Newspaper, 2012. Derrybrien Windfarm Purrs Away Despite Huge Landslide. Published Online 20/01/2006. http://www.independent.ie/business/irish/derrybrien-windfarm-purrs-away-despite-huge-landslide-26405150.html.

Jorgenson, M.T., 2013. Thermokarst terrains. In: Shroder, J.F., others (Eds.), Treatise on Geomorphology, vol. 8. Academic Press, San Diego, pp. 313−324.

JNCC, 2011. Towards an Assessment of the State of UK Peatlands. Joint Nature Conservation Committee report No 445. ISSN 0963 8901.

Kinahan, G.H., 1897. Peat bogs and debacles. Trans. Inst. of Civ. Eng. Irel. 26, 98−123.

Kirk, K.J., 2001. Instability of Blanket Bog Slopes on Cuilcagh Mountain, N.W. Ireland (Unpublished Ph.D. thesis). University of Huddersfield, U.K.

Landva, A.O., Korpijaakko, E.O., Pheeney, P.E., 1983. Geotechnical classification of peats and organic soils. In: Jarrett, P.M. (Ed.), Testing of Peats and Organic Soils, vol. 820. ASTM Special Technical Publication, Philadelphia, pp. 37−51.

Landva, A.O., Pheeney, P.E., 1980. Peat fabric and structure. Can. Geotech. J. 17, 416−435.

Landva, A.O., La Rochelle, P., 1983. Compressibility and shear characteristic of Radforth peats. In: Jarrett, P.M. (Ed.), Testing of Peats and Organic Soils, vol. 820. ASTM Special Technical Publication, Philadelphia, pp. 157−191.

Large, A.R.G., 1991. The Slievenakilla bog burst: investigations into peat loss and recovery on an upland blanket bog. Ir. Nat. J. 23, 354−359.

Latimer, J., 1897. Some notes on the recent bog-slip in the Co. Kerry. Trans. Inst. Civ. Eng. Irel. 26, 94−97.

Lewkowicz, A.G., 1990. Morphology, frequency and magnitude of active−layer detachment slides, Fosheim Peninsula, Ellesmere Island, N.W.T. Permaforost—Canada. In: Proceedings of the Fifth Canadian Permafrost Conference, Quebec, June 1990, pp. 111−118.

Lindsay, R.A., Bragg, O.M., 2005. Wind Farms and Blanket Peat: The Bog Slide of 16th October 2003 at Derrybrien, Co. Galway, Ireland. The Derrybrien Development Cooperatve Ltd, Galway.

Long, M., 2005. Review of peat strength, peat characterisation and constitutive modelling of peat with reference to landslides. Studua Geotech. et Mech. XXVII (3−4), 67−90.

Long, M., Boylan, N., 2012. In-situ testing of peat—a review and update on recent developments. Geotech. Eng. J. SEAGS & AGSSEA 43 (4), 41−55.

MacFarlane, I.C. (Ed.), 1969. Muskeg Engineering Handbook. University of Toronto Press, Toronto, p. 297.

MacFarlane, I.C., Radford, N.W., 1965. A study of the physical behaviour of derivatives under pressure. In: Proceedings 10th Muskeg Research Conference. NRC of Canada. Montreal, Canada.

McCahon, C.P., Carling, P.A., Pascoe, D., 1987. Chemical and ecological effects of a Pennine peat-slide. Environ. Pollut. 45, 275−289.

McEwen, L.J., Withers, C.W.J., 1989. Historical records and geomorphological events: the 1771 "eruption" of Solway Moss. Scott. Geogr. Mag. 105 (3), 149−157.

Mills, A.J., 2002. Peat Slides: Morphology, Mechanisms and Recovery (Unpublished Ph.D. thesis). University of Durham, UK.

Molyneux, W., 1697. Kapanihane bog flow, near Charleville, county Limerick. Philos. Trans. R. Dublin Soc. XIX, 714−716 (Oct.).

Munro, R., 2004. Dealing with Bearing Capacity Problems on Low Volume Roads Constructed on Peat: Including Case Histories from Roads Projects within the ROADEX Partner Districts.

Roadex II Project Report. http://www.roadex.org/roadex_new_site_wp/wp-content/uploads/2014/01/2_5-Roads-on-Peat_l.pdf (accessed May 2014.).

Munro, R., MacCulloch, F., 2006. Managing Peat Related Problems on Low Volume Roads—Executive Summary. Roadex II Project Report. http://www.roadex.org/roadex_new_site_wp/wp-content/uploads/2014/01/Guidelines-for-the-Risk-Management-of-Peat-Slips.pdf (accessed May 2014).

Myślińska, E., 2003. Classification of organic soils for engineering geology. Geol. Q. 47 (1), 39—42.

Ousley, R., 1788. An account of the moving of a bog, and the formation of a lake, in the County of Galway, Ireland. Trans. R. Ir. Acad. B2, 3—6.

Pearsall, W.H., 1950. Mountains and Moorland. Collins, London.

Praeger, L., 1897. Bog-bursts, with special reference to the recent disaster in Co. Kerry. Ir. Nat. 6, 141—162.

Selby, M.J., 1993. Hillslope Materials and Processes, second ed. Oxford University Press, Oxford.

Selkirk, J.M., 1996. Peat slides on subantarctic Macquarie Island. Z. Geomorphol. Suppl. 105, 61—72.

Shroder Jr., J.F., 1976. Mass movement on Nyika plateau, Malawi. Z. Geomorphol. 20 (1), 56—77.

Skempton, A.W., Petley, D.J., 1970. Ignition loss and other properties of peats and clays from Avonmouth, King's Lynn and Cranberry Moss. Géotechnique 20, 343—356.

Sollas, W.J., Praeger, R.L., Dixon, A.F., Delap, A., 1897. Report of the committee appointed by the Royal Dublin Society to investigate the recent bog-flow in Kerry. Sci. Proc. R. Dublin Soc. VIII, 475—510.

Tallis, J.H., Meade, R., Hulme, P.D., 1997. Blanket Mire Degradation: Causes, Consequences and Challenges. Proceedings. Mires Research Group. Aberdeen, 222 pp.

Tomlinson, R.W., Gardiner, T., 1982. Seven bog slides in the Slieve-an-Orra Hills, county Antrim. J. Earth Sci. R. Dublin Soc. 5, 1—9.

Tranter, D., 1999. Case study: Wingecarribee Swamp, water wet or dry? In: Proceedings of the Water and Wetlands Management Conference, November 1998. Nature Conservation Council of NSW, New South Wales, pp. 90—97.

Vidal, H., 1966. Die Moorbruchkatastrophe bei Schönberg/Oberbayern am 13./14.6.1960. Z. Dtsch. Geol. Ges. Jahrgang 1963 (115), 770—782.

von Post, L., 1924. Das genetische System der organogenen Bildungen Schwedens. Comité International de Pédologie IV Commission 22.

Warburton, J., Higgitt, D.L., Mills, A.J., 2003. Anatomy of a Pennine peat slide, Northern England. Earth Surf. Processes Landforms 28, 457—473.

Warburton, J., Evans, M.G., 2011. Geomorphic, sedimentary, and potential palaeoenvironmental significance of peat blocks in alluvial river systems. Geomorphology 130, 101—114.

Warburton, J., Holden, J., Mills, A.J., 2004. Hydrological controls of surficial mass movements in peat. Earth Sci. Rev. 67 (1—2), 139—156.

Wilson, P., Griffiths, D., Carter, C., 1996. Characteristics, impacts and causes of the Carntogher bog-flow, Sperrin Mountains, Northern Ireland. Scott. Geogr. Mag. 112, 39—46.

Chapter 7

Rock–Snow–Ice Avalanches

Rosanna Sosio
Università degli Studi di Milano-Bicocca, Italy

ABSTRACT

Rock avalanches which occur in glacial environments are controlled in their event dynamics and mode of propagation by the interplay between the detached rock and the icy component during all phases of motion, from initiation to final deposition. Because of the presence of ice and snow, the flow mobility is enhanced with respect to rock avalanches of comparable magnitude evolving in nonglacial settings by up to 25–30%. The high mobility, together with other possible secondary effects, caused by a change in flow behavior during propagation when glacial rock avalanches impact lakes or entrain and melt ice and snow at the flow base, determine glacial rock avalanches destructiveness. In recent decades, many among the most disastrous rock avalanches have occurred in glacial environments. In the future, possible increases in failure events occurring in formerly glaciated and permafrost areas are likely because of ongoing changes in climatic conditions.

Several factors converge to determine the mode of propagation and the high mobility observed for ice-rock avalanche events: (1) the debris-glacier interface provides a low-friction surface; (2) the basal topography may favor propagation due to funneling or air launching of the debris by moraines; (3) the detached material usually contains ice and snow in quantities which can be further increased by entrainment and reduces friction within the moving mass; (4) the snow and ice at the base of the flow can supply meltwater due to frictional heating, or compression of snow along the glacier surface increasing saturation and further reducing the flow resistance. The role of each factor has been analyzed based on post-event documentation, laboratory experiments, and numerical modeling. Despite a certain degree of the uncertainty of the results, which deserve more investigation, the presence of an icy basal material has been found to be particularly relevant at determining the flow mobility, either due to the smooth low-friction surface provided and lubrication or liquefaction effects of the propagating material due to melting processes.

7.1 INTRODUCTION

Rock avalanches are large, extremely rapid, flow-like movements of fragmented rock that originate as failures of intact rock mass but disaggregate in the course of

Landslide Hazards, Risks, and Disasters. http://dx.doi.org/10.1016/B978-0-12-396452-6.00007-0

FIGURE 7.1   Rock−snow−ice avalanche on Sherman Glacier triggered by the Alaska Earthquake (March 27, 1964). The source was from the area marked by the fresh scar on shattered peak (middle distance). The debris traveled for 5.6 km onto the Sherman Glacier until it stopped on a slope of 1−2°. The deposit, 1−2 m thick on average, displayed flow lines and terminal digitate lobes. The steep margin, about 20 m above the clear ice, was due to more rapid melting of the exposed glacier than the ice protected by the debris. No marginal dust layer was present. *U.S. Geological Survey, photo by A. Post, August 25, 1965. Image file: htmllib/batch81/batch81j/batch81z/ake00237.jpg.*

failure and runout (e.g., Figure 7.1; Hungr et al., 2001). Among rock avalanches, those in high-mountain, glacierized environments deserve increasing attention for the interactions between rock avalanches themselves and glaciers during all phases of motion, from initiation to final deposition. This has severe consequences on the event dynamics, mode of propagation, magnitude, and possible secondary effects, or chain reactions. Rock avalanches in glacierized environments are characterized by their sudden character and the hazard they pose, similar to their nonglacial counterparts. In addition, the class of rock avalanches known as rock−snow−ice (RSI) avalanches are of great interest because: (1) they are particularly mobile and destructive, as demonstrated by several among the most disastrous historical events that have recently occurred (Table 7.1); (2) their occurrence opens a series of issues about possible relationships with climate changes and ongoing deglaciation, with perhaps increases of event frequency in the near future; (3) they have strong geomorphic impact; (4) their significance is possibly still increasing with the recognition of new historical and prehistorical catastrophic rock-avalanche deposits in glacierized areas worldwide, as well as with the further increases of the elements at risk in mountain areas due to increasing tourism development and infrastructure installation.

TABLE 7.1 Description of Several among the Most Notable RSI Avalanches Occurred in the Last Decades

Event	Location	Date	Area Interested	Volume (× 10⁶ m³)	Evolution	Runout, L (km)	H (km)	Travel Angle (−)	Human, Economic Losses	Geomorphic Impact	References
Nevado Huascaran (6768 m l.s.l.)	Cordillera Blanca, Perú	January 10, 1962	Glacier 511 (RSIA) along the Rio Santa up to Pacific Ocean (DF)	13	RSIA-AF-DF	15.5 (RSI-A); 52 (total)	3.6	0.23	4000 casualties; Ranrahirca and Yungay towns were overwhelmed	The distal debris flow ran all the way to the Rio Santa	Morales (1966), Evans and Clague (1994)
		May 31, 1970		50–100	RSIA-AF-DF	15.6 (RSI-A), 175 (total)	3.85 (RSI-A), 6 (total)	0.25	18200 casualties; damage to communication facilities	Up to 10 m of debris deposited in the channel of Rio Santa	Ericksen et al. (1970), Evans and Clague (1994)
Pandemonium Creek	Coast Mountains, British Columbia	Summer 1959	South Atnarko River valley	5–7	RA-DF	8.6	2	0.23	–	Displacement waves along Knot Lakes	Evans et al. (1989)
Shattered Peak (1207 m a.s.l.), Sherman landslide	Chugach Mountains, Alaska	March 27, 1964	Andreas and Sherman Glaciers	13.7	RSIA	5.6	0.6	0.11	–	Aerial deposit extent of 8.6 km²	Shreve (1966), McSaveney (1975)

Continued

TABLE 7.1 Description of Several among the Most Notable RSI Avalanches Occurred in the Last Decades—cont'd

Event	Location	Date	Area Interested	Volume ($\times 10^6$ m^3)	Evolution	Runout, L (km)	H (km)	Travel Angle (–)	Human, Economic Losses	Geomorphic Impact	References
Mt Dzhimarai-khokh (4780 m a.s.l.), Kolka landslide	Kazbek massif in North Ossetia, Russia—Georgia	July 3–6, 1902	Kolka—Karmadon Glacier	75–110	GS-AF	14	1.7	0.12	36 casualties	—	Stoeber (1903)
		September 20, 2002	Kolka—Karmadon Glacier	100–140	RSIA-GS-AF-DF	19 (RSIA), 36 (total)	4.8	0.07	140 casualties	Almost complete detachment of the glacier from its bed. Area affected 12.5 km^2	Evans et al. (2009a)
Little Tahoma Peak	Mt Ranier, USA	December 14, 1963	Emmons Glacier and the White River valley	11	RSIA-DF	6.9	1.89	0.27	—	Aerial deposit extent of 5.1 km^2	Fahnestock (1978)
Lituya Mountain	Fairweather Mountain Range of Glacier Bay National Park, Alaska-British Columbia	June 11, 2012	A tributary of John Hopkins Glacier	18[1]	RSIA	9	2.4	0.27	—	Aerial deposit extent of 7 km^2, significant air blast	Geertsema (2012)

Triolet Glacier	Mont Blanc Massif (Italy)	September 12, 1717	Upper Ferret Valley	16–20	RSIA	7.2	1.86	0.26	7 casualties, 2 small settlements destroyed, loss of livestock	Aerial deposit extent of 0.9 km². The Doire River has been dammed for a short period and formed a lake	Eisbacher and Clague (1984), Deline (2009)
Mt Cook (3754 m a.s.l.)	Aoraki Mt Cook National Park, Southern Alps, NZ	December 14, 1991	Grand Plateaux ice field, Hochstetter Glacier, and Tasman Glacier	9.4–14.2	RSIA	6.9	0.76	0.11	–	The landslide reduced the Mt Cook height by 10 m	McSaveney (2002)

¹Volume suggested on the base of an average deposit depth of 3 m.

Despite being quite low frequency events, which commonly occur in remote areas, rock avalanches propagating onto glaciers have an enormous destructive potential and could cause high casualties if the process impacts on populated regions. With respect to other rock-avalanche types, those moving on glaciers are generally more mobile, at comparable magnitude. The involvement of snow and ice within the propagating material (volume up to 65−70 percent, Schneider et al., 2011a), or as substrate at the contact with the basal layer, strongly influences the flow, with direct consequences on the hazard they pose. Such catastrophic processes have been the cause of some of the worst natural disasters of this century (e.g., Table 7.1; Evans and Clague, 1993). Particularly destructive and far reaching are those events that change their flow behavior during propagation when they impact lakes and trigger outburst floods (Evans and Clague, 1994; Clague and Evans, 2000), or entrain and melt ice and snow at the flow base (Petrakov et al., 2008). Among the most disastrous historical events in recent decades, the two successive rock avalanches that occurred at Nevado Huascarán in the Cordillera Blanca, Perù (January 10, 1962 and May 31, 1970) are notable (Figure 7.2). Both initiated from the western face of the north summit of Mt Huascaran and involved large amounts of ice and snow. The material accelerated onto Glacier 511, reaching velocities as high as 278 m/s (Pflaker and Ericksen, 1978; Morales, 1966) and then traveled with low friction for approximately 16 km down the Rio Santa as pulverized and fluidized material entraining moraine debris (Petrakov et al., 2008). Together, the events caused a total loss of about 7,000 to 25,000 lives (Pflaker and Ericksen, 1978; Koerner, 1983; Evans et al., 2009a). The 1970 event, in particular, was among the largest RSI avalanches of the last decades with a volume of approximately $50-100 \times 10^6 \text{ m}^3$ (Pflaker and Ericksen, 1978). More recently, concern arose on September 2001, when a supraglacial lake, named Effimero, formed at the foot of Monte Rosa east face in the European Alps as a consequence of the surging activity of the Belvedere Glacier. The lake's last 2 years, had a variable depth of 15−35 m and reached a maximum volume of $3 \times 10^6 \text{ m}^3$ (Haeberli et al., 2002; Kääb et al., 2004; Tamburini and Mortara, 2005). If a mass movement entered the lake, the impact could trigger a secondary event (flood or debris flow). Two major RSI avalanches occurred in 2005 and 2007 and entered the (almost empty) depression of the former lake Effimero (Fischer et al., 2006; Huggel et al., 2010).

The last century has been marked by warming of atmospheric temperature (Houghton et al., 1996), to which perennially frozen and glacierized regions react very sensitively (Oerlemans and Fortuin, 1992; Haeberli and Beniston, 1998). Among other effects, climate warming is causing rapid glacier thinning and retreat and significant permafrost changes in mountainous regions (Table 7.2). From the maximum glacier extent recorded in many parts of the world during the last few centuries, a substantial reduction of glacial cover has been documented in North America (Hewitt et al., 2008), southeast Alaska, southwest Yukon, British Columbia (Clague and Evans, 1994), in European

FIGURE 7.2 Aerial view showing mountain range including Nevada Huascaran and the landslide event that was triggered on North Peak by May 31, 1970 earthquake. The event started as a rock avalanche with large amounts of snow and ice (up to 30 percent; Pflaker and Ericksen, 1978) and traveled along Glacier 511 for 15.6 km at extremely high velocities (see also Table 7.4) entraining ice and morainal material. By the time it reached the town of Yungay, the rock—snow—ice avalanche is estimated to have consisted of about 280 million cubic meters of water, mud, and rocks. The town of Yungay was overwhelmed by the debris and many thousands of people were killed. The mass continued as a debris flow along the Rio Santa all the way to Pacific Ocean. *U.S. Geological Survey, photo by Servicio Aerofotografico National, Perù. Image file: htmllib/batch89/batch89j/batch89z/pla00008.jpg.*

Alps (Paul et al., 2004; Barletta et al., 2006), in the Karakoram and in most other Himalayan and Inner Asian mountain systems (Shroder et al., 1993; Calkin, 1995). In the European Alps, permafrost warmed up by about 0.5—0.8 °C in the upper tens of meters during the last century (Harris and Haeberli, 2003; Harris et al., 2003), whereas the lower elevation limit of permafrost is estimated to have risen vertically by about 1 m/year since the Little Ice Age maximum (Frauenfelder, 2005). These changes in glacier extent and permafrost distribution altered alpine rockwall equilibria so that instabilities are increasingly expected in steep high-mountain walls, which are either affected by permafrost covered by firn fields, or no longer supported by

TABLE 7.2 Distributed Data of Glacial Retreat

	Glacier Ice Loss			Period	References
	Volume	Period	Extension		
New Zealand Alps	49%	Last century		1894–1990	Allen et al. (2009)
Alaska	61%	1850–1970	−25.9 m/year [*]		Hoelzle et al. (2007), [*] Oerlemans (1994)
	17 km³/year	1970–2000			Larsen et al. (2007)
Southern Alaska	20 km³/year	1850–2000			Larsen et al. (2005a,b)
Rocky Mountains	30%	Last century	−15.2 m/year [*]	1890–1974	Hewitt et al. (2008), [*] Oerlemans (1994)
	17 km³/year	1970–2000			Larsen et al. (2007)
European Alps	48%	1850–1970	−15.6 m/year [*]	1850–1988	Hoelzle et al. (2007), [*] Oerlemans (1994)
Himalaya			21%[1]	1962–2006	Kulkarni (2006)
			−9.9 m/year	1874–1980	Oerlemans (1994)
Kenya			−4.8 m/year	1893–1987	Oerlemans (1994)

[1]The value indicates a percent reduction in the aerial extent.
* It indicates the correspondence of the data reported in the row and its reference.

hanging glaciers (Harris, 2005; Huggel, 2008). The climate factor is related to triggering of small rock instabilities (10^4-10^5 m^3), such as the rockfalls that occurred in Europe (summer of 2003; Gruber et al., 2004) and in New Zealand (summer of 2007—2008; Allen et al., 2009). More recently, the observation that a number of events occurred after periods (i.e., several days—several weeks) of warmer than average temperature (Huggel et al., 2010), within areas of degrading permafrost (e.g., the event at Brenva Glacier; Barla and Barla, 2001), or retreating glaciers (e.g., several events in British Columbia; Geertsema et al., 2006) supports climate warming as a factor promoting rockwall failures of a wide range of magnitude which may cause rock avalanches (e.g., Mt Stellar about 5×10^7 m^3, Alaska; Huggel, 2009).

Rock avalanches impacting on glaciers represent an important geomorphological factor in alpine environments and on glacier behavior (Figure 7.1). They are capable of modifying the aspect of vast areas in a matter of minutes, and their deposits form debris-covered, ice-cored pedestals tens of meters high above the surrounding bare ice surface (Post, 1967; Shugar and Clague, 2011; Reznichenko et al., 2011). Very large events emplace conspicuous amounts of debris over time with consequences for the geomorphology of active orogens (Hewitt, 1988a; Korup and Clague, 2009), sediment delivery (Hovius et al., 2000; Korup et al., 2004), drainage systems (Korup, 2004; Hewitt, 2006), and moraine formation (Reznichenko et al., 2011). Debris sheets deposited over glacial surfaces can change the glacial mass balance, dynamics, and evolution, protecting the underlying glacier from ablation, affecting basal friction, and possibly triggering surges or glacier advances (Hewitt, 1998; Deline, 2008). Glaciers themselves move the debris blanket and supraglacial rock-avalanche deposits along the glacier or to its margins, eventually promoting the formation of terminal moraines (Kirkbride, 1995; Reznichenko et al., 2011). Glacial transport of the supraglacial debris is a major contributor to mountain denudation and sediment transfer (Korup et al., 2004), whereas meltwater streams may entrain and transport down valley much of the finer fraction of the rock-avalanche debris (McSaveney, 1975; Davies, 2013).

The link between glaciers and rock avalanches is so strong that dozens of catastrophic landslides have been misinterpreted as glacial deposits and subsequently reinterpreted (Hewitt, 1988b; Porter and Orombelli, 1980; McColl and Davies, 2011), either on the basis of sedimentologic and morphologic arguments (Heim, 1932; Shulmeister et al., 2009; Reznichenko et al., 2011), or taking advantage of the wider availability of high-resolution satellite images (Strom and Abdrakhmatov, 2004). In the perspective of Quaternary landscape evolution, the abundance of previously unrecognized landslide deposits increases recognition of the role played by catastrophic processes in shaping mountain landscapes (Hewitt, 1999; Fort, 2000), whereas the newly reinterpreted deposits increase the relevance of glacial rock-avalanche processes in terms of frequency and distribution (Whitehouse, 1983). The misinterpretation of the deposits also casts doubt on previous reconstructions of regional glacial chronologies (Hewitt, 1999) and paleoclimate conditions and evolution (Larsen et al., 2005a,b), in the cases where the effects produced by large rock avalanches (i.e., terminal moraine formation, glacial fluctuation, and surging) have been historically attributed to climatic influences (Oerlemans, 2005).

In the following, we define RSI avalanches as rock-avalanche events where ice, snow, and firn are involved in amounts that are sufficient to influence the runout dynamics (e.g., Figure 7.1) in comparison with other, relatively dry, rock-avalanche types. Events with volumes larger than $1 \times 10^6 \, \text{m}^3$ and runouts longer than 1 km are considered to emphasize the events that show excess travel distance relative to the single-block model (Hsü, 1975). The nature of the movement can possibly evolve during the event from sliding to dispersive, avalanche-like motion into highly mobile debris flow after complete lique-faction. Because of their extreme mobility and destructiveness, events un-dergoing flow transformations (e.g., Huascaran events, Pflaker and Ericksen, 1978) are generally termed "catastrophic glacier multiphase mass movements" (Figure 7.2; Petrakov et al., 2008). Pure ice avalanches may also occur as sudden failures from the terminus of glaciers and move rapidly downslope (Hanke, 1966; Alean, 1985). These events are less frequent, and usually only briefly preserved, but their inclusion can be useful as extreme end-member of RSI avalanches.

7.2 RAPID MASS MOVEMENTS ON GLACIERS

7.2.1 Frequency and Distribution

Large, rapid, mass movements are infrequent but not uncommon in high-mountain, glacierized environments and on rock slopes in permafrost condi-tions. Cases of RSI avalanches have been reported in the past decades worldwide either as isolated episodes, swarms of events widespread in space, or recurrent episodes repeated over time. Single episodes are distributed over most glacierized regions of the world. Examples are documented in the Eu-ropean Alps (e.g., Alean, 1984; Noetzli et al., 2006; Barla et al., 2000; Sosio et al., 2008; Deline and Kirkbride, 2009; Fischer et al., 2006), in the Caucasus (Haeberli et al., 2004), in Alaska, USA (Huggel et al., 2007), in British Columbia, Canada (Evans et al., 1989; Evans and Clague, 1988; Lipovsky et al., 2008; Geertsema et al., 2006), in the Karakoram Himalaya (Hewitt, 1988b, 2009), and in New Zealand, Southern Alps (Korup, 2005; Allen et al., 2009). A compilation of events is provided by Schneider et al. (2011a). In recent times, events occurred from eastern flanks of Lituya Mountain in Glacier Bay National Park (Alaska, June 11, 2012), and on the southern flank of Mt Dixon in the Mt Cook National Park (New Zealand, January 23, 2013, see The AGU Blogsphere, 2013). Traveling for about 9 km, partly over a tributary of Johns Hopkins Glacier, the event initiated at Lituya Mountain is among the longest runout landslides on a glacier that have occurred in Alaska and Canada in recent times (Geertsema, 2012). The seismic signals of 3.4 and

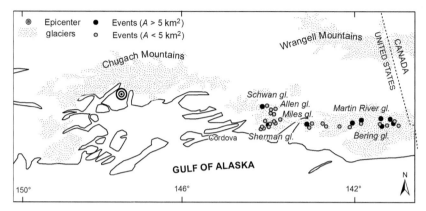

FIGURE 7.3   Cluster of rock–snow–ice avalanches triggered in south-central Alaska the by the March 27 1964 earthquake. *Figure modified after Post (1967).*

3.7 M, which were recorded by US and Canadian earthquake agencies, respectively, alerted distant scientists about the failure occurrence and allowed the location of the RSI avalanche to be identified.

Notable examples of swarms of rock avalanches have been documented particularly in seismically active environments. Earthquakes may trigger clusters of up to tens of events over extended areas, as observed in the Karakoram (Hewitt et al., 2011), Alaska (Figure 7.3; Post, 1967; Jibson et al., 2006), and British Columbia (Evans and Clague, 1999). Clusters of events can be also associated with melting of permafrost (e.g., Buckinghorse River area, Geertsema et al., 2006). Repeated failures are common in areas prone to pervasive degradation or rock weathering. Major rockfalls and rock avalanches themselves may favor further instabilities because of the consequent changes in surface geometry (Haeberli et al., 2007). In other cases, destabilizing factors may reactivate or persist for some time (e.g., geothermal gradient, Evans et al., 2009b). Among others, rockwalls which have undergone repeated failures over time include Mt Blanc (Deline, 2001), Mt Cook (McSaveney, 2002), Mt Iliamna (Figure 7.4; Huggel et al., 2007), and Monte Rosa (Fischer et al., 2006).

Despite the increasing number of documented events, frequency–magnitude relationships for RSI avalanches are sparse. An apparent increase in rock-avalanche frequency in (formerly) glacierized high mountains has been suggested for the European Alps (Fischer, 2009), New Zealand (Allen et al., 2011), Alaska (Huggel, 2009), and British Columbia (Geertsema et al., 2006). Moreover, expected frequencies and magnitudes could possibly change in the near future, both due to the implementation of existing (but still incomplete) inventories, and to a possible increase related to the ongoing deglaciation phase (e.g., Fischer, 2009; Davies et al., 2001). Underestimation of the number of events occurs for many reasons; among them are their occurrence in rugged,

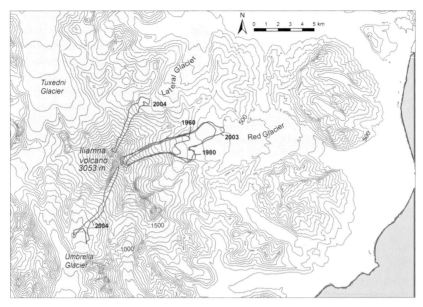

FIGURE 7.4 Multiple rock—snow—ice avalanches on Mt Iliamna (Alaska). The events replicate similar patterns in terms of failure and propagation dynamics, including entrainment and deposition. *Figure modified after Huggel et al. (2007).*

infrequently visited terrain and the circumstance that, by propagating onto glaciers, RSI avalanches are subjected to rapid burial of their deposits by snow cover or dispersion by ice flow, ablation, disappearing into crevasses or being misinterpreted as moraines. The recent increase of the number of deposits associated with historic RSI avalanches suggests that even more events may occur both in remote mountains (Eisbacher and Clague, 1984) and in long-settled regions (Whitehouse and Griffiths, 1983). The number of historic and prehistoric records may possibly increase from research in tectonically active regions (Topping, 1993) and from new recognition techniques (e.g., satellite data, sedimentologic and stratigraphic evidences, e.g., Krieger, 1977). On the other hand, future events have even more chances to be detected in areas where regional seismic monitoring networks are available (e.g., on Mt Cook, McSaveney, 2002; Mt Iliamna, Caplan-Auerbach et al., 2004; eastern Karakoram, Ekström and Stark, 2013) or through systematic observations of rockwall stability (e.g., on Mont Blanc Massif, Ravanel et al., 2010; Ravanel and Deline, 2011).

7.2.2 Causes

Many factors cause rockslides and avalanches in high mountain regions. Although rock avalanches occur in all types of rocks, topographic and geological environments that are most prone are: (1) weakly cemented, weathered, intensely fractured, or closely jointed rocks; and (2) well-indurated bedrock having prominent discontinuities or overhanging slopes. Weakly cemented rocks producing rock avalanches include shale, siltstone, sandstone, and conglomerate. Hard rocks that produce rock avalanches are usually intensely fractured or extremely weathered. Apart from lithological conditions, topographic environments that favor failures are slopes steeper than 40°, narrow spurs, ledges, and ridge crests (Table 7.3). Most of the instabilities are related to planes of local weakness such as faults, master joints, bedding planes, or foliation surfaces along which the rock mass is destabilized by a triggering factor (Figure 7.5). Given potentially unstable geological, glaciological, and topographic conditions, triggering factors may include strong earthquakes, intense rainfalls, and increasing temperatures that lead to glacial retreat and permafrost degradation. Locally, a further trigger is the interaction with heat fluxes as observed in ice-capped volcanoes and in areas of active geothermal flux.

Tectonic activity, including uplift (McSaveney, 2002) and seismic shaking (Post, 1967; Jibson et al., 2006), are typical of young, high and steep mountain ranges and favor large and widespread instabilities in rock (Post, 1967) and ice (Van der Woerd et al., 2004). A large number of RSI avalanches are triggered by large earthquakes (Richter magnitude, $M > 6$; Keefer, 1984; Malamud et al., 2004). The numbers of failures and the area affected by landslides are influenced by earthquake magnitude, and locally depend on regional differences in seismic attenuation, lithology, slope, and discontinuity orientation (Sepulveda et al., 2005), as well as on the angle and direction of incidence of the seismic waves (Ashford and Sitar, 2002). Steep slopes and ridge crests are geomorphic features with strong topographic amplification of the ground motion (Geli et al., 1988; Buech et al., 2010). Slopes subjected to seismic shaking may stay in near-critical stability conditions for years, before the full failure occurs (Dunning et al., 2007).

Rainfall is a minor triggering factor that rarely causes RSI avalanches. Intense rainfall triggered the 1999 rock avalanche at Howson Range at Fubar Glacier (Schwab et al., 2003), whereas a summer cloudburst probably triggered the 1999 rock avalanche at Kendall Glacier (Couture and Evans, 2002).

Glacier retreat since the last glacial maximum causes a decrease in ice thicknesses and a reduction in glacierized areas. These changes favor rock slope instabilities through: (1) a reduction of lateral support provided to adjacent steep rocky walls (O'Connor and Costa, 1993; Deline, 2009; Haeberli and Hohmann, 2008); (2) steepening of rockwalls (Augustinus, 1995); and (3) stress redistributions and debuttressing (Evans and Clague, 1999; Holm et al., 2004; but see McColl and Davies, 2013; McColl et al., 2010). Mechanical and thermal weathering acting on previously insulated surfaces can further modify periglacial slopes determining rockwall weakening and overstepping (Augustinus, 1995; Ballantyne, 2002). The Mt Fletcher RSI avalanches that occurred in 1992 (New Zealand, volume of $5-10 \times 10^6 \, m^3$) initiated from oversteepened rockwalls associated with the retreat of the Maud Glacier,

TABLE 7.3 Description of the Zone of Detachment for Several Recent RSI Avalanches

	Volume (× 106 m³)	Date	Source Areas					Permafrost	References
			Failure Thickness (m)	Slope (°)	Lithology	Aspect	Elevation Range (m a.s.l.)	Lower Limit (m a.s.l.)	
Vampire (New Zealand)	0.08–0.2	January, 2008	20–30 ± 10	65–73 (max)	Torlesse greywacke rock, moderately stratified	SE	2,380–2,560	2,000	Cox et al. (2008), Allen et al. (2009)
Mt Cook (New Zealand)	60–80	December 14, 1991	60	50–60	Closely jointed greywacke and argillite	E/SE	2,900–3,700	2,000	McSanevey (2002), Allen et al. (2009)
Mt Dixon (New Zealand)	Unknown	January 2013	Unknown	Unknown	Unknown	SW	Unknown	Unknown	The AGU Blogsphere, 2013
Mt Steller (Alaska)	40–60	September 14, 2005	Several tens of meters	45	Tertiary sedimentary rocks. Dip subparallel to the slope	S	2,500–3,100	Zone of cold permafrost	Huggel et al. (2010)
Mt Steller (Alaska)	1–1.15/1.4	July 2008. Two events.	Unknown	Unknown	Sedimentary rock extremely jointed	N/NE	2,350	Unknown	Huggel et al. (2010)
Mt Miller	16–28	August 6, 2008	Several tens of meters	Unknown	Intact basalt covered by 50–80 m glacial ice	N/NE	1,600–2,200	Zone of warming permafrost	Huggel et al. (2010)

Monte Rosa (Italy)	0.3/2.5	August 25, 2005; April 27, 2007	Unknown	>55 (max)	Layers of orthogneiss and paragneiss (pennininc Monte Rosa nap)	E	3,400–4,100	3,100–3,600	Fischer et al. (2006)
Monte Bianco (Italy)	0.1/0.5	August 1, 2007; September 30, 2007; September 10, 2008	From meters to tens of meters	65–80	Granitic intrusion in the gneissic basement (micashists and gneiss)	S and E	2,800–3,800	2,800–3,000	Ravanel et al. (2010)
Sperone della Brenva, (Mt Blanc, Italy)	2	January 19, 1997	Unknown	Unknown		W	3,400–3,725	Unknown	Barla and Barla (2001)
Lituya Mountain (Alaska)	Unknown	June 11, 2012	Unknown	>55	Layered hornblende-pyroxene gabbro	E	2,500–3,300	Below the source-area height	Geertsema (2012)
Mt Munday (British Columbia)	5	June 1997	Unknown	45–50	Gneiss	S	3,180–2,600	Unknown	Delaney and Evans (2008)
Punta Thurwieser (Italy)	2–2.5	September 18, 2004	10–20	70	Dolostone and black limestone with weathered and oxidized joint planes	S	3,600	Unknown	Sosio et al. (2008)

FIGURE 7.5   Source area on Punta Thurwieser (Valfurfa, Central Italian Alps). The slope failure started with rockfall activity, which involved a relatively thin slab of rock. The detachment completed in about 90 s and left a subvertical scarp 50 m high.

which had thinned approximately 250 m since Little Age Maximum, with acceleration of thinning after the middle of the nineteenth century (McSaveney, 1993).

Rising mean annual atmospheric temperatures induce changes in rock mass temperature and ongoing, progressive permafrost degradation. Atmospheric warming can affect the subsurface either by conduction (Noetzli et al., 2007) or by the development of thaw corridors along fractures. The latter process enables water to percolate much deeper into the bedrock and destabilize much larger volumes than pure conduction (Hasler et al., 2011). Permafrost degradation is faster in densely fractured rocks that promote active water drainage.

The observation that most RSI avalanches occur during summer and early autumn (Table 7.3), and the presence of massive ice along scars or within deposits (Dramis et al., 1995; Barla and Barla, 2001), supports the role of climate warming and permafrost degradation in promoting the instabilities that lead to these events. The ongoing permafrost degradation is testified to by the presence of water flowing on exposed bedrock in the detachment zone up to a few days after the failure (e.g., Mt Steller, Alaska, Huggel et al., 2010) or within boreholes and station tunnels at mountain stations (e.g., Mont Blanc massif, Gruber and Haeberli, 2007). Degrading permafrost leads to a rapid reduction of the strength of both ice-rich sediments and frozen jointed bedrock (Davies et al., 2001). In laboratory experiments, frozen rock joints show a temperature-dependent reduction in shear strength, with a minimum a little below 0 °C (Davies et al., 2001). This behavior is expected to favor instabilities in warming permafrost areas (Deline, 2001). Subsurface temperature responds rapidly (i.e., annual scale; Hasler et al., 2011; Gruber et al., 2004) to atmospheric warming, especially along steep slopes with reduced or absent snow cover (Allen et al., 2009; Gruber et al., 2004). Thawing of ice-filled fractures causes loss of ice/rock adhesion (Gruber and Haeberli, 2007). Melted water percolating at depth can penetrate into bedrock along joints and increases instability of rocky walls by the development of thaw corridors along fractures (Harris, 2005; Noetzli and Gruber, 2009). Rapid refreezing of melted water can strongly increase pore pressures within these joints (Huggel et al., 2010), whereas moving water can progressively widen and deepen passages in thawing rock (Gruber and Haeberli, 2007). Water percolation in highly fractured rock can contribute rather uniformly to heat transfer and lead to the development of an active layer (i.e., the top layer of thawing permafrost). Moving water can cause discrete thaw zones that penetrate into surrounding permafrost, and the active layer thus extends deeper and faster than through conduction alone, contributing to rapid destabilization of much larger volumes of rock than would be expected in a purely conductive system. The active layer forms during summer and then thickens, being deepest in September (Gruber et al., 2004), when the largest events appear to occur (Table 7.3). Consequences on slope stability can be inferred for zones of permafrost approaching marginal temperatures of c. 0 °C (Allen et al., 2008), or in zones subjected to different exposures and thermal exchanges (Noetzli et al., 2007). Permafrost bodies can also be present below slopes with positive, mean-annual, ground-surface temperatures, when induced by a colder surface nearby (Noetzli et al., 2007). Short, warm extremes or particularly hot summer seasons may have an effect on temperature distribution, meters below the surface. These are assumed to be responsible for several recent events on the eastern Eiger flank in the Swiss Alps (July 2006; Oppikofer et al., 2008), on Mt Steller, Alaska (on September 2005; Huggel et al., 2008), and on the east face of Monte Rosa, Italy (August 2005, Fischer et al., 2006). Deeper mountain permafrost (tens of meters depth) requires decades to warm. Thus, giving the ongoing warming trends, large instabilities will be favored on long timescales. Typical thermal patterns are recognized as promoting factors for instabilities at different locations (Alaska, Europe, and New Zealand). Huggel et al. (2010) found that several large slope failures were preceded by days to weeks of unusually warm temperatures, often followed by sudden drops in temperature, typically below freezing, lasting hours to days before failure.

Ice-capped active volcanoes are particularly prone to failures, which can be catastrophic (e.g., Nevado del Ruiz, Colombia in 1985; Pierson et al., 1990). More frequent than eruptions, volcanic unrest promotes instabilities through fumarolic activities, increasing heat flux within the edifice, and possible magmatic intrusions (Caplan-Auerbach and Huggel, 2007). Failure can occur at the glacier−rock interface by loss of cohesion associated with ice melting due to emission of volcanic gases through fractured rocks. Alternatively, failures can occur within unstable, hydrothermally altered rocks underlying the glacier (Huggel et al., 2007). The presence of rock weakened by alteration, as well as areas characterized by geothermal energy fluxes, promotes repeated failures, which seldom initiate in the same location. As an example, Iliamna volcano (Alaska) has been subject to failures cycles (every 3−6 years; Caplan-Auerbach and Huggel, 2007). Prehistoric failures of few millions of cubic meters involved the same sites and similar geometries (Waythomas et al., 2000) and evolution (Huggel et al., 2007). These repeated failures suggest possible mechanisms of mass accumulation up to a critical level where eventually failure occurs (Post and La Chapelle, 1971).

Failure of steep glaciers generally occurs during summer and results from loss of tensile strength in the ice mass through progressive fragmentation associated with crevasse development, melting of parts of the glacier, and reduction of the frictional resistance at the ice−rock interface due to increased water pressures (Rothlisberger, 1978). Most of the largest events occur as the ice breaks off hanging glaciers with ice near the bedrock at the pressure melting point (Alean, 1985). Many of the failures can be associated with: (1) earthquakes; (2) warm periods, with abundant meltwater at the sole of the glacier; and (3) significant changes in glacier mass balance or activity, for example during periods when the glacier shows "surge-like" behavior. Once the snouts of glaciers fail as ice avalanches, they in some case transform into debris flows or ice-rock avalanches (Haeberli et al., 2004). The Kolka event (Evans et al., 2009b) is an extreme manifestation of glacier instability, involving the complete detachment of the glacier mass from its bed, extreme mobility and velocities as high as 65−80 m/s.

7.2.3 Evolution

RSI avalanches usually initiate as the detachment of a large mass of rock in the form of slides, topples, and falls. The failure most often involves a relatively thin slab of rock (e.g., RSI avalanche at Aoraki/Mt Cook in 1991; McSaveney,

2002) or as wedge failure on joined surfaces (e.g., RSI avalanche at Mt Steller on 2005, Huggel et al., 2008). When the rupture occurs within the glacier, or at the rock—glacier interface (e.g., RSI avalanches at Mt Iliamna) analysis of the seismic signals constrained the failure mechanism as a ramp failure type, which involves a series of discrete slip events (Caplan-Auerbach and Huggel, 2007).

Precursory signals, such as gravitational creep and fracturing, are common and they are documented either as frequent, small-volume rockfalls and ice falls (e.g., preceding the Monte Rosa events, Fischer et al., 2006), sporadic, but large rock slope failures (e.g., preceding the Bualtar glacier RSI avalanches, Hewitt, 1988b), or glacier avalanches before the main event. Although not ubiquitous, precursory rockfall activity can take place months before the main event (e.g., during the period from August 1996 to January 1997 for the Sperone della Brenva; Barla and Barla, 2001) and then it can continue over months. As extreme cases, slopes may generate countless rockfalls before failing catastrophically (e.g., at Mt Fletcher, New Zealand, McSaveney, 2002) or may continue for decades after the major failure (e.g., Bualtar Glacier in Karakoram, Hewitt, 2006). The load of rock and debris falling onto the glacier surface can cause differential vertical displacements of the ice mass and increments in the sliding rate of the glacier which possibly result in ice avalanches (Post, 1967; Hewitt, 1998). Evans et al. (2009b), suggest that the failure of the Kolka Glacier was triggered by a series of ice-rockfalls from northern face of Mt Dzhimarai-khokh onto the rear part of the glacier. The triggering collapses started two months before the main failure took place and they loaded the glacier by a volume of about $18 \times 10^6 \, \mathrm{m}^3$ (Evans et al., 2009b). The detachment of the Kolka RSI avalanche ($15 \times 10^6 \, \mathrm{m}^3$ of ice and rock failed on an average slope of 9°) was induced by the disruption of the internal drainage of the glacier (Fountain et al., 2005), which developed excess water pressure at its base. Small-volume rockfalls may also follow the major events, until the slope equilibrium has been recovered.

Successive failures can possibly occur in steep glaciers that, after glacial retreat, leave behind a bare surface (e.g., Monte Rosa east face, Fischer et al., 2006). More rarely, multiple successive pulses occur in the same area either related to different stages of retrogressive collapses or to pervasive unstable conditions (e.g., events at Mt Steller; Huggel et al., 2008, 2010).

The detached rock mass initially slides under gravity and it then progressively collapses into joint-controlled blocks and comminutes (e.g., Davies and McSaveney, 2009) under the high stresses of the fall, generating fragments of all sizes to submicron diameter (Shugar and Clague, 2011; Reznichenko et al., 2011). Large quantities of fine particles are expelled from the surface to form a growing aerial dust cloud, which can be visible kilometers away from the source and from the path followed by the landslide (Hewitt, 1988b; Jibson et al., 2006). Geertsema (2012) estimated dust to be at least 500 m above the landslide deposit. Depending on the amount of ice and snow included in the

FIGURE 7.6 Aerial picture of rock avalanche on Punta Thurwieser (Valfurfa, Central Italian Alps) on September 18, 2004. The rock avalanche involved only minor amounts of snow and ice. The material propagated onto the Zebrù Glacier in its uppermost path and then continued, confined within the lateral moraines. *Photo by M. Ceriani.*

detached mass, the RSI avalanche can flow like an airborne powder avalanche. Dust production is documented for large events (e.g., $13.7 \times 10^6 \, \text{m}^3$ volume for the 1991 event at Mt Cook; McSaveney, 2002) and for relatively small-magnitude events (e.g., $2-3 \times 10^6 \, \text{m}^3$ volume for the event at Sperone della Brenva; Barla and Barla, 2001; at Punta Thurwieser; Sosio et al., 2008; Figure 7.6) and is more common during the initial phases of motion.

Associated with the detachment and the initial phases of the slide motion, blasts of compressed air can also occur. The air pressure wave scatters gravel and pebbles above the zone affected by the propagation of the main RSI avalanche body. Geertsema (2012) reported the presence of fist-sized pebbles above the zone covered by the dust. During the event at Nevado Huascaran in 1970, a unique case of a stone hailstorm was observed, with rocks over 1 Mt in weight which were scattered for up to 4 km requiring initial velocities higher than 230 m/s (Stadelmann, 1983). The air blast can blow down trees tens to hundreds of meters from the margins of the RSI avalanche. Similar pressure waves have been observed for other rock avalanche types (e.g., Mt St Helens collapse, Voight et al., 1983; Val Pola rock avalanche, Crosta et al., 2004; Yigong rock avalanche, Xu et al., 2012).

Once RSI avalanches impact the glacial ice, they transform their style of motion to flow (Schwab et al., 2003) eventually entraining variable volumes of ice and snow (McSaveney, 1975, 2002; The Seattle Times, 2011). While descending very steep slopes, and by the time they reach the glacier surface, such masses pervasively and intensively fragment and attain velocities generally higher than 60—70 m/s (Sosio et al., 2008), perhaps approaching 100 m/s (Shreve, 1966; Huggel et al., 2008). Similar velocities can be locally reached by nonglacial rock avalanches (e.g., the runup height observed for the Val Pola rock avalanche required velocities as high as 78—108 m/s; Crosta et al., 2004). High velocities result from the initial free fall and reduce slowly during propagation on low-friction surfaces provided by glacial ice (Hewitt, 1988b; Table 7.4). The high velocities allow the debris to rise hundreds of meters up topographic obstacles, and to travel very long distances onto the generally very low slope gradients typical of large glacier surfaces (Post, 1967). Average velocities of 40—50 m/s are inferred from the total duration of seismic signals produced by the propagation (McSaveney, 2002; Schneider et al., 2010).

The overall mobility is significantly higher than that of pure rock avalanches, as shown by empirical observations (Evans and Clague, 1988) and numerical simulations (Sosio et al., 2012). Maximum runout distances are commonly 5 to 10 times the fall height; nevertheless, these may vary considerably, depending on event volume and emplacement morphology. A number of elements contribute to the high mobility of RSI avalanches, among them their large size, travel on the glacier surface and the amount of ice involved (up to 80—90 percent). These conditions lead the mixture of ice, debris, and water to travel for tens of kilometers on glaciers reaching velocities as high as 70—90 m/s (Table 7.4) with apparent friction angles as low as 6° for the main part of the avalanche. The travel angles can be even lower if the event transforms into a debris/mud flow (Huggel et al., 2005).

During propagation, amounts of ice, snow, and substrate material can be entrained to mix with the initially detached material in a thin zone at the base of the flow. This can substantially change the character and mechanism of the flow by filling the interstices between rocks with fluid and reducing the friction coefficient (Pflaker and Ericksen, 1978). The mass eventually becomes quasi-saturated. In the field, this effect is documented by the sudden disappearance of the dust cover usually observed over the glacier (e.g., Figure 7.7). The same effect was observed at the Mt Dixon rock avalanche (Aoraki/Mt Cook National Park, New Zealand) on January 28, 2013. A video (The AGU Blogsphere, 2013) documents the event evolution: In the first phases of motion, the RSI avalanche shows a dusty flow. After about 50 s from the detachment, the style of motion changes to viscous and decreases its velocity. Amounts of snow and ice are eroded at the flow front and dust is not generated anymore.

TABLE 7.4 Flow Velocity Estimates for the Selected Events, Compared to Modeling Results. The Mode of Calculation and the Corresponding Referees are Indicated

Event	Date	Volume (×10⁶ m³)	Mode of Computation	Velocity Mean (m/s)	On Glacier	Local	Maximum	Acceleration Maximum (m/s²)	References
Bualtar I	July 29, 1986	10	Sliding block model formulas	62			124	4.1	Hewitt (1988b)
Bualtar II	July 29, 1986		Local runup		>44				Hewitt (1988b)
Huascaran	January 10, 1962	13	Unreported	47					Evans and Clague (1994)
Huascaran	May 31, 1970	50–100	Runup height	75			278		Pflaker and Ericksen (1978)
Triolet Glacier, Italy	September 12, 1717	16–20	Runup height			35–44			Porter and Orombelli (1980)
Iliamna	July 25, 2003	5	Seismic signal	41			80		Caplan-Auerbach and Huggel (2007), Schneider et al. (2010)
Sherman	March 27, 1964	13.7	Free fall conditions / Traveling duration / Flow lines curvature—kinetic arguments	26	67–22		90		Marangunic and Bull (1968), McSaveney (1975)

Location	Date		Method					References
Punta Thurwieser	September 18, 2004	2–2.5	Computed from a movie	36–38	70–80	60–65		Sosio et al. (2008)
Black Rapid east/west	November 3, 2002	5–7/9–14	Runup height		45			Jibson et al. (2006)
Mt Steller	September 14, 2005	10–20	Seismic signal				100	Huggel et al. (2008)
Becca di Luseney	June 8, 1952	1.1	Runup height		45			Dutto and Mortara (1991)
Mt Cook	December 14, 1991	9.4–14.2	Seismic signal duration	58				McSaveney (2002), Schneider et al. (2010)
Pandemonium, Canada	Summer 1959	5–6	Superelevation at valley bends	30		81–100		Evans et al. (1989)
Kolka, Russia	July, 1902	100	Seismic record from the impacting mass			50–80		Huggel et al. (2005)
Kolka, Russia	September 20, 2002	100	Superelevation at valley bends		70–90		0.8	Huggel et al. (2005)
Little Tahoma Peak, USA	December 14, 1963	11	Runup height			29–42		Fahnestock (1978)
			Flow lines curvature		60			Sheridan et al. (2005)
Val Pola, Italy	July 28, 1987	40	Runup height			76–108		Crosta et al. (2004)

Continued

TABLE 7.4 Flow Velocity Estimates for the Selected Events, Compared to Modeling Results. The Mode of Calculation and the Corresponding Referees are Indicated—cont'd

Event	Date	Volume (× 10⁶ m³)	Mode of Computation	Mean (m/s)	On Glacier	Velocity Local	Maximum	Acceleration Maximum (m/s²)	References
Goldeau, Switzerland	1806	30–40	Runup height						Heim (1932)
Mt St Helens, USA	May 18, 1980	2800	Timed photographs	39		70			Voight et al. (1983)
						90			Moore and Rice (1984)
Mt Clayley, Canada	1984	0.75	Superelevation at valley bends			42–70			Evans et al. (2001)

FIGURE 7.7   Aerial view of the Zebrù Glacier, which is partially covered by the debris produced by the Punta Thurwieser rock avalanche. At the photo bottom, the glacier is mantled almost continuously by a fan shaped, thin sheet of debris deposited by the rock avalanche. At the photo top, the glacier is covered by the more irregular and thicker accumulation of glacial debris. Note a thin layer of dust that covers the uppermost part of the rock-avalanche deposit. *Photo by M. Ceriani.*

Rock avalanches propagating onto glaciers are mostly unimpeded by topography. In those cases, the rock mass moving across the ice surface spreads both laterally and longitudinally. The mass stops as relatively thin sheets of crushed, pulverized debris (Figures 7.1 and 7.7) after the complete melting of the snow and ice component (Deline, 2008). Rarely, deposits more than 2–10 m thick have been measured (Hewitt et al., 2008; Reznichenko et al., 2011) whereas angular megaclasts may exceed 3–5 m in diameter. These deposit thicknesses are generally much lower than those typical for rock avalanches on soil/rock surfaces. In other cases, particularly for smaller events, the propagation can be confined laterally (e.g., within moraines at Punta Thurwieser, Figure 7.6). Propagation across rugged terrain can cause the mass to divide into several lobes. When weakly confined or unconfined, RSI deposits are generally tongue-shaped with digitate margins and with minor surface relief, which are features similar to wet snow avalanches.

RSI avalanches may eventually undergo a phase transition (Petrakov et al., 2008), which notably increases their destructiveness, as occurred during the Kolka event (Russian Caucasus, September 20, 2002; Haeberli et al., 2003; Kotlyakov et al., 2004) and at Nevado Huascaran, both in 1962 and 1970 (Figure 7.2, Cordillera Blanca, Perù; Pflaker and Ericksen, 1978). Multiple phase events initiated as glacier instabilities are favored by particular conditions such as glacier-surge behavior, the presence of geothermal fluxes, external loading by rockfalls onto the glacier, and possibly other causes (Huggel et al., 2005; Evans et al., 2009b). Once failure has occurred, the RSI has the potential of moving very long distances over moderately inclined slopes with great velocities, further entraining ice and snow and increasing saturation. The transformation into a more mobile debris flow can be favored either by the high content of ice (e.g., Kolka—Karmadon event) or by fine-grained, water-saturated debris (e.g., Huascarán event).

7.3 RSI AVALANCHE PROPAGATION

Rock avalanche mobility is generally described by the H/L elsewhere - ratio between the fall height, H, and the runout, L, (Figure 7.8). The longer runout distance of RSI avalanches was described by Evans and Clague (1988). Among other causes enhancing mobility of RSI avalanches, more notable are: (1) the debris—glacier interface provides a low-friction surface for the propagation (Evans and Clague, 1988); (2) the topographic effects due to funneling or air-launching of the debris by moraines (Evans et al., 1989; Nicoletti and Sorriso-Valvo, 1991; Shreve, 1966); (3) the propagating material usually contains ice and snow which can be further increased by entrainment and reduces friction within the moving mass (Ericksen et al., 1970); (4) the snow and ice at the base of the flow can supply meltwater due to frictional heating (Erismann and Abele, 2001), or compression of snow along the glacier surface (Geertsema et al., 2006) increasing saturation and further reducing the flow resistance. As an extreme consequence, ice and snow can fluidize the mass (McSaveney, 1978) eventually transforming the mode of propagation from almost dry rock avalanches into debris flows (e.g., Pflaker and Ericksen, 1978).

Due to the sudden initiation of rock avalanches and their rapid motion, any direct quantification of the effects on overall mobility due to ice and snow interaction along propagation is not feasible. Post-event documentation indicates that RSI avalanches travel downslope about 25 percent (Evans and Clague, 1988; Deline, 2001) to 30 percent (Bottino et al., 2002) farther compared to nonglacial events of similar magnitude. Although it is evident that glacial environments increase the runout distance of rapid mass movements, different processes often combine together, and individual ones are difficult to

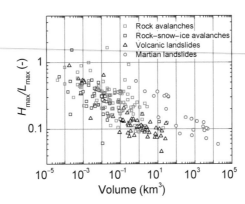

FIGURE 7.8 Bilogarithmic plot of the relative runout (H/L) versus volume (V) of rock avalanches from different settings. Note the large scatter of ice-rock avalanche data.

identify. The role of each factor will be discussed based on the analyses of postevent documentation, laboratory experiments, and numerical modeling.

Schneider et al. (2011a) performed comprehensive analyses of these factors on a data set consisting of 64 RSI avalanche events. The authors selected the events from among the most recent and best documented events worldwide. The selection is based on event volume (i.e., larger than 1×10^6 m^3 to observe excess travel distance related to volume effects), runout (i.e., horizontal length longer than 1 km), and time of occurrence (i.e., primarily 20th and 21st centuries). The events are characterized by their degree of interaction with ice and snow either quantitatively (e.g., amount of ice/snow into the moving mass, length of path trajectory run onto glaciers), or qualitatively (e.g., partial or complete liquefaction, when most grains were supposed to have been supported by the liquid phase), in order to find any relationship with the observed mobility (Schneider et al., 2011a). Typical deposit features develop during the propagation of RSI avalanches which are controlled by the emplacement conditions encountered, namely the high velocities or the interaction of the propagating mass with a low basal friction substrate, and reflect the flow conditions during emplacement.

Schneider et al. (2011b) performed laboratory experiments in partially-filled drums rotating at constant velocity, varying the proportion of gravel and ice, to test: (1) the reduction of material friction; and (2) the rise of water pressure and consequent reduction on shear resistance when ice is a part of the moving mass. Two drums of different sizes were equipped to measure flow depth, normal and shear forces, pore-water pressure, and temperature (Kaitna and Rickenmann, 2007; Hsü et al., 2007). The dimensionless forces estimated for the flow conditions in the experiments scale with those observed in nature. The friction coefficients found for mixtures of gravel and ice can serve as maximum values, as the experiments neglect several processes that can further reduce friction in nature. Among them are: dynamic grain fragmentation (Crosta et al., 2007; Davies et al., 2010), ice melting favored by the large pressure, sliding above low-friction surfaces on glaciers, and other possible scale effect related to the limited size of the sample, which does not mimic volume effects observed in real events.

Sosio et al. (2012) investigated the mobility of RSI avalanches through numerical modeling. They back analyzed 18 among the best documented case histories worldwide during recent decades. The events develop in a variety of settings and conditions, and vary with respect to their morphological constraints, materials, contact surfaces, and styles of failure, amounts of ice and snow entrained; their volumes range from M m^3 up to tens of M m^3.

7.3.1 Topographic Effects

Gravity is the main driving factor for mass propagation and spreading (Alean, 1985; Nicoletti and Sorriso-Valvo, 1991). Rapid mass movements become less sensitive to topographic features with increasing volume and velocity (Christen et al., 2010), and topographic effects may be highly variable along the path. RSI avalanches are characterized by extremely high velocities, which are among the highest observed for mass movements (Table 7.4). These result from the high relief and the low energy dissipation by friction. The runout paths can be either unconstrained or affected by more or less abrupt changes in slope, channeling by lateral moraines or in glacial troughs, interfluve overflows, caroming or swash effects (Hewitt, 2002; Hewitt et al., 2008).

The runup often observed demonstrates that the effect of topography can be minor during the first phases of motion (e.g., after descending the Shattered Peak, part of the material of the Sherman RSI avalanche overpassed the 140 m high Spur Ridge, Marangunic and Bull, 1968; McSaveney, 1975). By contrast, topographic influence becomes increasingly more relevant, as the RSI avalanche progressively reduces in velocity and its thickness. In these cases, when flowing onto unconstrained glacier surfaces with low gradient and almost flat topography, the effect of local micro-topography on propagation is enhanced (e.g., Mt Munday rock avalanche along the Icy Valley Glacier; Delaney and Evans, 2008). An increase in basal resistance or momentum adsorption, which is caused by irregularities or obstacles (Abdrakhmatov and Strom, 2006), as well as of the surrounding medium (Dufresne and Davies, 2009), are evidenced by the presence of transverse ridges (compressional features that form in the decelerating parts of the flow).

The effects exerted by lateral constraint on propagation efficiency are controversial. Channeling by lateral moraines or in glacial troughs, for example, is suggested to increase runout, leading to velocity increases (Evans and Clague, 1988) and reducing mechanical energy dissipation by higher flow depths (Nicoletti and Sorriso-Valvo, 1991). This can enhance the mobility by a funneling effect as documented at Little Tahoma Peak (Fahnestock, 1978) and at Nevado Huascaran (Pflaker and Ericksen, 1978). Analogously, to estimate quantitatively the influence of topography-induced deflections on the moving mass, Schneider et al. (2011a) defined a path deviation index as the (horizontal and vertical) deviation of the path length with respect to a direct connection between its highest and lowest points. The index does not provide any clear relationship between path deviation indices and apparent friction angles, as expected because of energy loss due to changes in flow direction (Nicoletti and Sorriso-Valvo, 1991; Okura et al., 2003). Frequent or abrupt changes of slope or direction affect the avalanche propagation by reducing the volume of the moving avalanche due to deposition of ice on flatter terraces or behind obstacles.

7.3.2 Motion on Low-Friction Glaciers

On striking the glacier, the motion of the distal debris in an RSI avalanche melts enough ice to smoothen a relatively rough surface. The mobility of RSI avalanches is thus enhanced (relative to motion over terrestrial surfaces) by the low-roughness basal surface and by possible further reduction of the basal friction by fluidization generated by frictional heating of ice (Evans and Clague, 1988).

A reduction of the basal friction and the predominance of inertia over granular dispersion in the style of flow is demonstrated by some morphological features observable on the deposits, such as the longitudinal grooves, well visible in Sherman as well in other landslides (Figure 7.1) and the alignment of elongated clasts along the flow direction (e.g., RSI avalanches onto Sherman and Black Rapids Glaciers; Shugar and Clague, 2011). Propagating on a basally weak substrate (Dufresne and Davies, 2009), shear is concentrated at the base of the avalanche, with minimal grain crushing at the periphery (e.g., Sherman Peak RSI avalanche, McSaveney, 1978).

RSI avalanches are characterized by large acceleration in the initial stages of the movements and by high maximum and mean velocities (Table 7.4). The low-friction basal surface provided by ice and snow enhances spreading, leaving much thinner deposits than on other surfaces, and allows propagation on gently sloping surfaces, especially in the absence of confinement. The Shattered Peak rock avalanche propagated onto Sherman Glacier for about 4 km at a gradient of $1°-3°$.

Despite the scatter in empirical data of historical events, Schneider et al. (2011a) found a positive correlation between mobility of RSI avalanches and the relative proportions of their paths upon glaciers, and this effect is almost independent of glacier debris cover, suggesting that debris cover and crevasses can perhaps increase glacier friction initially, but this effect becomes negligible once entrainment begins (Schneider et al., 2011a). The empirical relationship predicts a reduction in the overall apparent friction angle of some $25-30$ percent when the propagation is entirely over glacier ice with respect to propagating onto a different surface (Figure 7.9).

Similarly, the numerical replication of the propagation highlights the influence of the various surface materials encountered along the propagation path (Sosio et al., 2008). To model the effect of glacial ice, in particular, requires very low-resistance parameters that allow the propagation onto generally flat surfaces (up to few degrees), typical of glaciers. The basal friction angles back-calculated from modeling, although sparse, reduce by about 20 percent when the RSI avalanche is entirely emplaced onto glacier ice (Figure 7.9).

7.3.3 Snow and Ice Content of the Granular Mass

All the way from initial failure to final halt, rock, ice, snow, and frozen debris can be incorporated into the rocky mass by: (1) direct detachment, either as ice filling in fractures and pore spaces within bedrock or through direct glacier failure; and (2) entrainment during propagation along glaciers or along debris-covered areas with seasonal frost or permafrost (potentially degraded). Available data indicate amounts of ice and snow in the deposits that vary from

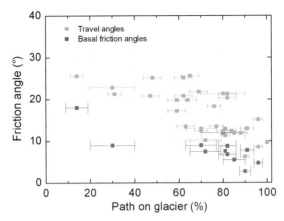

FIGURE 7.9 Dependence of the friction angle on the length of the path traveled onto glacier surface (perhaps covered by debris) expressed as percentage. Empirical (i.e., travel angles; Schneider et al., 2011a), and modeling (i.e., basal friction angles obtained through back analyses; Sosio et al., 2012) data are compared. Error bars indicate uncertainty in the evaluation of the length of the path traveled onto glaciers. No distinctions are made about the presence of debris covering the glacier surface.

being almost negligible (e.g., Thurwieser rock avalanche, Sosio et al., 2008) to representing up to 60–75 percent (e.g., events at Illiamna, Steller, Brenva Glaciers; Huggel et al., 2007; Deline, 2001; Deline and Kirkbride, 2009). The presence of ice and snow as part of the propagating mass is expected to reduce basal resistance due to the lower friction relative to rock and to possible changes in flow conditions (i.e., more fluid lubrication).

Evaluating the role exerted by ice content in deposits suffers from the sparse data available and from difficulties in quantifying ice/snow amounts and their possible changes along the path. As a result, travel angles evaluated in the field do not clearly show any dependency on ice content (Schneider et al., 2011b).

Laboratory experiments predict a reduction in bulk friction coefficient by 20 percent for dry granular ice compared with dry granular rock (Schneider et al., 2011a; Figure 7.10). The authors observed a stronger dependence on ice content with increasing size and angularity of grains, with reductions in the bulk friction angle of 27 percent (fine angular grains), 13 percent (fine rounded grains), and 50 percent (coarse angular grains) for pure granular ice compared to pure grains.

Basal friction angles obtained by back analyses of historical events through numerical modeling show a reduction as the amount of ice and snow in the propagating material increases, and such a decrease seems linear, as also observed from rotating drum experiments (Figure 7.10; Sosio et al., 2012). RSI avalanches show an 8 percent reduction of the basal friction angle for every 10 percent increase in ice content leading to basal friction angles for pure ice

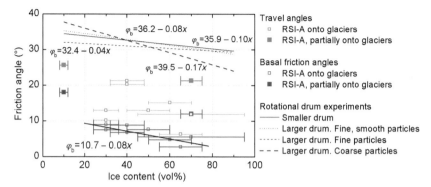

FIGURE 7.10   Dependence of the friction angle on ice content in the propagating material. Empirical (i.e., travel angles; Schneider et al., 2011a), laboratory (i.e., relationships obtained through rotating drum experiments; Schneider et al., 2011b), and modeling (i.e., basal friction angles obtained through back analyses; Sosio et al., 2012) data are compared. Error bars indicate uncertainty in the evaluation of the ice content within the whole material.

materials that are about 25 percent that of pure rock. Nevertheless, this rule needs to be refined with more data and suffers from uncertainty in the amounts of ice involved. No differences in the modeling results are observed by reducing the material density so as to account for the presence of snow and ice in the propagating mass.

7.3.4 Melting of Ice and Snow Due to Frictional Heating

For rock avalanches, De Blasio and Elverhoi (2008) suggest that, in cases of very smooth surfaces, heat produced at the shear layer by frictional resistance during sliding could change the material properties at the interface by the formation of a thin layer of lubricating melt in the shear zone. For rock avalanches propagating onto glaciers, frictional heating is still more efficient at melting the underlying ice (Huggel et al., 2005) by generating water that contributes to the lubrication of the propagating mass onto ice. Available water can either reduce the frictional resistance by increasing pore-water pressure (Evans and Clague, 1988) or by changing the basal rheology from purely frictional to viscous-frictional as the water interacts with fines produced at the base due to rock fragmentation (De Blasio, 2009). It is possible that large quantities of water can transform ice-rock avalanches into multiphase flows (Petrakov et al., 2008).

In the field, evidence of ice melting has been observed at the base of snow avalanches (Bartelt et al., 2006; De Blasio and Elverhoi, 2008) and ice-rock avalanches, either as the presence of refrozen ice within the debris (Marangunic and Bull, 1968), as a thin coat of wet mud at the deposit margins (McSaveney, 2002), up to a complete saturation of the mass (Hewitt, 1988b). Longitudinal flow bands (i.e., bands of debris of distinct lithology commonly extending over the entire length of the flow and separated from adjacent flow bands by grooves and furrows) are a characteristic of RSI avalanches emplaced onto nondry substrates. Flow bands develop in weak source material or fluidized substrate, being favored by high-velocity emplacement with a prominent component in the longitudinal flow direction (Dufresne and Davies, 2009), and may reflect ice and snow melting during propagation over glaciers. Ice and snow melt and mix with fragmenting rock, fluidizing it and allowing the debris sheet to extend along the travel direction of the landslide. The transformation of a dusty rock avalanche into a dense, muddy flow occurred within tens of seconds during propagation onto glaciers as documented by video tapes recorded at Nisqually Glacier, Mt Rainier on June 24, 2011 (Geertsema, 2012), and from Mt Dixon, in Aoraki Mt Cook, New Zealand on January 28, 2013 (The AGU Blogsphere, 2013). Based on evidence for quicklime (a calcined carbonate frictionite) at the Bualtar rock-ice avalanche in the Karakoram Himalaya, Hewitt (1988a,b) argued that strong frictional heating can only occur when the mass has attained a critical velocity, but before it disintegrates when the heat would diffuse rapidly (usually at very high velocities on irregular terrain with turbulent flow).

Laboratory experiments produced evidence of flow transformation in samples with ice contents larger than 40 percent (Schneider et al., 2011b). After the initial stages of the experiments (c. 5−15 min) the authors observed water supply by ice and snow melting that formed an intergranular water film around the grain surface (Casassa et al., 1991; Iverson et al., 2004). The intergranular force decreases as soon as partial saturation started at the tail of the moving mass, finally resulting in a partial or complete liquefaction (after c. 30 min, Schneider et al., 2011b) and in a reduction of the friction coefficient by more than 50 percent from a dry to fully saturated flow. The initial, dry, granular flow develops into hyperconcentrated flow, which is sustained by the liquid phase. Small quantities of fines are produced during the experiments through fragmentation, which are expected to increase the fluid viscosity (De Blasio, 2011b; De Blasio, in press) and, eventually, to support the development into debris flow or hyperconcentrated flows (Schneider et al., 2011b).

Erismann and Abele (2001) calculated that a mass with a thickness of 100 m and 1 km travel distance could melt a layer of ice as thick as 1.6 m. An equivalent water volume of 70 kg/m was estimated for the 1991 Mt Cook rock avalanche (McSaveney, 2002). Supported by the results of numerical modeling, which provide the evolution of the flow thickness and velocity during the propagation, Sosio et al. (2012) evaluated the efficiency of melting for the Shattered Peak rock avalanche traveling over the Sherman Glacier. They estimated that 86.2 ± 5.9 kg/m of equivalent water may have been produced by frictional heating during the 1964 Sherman rock avalanche, corresponding to an average melting depth of 9−10 cm. The melting rate (0.08 mm/s) is highest during the initial tens of seconds that correspond to the largest flow thicknesses. The layer of snow melted cumulatively during the whole propagation can vary locally from a few to tens of centimeters, being higher proximally. According to the authors, pore pressure development due to the rapid diffusion of the equivalent melted water within the basal layer is expected to cause an overall reduction in the basal friction angle by about 20–35 percent. The pore pressure developed within a 40 and 20-cm-thick shear layer in tens of seconds after the flow propagation (Schneider et al., 2011b; Sosio et al., 2012). The reduction of friction would be locally higher at the flow front and during the first phases of motion (Sosio et al., 2012). The downward dissipation of heat, responsible for ice melting, is proportional to the flow velocity and thickness, and to Coulomb frictional shear resistance (Colbeck, 1995; De Blasio and Elverhoi, 2008). A dependence of melting efficiency with the flow velocity is confirmed by laboratory experiments (Schneider et al., 2011b).

7.3.5 Snow and Ice Entrainment

Large and rapid rock mass movements can cause basal and frontal erosion leading to deposit volumes significantly larger than the detached volumes, especially when propagation occurs onto low resistance and weak substrates as glacial surfaces (Sovilla et al., 2006, 2007). Traveling over glaciers, RSI avalanches can increase their volume considerably, up to twice the detached volume (e.g., events from Mt Steller, Shattered Peak, Illiamna). Distal rims commonly occur, which are produced by snow entrainment and bulldozing of debris over dirty snow (Marangunic and Bull, 1968; The AGU Blogsphere, 2013; The Seattle Times, 2011).

In the field, the sparse available data indicate that the runout increases with the erosion of larger amounts of snow and ice (Sovilla et al., 2006). Numerical modeling shows that entrainment strongly influences both the flow mobility and the propagation direction (Sosio et al., 2012). Modeling the Sherman Glacier rock avalanche, entrainment retards the increase in the flow velocity and thus affects the propagation, particularly in relation to slope changes or topographic obstacles (Sovilla et al., 2007). A more direct quantification of the mobility increase due to entrainment is not presently feasible.

7.4 IMPLICATIONS FOR HAZARD ASSESSMENT

Current knowledge of RSI avalanches is largely based on postevent documentation, which provides event descriptions (i.e., detached volume, entrainment, runout, and snow/ice involvement). Failure events in rugged mountains are rarely documented or witnessed; exceptions include the events at Mt Cook (McSaveney, 2002), and Mt Dixon (The AGU Blogsphere, 2013). In most cases, these are identified from their deposits (Post, 1967; Jibson et al., 2006) or from the seismic tremor that they produce (Delaney and Evans, 2008; Geertsema, 2012; Ekström and Stark, 2013).

Nevertheless, RSI avalanches have caused some of the most tragic catastrophes related to glacial hazards known in history, when low-frequency/high magnitude events impacted densely populated regions or mountain infra-structure (Evans et al., 2009a,b; Haeberli et al., 1997; Deline, 2009). In spite of the significant damage produced by these events, and of the certainty that similar events will occur in the future, the hazards associated with their occurrence have only seldom been evaluated because of their low frequency, and are limited to specific areas of known instability. Analogously, standard engineering countermeasures are largely inefficient for large long-runout failures (Crosta et al., 2006). Determining the hazard of RSI avalanches re-quires quantitative estimates of the occurrence probability in space and time for a given event size, its runout mobility, and depositional area.

7.4.1 Probability of Occurrence in Time

The expected probability of occurrence of an event of a given magnitude is usually approximated by statistical distributions of documented time series using historical data (Abele, 1974; Whitehouse and Griffiths, 1983). This re-quires a reasonably complete inventory of dated events of known size. For RSI avalanches, however, time series rarely capture the full range of process magnitudes: they are of short duration (i.e., recent decades), lack systematic observations, and mostly rely on localized events. This record fails to recog-nize or undersample events occurring in remote areas (Hewitt et al., 2008), or of moderate size, which have relatively short residence times in the landscape before being obliterated by glacier dynamics. Following climate evolution, glacial and periglacial areas are currently undergoing dramatic changes, and may do so even more in the near future (Haeberli and Hohmann, 2008; Hewitt et al., 2008; O'Connor and Costa, 1993; Clague and Evans, 1994). Variations in the frequency and magnitude of landslides (i.e., rockfalls, RSI avalanches) are likely in alpine regions where glaciers have been retreating for decades and permafrost has been degrading (Jakob and Lambert, 2009; Haeberli et al., 2004; Kääb et al., 2005; Geertsema et al., 2006). Analogously, the frequency of warm air-temperature events increased during the twentieth century (Alexander et al., 2006) and further increases in warm extremes are still likely (Beniston et al., 2007).

Recurring events are documented for mountain regions particularly prone to failures. Repeated rockfalls and rock avalanches were recorded at Monte Rosa (Fischer et al., 2006), Monte Bianco (Deline, 2001) and Mt Iliamna (Huggel et al., 2007), which may reactivate in the future. In some cases, recurrent mass movements replicate similar patterns in terms of failure and propagation dy-namics, including entrainment and deposition (Huggel et al., 2007); or result in highly destructive events, as occurred at the Genaldon River Valley, which experienced repeated catastrophic failures from the Kolka Glacier, in September 2002 (Evans et al., 2009b) and in July 1902 (Stoeber, 1903). Hazard analysis in these cases can be assessed either by monitoring activity or by prediction of possible scenarios (Caplan-Auerbach and Huggel, 2007).

Seismogram analyses have become established for identifying and interpreting instabilities (Weaver et al., 1990; Suriñach et al., 2001) and each type of mass movement exhibits a characteristic spindle-shape seismogram. An exceptionally rich sequence of data are provided by the Mt Iliamna station, where a seismic network was installed in 1997 and has recorded about one ice-rock avalanche larger than 1×10^6 m^3 every 3 years (Caplan-Auerbach and Huggel, 2007). The signals identify the source zone, and estimate avalanche velocities, providing a better constraint on the failure mechanism (Caplan-Auerbach et al., 2004). In addition, the seismograms show a signal sequence up to 2 h prior to failure that characterizes the events initiating in the ice or at the ice—rock interface and provide an early indication of possible mass instabilities (Caplan-Auerbach et al., 2004).

7.4.2 Zone of Possible Initiation

Zones of recent glacier retreat appear particularly prone to failure, and a spatial shifting of the detachment zone is observed following the decrease in glacier extent in some cases (Fischer et al., 2006). The source areas of ice avalanches commonly correspond to zones of ablation of steep hanging glaciers (Alean, 1985), as observed on the Monte Rosa east face (Fischer et al., 2006). Failures of hanging glaciers are favored by high ice temperatures and percolating meltwater at the glacier bed, which destabilize the front of steep glaciers in warm-based conditions or at the transition from cold- to warm-based conditions.

Many unstable zones have been observed to be close to the lower limits of permafrost and in zones of warm permafrost (i.e., temperatures of -5 °C or higher) suggesting that permafrost degradation can be a significant factor in rock slope destabilization (Davies et al., 2001; Gruber and Haeberli, 2007; Gruber et al., 2004; Harris et al., 2003). In high-mountain topographies, the heat flux is driven by differential temperature at the surface between opposite mountainsides with an almost negligible contribution of geothermal gradient at the higher elevations (Noetzli et al., 2007). Thawing and migrating permafrost can lead to failure in the warmer side of ridges or peaks. This was observed at the Punta Thurwieser (Sosio et al., 2008), where the failure initiated at about 3600 m a.s.l. from a 70° steep and south-facing rockfall. Sun-facing instabilities are not always the case, and failures eventually occur in zones of cold permafrost that are encountered at higher elevation or in slopes facing away from the sun (Noetzli et al., 2003; Fischer et al., 2006). Nevertheless, the average altitude of scars on slopes facing away from the sun is lower than the ones facing the sun (Table 7.3), according to temperature distributions at the surface and within the rockwalls at a variety of exposures (Noetzli et al., 2007; Ravanel et al., 2010).

Krautblatter et al. (2013) developed a rock—ice mechanical model combining mechanical, geotechnical, and geomorphological stability analysis. The model considers two possible mechanisms acting over different time periods, which explains why all magnitudes of rock slope failures can be prepared by permafrost degradation. The first mechanism is the gradual and presumably cyclic warming/thawing of Alpine rockwalls consequent on the warming-impulse, relaxation time in the Late-glacial/Holocene period (Noetzli and Gruber, 2009). In that case, warming causes slow deformations along critical paths and influences the total friction and the fracture toughness of rock bridges at the microscale (Model I). This model explains high magnitude failures (>20 m thick detachment layer) that occur over months to millennia. The second mechanism is driven by the gradual decrease in strength that causes long-term fracture propagation and acceleration of rock slope failure along existing shear planes. During fast deformation, warming influences the creep of ice, the propensity of rock—ice detachments and total friction (Model II). This model explains the rapid response of rockslides of smaller magnitudes to warming that occurs along existing shear planes over days to months (short reaction time).

7.4.3 Runout Prediction

When a potential source of instability is identified, hazard mapping through runout analyses is necessary to define which areas could be threatened by landslide propagation. Different numerical models have addressed the dynamics of landslides traveling on glaciers, based on empirical relationships (Noetzli et al., 2006) assuming a certain rheology (i.e., frictional with finite pore pressure and Voellmy; Sosio et al., 2008) or introducing a more-or-less ad hoc decay of the friction coefficient (McSaveney, 2002; De Blasio, 2009, 2011a). Input data required for forward modeling include: (1) the event magnitude; (2) the path topography; and (3) a rheological approximation of the flow behavior.

Estimating the event location, volume, and the local thickness distribution in the source area is a critical and difficult task. The sliding surface in the source area is fundamental in controlling the initial distribution of the detached debris and strongly influences its subsequent propagation. The failure morphology can direct the debris initial motion preventing or encouraging a strong lateral dispersion. Failures occurring along bedding and joints may be constrained by the thickness of the bedding and spacing of the joints (Cox et al., 2008). The local thickness of the detached material controls directions of propagation, particularly for crest failures (a common feature where detachment is triggered by seismic tremor), or for detachments occurring in other zones of unconstrained topography, whereas the volume and initial location of the center of mass control the maximum runout.

The various surface materials encountered along the propagation path influence the event evolution. In order to capture and differentiate the effects of the surface materials, it is fundamental to characterize differently each phase of the propagation and to distinguish the basal surfaces by their frictional properties and entrainment susceptibility to improve the runout prediction (Hungr and Evans, 1996; McDougall and Hungr, 2004). In particular, the presence of glacial ice plays a substantial role at increasing the mobility of the rock avalanche (i.e., by some 25–30 percent when the propagation is entirely emplaced over glacier ice, Schneider et al., 2011a) and must be accounted for in the modeling. Avalanches that first descend over particularly steep terrain seem to have relatively large average slope angles (i.e., comparatively short runouts, Alean, 1985). This can be interpreted as a greater reduction of the initial potential energy at greater velocities. The Voellmy model is more effective than the frictional model at replicating the flow behavior over glacial ice, where the resistance has to be reduced to properly enhance the flow mobility. Values of $\mu = 0.05$ and $\xi = 1000$ m/s provided good results in several cases (Sherman Glacier, Pandemonium Creek (Hungr, unpublished; Hungr and Evans, 1996); Huascaran (Pirulli et al., 2004); Kolka (McDougall, 2006)). In the Voellmy model the frictional term depends on total stresses, whereas the turbulent term provides a correction that mimics the influence of velocity-dependent effective stress change (Hungr and Evans, 2004). The use of a Voellmy rheology, although empirical, is physically justified by the observation that, in a dense grain suspension in a fluid rapidly sheared at constant volume under inertial conditions, both the effective normal stress and the shear stress are proportional to the square of the shear strain rate (Bagnold, 1954; Hungr and Evans, 2004). These conditions probably apply to RSI avalanches until they start to slow down, and/or until they entrain enough water to become macroviscous.

For the purposes of hazard analysis, adopting predefined values of H/L to predict the mobility of RSI avalanches propagating in glacial settings is unreliable because of the spread of the H/L values as a function of the event volume (Figure 7.8), which is somewhat larger than observed for rock avalanches propagating in different settings. The large scatter of the travel angles observed for historical events suggests that potential energy loss cannot simply be explained by uniform loss in a conservative vector field, but it depends on other factors (Straub, 1997). Among them, most notable are the path constrictions (Strom, 2006; Schneider et al., 2011a), and the presence of amounts of ice and snow as contact surfaces (Schneider et al., 2011a), or within the propagating mass (Schneider et al., 2011b; Erismann and Abele, 2001). All these factors contribute to substantially reducing the basal resistance, both with respect to "real" material properties as well as to empirical parameters that describe nonglacial rock avalanches of comparable volumes. Due to the sudden onset of the phenomena, and to the infeasibility of using measured material properties in modeling process, a gap remains in the selection of the input values introducing large uncertainty into modeling prediction. Basal friction angles resulting from back analyses are less scattered than corresponding travel angles (Figures 7.9 and 7.10), and they describe the basal resistance well when assuming a frictional rheology. These are inversely

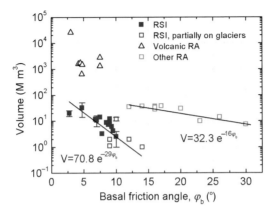

FIGURE 7.11 Event volume versus basal friction angle obtained through numerical modeling (Hungr and Evans, 1996; McDougall, 2006; Sosio et al., 2012). Error bars indicate volume changes related to entrainment during landslide propagation. Basal friction angles relate with event volume following exponential relationships, which depend on the event environment. Regression lines are indicated. RSI, rock–snow–ice.

related to the event magnitude according to $V = a \exp^{b_\varphi}$ where a and b are empirical coefficients (Figure 7.11) that decrease exponentially with event volumes (Sosio et al., 2012), and φ is the basal friction angle. The relationship between event volumes and the best fitting basal friction angles suggests possible friction angle values for forward modeling, once the event volume is known.

Particular concern and specific hazard analysis are required for those events that result as secondary effects of major RSI avalanches, such as: (1) cascading processes, including outburst flows resulting from the impact of landslides onto lakes (Evans and Clague, 1994); and (2) events that develop multiple flow behavior such as debris flows (e.g., Kolka–Karmadon).

7.5 CONCLUSIONS

RSI avalanches have attracted the attention of scientists, especially in recent years due to the Kolka–Karmadon event in 2002. From a theoretical point of view, the question why large landslides reach runout distances exceeding those expected from conventional frictional models (Heim, 1932; Hsü, 1975; Legros, 2002; Davies and McSaveney, 2012) can be applied to events occurring in glacial settings that are more mobile than their nonglacial counterparts (Alean, 1985; Evans and Clague, 1988). Moreover, rock avalanches propagating onto glaciers are mostly unimpeded by topography, being weakly confined or unconfined, and they offer the possibility of observing morphological features over a flat surface. From a practical point of view, the understanding of their flow behavior and a-priori delineation of the area vulnerable to their propagation are fundamental for hazard analyses. The most disastrous rapid mass movements from glacial environments have commonly shown a combination of features, such as large volumes, flow paths over glaciers or weak substrate, high ice and water contents, strong material entrainment, and flow transformations or chain reactions. Glacial and peri-glacial, high-mountain areas currently show increasing mass–movement activity due to the warmer temperature trends (Geertsema et al., 2006; Gruber et al., 2004; Haeberli and Hohmann, 2008).

Ongoing changes in climatic conditions raise questions about possible future increases in failure events occurring in formerly glaciated and permafrost areas. Actually, despite the number of events observed over the last decades, a real frequency increase is only postulated and not yet proven, and how climate change could affect landslide activity, and on which timescale, remain largely unresolved (Huggel, 2009).

Many studies have addressed evaluation of the hazard posed by RSI avalanches occurring in glacierized environments, including: (1) failure initiation mechanisms; (2) factors explaining the extremely high-observed mobility observed during propagation; and (3) modeling strategies for the prediction of the propagation of possible future events.

Krautblatter et al. (2013) developed a rock–ice mechanical model combining mechanical, geotechnical, and geomorphological stability analysis, which models the changes in rockwall strength by means of their fractures. In the context of an evolution toward a new equilibrium with changing climate conditions research is both ongoing and needed at defining:

1. Near-surface temperatures within steep rock faces (Allen et al., 2009; Gruber et al., 2003). A basis of systematic temperature measurements and models exists for near-vertical, unfractured rock, which allows simulation of temperature time series for various locations and depths (Gruber et al., 2004; Gruber, 2005).
2. Three-dimensional permafrost distributions (Allen et al., 2008; Noetzli et al., 2007) and zone of possible thawing (Wegmann et al., 1998; Noetzli et al., 2007). Transient effects as well as the influence exerted by the snow cover, the ice contained in pore spaces, and the percolating water greatly influence the location and extent of permafrost (Gruber and Haeberli, 2007).
3. Fracture distribution and monitoring at critical locations (Gruber and Haeberli, 2007).

Several factors converge to determine the high mobility observed for ice-rock avalanche events. Among them, the role of ice is particularly relevant when it constitutes the basal material and can be either due to the smooth surface or to lubrication or liquefaction of the propagating material due to melting (Schneider et al., 2011a).

Basal friction angles used to characterize the flow rheology in forward modeling cannot be directly inferred from the empirical indexes commonly used to describe the exceptional mobility of granular flows, particularly for RSI avalanche events that interact with the glacial ice and snow at every phase of their motion with variable effect on the overall mobility. From the perspective of hazard assessment, a negative exponential law relating the event volume and the probable basal friction angle has been proposed which results from the back analyses of several events of ice-rock avalanches propagating onto glaciers (Sosio et al., 2012).

REFERENCES

Abdrakhmatov, K.E., Strom, A.L., 2006. Dissected rockslide and rock avalanche deposits; Tien Shan, Kyrgyzstan. In: Evans, S.G., Scarascia-Mugnozza, G., Hermanns, R., Strom, A. (Eds.), Massive Rock Slope Failures. NATO Science Series. Kluwer Academic, Dordrecht, The Netherlands, pp. 573−595.

Abele, G., 1974. Bergstürze in den Alpen-ihre Verbreitung, Morphologie und Folgeerscheinungen. Wiss. Alpenl. Ver. Hefte 25, 1−165.

Alean, J.C., 1984. Ice avalanches and a landslide on Grosser Aletschgletscher. Z. Gletscherkd. Glazialgeol. 20, 9−25.

Alean, J.C., 1985. Ice avalanches: some empirical information about their formation and reach. J. Glaciol. 31, 324−333.

Alexander, L.V., et al., 2006. Global observed changes in daily climate extremes of temperature and precipitation. J. Geophys. Res. 111, D05109 http://dx.doi.org/10.1029/2005JD006290.

Allen, S., Owens, I., Huggel, C., 2008. A first estimate of mountain permafrost distribution in the Mount Cook region of New Zealand's Southern Alps. In: Proceedings of the 9th International Conference on Permafrost, June 28−July 3, 2008, Fairbanks, Alaska.

Allen, S.K., Gruber, S., Owens, I.F., 2009. Exploring steep permafrost bedrock and its relationship with recent slope failures in the Southern Alps of New Zealand. Permafrost Periglac. Process. 20, 345−356.

Allen, S.K., Simon, I., Cox, C., Ian, I., Owens, F., 2011. Rock avalanches and other landslides in the central Southern Alps of New Zealand: a regional study considering possible climate change impacts. Landslides 8, 33−48. http://dx.doi.org/10.1007/s10346-010-0222-z.

Ashford, S., Sitar, N., 2002. Simplified method for evaluating seismic stability of steep slopes. J. Geotech. Geoenviron. Eng. 128 (2), 119−128.

Augustinus, P.C., 1995. Glacial valley cross-profile development: the influence of in situ rock stress and rock mass strength, with examples from the Southern Alps, New Zealand. Geomorphology 14, 87−97.

Bagnold, R.A., 1954. Experiments on a gravity-free dispersion of large solid spheres in a Newtonian fluid under shear. Proc. R. Soc. Lond. Ser. A 225, 49−63.

Ballantyne, C.K., 2002. Paraglacial geomorphology. Quat. Sci. Rev. 21, 1935−2017.

Barla, G., Barla, M., 2001. Investigation and modelling of the Brenva Glacier rock avalanche on the Mount Blanc Range. In: Proceedings of the ISRM Regional Symposium Eurock 2001, Espoo, Finlandia, 3-7 Giugno, pp. 35−40.

Barla, G., Dutto, F., Mortara, G., 2000. Brenva glacier rock avalanche of 18 January 1997 on the mount blanc range, northwest Italy. Landslide News 13, 2−5.

Barletta, V.R., Ferrari, C., Diolaiuti, G., Carnielli, T., Sabadini, R., Smiraglia, C., 2006. Glacier shrinkage and modeled uplift of the Alps. Geophys. Res. Lett. 33, L14307. http://dx.doi.org/10.1029/2006GL026490.

Bartelt, P., Buser, O., Platzer, K., 2006. Fluctuation-dissipation relations for granular snow avalanches. J. Glaciol. 52 (179), 631–643.

Beniston, M., et al., 2007. Future extreme events in European climate; an exploration of regional climate model projections. Clim. Change 81, 71–95.

Bottino, G., Chiarle, M., Joly, A., Mortara, G., 2002. Modelling rock avalanches and their relation to permafrost degradation in glacial environments. Permafrost Periglac. Process. 13, 283–288.

Buech, F., Davies, T.R.H., Pettinga, J.R., 2010. The little Red Hill seismic experimental study: topographic effects on ground motion at a bedrock-dominated mountain edifice. Bull. Seismol. Soc. Am. 100, 2219–2229.

Calkin, P.E., 1995. Global glacial chronologies and causes of glaciation. In: Menzies, J. (Ed.), Modern Glacial Environments. Butterman-Heinemann, Oxford, pp. 9–75.

Caplan-Auerbach, J., Huggel, C., 2007. Precursory seismicity associated with frequent, large avalanches on Iliamna Volcano. Alaska J. Glaciol. 53 (180), 128–140.

Caplan-Auerbach, J., Prejean, S.G., Power, J.A., 2004. Seismic recordings of ice and debris avalanches on Iliamna Volcano (Alaska). Acta Vulcanol. 16, 9–20.

Casassa, G., Narita, H., Maeno, N., 1991. Shear cell experiments of snow and ice friction. J. Appl. Phys. 69 (6), 3745–3756.

Christen, M., Bartelt, P., Kowalski, J., 2010. Back calculation of the In den Arelen avalanche with RAMMS: interpretation of model results. Ann. Glaciol. 51 (54), 161–168.

Clague, J.J., Evans, S.G., 1994. Historic retreat of Grand Pacific and Melbern glaciers, Saint Elias Mountains, Canada: an analogue for decay of the Cordilleran ice sheet at the end of the Pleistocene. J. Glaciol. 39, 619–624.

Clague, J.J., Evans, S.G., 2000. A review of catastrophic drainage of moraine-dammed lakes in British Columbia. Quat. Sci. Rev. 19, 1763–1783.

Colbeck, S.C., 1995. Pressure melting and ice-skating. Am. J. Phys. 63, 888–890.

Couture, R., Evans, S.G., 2002. A rock topple–rock avalanche, near Goat Mountain, Caribou Mountains, British Columbia, Canada. In: Proceedings, 9th Congress, International Association of Engineering Geology and the Environment, Durban, South Africa.

Cox, S.C., Allen, S.K., Ferris, B.G., 2008. Vampire Rock Avalanches, Aoraki/Mount Cook National Park, New Zealand. GNS Science Report 2008/10, 34 pp.

Crosta, G.B., Chen, H., Frattini, P., 2006. Forecasting hazard scenarios and implications for the evaluation of countermeasure efficiency for large debris avalanches. Eng. Geol. 83, 236–253.

Crosta, G.B., Chen, H., Lee, C.F., 2004. Replay of the 1987 Val Pola landslide, Italian Alps. Geomorphology 60 (1–2), 127–146.

Crosta, G.B., Frattini, P., Fusi, N., 2007. Fragmentation in the Val Pola rock avalanche, Italian Alps. J. Geophys. Res. 112, F01006.

Davies, M.C.R., Hamza, O., Harris, C., 2001. The effect of rise in mean annual temperature on the stability of rock slopes containing ice-filled discontinuities. Permafrost Periglac. Process. 12, 137–144.

Davies, T.R.H., 2013. Fluvial processes in Proglacial environments. In: Shroder, J.F. (Ed.), Treatise on Geomorphology, vol. 8. Academic Press, San Diego, pp. 141–150.

Davies, T.R.H., McSaveney, M.J., 2009. The role of rock fragmentation in the motion of large landslides. Eng. Geol. 109, 67–79.

Davies, T.R.H., McSaveney, M., Kelfoun, K., 2010. Runout of the Socompa volcanic debris avalanche, Chile: a mechanical explanation for low basal shear resistance. Bull. Volcanol. 72 (8), 933−944. Doi: 201010.1007/s00445-010-0372-9.

Davies, T.R.H., McSaveney, M.J., 2012. Mobility of long-runout rock avalanches. In: Clague, J.J., Stead, D. (Eds.), Landslides: Types, Mechanisms and Modeling. Cambridge University Press, pp. 50−58. ISBN-13: 9781107002067.

De Blasio, F.V., 2009. Rheology of a wet, fragmenting granular flows and the riddle of the anomalous friction of large rock avalanches. Gran. Matter 11, 179−184.

De Blasio, F.V., 2011a. Introduction to the Physics of Landslides. Springer Verlag, Berlin, 420 pp.

De Blasio, F.V., 2011b. Landslides in Valles Marineris (Mars): a possible role of basal lubrication by sub-surface ice. Planet. Space Sci. 59 (11), 1384−1392.

De Blasio, F.V., Elverhoi, A., 2008. A model for frictional melt production beneath large rock avalanches. J. Geophys. Res. 113 (F02014), 1−13.

De Blasio, F.V. Friction and dynamics of rock avalanches travelling on glaciers. Geomorphology, 213, pp. 88−98.

Delaney, K.B., Evans, S.G., 2008. Application of digital cartographic techniques in the characterization and analysis of catastrophic landslides. The 1997 Mount Munday rock avalanche, British Columbia. In: Locat, D., Perret, D., Turmel, D., Demers, D., Leroueil, S. (Eds.), Proceedings of the 4th Canadian Conference on Geohazards: From J. Causes to Management. Presse de l'Université Laval, Québec, pp. 141−146.

Deline, P., 2001. Recent Brenva rock avalanches (Valley of Aosta): new chapter in an old story? Suppl. Geogr. Fis. Dinam. Quat. V, 55−63.

Deline, P., April 2008. Paraglacial Control on Rock Avalanches Occurred during the Recent Holocene in the Mont Blanc Massif. EGU General Assembly, Vienna, Austria, pp. 13−18.

Deline, P., 2009. Interactions between rock avalanches and glaciers in the Mont Blanc massif during the late Holocene. Quat. Sci. Rev. 28, 1070−1083.

Deline, P., Kirkbride, M.P., 2009. Rock avalanches on a glacier and morainic complex in Haut Val Ferret (Mont Blanc Massif, Italy). Geomorphology 103, 80−92.

Dramis, F., Govi, M., Guglielmin, M., Mortara, G., 1995. Mountain permafrost and slope instability in the Italian Alps: the Val Pola landslide. Permafrost Periglac. Process. 6 (1), 73−81.

Dufresne, A., Davies, T.R., 2009. Longitudinal ridges in mass movement deposits. Geomorphology 105, 171−181.

Dunning, S.A., Mitchell, W.A., Rosser, N.J., Petley, D.N., 2007. The Hattian Bala rock avalanche and associated landslides triggered by the Kashmir Earthquake of 8 October 2005. Eng. Geol. 93, 130−144.

Dutto, F., Mortara, G., 1991. Grandi frane storiche con percorso su ghiacciaio in Valle d'Aosta. Rev. Valdôtaine Hist. Nat. 45, 21−35.

Eisbacher, G.H., Clague, J.J., 1984. Destructive mass movements in high mountains: hazard and management. Geol. Surv. Can. Pap. 84-16, 230.

Ekström, G., Stark, C.P., 2013. Simple scaling of catastrophic landslide dynamics. Science 339, 1416−1419.

Ericksen, G.E., Pflaker, G., Fernández Concha, J., 1970. Preliminary report on the geologic events associated with the May 31, 1970, Peru earthquake. U.S. Geol. Surv. Circ. 639, 25.

Erismann, T.H., Abele, G., 2001. Dynamics of Rockslides and Rockfall. Springer, New York.

Evans, S.G., Clague, J.J., 1988. Catastrophic rock avalanches in glacial environments. In: Proceedings, 5th International Symposium on Landslides, Lausanne, pp. 1153−1158.

Evans, S.G., Clague, J.J., 1993. Glacier-related hazards and climate change. In: Bras, R. (Ed.), The World at Risk: Natural Hazards and Climate Change, American Institute of Physics Conference Proceedings, vol. 277, pp. 48—60.

Evans, S.G., Clague, J.J., 1994. Recent climatic change and catastrophic geomorphic processes in mountain environments. Geomorphology 10, 107—128.

Evans, S.G., Clague, J.J., 1999. Rock avalanches on glaciers in the coast and St. Elias Mountains, British Columbia. Slope stability and landslides. In: Proceedings, 13th Annual Geotechnical Society Symposium, Vancouver, B.C, pp. 115—123.

Evans, S.G., Hungr, O., Clague, J.J., 2001. Dynamics of the 1984 rock avalanche and associated distal debris flow on Mount Cayley, British Columbia, Canada; implications for landslide hazard assessment on dissected volcanoes. Eng. Geol. 61, 29—51.

Evans, S.G., Bishop, N.F., Smoll, L.F., Murillo, P.V., Delaney, K.B., Oliver-Smith, A., 2009a. A re-examination of the mechanism and human impact of catastrophic mass flows originating on Nevado Huascarán, Cordillera Blanca, Peru in 1962 and 1970. Eng. Geol. 108, 96—118.

Evans, S.G., Tutubalina, O.V., Drobyshev, V.N., Chernomorets, S.S., McDougall, S., Petrakov, D.A., Hungr, O., 2009b. Catastrophic detachment and high-velocity long-runout flow of Kolka Glacier, Caucasus Mountains, Russia in 2002. Geomorphology 105, 314—321.

Evans, S.G., Clague, J.J., Woodsworth, G.J., Hungr, O., 1989. The Pandemonium Creek Rock Avalanche, British Columbia. Can. Geotech. J. 26, 427—446.

Fahnestock, R.K., 1978. Little Tahoma Peak rockfalls and avalanches, Mount Rainier, Washington, USA. In: Voight, B. (Ed.), Rockslides and Avalanches. Elsevier, Amsterdam, The Netherlands, pp. 181—196.

Fischer, L., 2009. Slope Instabilities on Perennially Frozen and Glacierised Rock Walls: Multi-Scale Observations, Analyses and Modelling. University of Zurich, Zurich, 81 pp.

Fischer, L., Kääb, A., Huggel, C., Noetzli, J., 2006. Geology, glacier retreat and permafrost degradation as controlling factor of slope instabilities in a high-mountain rock wall: the Monte Rosa east face. NHESS 6, 761—772.

Fort, M., 2000. Glaciers and mass wasting processes: their influence on the shaping of Kali Gandaki valley, Nepal. Quat. Int. 65/66, 101—119.

Fountain, A.G., Jacobel, R.W., Schlichting, R., Jansson, P., 2005. Fractures as the main pathways of water flow in temperate glaciers. Nature 433, 618—621.

Frauenfelder, R., 2005. Regional-Scale Modelling of the Occurrence and Dynamics of Rock Glaciers and the Distribution of Paleo-permafrost. Schriftenreihe Physische Geographie, Glaziologie und Geomorphodynamik. University of Zurich.

Geertsema, M., 2012. Initial observations of the 11 June 2012 rock/ice avalanche, Lituya Mountain, Alaska. In: Proceedings of the Cold Regions Landslide Network, Harbin, China First Meeting.

Geertsema, M., Clague, J.J., Schwab, J.W., Evans, S.G., 2006. An overview of recent large catastrophic landslides in northern British Columbia, Canada. Eng. Geol. 3, 120—143.

Geli, L.G., Bard, P., Jullien, B., 1988. The effect of topography on earthquake ground motion: a review and new results. Bull. Seismol. Soc. Am. 78, 42—63.

Gruber, S., 2005. Mountain permafrost: transient spatial modelling, model verification and the use of remote sensing (Ph.D. thesis). Universität Zürich, Zürich.

Gruber, S., Haeberli, W., 2007. Permafrost in steep bedrock slopes and its temperature-related destabilization following climate change. J. Geophys. Res. 112, F02S18.

Gruber, S., Hoelzle, M., Haeberli, W., 2004. Permafrost thaw and destabilization of Alpine rock walls in the hot summer of 2003. Geophys. Res. Lett. 31, L13504. http://dx.doi.org/10.1029/2006JF000547.

Gruber, S., Peter, M., Hoelzle, M., Woodhatch, I., Haeberli, W., 2003. Surface Temperatures in Steep Alpine Rock Faces—A Strategy for Regional Scale Measurement and Modelling. Paper Presented at 8th International Conference on Permafrost, Int. Permafrost Assoc., Zurich, Switzerland.

Haeberli, W., Beniston, M., 1998. Climate change and its impacts on glaciers and permafrost in the Alps. Ambio 27 (4), 258−265.

Haeberli, W., Hoelzle, M., Paul, F., Zemp, M., 2007. Integrated monitoring of mountain glaciers as key indicators of global climate change: the European Alps. Ann. Glaciol. 46, 150−160.

Haeberli, W., Huggel, C., Kääb, A., Oswald, S., Polkvoj, A., Zotikov, I., Osokin, N., 2004. The Kolka−Karmadon rock/ice slide of 20 September 2002 − an extraordinary event of historical dimensions in North Ossetia (Russian Caucasus). J. Glaciol. 50, 533−546.

Haeberli, W., Huggel, C., Kääb, A., Polkvoj, A., Zotikov, I., Osokin, N., 2003. Permafrost conditions in the starting zone of the Kolka−Karmadon rock/ice slide of 20 September 2002 in North Osetia (Russian Caucasus). In: Haeberli, W., Brandova, D. (Eds.), Extended Abstracts on Current Research and Newly Available Information 8th Int. Conf. on Permafrost, University of Zurich, Switzerland, pp. 21−25.

Haeberli, W., Kääb, A., Paul, F., Chiarle, M., Mortara, G., Mazza, A., Deline, P., Richardson, S., 2002. A surge-type movement at Ghiacciaio del Belvedere and a developing slope instability in the east face of Monte Rosa, Macugnaga, Italian Alps. Norw. J. Geogr. 56, 104−111.

Haeberli, W., Wegmann, M., Vonder-Muhll, D., 1997. Slope stability problems related to glacier shrinkage and permafrost degradation in the Alps. Eclogae Geol. Helv. 90, 407−414.

Haeberli, W., Hohmann, R., 2008. Climate, glaciers and permafrost in the Swiss Alps 2050, scenarios, consequences and recommendations. In: Kane, D.L., Hinkel, K.M. (Eds.), Proc. 9th Int. Conf. on Permafrost. University of Alaska, Fairbanks, pp. 607−612.

Hanke, H., 1966. Gletscherkatastrophen. Die Bergsteiger 33, 433−556.

Harris, C., 2005. Climate change, mountain permafrost degradation and geotechnical hazard. In: Huber, U.M., Bugmann, H.K.M., Reasoner, M.A. (Eds.), Global Change and Mountain Regions. An Overview of Current Knowledge. Springer, Dordrecht, pp. 215−224.

Harris, C., Haeberli, W., 2003. Warming permafrost in the mountains of Europe. World Meteorol. Organ. Bull. 52 (3), 252−257.

Harris, C., Vonder Muhll, C., Isaksen, K., Haeberli, W., Sollid, J.L., King, L., Holmlund, P., Dramis, F., Gugliemin, M., Palacios, D., 2003. Warming permafrost in European mountains. Glob. Planet. Change 39, 215−225.

Hasler, A., Gruber, S., Font, M., Dubois, A., 2011. Advective heat transport in frozen rock clefts − conceptual model, laboratory experiments and numerical simulation. Permafrost Periglac. Process. 22 (4), 378−389.

Heim, A., 1932. Bergsturz und Menschenleben. Beiblatt zur Vierteljahrsschrift. Fretz and Wasmuth Verlag, Zürich, 218 pp.

Hewitt, K., 1988a. Styles of rock avalanche depositional complex in very rugged terrain, Karakoram Himalaya, Pakistan. In: Evans, S.G. (Ed.), Catastrophic Landslides: Processes, Events, Environments, Geological Society of America (1998). Review in Engineering Geology No 14. Geological Society of America.

Hewitt, K., 1988b. Catastrophic landslide deposits in the Karakoram Himalaya. Science 242, 64−67.

Hewitt, K., 1998. Catastrophic landslides and their effects on the Upper Indus streams, Karakoram Himalaya, northern Pakistan. Geomorphology 26, 47−80.

Hewitt, K., 1999. Quaternary moraines vs. catastrophic rock avalanches in the Karakoram Himalaya, Northern Pakistan. Quat. Res. 51, 220−237.

Hewitt, K., 2002. Styles of rock avalanche depositional complex in very rugged terrain, Karakoram Himalaya, Pakistan. In: Evans, S.G. (Ed.), Catastrophic Landslides: Effects, Occurrence and Mechanisms, Reviews in Engineering Geology. Geological Society of America, Boulder, CO, pp. 345−378.

Hewitt, K., 2006. Disturbance regime landscapes: mountain drainage systems interrupted by large rockslides. Prog. Phys. Geogr. 30, 365−393.

Hewitt, K., 2009. Rock avalanches that travel onto glaciers and related developments, Karakoram Himalaya, Inner Asia. Geomorphology 103 (1), 66−79.

Hewitt, K., Clague, J.J., Orwin, J.F., 2008. Legacies of catastrophic rock slope failures in mountain landscapes. Earth Sci. Rev. 87, 1−38.

Hewitt, K., Gosse, J., Clague, J.J., 2011. Rock avalanches and the pace of late Quaternary development of river valleys in the Karakoram Himalaya. Geol. Soc. Am. Bull. 123 (9−10), 1836−1850.

Hoelzle, M., Chinn, T., Stumm, D., Paul, F., Zemp, M., Haeberli, W., 2007. The application of glacier inventory data for estimating past climate change effects on mountain glaciers: a comparison between the European Alps and the Southern Alps of New Zealand. Glob. Planet. Change 56 (1−2), 69−82.

Holm, K., Bovis, M.J., Jakob, M., 2004. The landslide response of alpine basins to post-Little Ice Age glacial thinning and retreat in southwestern British Columbia. Geomorphology 57, 201−216.

Houghton, J.T., Meira Filho, L.G., Callander, B.A., Harris, N., Kattenberg, A., Maskell, K., 1996. Climate Change 1995: The Science of Climate Change. Cambridge University Press, Cambridge, p. 572.

Hovius, N., Stark, C.P., Chu, H.T., Lin, J.C., 2000. Supply and removal of sediment in a landslide-dominated mountain belt: Central Range, Taiwan. J. Geol. 108, 73−89.

Hsü, K.J., 1975. Catastrophic debris streams (sturzströms) generated by rock falls. Geol. Soc. Am. Bull. 86, 129−140.

Hsü, L., Dietrich, W.E., Sklar, L.S., 2007. Normal stresses, longitudinal profiles, and bedrock surface erosion by debris flows: initial findings from a large, vertically rotating drum. In: Chen, C.L., Major, J.J. (Eds.), Debris-Flow Hazards Mitigation: Mechanics, Prediction, and Assessment: Proceedings 4th International DFHM Conference. Millpress, Amsterdam, Chengdu, China, p. 11.

Huggel, C., 2008. Recent extreme slope failures in glacial environments: effects of thermal perturbation. Quat. Sci. Rev. 28 (11−12), 1119−1130. http://dx.doi.org/10.1016/j.quascirev. 2008.06.007.

Huggel, C., 2009. Recent extreme slope failures in glacial environments: effects of thermal perturbation. Quat. Sci. Rev. 28, 1119−1130.

Huggel, C., Caplan-Auerbach, J., Gruber, S., Molina, B., Wessels, R., 2008. The 2005 Mt. Steller, Alaska, rock−ice avalanche, a large slope failure in cold permafrost. In: Kane, D.L., Hinkel, K.M. (Eds.), Proc. 9th Int. Conf. on Permafrost. University of Alaska, Fairbanks, pp. 747−752.

Huggel, C., Caplan-Auerbach, J., Waythomas, C.F., Wessels, R.L., 2007. Monitoring and modeling ice-rock avalanches from ice-capped volcanoes: a case study of frequent large avalanches on Iliamna Volcano, Alaska. J. Volcanol. Geotherm. Res. 168, 114−136.

Huggel, C., Salzmann, N., Allen, S., Caplan-Auerbach, J., Fischer, L., Haeberli, W., Larsen, C., Schneider, D., Wessels, R., 2010. Recent and future warm extreme events and high-mountain slope stability. Philoso. Trans. R. Soc. A 368, 2435–2459.

Huggel, C., Zgraggen-Oswald, S., Haeberli, W., Kääb, A., Polkvoj, A., Galushkin, I., Evans, S., 2005. The 2002 rock/ice avalanche at Kolka/Karmadon, Russian Caucasus: assessment of extraordinary avalanche formation and mobility, and application of QuickBird satellite imagery. NHESS 5 (2), 173–187.

Hungr, O., Evans, S.G., 1996. Rock avalanche runout prediction using a dynamic model. In: Senneset, K. (Ed.), Procs. 7th. International Symposium on Landslides, Trondheim, Norway, pp. 233–238.

Hungr, O., Evans, S.G., 2004. Entrainment of debris in rock avalanches; an analysis of a long runout mechanism. GSA Bull. 116 (9/10), 1240–1252.

Hungr, O., Evans, S.G., Bovis, M.J., Hutchinson, J.N., 2001. A review of the classification of landslides of the flow type. Environ. Eng. Geosci. 7, 221–238.

Iverson, R.M., Logan, M., Denlinger, R.P., 2004. Granular avalanches across irregular three-dimensional terrain: 2. Experimental tests. J. Geophys. Res. 109 (F01015), 1–16.

Jakob, M., Lambert, S., 2009. Climate change effects on landslides along the southwest coast of British Columbia. Geomorphology 107, 275–284.

Jibson, R.W., Harp, E.L., Schulz, W., Keefer, D.K., 2006. Large rock avalanches triggered by the M 7.9 Denali Fault, Alaska, earthquake of 3 November 2002. Eng. Geol. 83, 144–160.

Kääb, A., Huggel, C., Barbero, S., Chiarle, M., Cordola, M., Epifani, F., Haeberli, W., Mortara, G., Semino, P., Tamburini, A., Viazzo, G., 2004. Glacier hazards at Belvedere glacier and the Monte Rosa east face, Italian Alps: processes and mitigation. In: International Symposium, Interpraevent 2004 – Riva/Trient.

Kääb, A., Reynolds, J.M., Haeberli, W., 2005. Glaciers and permafrost hazards in high mountains. In: Huber, U.M., Burgmann, H.K.H., Reasoner, M.A. (Eds.), Global Change and Mountain Regions – An Overview of Current Knowledge. The Netherlands Springer, Dordrecht, pp. 225–234.

Kaitna, R., Rickenmann, D., 2007. A new experimental facility for laboratory debris flow investigation. J. Hydraul. Res. 45 (6), 797–810.

Keefer, D.K., 1984. Landslides caused by earthquakes. Geol. Soc. Am. Bull. 95 (4), 406–421.

Kirkbride, M.P., 1995. Processes of transportation. In: Menzies, J. (Ed.), Modern Glacial Environments. Butterman-Heinemann, Oxford, pp. 261–292.

Koerner, H.J., 1983. Zur mechanic der bergsturzstrome vom Huascarán, Peru. In: Patzelt, G. (Ed.), Die Berg und Gletschersturze vom Huascarán, Cordillera Blanca, Peru, Hochgebirgsforschung Heft, vol. 6. UniversitatsverlagWagner, Innsbruck, pp. 71–110.

Korup, O., 2004. Landslide-induced river channel avulsions in mountain catchments of southwest New Zealand. Geomorphology 63, 57–80.

Korup, O., Clague, J.J., 2009. Natural hazards, extreme events, and mountain topography. Quat. Sci. Rev. 28, 977–990.

Korup, O., McSaveney, M.J., Davies, T.R.H., 2004. Sediment generation and delivery from large historic landslides in the Southern Alps, New Zealand. Geomorphology 61, 189–207.

Korup, O., 2005. Distribution of landslides in southwest New Zealand. Landslides 2, 43–51.

Kotlyakov, V.M., Rototaeva, O.V., Nosenko, G.A., 2004. The September 2002 Kolka glacier catastrophe in North Ossetia, Russian Federation: evidence and analysis. Mt. Res. Dev. 24 (1), 78–83.

Krautblatter, M., Funk, D., Gunzel, F.K., 2013. Why permafrost rocks become unstable: a rock-ice-mechanical model in time and space. Earth Surf. Proc. Landforms 38 (8), 876–887. http://dx.doi.org/10.1002/esp.3374.

Krieger, M.H., 1977. Large Landslides, Composed of Megabreccia, Interbedded in Miocene Basin Deposits, Southeastern Arizona. Geological Survey Professional Paper 1008, Washington, DC.

Kulkarni, A.V., 2006. Glacial retreat in Himalaya using Indian Remote Sensing satellite data In: Agriculture and Hydrology Applications of Remote Sensing, edited by Robert, J. Kuligowski, Jai, S. Parihar, Genya, Saito, Proc. of SPIE Vol. 6411, 641117.

Larsen, C.F., Motyka, R.J., Arendt, A.A., Echelmeyer, K.A., Geissler, P.E., 2007. Glacier changes in southeast Alaska and northwest British Columbia and contribution to sea level rise. J. Geophys. Res. 112, F01007. http://dx.doi.org/10.1029/2006JF000586.

Larsen, C.F., Motyka, R.J., Freymueller, J.T., Echelmeyer, K.A., Ivins, E.R., 2005a. Rapid viscoelastic uplift in southeast Alaska caused by post-Little Ice Age glacial retreat. Earth Planet. Sci. Lett. 237, 548—560.

Larsen, S.H., Davies, T.R.H., McSaveney, M.J., 2005b. A possible coseismic landslide origin of late Holocene moraines of the Southern Alps, New Zealand. N. Z. J. Geol. Geophys. 48, 311—314.

Legros, F., 2002. The mobility of long-runout landslides. Eng. Geol. 63, 301—331.

Lipovsky, P.S., Evans, S.G., Clague, J.J., Hopkinson, C., Couture, R., Bobrowsky, P., Ekström, G., Demuth, M.N., Delaney, K.B., Roberts, N.J., 2008. The July 2007 rock and ice avalanches at Mount Steele, St. Elias Mountains, Yukon, Canada. Landslides 5, 445—455.

Malamud, B., Turcotte, D.L., Guzzetti, F., Reichenbach, P., 2004. Landslide inventories and their statistical properties. Earth Surf. Proc. Land. 29, 687—711.

Marangunic, C., Bull, C., 1968. The Landslide on the Sherman Glacier. The Great Alaska Earthquake of 1964 e Hydrology. National Academy of Sciences, Washington, DC, 383—394.

McColl, S.T., Davies, T.R.H., 2011. Evidence for a rock-avalanche origin for 'The Hillocks' "moraine", Otago, New Zealand. Geomorphology 127 (3—4), 216—224.

McColl, S.T., Davies, T.R.H., 2013. Large ice-contact slope movements and glacial buttressing, deformation and erosion. Earth Surf. Proc. Land. 38, 1102—1115.

McColl, S.T., Davies, T.R.H., McSaveney, M.J., 2010. Glacier retreat and rock-slope stability: Debunking debuttressing. In: Williams, A.L., Pinches, G.M., Chin, C.Y., McMorran, T.J., Massey, C.I. (Eds.), Geologically Active: Proceedings of the 11th IAEG Congress. Auckland, New Zealand.

McDougall, S., 2006. A new Continuum dynamic model for the analysis of extremely rapid landslide motion across complex 3d terrain (Ph.D. thesis). University of British Columbia, Vancouver, Canada.

McDougall, S., Hungr, O., 2004. A model for the analysis of rapid landslide motion across three-dimensional terrain. Can. Geotech. J. 41, 1084—1097.

McSaveney, M.J., 1975. The Sherman Glacier rock avalanche of 1964: its emplacement and subsequent effects on the glacier beneath it (Ph.D. thesis). Ohio State University, Columbus, OH.

McSaveney, M.J., 1978. Sherman glacier rock avalanche, Alaska, USA. In: Voight, B. (Ed.), Rockslides and Avalanches, Natural Phenomena. Elsevier, Amsterdam, pp. 197—258.

McSaveney, M.J., 1993. Rock avalanches of 2 May and 16 September 1992, Mount Fletcher, New Zealand. Landslide News 7, 2—4.

McSaveney, M.J., 2002. Recent rockfalls and rock avalanches in Mount Cook National Park, New Zealand. In: Evans, S.G., DeGraff, J.V. (Eds.), Catastrophic Landslides: Effects, Occurrence, and Mechanisms, Geol. Soc. Am. Rev. Eng. Geol., vol. 15, pp. 35—70.

Moore, J.G., Rice, C.J., 1984. Chronology and Character of the May 18, 1980, Explosive Eruption of Mount St. Helens, in Explosive Volcanism: Inception, Evolution, Hazards. National Academy Press, Washington, D.C., pp. 133—142. America Special Paper 229, pp. 23—36.

Morales, B., 1966. The Huascaran avalanche in the Santa Valley, Peru. Int. Assoc. Sci. Hydrol. Publ. 69, 304−315.

Nicoletti, P.G., Sorriso-Valvo, M., 1991. Geomorphic controls of the shape and mobility of rock avalanches. Geol. Soc. Am. Bull. 103, 1365−1373.

Noetzli, J., Gruber, S., 2009. Transient thermal effects in Alpine permafrost. The Cryosphere 3, 85−99.

Noetzli, J., Gruber, S., Kohl, T., Salzman, N., Haeberli, W., 2007. Three-dimensional distribution and evolution of permafrost temperatures in idealized high-mountain topography. J. Geophys. Res. 112, F02S13. http://dx.doi.org/10.1029/2006JF000545.

Noetzli, J., Hoelzle, M., Haeberli, W., 2003. Mountain permafrost and recent Alpine rock-fall events: a GIS-based approach to determine critical factors. In: Phillips, M., Springman, S., Arenson, L. (Eds.), 8th International Conference on Permafrost, Proceedings. 2, Zurich, Swets & Zeitlinger, Lisse, pp. 827−832.

Noetzli, J., Huggel, C., Hoelzle, M., Haeberli, W., 2006. GIS-based modelling of rock-ice avalanches from Alpine permafrost areas. Comput. Geosci. 10, 161−178. http://dx.doi.org/10.1007/s10596-005-9017-z.

O'Connor, J.E., Costa, J.E., 1993. Geologic and hydrologic hazards in glacierized basins in North America resulting from 19th and 20th Century global warming. Nat. Hazards 8, 121−140.

Oerlemans, J., 1994. Quantifying global warming from the retreat of glaciers. Science 264 (5156), 243−245.

Oerlemans, J., 2005. Extracting a climate signal from 169 glacier records. Science 308, 675−677.

Oerlemans, J., Fortuin, J.P.F., 1992. Sensitivity of glaciers and small ice caps to greenhouse warming. Science 258, 115−117.

Okura, Y., Kitahara, H., Kawanami, A., Kurokawa, U., 2003. Topography and volume effects on travel distance of surface failure. Eng. Geol. 67 (3−4), 243−254.

Oppikofer, T., Jaboyedoff, M., Keusen, H.R., 2008. Collapse at the eastern Eiger flank in the Swiss Alps. Nat. Geosci. 1, 531−535.

Paul, F., Kääb, A., Maisch, M., Kellenberger, T.W., Haeberli, W., 2004. Rapid disintegration of Alpine glaciers observed with satellite data. Geophys. Res. Lett. 31, L21402. http://dx.doi.org/10.1029/2004GL020816.

Petrakov, D.A., Chernomorets, S.S., Evans, S.G., Tutubalina, O.V., 2008. Catastrophic glacial multi-phase mass movements: a special type of glacial hazard. Adv. Geosci. 14, 211−218.

Pflaker, G., Ericksen, G.E., 1978. Nevados Huascaran avalanche, Peru. In: Voight, B. (Ed.), Rockslides and Avalanches, 1 Natural Phenomena. Elsevier, NY, pp. 277−314.

Pierson, T.C., Janda, R.J., Thouret, J.C., Borrero, C.A., 1990. Perturbation and melting of snow and ice by the 13 November 1985 eruption of Nevado del Ruiz, Colombia, and mobilization, flow and deposition of lahars. J. Volcanol. Geotherm. Res. 41, 17−66.

Pirulli, M., Scavia, C., Hungr, O., 2004. Determination of rock avalanche run-out parameters through back analyses. In: Lacerda, W.A., Ehrlich, M., Fontoura, S.A.B., Sayão, A.S.F. (Eds.), Proceedings of the 9th International Symposium on Landslides, Rio de Janeiro. Balkema, London, pp. 1361−1366.

Porter, S.C., Orombelli, G., 1980. Catastrophic rockfall of September 12, 1717 on the Italian flank of the Mont Blanc massif. Z. Geomorphol. N.F. 24, 200−218.

Post, A., 1967. Effects of the March 1964 Alaska earthquake on glaciers. U.S. Geol. Surv. Prof. Pap. 544-D, D1−D42.

Post, A., La Chapelle, E.R., 1971. Glacier Ice, 110. The Mountaineers, Seattle.

Ravanel, L., Deline, P., 2011. Climate influence on rockfalls in high-Alpine steep rockwalls: the north side of the Aiguilles de Chamonix (Mont Blanc massif) since the end of the Little Ice Age. Holocene 21, 357—365. http://dx.doi.org/10.1177/0959683610374887.

Ravanel, L., Allignol, F., Deline, P., Ravello, M., 2010. Rock falls in the Mont Blanc massif in 2007 and 2008. Landslides 7, 493—501.

Reznichenko, N.V., Davies, T.R.H., Alexander, D.J., 2011. Effects of rock avalanches on glacier behaviour and moraine formation. Geomorphology 132, 327—338.

Rothlisberger, H., 1978. Eislawinen und AusbrUche von Gletscherseen. Jahrbuch der Schweizerischen Naturforschenden Gesellschafl. Wiss. Teil, 170—212.

Schneider, D., Bartelt, P., Caplan-Auerbach, J., Christen, M., Huggel, C., McArdell, B.W., 2010. Insights into rock-ice avalanche dynamics by combined analysis of seismic recordings and a numerical avalanche model. J. Geophys. Res. 115 (F04026), 1—20.

Schneider, D., Huggel, C., Haeberli, W., Kaitna, R., 2011a. Unravelling driving factors for large rock-ice avalanche mobility. Earth Surf. Process. Landforms 36, 1948—1966.

Schneider, D., Kaitna, R., Dietrich, W.E., Hsu, L., Huggel, C., McArdell, B.W., 2011b. Frictional behavior of granular gravel-ice mixtures in vertically rotating drum experiments and implications for rock-ice avalanches. Cold Reg. Sci. Technol. 69 (1), 70—90.

Schwab, J.W., Geertsema, M., Evans, S.G., 2003. Catastrophic rock avalanches, west central B.C., Canada. In: Proc. 3rd Canadian Conference on Geotechnique and Natural Hazards. Edmonton, AB, pp. 252—259.

Sepulveda, S.A., Murphy, W., Jibson, R.W., Petley, D.N., 2005. Seismically induced rock slope failures resulting from topographic amplification of strong ground motions: the case of Pacoima Canyon, California. Eng. Geol. 80, 336—348.

Sheridan, M.F., Stintona, A.J., Patrab, A., Pitmanc, E.B., Bauerb, A., Nichita, C.C., 2005. Evaluating Titan2D mass-flow model using the 1963 Little Tahoma Peak avalanches, Mount Rainier, Washington. J. Volcanol. Geotherm. Res. 139 (1—2), 89—102.

Shreve, R.L., 1966. Sherman landslide, Alaska. Science 154 (3757), 1639—1643.

Shroder, J.F., Owen, L., Derbyshire, E., 1993. Quaternary glaciation of the Karakoram and Nanga Parbat Himalaya. In: Shroder Jr., J.F. (Ed.), Himalaya to the Sea. Routledge, London, pp. 132—158.

Shugar, D.H., Clague, J.J., 2011. The sedimentology and geomorphology of rock avalanche deposits on glaciers. Sedimentology 58 (7), 1762—1783.

Shulmeister, J., Davies, T.R., Evans, D.J.A., Hyatt, O.M., Tovar, D.S., 2009. Catastrophic landslides, glacier behaviour and moraine formation — a view from an active plate margin. Quat. Sci. Rev. 28, 1085—1096.

Sosio, R., Crosta, G., Hungr, O., 2008. Complete dynamic modeling calibration for the Thurwieser rock avalanche (Italian Central Alps). Eng. Geol. 100, 11—26.

Sosio, R., Crosta, G.B., Chen, J.H., Hungr, O., 2012. Modelling rock avalanche propagation onto glaciers. Quat. Sci. Rev. 47, 23—40.

Sovilla, B., Burlando, P., Bartelt, P., 2006. Field experiments and numerical modeling of mass entrainment in snow avalanches. J. Geophys. Res. 111, F03007.

Sovilla, B., Margreth, S., Bartelt, P., 2007. On snow entrainment in avalanche dynamics calculations. Cold Reg. Sci. Technol. 47, 69—79.

Stadelmann, J., 1983. Zur dokumentation der bergsturzereignisse vom Huascarán. In: Patzelt, G. (Ed.), Die Berg — und Gletschersturze vom Huascarán, Cordillera Blanca, Peru, Hochgebirgsforschung Heft, vol. 6. Universitatsverlag Wagner, Innsbruck, pp. 51—70.

Stoeber, E.A., 1903. Glacier collapses in the headwaters of the Genal-don River in the Caucasus. Terskiy Sbornik Ekaterinoslavskoe Nauchnoe Obshestvo 2 (7), 72—81 (in Russian).

Straub, S., 1997. Predictability of long runout landslide motion: implications from granular flow mechanics. Geol. Rundsch. 86 (2), 415−425.

Strom, A.L., 2006. Morphology and internal structure of rockslides and rock avalanches; grounds and constraints for their modelling. In: Evans, S.G., Scarascia-Mugnozza, G., Strom, A.L., Hermanns, R.L. (Eds.), Landslides from Massive Rock Slope Failure. NATO Science Series: IV, Earth and Environmental Sciences. Springer, Dordrecht, pp. 305−326.

Strom, A.L., Abdrakhmatov, K., 2004. Rock Avalanches and Rockslide Dams of the Northern Kyrgyzstan: Field Excursion Guidebook. NATO Advanced Res. Workshop. Bishkek, Kyrgyzstan, 36 pp.

Suriñach, E., Furdada, G., Sabot, F., Biescas, B., Vilaplana, J.M., 2001. On the characterization of seismic signals generated by snow avalanches for monitoring purposes. Ann. Glaciol. 32, 268−274.

Tamburini, A., Mortara, G., 2005. The case of the "Effimero" Lake at Monte Rosa (Italian Western Alps): studies, field surveys, monitoring. In: Progress in Surface and Subsurface Water Studies at Plot and Small Basin Scale. Proceedings of the 10th ERB Conference, Turin, 13−17 October 2004, Unesco, IHP-VI Technical Documents in Hydrology, vol. 77, pp. 179−184.

The AGU blogsphere, 2013. http://blogs.agu.org/landslideblog/2013/02/24/an-intrepretation-of-the-mount-dixon-rock-avalanche-in-new-zealand-using-satellite-imagery/.

The Seattle Times, 2011. http://seattletimes.nwsource.com/html/localnews/2015453613_rainier29m.html.

Topping, D.A., 1993. Paleogeographic reconstruction of the Death Valley extended region: evidence for Miocene large rock-avalanche deposits in the Amargosa Chaos basin, California. Geol. Soc. Am. Bull. 105, 1190−1213.

Van der Woerd, J., Owen, L.A., Tapponnier, P., Xu, X.W., Kervyn, F., Finkel, R.C., Barnard Giant, P.L., 2004. ∼M8 earthquake-triggered ice avalanches in the eastern Kunlun Shan, northern Tibet: characteristics, nature and dynamics. Geol. Soc. Am. Bull. 116, 394−406.

Voight, B., Janda, R.J., Glicken, H., Douglass, P.M., 1983. Nature and mechanics of the Mount St. Helens rockslide-avalanche of 18 May 1980. Geotechnique 33 (3), 243−273.

Waythomas, C.F., Miller, T.P., Beget, J.E., 2000. Record of Late Holocene debris avalanches and lahars at Iliamna volcano, Alaska. J. Volcanol. Geotherm. Res. 104, 97−130.

Weaver, C.S., Norris, R.D., Jonientz-Trisler, C., 1990. Results of seismological monitoring in the Cascade Range, 1962−1989: earthquakes, eruptions, avalanches and other curiosities. Geosci. Can. 17, 158−162.

Wegmann, M., Gudmundsson, G.H., Haeberli, W., 1998. Permafrost changes in rock walls and the retreat of Alpine glaciers: a thermal modelling approach. Permafrost Periglac. Process. 9, 23−33.

Whitehouse, I.E., 1983. Distribution of large rock avalanche deposits in the central Southern Alps, New Zealand. N. Z. J. Geol. Geophys. 26, 271−279.

Whitehouse, I.E., Griffiths, G.A., 1983. Frequency and hazard of large rock avalanches in the central Southern Alps, New Zealand. Geology 11, 331−334.

Xu, Q., Shang, Y., van Asch, T., Wang, S., Zhang, Z., Dong, X., 2012. Observations from the large, rapid Yigong rock slide − debris avalanche, southeast Tibet. Can. Geotech. J. 49, 589−606.

Chapter 8

Multiple Landslide-Damming Episodes

Oliver Korup [1] and Gonghui Wang [2]

[1] *Institute of Earth and Environmental Science, University of Potsdam, Potsdam, Germany,*
[2] *Disaster Prevention Research Institute, Kyoto University, Kyoto, Japan*

ABSTRACT

The natural blockage of river channels by landslide debris gives rise to a range of potentially adverse processes including complete or partial damming, backwater inundation and sedimentation, catastrophic outbursts of impounded lake waters, and downstream river aggradation following dam failure. Most reports portray landslide dams as short-lived and hazardous landforms that often require immediate mitigation based on engineering solutions. Despite a growing body of research thanks to increasingly detailed remote sensing and field data, we often lack readily accessible tools for reliably predicting the stability, or timing of failure, of landslide dams. In this chapter, we augment a number of detailed reviews on landslide dams by focusing on recent insights into the synchronous and regional-scale emplacement of several dozens to hundreds of landslide dams and their consequences on water and sediment fluxes. Such multiple landslide-damming episodes have been rarely addressed in the literature so far, though they create a number of distinct challenges for both research and hazard mitigation. These challenges include rapid appraisals of how to distinguish from among multiple coevally formed dams that are most hazardous; suitable engineering solutions for stabilizing landslide dams; management of massive sediment pulses—and their risks—in the wake of landslide-dam failures; and the paleoseismological interpretation of clusters of coeval landslide dams.

8.1 INTRODUCTION

Landslide dams result from the blockage of stream channels by hillslope-derivedmass-wasting debris. Landslide dams have been mainly reported from hilly to mountainous terrain, where strong rainstorms, snowmelt events, and earthquake shaking provide suitable conditions for prolific mass wasting (e.g., Costa and Schuster, 1988; Korup and Tweed, 2007; O'Connor and Beebee, 2009). Landslide dams can be formed either gradually by slow-moving

Landslide Hazards, Risks, and Disasters. http://dx.doi.org/10.1016/B978-0-12-396452-6.00008-2
241

landslides or, more often, catastrophically, that is, within a matter of seconds or minutes. Accordingly, landslide dams dominantly comprise soil, debris, or disrupted rock materials. In high-altitude and high-latitude terrain, however, some landslide dams may also contain significant quantities of ice that were either entrained during runout or formed after emplacement of the dam under permafrost conditions.

Lakes may form behind landslide dams that remain intact long enough to interrupt or attenuate river flows. Landslide dams may be prone to sudden failure and subsequent catastrophic outburst of water masses from impounded lakes. Common modes of failure include overtopping of the dam with subsequent breaching, internal erosion by groundwater seepage, and wave overtopping induced by rapid mass movements entering the lake (Costa and Schuster, 1988; Peng and Zhang, 2012). Such outburst events may entrain substantial amounts of debris from the failing dam, any backwater sediments, and the downstream valley floor, thus giving rise to fast-flowing mixtures with high sediment–water ratios such as hyperconcentrated flows, and sometimes even debris flows. The subsequent catastrophic flooding of valley floors may cause substantial channel changes for up to several tens to hundreds of kilometers, depending upon parameters such as the volume of water released, peak discharge, sediment availability along the flow path, and downstream wave attenuation by valley-floor topography (Korup, 2012). Following dam failure, the partly eroded landslide dam remains as a knick point in the channel longitudinal profile. This knick point continues to cause reach-scale fluvial erosion and aggradation as a means of channel adjustment ranging from years to millennia after the outburst event.

Landslide dams are often portrayed as short-lived landforms. Indeed, most of the dams that have reportedly failed did so within a year after their formation (Costa and Schuster, 1988; Peng and Zhang, 2012). Yet, the sample number used to uphold this claim is rather small considering the number of landslide dams reported, and the greatest uncertainty remains with the most short-lived dams (Figure 8.1). While the number of breached landslide dams reported throughout the world greatly exceeds those of intact dams, comparatively few data are available on the maximum life span of landslide dams. Clearly, more insights into the longevity of dams are essential, given that a number of landslide-dammed lakes have remained in place for up to many thousands of years (Korup and Tweed, 2007). In those cases, inflow and outflow may have achieved an equilibrium, or the lake may have formed a seemingly stable outlet. Also, a number of lakes have been completely infilled with sediment before the dam showed any significant structural damage (Casagli and Ermini, 1999). Many of these seemingly more robust landslide dams also occur in humid and tectonically active uplands or mountain belts that are prone to episodes of strong seismic ground motion and high-intensity rainfalls (Korup, 2002).

In any case, landslide-dammed lakes inundate large tracts of valley floors (inclusive of settlements and infrastructure), and attenuate river water and

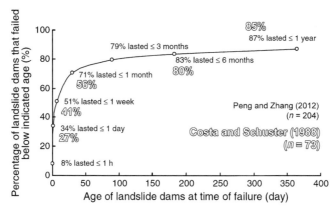

FIGURE 8.1   Cumulative distribution of the age of landslide dams at the time of failure, based on $n = 204$ cases (Peng and Zhang, 2012). This data set is nearly three times larger than the widely cited data set by Costa and Schuster (1988). While the reported percentages are consistent for longer time intervals, some significant differences concern the shorter life spans of dams, where time for mitigation options may be limited. *Modified after Peng and Zhang (2012).*

sediment flows, thus shifting the locus of fluvial erosion and aggradation. Reaches downstream of the dam will respond to sediment starvation and begin to incise, whereas reaches upstream of the dam will be prone to backwater aggradation, forming a deltaic wedge that progrades into the landslide-dammed lake. If large and long-lived enough, such natural reservoirs may also provide benefits in that they may act as natural flood control and sediment retention basins; the 2,000-year-old Waikaremoana landslide dam in New Zealand, for example, has been utilized for hydropower generation. Landslide dams thus share some, but not all, of the characteristics of artificial earth-fill dams (O'Connor and Beebee, 2009). By contrast, a landslide dam that fails releases large volumes of water to the downstream river causing potentially severe, extensive and long-lasting sedimentation, avulsion, and flooding (Davies and Korup, 2007).

The more significant, and potentially crucial, differences between natural landslide-dammed lakes and artificial earth-fill dams concern their internal sedimentology. Depending on landslide type, the stratigraphy of the dam may feature crudely stratified layers, surface armoring, or largely unstratified deposits of chiefly poorly sorted clasts (Dunning and Armitage, 2012). The internal structure of landslide dams is among the main unknowns in many studies, and reflects much of the lack of understanding regarding the assessment and prediction of landslide-dam stability and longevity.

Nevertheless, much of the study on landslide dams has focused on detailed case studies that serve to emphasize the broad variety of individual dam types, formation, and failure histories, as well as the geomorphic consequences. A growing number of published inventories have gone beyond individual case studies and permit preliminary insights into the regional pattern of river blockage by hill slope processes. One limitation to this approach is that a regional sample of landslide dams is likely to be composed of many dams of differing formation ages. However, a number of recent landsliding episodes triggered by strong earthquake shaking and high-intensity rainstorms have provided the opportunity to compile substantially complete landslide inventories that include many river-blocking slope failures. Given the breadth and focus of previous work on landslide dams (including several thorough reviews listed below), we focus in this contribution on recent research devoted to the study of such regional earthquake- and rainstorm-triggered landslide episodes that also triggered numerous landslide dams. We review three such recent episodes, and discuss the novel insights that studying the synchronous formation of multiple landslide dams may offer against the backdrop of dozens of previous case studies devoted to individual river blockage.

8.2 PREVIOUS WORK ON LANDSLIDE DAMS

Several reviews have summarized the formation, failure, and consequences arising from natural river blockage by landslides: that of Costa and Schuster (1988) remains one of the first and most authoritative treatments of the topic. Their catalog of nearly 500 landslide dams around the world (Costa and Schuster, 1991) has been widely used and analyzed. Cenderelli (2000) reviewed floods from natural and artificial dam failures, and Korup (2002) also provided a review in the context of geomorphic hill slope-channel coupling, and the persistence of landslide-dammed lakes in tectonically active terrain. Korup and Tweed (2007) focused on the longer-term, that is, Quaternary, geomorphic implications of landslide, and other natural, dams for river adjustment and landscape evolution. O'Connor and Beebee (2009) provided a thorough review of all aspects of the formation, stability, and failure of landslide dams, and included a detailed database containing well-documented case studies of natural and artificial dams. Peng and Zhang (2012) presented one of the largest landslide-dam inventories published so far, containing nearly 1240 cases that were partly merged from previous inventories together with further analyses of some of the pertinent breaching parameters. Korup (2012) reviewed worldwide data on extreme sediment yields and concluded that landslide (and other natural) dam breaks were some of the most effective sediment transport events in mountain rivers. This study supported earlier work highlighting the importance of non-meteorological causes of the Earth's largest floods during the Quaternary (O'Connor and Costa, 2004).

Summarizing these reviews, our knowledge about the formation, stability, failure, and consequences of landslide dams is mainly derived from two sources, that is, (1) an array of detailed field studies and (2) a growing number of detailed landslide-dam inventories.

8.3 LANDSLIDE-DAM EPISODES: LESSONS FROM CASE STUDIES

Most case studies focus on the processes involved with a single or several landslide dams. However, more recent work focuses on the analysis of many tens to hundreds of landslide dams triggered more or less coevally by a regional event such as a strong earthquake or a high-intensity rainstorm. Here, we briefly review some of the key insights regarding the forming conditions, dam stability, and longevity of impact from this work.

8.3.1 Wenchuan Earthquake (M_w 7.9), China, 2008

To date, the 2008 Wenchuan earthquake, Sichuan, China, remains the most thoroughly studied source of coseismic slope instability and associated river blockage. Nearly 40 scientific papers have been published on landslide dams triggered by this earthquake alone (Elsevier SCOPUS database, last accessed December 2013). Detailed mapping from pre- and post-earthquake remote-sensing data indicate that at least 828 landslide dams were triggered by the earthquake (Fan et al., 2012; Table 8.1; Figure 8.2), whereas many earlier papers had greatly underestimated this number. For example, Cui et al. (2011) had proposed that only 257 landslide dams were formed as a result of the earthquake. Similarly, published estimates regarding the total number and volume of landslides triggered by the Wenchuan earthquake differ widely, and range from approximately 11,000 to at least 60,000 (Gorum et al., 2011; Figure 8.3), although the exact number remains disputed. This uncertainty mainly reflects several key methodological difficulties and uncertainties that remain despite widely accessible remote-sensing technology in the early twenty-first century. These difficulties include, among others, topographic shadow effects and cloud cover that mask landslide occurrence, but mainly the problem of coalescing landslide scars that makes discerning individual slope failures more or less intractable. This in turn limits the reliability of the bulk sediment budgets that are mainly derived from the statistical scaling of individually mapped landslide planform areas to landslide volumes (Parker et al., 2011).

The abundance of landslide dams associated with the Wenchuan earthquake is unprecedented in historic reports, and allowed a number of novel quantitative insights regarding the short-term impact on postseismic water and sediment fluxes in the Longmen Shan mountain range (Fan et al., 2012). Most importantly, the synchronous formation of the landslide dams allowed a direct comparison of the longevity of hundreds of landslide dams of comparable formation age. For instance, this allowed comparisons of the ways in which non-river-blocking landslides differed from blocking ones. Gorum et al. (2011) found that, in all the major lithologies, the spatial density of river-blocking landslides (km^{-2}) was similar to that of slope failures that did not dam rivers. Notable exceptions were areas locally underlain by mudstone/shalestone and carbonate lithologies that featured four and 1.5 times more river-blocking

TABLE 8.1 Selected Characteristics of Recent Episodes of Multiple Landslide Damming Discussed in this Review

	M_w 7.8 Murchison Earthquake, New Zealand	M_w 7.9 Wenchuan Earthquake, Sichuan, China	Typhoon Talas (No. 12), Japan
Date	June 17, 1929	May 12, 2008	September 2–5, 2011
Landslide-affected area (km^2)	~4500–7,000	~20,000	not determined
Number of landslides	~10,000	>60,100	>3000
Number of landslide dams	>38	>828[1]	>17
Number of landslide-dammed lakes remaining	20	~20	5
Total landslide volume (m^3)	$(1.3–2.0) \times 10^9$	$(5–15) \times 10^9$	$>95 \times 10^6$
Range of river-blocking landslide volumes (m^3)	$(\sim 0.5–18) \times 10^6$	$\sim 2 \times 10^3 – 0.9 \times 10^9$	$(0.9–14) \times 10^6$
Fatalities	17 (14[2])	~88,000	97[2]
Key references	Pearce and Watson (1986) and Hancox et al. (2002)	Gorum et al. (2011) and Fan et al. (2012)	Uchida et al. (2012) and Chigira et al. (2013)

Note that many of these characteristics, particularly the total number, area, and volumes of landslides are estimates and differ between various studies. The number of landslide dams associated with each event is a minimum estimate, and does not include very short-lived blockages.
[1] Five hundred and one full blockages and 327 partial blockages.
[2] From landslides exclusively.

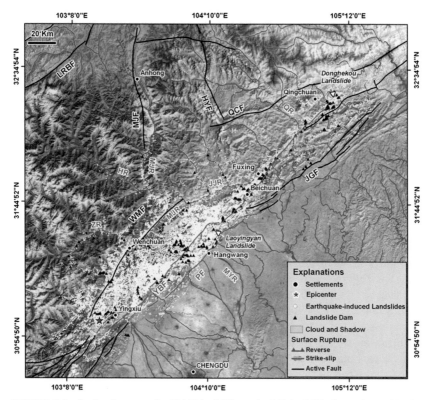

FIGURE 8.2   Regional pattern of >60,100 landslides and >250 landslide dams triggered by the 2008 Wenchuan earthquake, China; MJR: Minjiang River; ZR: Zagunao River; HR: Heishuehe River; MYR: Mianyuan River; JJR: Jianjiang River; and QR: Qingzhu River. *After Gorum et al. (2011).*

FIGURE 8.3   Earthquake-triggered landslide scars dot the steep and dissected valleys of the Longmen Shan, Sichuan Province, China. More than 60,000 slope failures were triggered by the M_w 7.9 Wenchuan earthquake on May 12, 2008; at least 828 of these landslides formed mostly short-lived dams.

than non-river-blocking landslides, respectively. Any possible lithological effect on river impoundment requires detailed site investigations, however, given that only some 60 percent of all landslide dams may have involved significant amounts of bedrock (Fan et al., 2012). Another important observation was that the spatial density of landslide dams followed an exponential decay with distance from the fault rupture, similar to that observed for all slope failures. Landslide dams occurred at distances of up to approximately 70 km from the fault rupture, that is, about half the maximum distance of all coseismic landslides, yet most landslide dams occurred within 10 km of the fault rupture (Gorum et al., 2011). According to Cui et al. (2011), mass wasting-induced river blockage was most abundant in areas that experienced Modified Mercalli Intensities of IX−XI.

Fan et al. (2012) estimated that at least 76 km of river network had been buried directly by landslide debris, excluding the backwater reaches that were subsequently subjected to impoundment and rapid sedimentation. In some catchments, landslides dotted river reaches like strings of pearls: Nearly half a dozen river reaches hosted up to 0.4−0.88 dams per kilometer of river length (Cui et al., 2011). According to Fan et al. (2012), about 14−18 percent of the total landslide volume mobilized by the earthquake was contained in the 828 river-blocking landslides that made up only approximately 1.4 percent of all approximately 60,000 mapped coseismic slope failures. Size−frequency estimates showed that mostly moderate-sized landslides were responsible for impounding rivers, creating a total backwater accommodation space of approximately 0.6×10^9 m^3. Yet little of this storage capacity was infilled with sediment, as >90 percent of the dams had failed within one year, thereby losing a very similar fraction of retention space (Figure 8.4). Several of these dam failures had been artificially induced by Chinese authorities to prevent

FIGURE 8.4 Remnants of former lakes that were impounded behind landslides triggered by the 2008 Wenchuan earthquake. (a) Upstream view of one of hundreds of smaller lakes that was artificially drained, leaving an elongated pool flanked by dissected backwater sediments. (b) An earthquake-damaged building partly buried beneath backwater sediments. (c) Boulder (∼3 m high) derived from catastrophic landslide-dam burst flow locked on the upstream side of the bridge. (d) Drained landslide-dammed lake floor exposing the infill of anthropogenic debris.

catastrophic outbursts (from the larger lakes in particular that had formed rapidly behind the debris barriers), and were threatening the earthquake-stricken areas downstream.

Several lessons were learned regarding such engineering mitigation of the Wenchuan landslide dams. In the case of Tangjiashan landslide dam, the largest $(8-9 \times 10^8 \text{ m}^3)$ formed during the earthquake, a spillway was constructed by June 8, 2008. However, abrupt breaching occurred on June 11, 2008, resulting in an estimated peak discharge of approximately 6,500 m³ s⁻¹ (Cui et al., 2011), and the former (spillway) bed was rapidly lowered by 25 m. After that, the local government further excavated the spillway, such that the water level was lowered by another 3 m. In 2011–2012, a number of engineering structures were built, including a concrete-lined floor on the upstream entrance of the spillway, bridges crossing the remnant landslide-dammed lake on the upper stream of the dam, and a drainage tunnel for routing excess floodwater. However, on July 9, 2013, flooding caused a 6-m increase in the lake level, causing yet another breach of the partly engineered landslide dam, culminating in a maximum discharge of approximately 5,000 m³ s⁻¹, eroding the concrete spillway and parts of the landslide dam by nearly another 10 m vertically, and depressing the lake level by >20 m. Clearly, some of the engineering solutions were only of ephemeral use, and did not significantly counteract the river's adjustment to the knick point formed by the Tangjiashan dam.

Overall, failure and partial erosion of the landslide dams liberated up to $0.2 \times 10^9 \text{ m}^3$ of sediment in the drainage network in the first two years following the earthquake (Fan et al., 2012), causing massive valley-floor aggradation (Figure 8.5), partly also in areas where large refugee camps and critical infrastructure had been erected. Only 23 landslide dams had remained intact 26 months after the earthquake, yet these deposits from mostly deep-seated failures retain 45 percent of the total original landslide-dam volume mobilized by the earthquake. Monsoon-fed rainstorms hitting the earthquake area are thought to be among the main reasons for the rapid decay of this landslide dam population, and the associated loss of retention spaces for incoming water and sediment. Tang et al. (2009) registered 72 debris flows that mostly reworked much of the ubiquitous landslide debris during a single storm in September 2008, that is, four months after the earthquake, and it is likely that many of the smaller dams failed by overtopping during that and later storms. Such postseismic sediment pulses may involve extremely high sediment yields, last several years to decades, and rapidly aggrade river beds by meters to tens of meters (e.g., Yanites et al., 2010, Figure 8.6). Such massive channel changes contribute to impeding or delaying postdisaster reconstruction efforts of much needed infrastructure repairs or renewals concerning access roads, bridges, power lines, etc., over extensive distances downstream. If the estimates by Fan et al. (2012) turn out to be correct to the first order, then the bulk (~80 percent) of coseismic landslide debris may remain sitting on the lower portions of denuded hill slopes or on valley floors, waiting to be sluiced

FIGURE 8.5 Massive river aggradation following the failure of Tangjiashan landslide dam, the largest river blockage triggered by the 2008 Wenchuan earthquake. (a) Satellite image *(adapted from GoogleEarth)* dated May 19, 2001, that is, 7 years before the earthquake, showing the location of a future landslide dam, hydropower station, and tunnel west of Beichuan City, which was completely destroyed during the earthquake. (b) and (c) Comparison of the river-bed level at a hydropower station before and four months after the earthquake indicate >15 m of catastrophic bed aggradation some 2–3 km downstream of the failed Tangjiashan landslide dam; the channel remained aggraded to >2 m above preearthquake levels some three years after the event at a tunnel approximately 5 km downstream of the landslide dam (d) and (e).

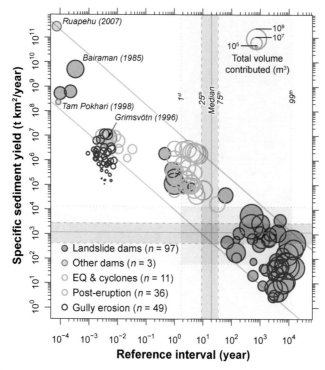

FIGURE 8.6 Specific sediment yields measured in the wake of landslide-dam failures compared to yields from other local to regional-scale disturbances. Note that longer reference intervals curtail these yields with increasing inclusion of time intervals without much geomorphic activity. Accordingly, red lines are temporal trajectories of high transient yields that decay with increasing reference intervals. Nevertheless, the specific sediment yields from dissected landslide dams may remain exceptionally high when compared to a global sample of sediment yields in mountain rivers. *Gray shaded areas show quantiles of this sample; after Korup (2012).*

out of the Longmen Shan mountains into the Sichuan Basin during future rainstorms. This sediment pulse is likely to lead to protracted setbacks to engineering efforts of restabilizing rivers not only in the earthquake-affected region but also farther downstream.

8.3.2 Murchison (Buller) Earthquake (M_w 7.8), New Zealand, 1929

The M_w 7.8 Murchison earthquake shook the mountainous northwest South Island of New Zealand in June 1929, and remains another prominent example that triggered widespread landsliding and multiple simultaneous river blockages. The total number and volume of coseismic landslides, however, remains elusive and based on first-order estimates (Table 8.1; Figure 8.7; Hancox et al., 2002). Some 20 landslide-dammed lakes dot the landscape as a striking legacy

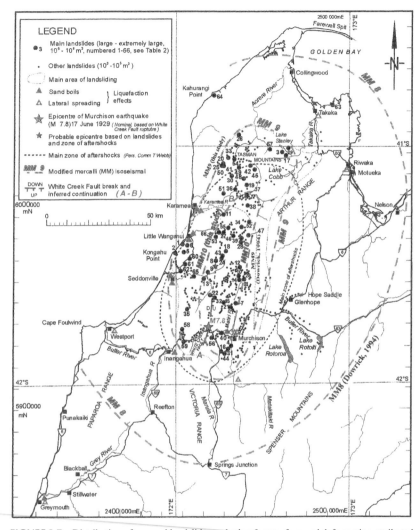

FIGURE 8.7 Distribution of mapped landslides and other forms of ground deformation attributed to the M_w 7.8 1929 Murchison earthquake, northwest South Island of New Zealand. Only the largest slope failures are shown, while the pattern of isoseismals, that is, lines of equal Mercalli Intensity, have been estimated from the regional landslide distribution. The pattern of existing landslide-dammed lakes has been used to map the Mercalli X contour. *After Hancox et al. (2002).*

of the earthquake (Table 8.1), although many of the original landslide scars also remain clearly visible in the otherwise densely forested terrain. This reported clustering of persistent landslide-dammed lakes triggered by a single event some 85 years ago seems to be somewhat rare. Crozier et al. (1995) reported distinct clusters of deep-seated landslides in the southern Taranaki region, North Island of New Zealand. Many of these mostly rotational failures involving Pliocene marine sediments are spread out over an area of approximately 1,000 km^2, had dammed or ponded small lakes, and were interpreted collectively as evidence of a prehistoric earthquake in this region. Similar reports remain rare, and further research is needed to elucidate whether other tectonically active landscapes host comparable evidence that may be used for linking clusters of coeval landslide-dammed lakes to earthquake triggers.

In a more recent reassessment of the Murchison earthquake, Hancox et al. (2002) estimated that the earthquake had triggered the formation of at least 38 landslide dams, which would imply a survival rate of intact landslide dams that significantly exceeded that of the Wenchuan river blockages. Individual lakes such as that impounded behind the Matakitaki landslide (18×10^6 m^3) survived for nearly 10 years before being emptied during a major flood event. Most landslide dams had failed within one year of the earthquake, although a few remain intact to the present day. Hancox et al. (2002) argued that the main reasons why so many dams remain intact despite seeing no engineering mitigation (in this area of low population density) were that their volumes greatly exceeded those of their impounded lakes. The same reasoning was encapsulated in an impoundment index proposed originally by Casagli and Ermini (1999) who argued that this index helped to distinguish between failed and intact landslide dams for a range of environments. Clearly, such distinctions must remain preliminary and a function of time, as landslide dams may fail catastrophically even millennia after their formation (Korup and Tweed, 2007).

Adams (1980) argued that a "smooth ellipse that just contains the landslide-dammed lakes of a single age is an indication of the area shaken to MM intensity X," and used this proposition to map out the Mercalli Intensity X contour from the landslide-dammed lakes attributed to the 1929 Murchison earthquake. This approach has been used for mapping the intensity isoseismals of a number of historic earthquakes in New Zealand (Hancox et al., 2002), and appears to be consistent with the observations from the Wenchuan earthquake.

Pearce and Watson (1986) were among the first to quantify in detail some of the geomorphic impacts of these earthquake-triggered landslides and the associated river blockages in the Matiri River catchment close to the epicenter, at least as far as this could be detected decades after the triggering event. They stressed the effective trapping of earthquake-generated sediment in their study area, which they argued was due to the coarseness of the debris and multiple occurrences of landslide-dammed lakes that sequestered much of the incoming bed load. Their overall estimate was that 50–75 percent of the landslide-delivered debris did not evacuate the approximately 10-km long headwater study reach—that also included five landslide-dammed lakes—within 50 years of the earthquake. This estimate implies that the sediment pulse would have prograded at an average rate of approximately 150 m year^{-1} Pearce and Watson (1986) emphasized that these rates of sediment evacuation from earthquake-impacted catchments were rather low when compared to those in tectonically active regions elsewhere.

8.3.3 Typhoon Talas, Japan, 2011

Multiple landslide dams may also be formed during heavy rainstorms, and this has been documented for the Kii Peninsula of southwest Japan. This area receives most of its high annual rainfall from tropical cyclones (typhoons) that make landfall in the steeply dissected mountains mainly consisting of highly tectonized and easily erodible rocks in Jurassic to Paleogene—lower Miocene accretionary complexes. There, Typhoon Talas in September 2011 brought >2,000 mm of rainfall, triggering thousands of landslides (Uchida et al., 2012). Chigira et al. (2013) including >70 catastrophic slope failures with individual volumes $>10^5$ m^3. Among these, at least 17 largely deep-seated landslides formed dams in the deeply incised river valleys (Figure 8.8; Inoue and Doshida, 2011). Many of these deep-seated rockslides detached from hill slopes that supported a dense tree cover, including significant portions of forestry areas. The dense spatial clustering of such large slope failures is similar to that documented for the 1929 Murchison earthquake.

In the Japan example, five partly drained landslide-dammed lakes from Typhoon Talas have remained in place (Chigira et al., 2013), although they have seen major engineering works since then. These engineering solutions include the reinforcement of artificial flood overflow and outlet channels through the dams and armoring of river beds downstream; controlled drainage

FIGURE 8.8 Aerial view of several large landslide dams triggered during Typhoon Talas, Kii Peninsula, Japan, 2011. Note landslide-dammed lakes in the left foreground, and also the conspicuous clustering of these large failures, whereas smaller slope failures are strikingly absent. Volumetric estimates derived from 1-m LiDAR scans (Chigira et al., 2013). *Photograph courtesy of Yukio Michikubo.*

FIGURE 8.9   Mitigation works on landslide-dammed lake formed during Typhoon Talas, Kii Peninsula, southwest Japan, in 2011. (a) Artificially lowered lake with an automatic pumping station in the foreground. (b) Stabilized flood-overflow channel cut into the artificially graded landslide dam surface; flow is from the left to the right.

of some lakes; and establishment of automated monitoring and early warning systems for the case of flood-induced lake-level rise (Figure 8.9). In several cases, these artificial drainage measures appear to be resistant to water erosion than the surrounding landslide debris such that any other water-induced erosion of the regraded dam faces need to be avoided at any cost (Figure 8.10). Heavy rainstorms contributed to the destruction of most of the engineering structures on the Akatani landslide dam in July 2013, thus clearly pointing out the need for more detailed research on landslide-dam internal stability.

Forthcoming studies on the landslide dams triggered by Typhoon Talas also herald the firm establishment of a number of research methods that were

FIGURE 8.10   Engineering solutions for stabilizing large ($>10^6$ m^3) landslide dams triggered by Typhoon Talas, Kii Peninsula, Japan. (a) Downstream view of grading works on a channelized long-runout landslide. (b) Upstream view of reinforced and artificially stepped flood outlet channel. (c) Concrete-lined overflow channel on top of artificially graded landslide-dam crest. (d) Downstream view of artificially armored outlet channel under construction.

considered novel only a few years back. Chigira et al. (2013) used 1-m resolution light detection and ranging (LiDAR; laser scanning) imagery of nearly a dozen of the large landslide sites for visualization, morphometric and structural characterization, inundation computations, and displacement analysis. Digital elevation models predating and postdating slope failure were available in some cases, and allowed unprecedented insights into the volumetric budget of these failures, a task that is commonly difficult to achieve from fieldwork and other remote sensing data alone. Yamada et al. (2012) were able to detect the seismic signals of 18 of the large landslides triggered by the typhoon in high-frequency wave forms, and established an empirical relationship between a seismic energy parameter and landslide size.

Finally, incidences of multiple typhoon-triggered river blockage are not uncommon in this region of Japan, judging from a number of documented historic events. Indeed, a storm very similar to Typhoon Talas affected the Kii Peninsula in August 1889. More than 33 landslide dams formed during this storm, and most had failed within a month (Inoue and Doshida, 2011). Clearly, the clustered coeval occurrence of landslide dams is not an exclusive effect of strong seismic ground motion.

8.4 DISCUSSION

The more or less synchronous triggering of multiple landslide dams creates geomorphic process cascades and natural hazards that may differ distinctly from the formation of a single landslide dam. From a mitigation perspective, multiple landslide dams require rapid screening and monitoring to decide which of the dams may be the most prone to sudden failure. Multiple landslide-dammed lakes may effectively block access to valleys in mountainous terrain, and contribute to deteriorating water quality in earthquake or storm disaster scenarios. A sudden outburst from one landslide-dammed lake may cause flows that could cause downstream lakes to fail due to increased overtopping, thus triggering cascades of catastrophic water and sediment waves.

The possibility for forming multiple landslide dams more or less simultaneously leads to a number of open research questions. The most pressing concerns immediate assessment of any resulting hazards and risk: Which dams are the most dangerous from a given population that may exceed several hundreds over a large region? In the case of the Wenchuan earthquake, some 34 landslide dams were considered to be the most hazardous, and the largest, that is, the Tangjiashan landslide, was deemed to be the most problematic (Cui et al., 2011). Mitigating several dozens of hydrogeomorphic hazards can be enormously challenging in an area where much critical infrastructure has suffered from severe seismic ground motion and ubiquitous landsliding. This raises the question of suitable means of assessing dam stability or detecting imminent failure. Near-surface geophysical methods may offer both such noninvasive means and valuable insights into the internal structure of landslide dams. For example, Wang et al. (2013) used near-surface seismic investigation and demonstrated that differences in shear-wave velocities revealed distinctly different facies in a dissected river-blocking deposit. Multifacies landslide-dam deposits have been reported from other locations (e.g., Wassmer et al., 2004; Weidinger et al., 2014), and may be crucial in governing the groundwater flow through—and hence affecting the stability of—the landslide dam.

The three case studies sketched above may be combined in a number of possible research themes: What can we learn from the occurrence of spatially clustered landslide-dammed lakes in a given region? What do we know about the fraction of river-damming landslides within a given landslide inventory? How do we assess the longevity of river blockage from a single coevally formed population of landslide dams?

The potential for triggering multiple landslide dams during heavy rainstorms, as highlighted by Typhoon Talas in Japan in 2011 and its predecessor in 1889, may confound interpretations regarding seismic triggers of clusters of coeval landslide dams. Hancox et al. (2002, p. 91) noted that the "principal arguments hinged on modern analogues, which indicate that seismicity and climate (heavy rainfall) are the only triggering factors that produce large clusters of individual landslides. [...] Heavy rainfall may produce isolated large deep-seated landslides, but apparently only moderate and large earthquakes produce coeval clusters of large deep-seated landslides." The observation that Typhoon Talas triggered dozens of such larger, deep-seated catastrophic landslides (Chigira et al., 2013) challenges this notion, and highlights the necessity of obtaining more detailed data to be able to test further whether clusters of coeval large deep-seated landslide deposits (and landslide dams) can be of seismic origin exclusively. Part of this problem clearly depends on somewhat arbitrary definitions of what constitutes a "large" or "deep-seated" landslide in the first place, and calls for further studies in this regard.

Lessons from engineering works on major landslide dams from both the earthquake-triggered examples in Wenchuan, and the rainfall-triggered landslide dams in southwestern Japan, were sobering, but strikingly similar. Simply lining potential spillways or breach channels with concrete may be of limited use if the surrounding dam structure remains highly erodible and thus susceptible to surface runoff, groundwater erosion, or overtopping. A recent showcase example was the $85 \times 10^6 \text{m}^3$ rockslide/avalanche dam that impounded nearly $70 \times 10^6 \text{ m}^3$ of water upstream the village of Hattian Bala in the wake of the 2005 Kashmir earthquake. Authorities decided to excavate a rock-covered spillway channel across the dam to mitigate the risk from catastrophic flooding. This engineering solution appeared to be successful, but the dam breached more than four years later following a combination of piping and heavy rainstorms, thus releasing more than half of the stored water volume (Konagai and Sattar, 2012). Similarly, some of the engineering works on the rainfall-triggered landslide dams in Kii Peninsula had been destroyed on several dams within only years of their construction, mostly during typhoon-induced rainfall that remobilized much of the artificially regraded, though unvegetated, landslide debris, while also scouring and undermining concrete spillway channels. At Kuridaira landslide dam, the highest (~ 100 m) formed during Typhoon Talas, nearly two-thirds of an approximately 580-m-long highly engineered concrete diversion channel was eroded during Typhoon Jelawat only about one year later (Sakurai, 2013). A common underlying problem with this and several similar case studies appears to be the construction of rigid concrete structures, mainly spillways, in poorly consolidated and highly permeable landslide debris that is prone to remobilization during rainstorms. Clearly, more reliable and noninvasive methods for elucidating the internal structure of dams will be desirable to better quantify their stability.

8.5 CONCLUSIONS

- We have briefly reviewed several historic case studies of regional landsliding episodes. We summarize that both earthquakes and heavy rainstorms may trigger up to hundreds of landslide dams more or less simultaneously.
- Such synchronous formation of multiple landslide dams may lead to substantial problems in natural disaster mitigation efforts, if landslide-dammed lakes (1) block access to areas impacted by earthquake or rainstorm damage; (2) pose substantial outburst hazards following potential breaks;

(3) induce massive aggradation of valley floors from landslide-derived debris; and (4) contribute to deteriorating water quality in emergency scenarios.

- Deciding with reliability which landslide-dammed lakes may pose the highest hazards and risk may require time-consuming consideration and critical decisions, and requires a trade-off between rapid and reliable investigation methods, as individual field checks of hundreds of dams over a large area may not be tractable during many disaster scenarios.

- Multiple coeval landslide dams contribute to substantially modulating water and sediment fluxes, at least for the first months to years following their formation. Dams that fail can drastically affect river behavior and land-use downstream, and cascading failures have correspondingly greater potential in this regard. Much research is still required on useful proxies of the longevity of a given landslide dam, as such lifetimes may range from a few minutes to thousands of years.

- Clusters of coeval lakes dammed by large landslides may not be uniquely diagnostic of strong seismic ground motion, but may result from high-intensity rainfall events instead. We thus caution against using the spatial pattern of landslide-dammed ages of equal age as proxies for mapping the trace of earthquake intensity isoseismals.

ACKNOWLEDGMENTS

This study was supported by the Potsdam Research Cluster for Georisk Analysis, Environmental Change, and Sustainability (PROGRESS), and a research grant by the Disaster Prevention Research Institute at the University of Kyoto, Japan.

REFERENCES

Adams, J., 1980. Earthquake-dammed lakes in New Zealand. Geology 9, 215–219.

Casagli, N., Ermini, L., 1999. Geomorphic analysis of landslide dams in the Northern Apennine. Transac.-Japanese Geomorphol. Union 20, 219–249.

Cenderelli, D.A., 2000. Floods from natural and artificial dam failures. In: Wohl, E. (Ed.), Inland Flood Hazards. Cambridge University Press, New York, pp. 73–103.

Chigira, M., Tsou, C.Y., Matsushi, Y., Hiraishi, N., Matsuzawa, M., 2013. Topographic precursors and geological structures of deep-seated catastrophic landslides caused by Typhoon Talas. Geomorphology 201, 479–493.

Costa, J.E., Schuster, R.L., 1988. The formation and failure of natural dams. Geol. Soc. Am. Bull. 100, 1054–1068.

Costa, J.E., Schuster, R.L., 1991. Documented Historical Landslide Dams from Around the World. U.S. Geological Survey Open-File Report 91-239, 486 pp.

Crozier, M.J., Deimel, M.S., Simon, J.S., 1995. Investigation of earthquake triggering for deep-seated landslides, Taranaki, New Zealand. Quat. Int. 25, 65–73.

Cui, P., Han, Y., Chao, D., Chen, X., 2011. Formation and treatment of landslide dams emplaced during the 2008 Wenchuan earthquake, Sichuan, China. In: Evans, S.G., Hermanns, R.L.,

Strom, A., Scarascia-Mugnozza, G. (Eds.), Natural and Artificial Rockslide Dams. Springer, Heidelberg, pp. 295–321.

Dunning, S.A., Armitage, P.J., 2012. The grain-size distribution of rock-avalanche deposits: implications for natural dam stability. In: Evans, S.G., Hermanns, R.L., Strom, A., Scarascia Mugnozza, G. (Eds.), Natural and Artificial Rockslide Dams. Springer, Heidelberg, pp. 479–498.

Davies, T.R.H., Korup, O., 2007. Persistent alluvial fanhead trenching resulting from large, infrequent sediment inputs. Earth Surface Processes and Landforms 32, 725–742.

Fan, X., van Westen, C.J., Korup, O., Gorum, T., Xu, Q., Dai, F., Huang, R., Wang, G., 2012. Transient water and sediment storage of the decaying landslide dams induced by the 2008 Wenchuan earthquake, China. Geomorphology 171–172, 58–68.

Gorum, T., Fan, X., van Westen, C.J., Huang, R.Q., Xu, Q., Tang, C., Wang, G., 2011. Distribution pattern of earthquake-induced landslides triggered by the 12 May 2008 Wenchuan earthquake. Geomorphology 133, 152–167.

Hancox, G.T., Perrin, N.D., Dellow, G.D., 2002. Recent studies of historical earthquake-induced landsliding, ground damage, and MM intensity in New Zealand. Bull. N.Z. Soc. Earthquake Eng. 35, 59–95.

Inoue, K., Doshida, S., 2011. Comparison of distribution of disasters occurring in 1889 and 2011 on Kii Peninsula. J. Jpn. Soc. Erosion Control Eng. 65, 42–46 (in Japanese).

Konagai, K., Sattar, A., 2012. Partial breaching of Hattian Bala landslide dam formed in the 8th October 2005 Kashmir earthquake, Pakistan. Landslides 9, 1–11.

Korup, O., 2002. Recent research on landslide dams—a literature review with special attention to New Zealand. Prog. Phys. Geogr. 26, 206–235.

Korup, O., 2012. Earth's portfolio of extreme sediment transport events. Earth-Sci. Rev. 112, 115–125.

Korup, O., Tweed, F., 2007. Ice, moraine, and landslide dams in mountainous terrain. Quat. Sci. Rev. 26, 3406–3422.

O'Connor, J.E., Beebee, R.A., 2009. Floods from natural rock-material dams. In: Burr, D.M., Carling, P.A., Baker, V.R. (Eds.), Megaflooding on Earth and Mars. Cambridge University Press, Cambridge, pp. 128–171.

O'Connor, J.E., Costa, J.E., 2004. The World's Largest Floods, Past and Present—Their Causes and Magnitudes. U.S. Geological Survey Circular 1254, 13 pp.

Parker, R.N., Densmore, A.L., Rosser, N.J., de Michele, M., Li, Y., Huang, R., Whadcoat, S., Petley, D.N., 2011. Mass wasting triggered by the 2008 Wenchuan earthquake is greater than orogenic growth. Nat. Geosci. 4, 449–452.

Pearce, A.J., Watson, A.J., 1986. Effects of earthquake-induced landslides on sediment budget and transport over a 50-yr period. Geology 14, 52–55.

Peng, M., Zhang, L.M., 2012. Breaching parameters of landslide dams. Landslides 9, 13–31.

Sakurai, W., 2013. On the countermeasures to the large-scale geohazards triggered by typhoon no.12 in September 2011, Japan. In: Deep-seated Landslides and Landslide Dams, Proceedings of Landslide Field Trip and Symposium. Kansai Branch of Japan Landslide Society, pp. 7–22.

Tang, C., Zhu, J., Li, W., Liang, J., 2009. Rainfall-triggered debris flow following the Wenchuan earthquake. Bull. Eng. Geol. Environ. 68, 187–194.

Uchida, T., Sato, T., Mizuno, M., Hayashi, S., Okamoto, A., 2012. Characteristics of landslides and the rainfall condition due to typhoon no. 12, 2011. Civil Eng. J. 54, 10–13 (in Japanese).

Wang, G., Huang, R., Kamai, T., Zhang, F., 2013. The internal structure of a rockslide dam induced by the 2008 Wenchuan (M_w 7.9) earthquake, China. Eng. Geol. 156, 28–36.

Wassmer, P., Schneider, J.L., Pollet, N., Schmitter-Voirin, C., 2004. Effects of the internal structure of a rock-avalanche dam on the drainage mechanism of its impoundment, Flims sturzstrom and Ilanz paleo-lake, Swiss Alps. Geomorphology 61, 3—17.

Weidinger, J.T., Korup, O., Munack, H., Altenberger, U., Dunning, S.A., Tippelt, G., Lottermoser, W., 2014. Giant rockslides from the inside. Earth Planet. Sci. Lett. 389, 62—73.

Yamada, M., Matsushi, Y., Chigira, M., Mori, J., 2012. Seismic recordings of landslides caused by typhoon Talas (2011), Japan. Geophys. Res. Lett. 39, L13301. http://dx.doi.org/10.1029/2012GL052174.

Yanites, B.J., Tucker, G.E., Mueller, K.J., Chen, Y.G., 2010. How rivers react to large earthquakes: evidence from central Taiwan. Geology 38, 639—642.

Chapter 9

Rock Avalanches onto Glaciers

Philip Deline [1], Kenneth Hewitt [2], Natalya Reznichenko [3] and
Dan Shugar [4]

[1] *EDYTEM Lab, Université de Savoie, CNRS, Le Bourget-du-Lac, France,* [2] *Geography and Environmental Studies, Wilfrid Laurier University, Waterloo, Ontario, Canada,* [3] *Department of Geography, Durham University, Durham, UK,* [4] *Department of Geography, University of Victoria, British Columbia, Canada*

ABSTRACT

The chapter looks mainly at massive rock slope failures that generate high-speed, long-runout rock avalanches onto glaciers in high mountains, from subpolar through tropical latitudes. Drastic modifications of mountain landscapes and destructive impacts occur, and initiate other, longer-term hazards. Worst-case calamities are where mass flows continue into inhabited areas below the glaciers. Travel over glaciers can change landslide dynamics and amplify the speed and length of runout. Conversely, landslide material deposited onto ice modifies glacier behavior, protecting ice from ablation, usually causing advances and hugely increasing glacier sediment delivery. Hitherto, recognition of the risks associated with these processes has been compromised by observational difficulties and theoretical disagreements, lack of evidence in many regions and widespread misclassification of rock avalanche deposits as moraines. The latter has also compromised glacial sequences and risk scenarios. We emphasize the need for improved understanding of processes and diagnostics, as prelude to risk assessments, including the roles of earthquakes and climate change.

9.1 INTRODUCTION

Landslide processes that affect glaciers range from small rockfalls and snow avalanches to massive rock slope failures and deep-seated gravitational creep[1]. The more frequent, smaller events may have a significant role as they affect virtually all mountain glaciers seasonally every year. This is reflected in the debris-covered ablation zones common in high mountain glaciers, and substantial delivery of coarse, angular debris to glacier margins. However, though infrequent, large volume and long-runout events can have powerful influences on glacier

---

[1] Photos by authors unless otherwise stated

dynamics, sediment assemblages, and landform development (Hewitt et al., 2011b; Reznichenko et al., 2012; Shugar et al., 2012). Travel over glaciers results in greater speed and reach of long-runout events than is the case with landslides that travel over other substrates. The mobility of the landslide debris is affected by travel over ice and by incorporation of snow, ice, meltwater, or wet sediment from the glacier (Evans and Clague, 1988). This may also result in changes of character and dynamics of mass flows. Rock avalanches may be transformed into debris avalanches—large debris-flow-like or complex mass flow events.

The chapter will focus mainly on massive rock slope failures that result in rock avalanches onto alpine glaciers in high mountains. Rock avalanches, involving the extremely rapid, flowlike movement of crushed and pulverized rock material, are among the few surface processes that can cause large, almost instantaneous modifications of mountain landscapes. Massive rock slope failures produce deep head-scarps, leaving a strong imprint on landforms in their travel zones and huge areas buried in debris. They are characterized by large volumes, generally not less than 1 Mm^3. Some are "megaslides" exceeding 1 km^3. In high mountains their motion involves great vertical falls (>1,000 m), long horizontal travel distances (>5 km), and high velocities (>25 m/s).

These landslides pose both direct and indirect hazards to society. No one is likely to survive in the path of a rock avalanche. However, they are sufficiently rare, and the exposure of humans on glaciers is so limited that few fatal cases are known. The greatest risks and worst calamities associated with these events are where runout continues below the glaciers into areas that have human settlements and infrastructure. A number of recent calamitous examples have occurred in the Andes and Inner Asian high mountains (Table 9.1). Indirect or secondary hazards can arise from the impact of the landslide on glacier activity, notably if it results in thickening and sudden advance, or where the landslide mass impounds meltwater along the glacier margins or streams in ice-free valleys below. The inundation of land or outburst floods from such impoundments can be serious threats.

We examine the conditions likely to create rock slope instability and trigger catastrophic failures, landslide behavior during travel over and beyond the glaciers, and the resulting depositional and erosional legacies. An outstanding question is the extent to which global climate change may increase these hazards or alter their geographical patterns. This may follow from greater weather extremes that trigger events, or degrading of mountain permafrost so as to destabilize rockwalls (Davies et al., 2001; Haeberli et al., 2002; Fischer et al., 2006; Huggel, 2008). Meanwhile, risks will inevitably increase because of greater human presence and activity in the mountains. However, these require risk assessments using probabilities based on evidence of older or prehistoric events, even though they occurred when settlements were absent or much more dispersed.

Full appreciation of the topic and the hazards involved has been compromised by three main factors: (1) difficulty in identifying deposits in remote mountain ranges before they are covered by snow, heavily modified, and

TABLE 9.1 Statistics of Documented Rock Avalanches onto Glacier with a Rock Volume >1 Mm³ since 1950

Rock Avalanche (Country Codes: ISO 3166-1 Alpha-3)	Year	V_R^1 (Mm³)	V_i^2 (Mm³)	H (km)	L (km)	μ^3	L_e^4 (km)	A^5 (km²)	Th^6 (m)	V_{i+s}^7 (Mm³)	P_{Gl}^8 (km)	P_{Marg}^9 (km)	Z^{10} (m)	References
Becca di Luseney (Valpelline, ITA)	1952	1.0	0.0	1.65	4.00	0.41	1.35	?	?	?	0.30	0.00	3,265	Dutto and Mortara (1991)
Tim Williams Glacier (Coast Mts, CAN)	1956	3.0	?	0.94	3.70	0.25	2.20	?	?	?	?	0.00	?	Evans and Clague (1999)
Pandemonium Creek (Coast Mts, CAN)	1959	5.5	0.0	2.00	8.60	0.23	5.35	1.50	?	50%	1.39	0.00	2,600	Evans and Clague (1999), Evans and Clague (1988), Schneider et al. (2011b)
Iliamna Red Glacier (Chigmit Range, USA)	1960	*<2.0	?	1.47	5.50	0.27	3.15	?	?	60%	5.10	0.00	2,125	Schneider et al. (2011b)
Nevados Huascarán (Cordillera Bl., PER)	1962	1.0	2.0	4.08	18.10	0.23	11.50	?	?	15.6	2.40	3.50	6,654	Schneider et al. (2011b), Plafker and Ericksen (1978), Evans et al. (2009a)
Little Tahoma Peak[12] (Mount Rainer, USA)	1963	10.7	0.0	1.88	7.25	0.26	4.25	5.00	1–20	20%	4.33	13.55	3,230	Crandell and Fahnestock (1965), Fahnestock (1978)
Schwan Glacier 1 (Chugach Mts, USA)[11]	1964	27.0	?	1.45	6.00	0.24	3.65	9.00	?	?	5.35	0.00	2,180	Evans and Clague (1988), Post (1967)
Sherman Glacier (Chugach Mts, USA)[11]	1964	10.1	2.0	1.16	6.00	0.19	4.15	8.25	1.6	>2.0	5.13	3.60	1,310	Post (1967), McSaveney (1978)

Continued

TABLE 9.1 Statistics of Documented Rock Avalanches onto Glacier with a Rock Volume >1 Mm^3 since 1950—cont'd

Rock Avalanche (Country Codes: ISO 3166-1 Alpha-3)	Year	V_R[1] (Mm^3)	V_i[2] (Mm^3)	H (km)	L (km)	μ[3]	L_e[4] (km)	A[5] (km^2)	Th[6] (m)	V_{i+s}[7] (Mm^3)	P_{Gl}[8] (km)	P_{Marg}[9] (km)	Z[10] (m)	References
Sioux Glacier 1 (*Chugach Mts, USA*)[11]	1964	7.0	?	1.35	4.50	0.30	2.30	3.10	2.0	2.3	3.60	0.70	1,525	Post (1967), Reid (1969)
Steller Glacier 1 (*Chugach Mts, USA*)[11]	1964	20.0	?	1.15	6.70	0.17	4.85	7.50	?	?	5.80	0.63	2,700	Evans and Clague (1988), Post (1967)
Allen Glacier 4 (*Chugach Mts, USA*)	?1965	23.0	?	1.23	7.70	0.16	5.75	7.50	?	?	7.03	0.45	1,925	Evans and Clague (1988), Post (1967)
Fairweather Glacier (*Chugach Mts, USA*)	?1965	26.0	?	3.35	10.50	0.32	5.10	8.50	?	?	7.90	0.00	4,050	Evans and Clague (1988), Post (1967)
Steinsholtsjokull Glacier (*Eyjafjallajökull, ISL*)	1967	15.0	<6.00	0.32	5.00	0.06	4.50	0.28	2–5	>20.0	1.20	1.10	710	Kjartansson (1967)
Nevados Huascarán (*Cordillera Bl., PER*)[11]	1970	6.5	1.0	3.92	18.20	0.22	11.90	22.00	?	58.2	2.40	0.00	6,654	Plafker and Ericksen (1978), Evans et al. (2009a)
Winthrop Glacier (*Mount Rainier, USA*)	1974	?	0.0	1.39	2.70	0.51	0.45	0.75	?	?	2.25	0.00	3,705	Scott and Vallance (1995)
Devastation Glacier (*Coast Mts, CAN*)	1975	13.0		1.22	7.00	0.17	5.05	?	?	?	2.50	0.00	2,010	Evans and Clague (1999), Evans and Clague (1988)
Iliamna Red Glacier (*Chigmit Range, USA*)	1978	*6.8	?	1.78	7.70	0.23	4.85	?	?	60%	7.00	?	2,310	Schneider et al. (2011b)

Ama Dablam (*Khumbu Himal, NPL*)	1979	>1.0	0.0	1.40	3.00	0.47	0.75	>0.75	?	?	?	2.10	1.80	5,600	Selby (1993)
Iliamna Red Glacier (*Chigmit Range, USA*)	1980	*11.2	?	1.68	7.80	0.22	5.10	?	?	60%	7.35	0.07[13]	2,193	Schneider et al. (2011b)
W Fork Robertson Glacier (*Alaska Range, USA*)	1981	2.2	?	0.70	3.50	0.20	2.35	1.50	2–5	?	3.00	0.14	2,100	Herreid et al. (2010)
Marvine Glacier (*St. Elias Mountains, USA*)	1983	1	?	0.86	3.15	0.27	1.75	?		70%	2.05	0.27	1,650	Schneider et al. (2011b)
Glacier de Bualtar 1 (*Karakorum, PAK*)	1986	10.0	0.0	1.49	4.80	0.31	2.40	4.10	2.5	?	3.25	0.00	4,450	Hewitt (1988)
Glacier de Bualtar 2 (*Karakorum, PAK*)	1986	7.0	0.0	1.43	3.70	0.39	1.40	3.30	2.1	?	2.15	0.00	4,450	Hewitt (1988)
Glacier de Bualtar 3 (*Karakorum, PAK*)	1986	3.0	0.0	1.30	2.40	0.54	0.30	1.20	2.5	?	0.85	0.50	4,450	Hewitt (1988)
Mount Meager (*Coast Mountains, CAN*)	1986	0.5–1.0	0.0	1.34	3.68	0.36	1.50	?	?	?	?	0.00	2,400	Evans and Clague (1988)
North Creek (*Coast Mountains, CAN*)	1986	1.0–2.0	0.0	0.75	2.85	0.26	1.65	?	?	?	1.00	0.75	1,980	Evans and Clague (1988)
Aroaki/Mount Cook (*Southern Alps, NZL*)	1991	9.4–14.2	0.4–0.6	2.72	7.50	0.36	3.10	7.00	2–10	>30.0	6.66	0.60	3,764	McSaveney (2002)
Mount Fletcher 1 (*Southern Alps, NZL*)	1992	>7.8	4%	1.44	3.80	0.38	1.00	1.75	?	?	2.29	1.20	2,450	McSaveney (1993)
Mount Fletcher 2 (*Southern Alps, NZL*)	1992	>5.0	0.0	1.44	3.80	0.38	1.45	1.75	?	?	2.36	0.50	2,450	McSaveney (1993)

Continued

TABLE 9.1 Statistics of Documented Rock Avalanches onto Glacier with a Rock Volume >1 Mm³ since 1950—cont'd

Rock Avalanche (Country Codes: ISO 3166-1 Alpha-3)	Year	V_R^1 (Mm³)	V_I^2 (Mm³)	H (km)	L (km)	μ^3	L_e^4 (km)	A^5 (km²)	Th^6 (m)	V_{i+s}^7 (Mm³)	P_{Gl}^8 (km)	P_{Marg}^9 (km)	Z^{10} (m)	References
Winthrop Glacier (Mount Rainier, USA)	1992	?	0.0	1.30	2.45	0.53	0.35	0.45	?	?	1.90	0.55	3,780	Scott and Vallance (1995)
Kshwan Glacier (Coast Mountains, CAN)	?1992	3.2	0.0	0.79	2.25	0.35	1.00	0.68	5	—	1.42	16.3	1,750	Evans and Clague (1999), Mauthner (1996)
Iliamna Red Glacier (Chigmit Range, USA)	1994	*10.2	?	1.80	10.0	0.18	7.10	?	?	60%	9.50	0.00	2,230	Schneider et al. (2011b)
Mount Munday (Coast Mountains, CAN)	1997	3.2	0.0	0.88	4.70	0.19	3.30	2.20	1.5	?	>3.50	0.00	3,000	Evans and Clague (1999), Evans and Clague (1998)
Brenva Glacier (Mont Blanc massif, ITA)	1997	2.0	0.0	2.33	5.75	0.40	0.20	?	1	>4.5	5.40	0.00	3,725	Deline (2001, 2009)
Iliamna Red Glacier (Chigmit Range, USA)	1997	*8.4	?	1.71	7.70	0.22	4.95	?	?	60%	7.10	0.00	2,230	Schneider et al. (2011b)
Howson Glacier 2 (Coast Mountains, CAN)	1999	0.9	?	1.30	2.70	0.48	0.60	?	?	?	1.50	0.00	?	Schwab (2002)
Iliamna Red Glacier (Chigmit Range, USA)	2000	*9.0	?	1.83	8.90	0.21	5.95	?	?	60%	8.40	0.00	2,310	Schneider et al. (2011b)
Tsar Mountain (Rockies Mountains, CAN)	2000	1.6	?	0.61	2.25	0.27	1.25	?	?	40%	1.70	0.00	2,778	Jiskoot (2011a)

Kolka-Karmadon (Caucasus, RUS)	2002	10–14.0	8.5–13	2.05	19.40	0.11	16.10	?	?	90.0	3.10	0.00	3,368	Schneider et al. (2011b), Huggel et al. (2005)
McGinnis Peak Glacier N (Alaska Range, USA)[11]	2002	18.4	?	1.65	11.00	0.15	8.45	10.21	2	2.0	10.50	0.00	2,740	Jibson et al. (2006)
McGinnis Peak Glacier S (Alaska Range, USA)[11]	2002	11.4	?	1.80	11.50	0.16	8.60	5.71	2	?	10.70	2.30	2,770	Jibson et al. (2006)
Black Rapids Glacier E (Alaska Range, USA)[11]	2002	9.3–14.0	?	0.98	4.10	0.24	2.50	4.64	2–3	?	3.60	0.00	2,130	Jibson et al. (2006), Shugar and Clague (2011)
Black Rapids Glacier M (Alaska Range, USA)[11]	2002	7.7–11.6	?	0.80	5.60	0.14	4.30	4.55	2–3	?	3.20	0.00	2,050	Jibson et al. (2006), Shugar and Clague (2011)
Black Rapids Glacier W (Alaska Range, USA)[11]	2002	4.9–7.4	?	0.73	3.40	0.21	2.20	3.24	2–3	?	2.20	0.00	2,070	Jibson et al. (2006), Shugar and Clague (2011)
West Fork Glacier N (Alaska range, USA)[11]	2002	4.1	?	0.76	3.30	0.23	2.10	1.37	3	?		0.85	1,980	Jibson et al. (2006)
West Fork Glacier S (Alaska Range, USA)[11]	2002	4.4	?	0.90	4.10	0.22	2.65	1.47	3	?		10.85	2,070	Jibson et al. (2006)
Iliamna Red Glacier (Chigmit Range, USA)	2003	*6.0	6.0–14	1.98	8.60	0.23	5.40	4.00	1–2	60%	8.00	0.00	2,256	Huggel et al. (2007)
Iliamna Umbrella Glacier (Chigmit Range, USA)	2004	*2.0	?	1.76	6.05	0.29	3.20	?	?	50%	5.55	0.00	2,406	Schneider et al. (2011b)
Punta Thurwieser (Ortles-Cevedale, ITA)	2004	2.5	0.0	1.30	2.70	0.48	0.60	?	?	10%	0.40	0.00	3,570	Schneider et al. (2011b), Sossio et al. (2008), Pirulli (2009)
Mount Steller (St Elias Mountains, USA)	2005	>40.0	>3.0	2.43	9.00	0.27	5.10	?	?	0.36	8.64	0.00	3,100	Huggel et al. (2008)

Continued

TABLE 9.1 Statistics of Documented Rock Avalanches onto Glacier with a Rock Volume >1 Mm³ since 1950—cont'd

Rock Avalanche (Country Codes: ISO 3166-1 Alpha-3)	Year	V_R^1 (Mm³)	V_i^2 (Mm³)	H (km)	L (km)	μ^3	L_e^4 (km)	A^5 (km²)	Th^6 (m)	V_{i+s}^7 (Mm³)	P_{Gl}^8 (km)	P_{Marg}^9 (km)	Z^{10} (m)	References
Mount Steele (*St Elias Mountains, CAN*)	2007	27.5–80.5		2.16	5.80	0.32	2.30	5.28	4–22	30%	4.05	0.00	4,650	Lipovski et al. (2008)
Morsárjökull (*Vatnajökull Ice Cap, ISL*)	2007	4.0	0.0	0.66	1.40	0.47	0.35	0.72	5.5	–	1.70	0.00	950	Decaulne et al. (2010)
Mount Miller (*St Elias Mountains, USA*)	2008	*22.0	?	0.91	4.50	0.20	3.05	?	?	?	3.65	3.50	2,200	Schneider et al. (2011b)
Mount Meager (*Coast Mts, CAN*)	2010	48.5	<0.2	2.18	12.70	0.17	9.20	9.00	?	<1%	<0.50	>12.0	2,554	Guthrie et al. (2012)
Lituya Mountain (*Coast Mts, USA*)	2012	>20.0		2.50	9.00	0.28	4.95	7–8.00	?	?	6.50	0.00	3,030	Geertsema (2012)
Mount Dixon (*Southern Alps, NZL*)	2013	1.5?	?	0.85	2.80	0.30	1.45	0.75	1–2	?	2.55	3.60	2,950	http://blogs.agu.org/landslideblog

Name: Catastrophic glacier multiphase event (*sensu* Petrakov et al., 2008).
? – Unknown data.
[1] *Detached (*: deposited) rock volume.*
[2] *Detached ice volume.*
[3] *μ = H/L.*
[4] *Excessive travel distance $L_e = L - H/\tan 32$.*
[5] *Surface area of the deposit.*
[6] *Mean thickness of final rock deposit (___: ice and rock deposit).*
[7] *Ice and snow volume of the deposit (or % of the total (rock + ice) deposited volume).*
[8] *Path onto glacier surface.*
[9] *Path on margins (outside of the glacier).*
[10] *Maximal elevation of the scar.*
[11] *Rock avalanche triggered by earthquake.*
[12] *Seven successive rock collapses (data: main collapse).*
[13] *Run-up on moraine proximal side.*

dispersed by glacier transport; (2) widespread misclassifying of rock avalanche deposits as moraines (Heim, 1932; Porter and Orombelli, 1980; Hewitt, 1999; Prager et al., 2009; McColl and Davies, 2011); and (3) recognition that some moraines result from supraglacial rock avalanche deposits rather than from climatic variation (Reznichenko et al., 2012). Rock avalanche deposits need to be distinguished from the many other coarse, poorly-sorted, or unsorted materials in mountain glacier environments. The problem is further complicated where rock avalanches descend over surge-type glaciers, as reported in the Alaska–Yukon ranges, Argentinian Andes, Caucasus, Pamirs, and Karakoram. In such cases they become involved in the already complex issues of interpreting surge-induced glacial deposits (Sharp, 1988; Evans and Rea, 2003). Further, inventories of rock avalanches on glaciers have been hindered by difficulties associated with observing them. Their unpredictability and frequent location in remote mountains have made observations rare. Their known numbers are certainly underestimates.

The first section of the chapter is dedicated to the processes that are involved in the detachment, the displacement, and the deposition of the rock avalanches onto and beyond glaciers, and the subsequent transformation of the deposits. Then the interactions between rock avalanches and glaciers are explored in terms of high mobility of the landslide material, debris supply to glaciers and changes in glacier dynamics, and character of the resulting moraine complexes. Finally, three regional and local case studies are presented from the Karakoram Himalaya, New Zealand Alps, and Argentinian Andes.

9.2 PROCESSES

9.2.1 Detachment Zone and Conditions

Several sets of factors affect the occurrence of a landslide: inherent factors (e.g., rock structure, slope form), preparatory factors (e.g., weathering, debuttressing, climate change), triggering factors (e.g., earthquake, rainstorm), and factors that may affect mobility (e.g., glacier surface) (Pacione, 1999). Here we consider two categories of causal factors in a rock failure: preparatory factors and triggers (see also Chapter 2 in this volume). The former operate over a lengthy period to reduce the stability of the slope, while the latter are external stimuli, operating over a short period, which rapidly increase the total stress on a slope or reduce the strength of the slope, causing it to fail (Lee and Jones, 2004). In some cases, no identifiable trigger is evident. In these cases it seems the stability of the slope has deteriorated over time, e.g., stress corrosion to reach a spontaneous failure threshold (Eberhardt et al., 2004), or the actual trigger factor remained unidentified (e.g., rock avalanches described by McSaveney (2002), Lipovsky et al. (2008)).

9.2.1.1 Preparatory Factors

Lithology and structure are two well-known control factors for rockfall and massive rock slope failure. Shear zones with crushed rock (e.g., mylonitoschists in the Mont Blanc massif; the active Raikot-Sassi and Stak fault systems of the Nanga Parbat-Haramosh massif, NW Himalaya) are particularly prone to rockfall and rock avalanching.

These conditions may combine with paraglacial dynamics; conditions that follow from former glacial action and deglaciation. During a paraglacial period, three processes can act to destabilize rock slopes:

1. glacial overdeepening may increase the preglacial rock slope stress and cause possible slope instability (McColl, 2012); due to uplift and glacier erosion, relative relief of mountainsides in Himalaya, Rockies, or European Alps can exceed 2,500–3,000 m, and with slope angles >45°;
2. shrinkage and downwasting of glaciers results in glacial debuttressing (Cossart et al., 2008), i.e., the loss of slope support provided by glacial ice; although whether this can prevent a failure being triggered, or simply constrains the movement of the failed mass, is unclear (McColl and Davies, 2013);
3. the removal of glacier ice can generate near-surface rock fractures parallel to the slope; this stress-release fracturing weakens the rock and prepares its instability (McColl, 2012).

The 1992 rock avalanches on Mount Fletcher (Southern Alps, NZ) illustrate these processes. They combine a 250-m post-Little Ice Age (LIA) debuttressed foot of a 1,000-m-high, 57° slope angle, mountain side, with near-surface fractures parallel to the slope, corresponding to the detachment zones of the rock avalanches (McSaveney, 1993). Generally, there is a strong association in the central Southern Alps of recent slope failures with glacial shrinkage over the past 100–150 years (Allen et al., 2011).

Moreover, glacier surfaces lowered by tens to hundreds of meters, especially in the ablation zone, may have an effect on rock temperature regime, as identified for >300 m in the frontal area of Grosser Aletschgletscher since the LIA termination, and for 250 m at Black Rapids Glacier since 1949 (Shugar et al., 2010). Once close to 0 °C, the rock temperature of the deglaciated rockwalls responds to local air temperature, depending on orientation, so that freeze-front propagation can fracture the rock (Wegmann et al., 1998; Nagai et al., 2013). At lower elevations, slope dewatering due to glacier retreat tends to destabilize the foot slope because of the reorientation of its stress field (Haeberli et al., 1997; McColl et al., 2010)—although it also reduces the pore-water pressure. On the other hand, surface lowering of glaciers and melting of ice aprons and small hanging glaciers can lead to permafrost development in rock slopes previously covered by ice (Haeberli et al., 1997; Fischer et al., 2006).

Permafrost degradation due to climate warming generates physical changes in the formerly frozen rock mass. Interstitial ice can melt, whereas heat advection and hydrostatic pressure can be produced at depth in a densely fractured rock mass by circulation of water from the melting of surface snow cover and/or cleft ice (Hasler et al., 2011; Wegmann et al., 1998). The upward warming of the permafrost base in high-elevation rockwalls combined with a frozen rock surface layer can produce high groundwater pressure (Wegmann et al., 1998). Deep degradation of rock permafrost may be delayed by decades, centuries, or millennia (Gruber and Haeberli, 2007). The 22 rock failures observed in the central region of the NZ Southern Alps since the mid-twentieth century have mainly affected rockwalls with warm permafrost, i.e., whose temperature is in the range −2 to 0 °C, and nearly all of the recent rock avalanches were associated with glaciers (Allen et al., 2011).

Because the majority of rock avalanches have occurred in glaciated mountain ranges that are tectonically active (Figure 9.1), or impacted by the removal of ice caps (McColl et al., 2012), seismicity can also lead to the gradual, or intermittent destabilization of susceptible slopes, acting as a pre-paratory as well as a common triggering factor. Repeated seismic shaking weakens a rock mass until it reaches the threshold when a rock avalanche may be triggered by the next seismic event or by other causes (Voight, 1978; Eisbacher and Clague, 1984).

9.2.1.2 Triggering Factors

Freeze-thaw action is conditioned by the temperature fluctuations around 0 °C and may occur at depth during or following permafrost degradation. Volumetric expansion of ice in the slope generally widens existing fractures and prepares the rock for failure (Gruber and Haeberli, 2007; Murton and Matsuoka, 2008). Whereas this freeze-thaw action was observed and empirically proven to cause smaller mass failure events such as rockfalls (Gruber et al., 2004; Matsuoka and Murton, 2008), it is uncertain whether it could lead to massive rock slope failures and rock avalanches (Whalley, 1984; Davies et al., 2001).

Strong ground shaking during earthquakes has triggered landslides in diverse topographic and geologic settings. In the Chugach Mountains of south-central Alaska, Uhlmann et al. (2013) estimate that half of the supraglacial rock ava-lanches and c. 73 percent of the RA volume emplaced onto glaciers were triggered by coseismic shaking. The March 27, 1964 M 9.2 earthquake located on the south coast of Alaska triggered 50 rock avalanches onto glaciers with a total deposit area $>0.5 \times 10^6 \, \text{m}^2$ (Post, 1967), whereas the 1979 St Elias earthquake triggered at least three rock avalanches onto glaciers in northern British Columbia (Evans and Clague, 1999). As suggested by a study of some 50 large coseismic landslides worldwide (Keefer, 1984), slopes affected generally have a relative relief $>150 \, \text{m}$ (median: 500 m) and a slope angle $>25°$ (median 40°). Keefer (2002) also established a correlation between earthquake magnitude and landslide-affected area for a large sample of events. However, the area affected by coseismic landslides varies greatly from event to event. The 1964 M 9.2 earthquake involved 269,000 km²; the 2002 Alaskan M 7.9 earthquake only 10,000 km², much less than expected (Jibson et al., 2006).

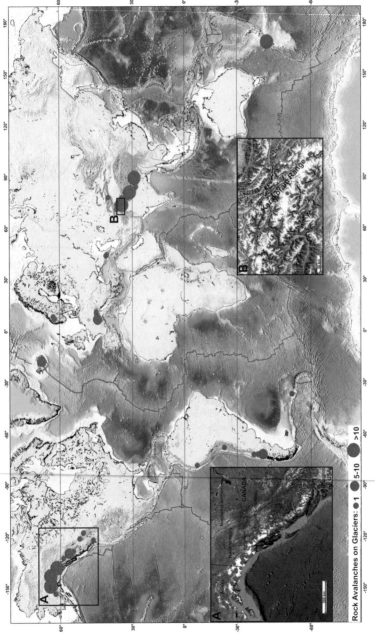

FIGURE 9.1 Distribution of main documented rock avalanches onto glacier worldwide. A: Northwestern North America; B: Karakoram Range.

The dimensions of the seismic event (e.g., magnitude, etc.) control the likely shear strength reduction in the slopes, and chances of failure (Voight, 1978). Malamud et al. (2004) suggested that relatively small (c. 10^4 m^3) landslides can be triggered by earthquakes with threshold magnitudes of only M 4.3 ± 0.4. A landslide with a volume of 10^6 m^3 would require an earthquake of M 6, while one with a volume of 10^9 m^3 would require M 8. Keefer (1999) similarly estimated that gigantic rock avalanches (in the order of 10^9 m^3) would require earthquakes of M > 8. Finally, seismic events may reactivate existing landslides as well as trigger new failures.

9.2.1.3 Glacier Basins and Rock Avalanches

Evidence from various mountain ranges suggests massive rock slope failures are as frequent or more frequent in glaciated basins than in surrounding, ice-free valleys (McSaveney, 2002; Geertsema et al., 2006; Hewitt, 2009). It follows that glacier basin environments increase either preparatory and triggering factors, or both. Several elements may contribute:

1. glacier basins, at least outside high latitudes, are located in high, steep, tectonically and seismically active and densely fractured mountain ranges, recent from the geological perspective and with high erosion rates;
2. glacier erosion tends to greatly steepen rock slopes;
3. glacier shrinkage causes debuttressing of slopes; and
4. isostatic rebound resulting from regional deglaciation may generate stresses and fracture in bedrock.

In addition, triggering factors within, as well as outside glacier basins arise from higher seismicity due to tectonics, frequent and large freeze-thaw cycles, and permafrost in rockwalls or its degradation.

Remote sensing methods have made rock avalanches on glaciers visible as never before, and increase the chance of event detection before removal of the deposit. In contrast, reported frequencies of past supraglacial rock avalanches certainly underestimate them. Deposits are mostly in uninhabited, perhaps rarely visited areas where glacionival processes quickly remove the evidence. Once they are incorporated into moraines, or reworked by glaciofluvial processes, it may be difficult or impossible to detect them. Of particular relevance are the results of a study by Ekstrom and Stark (2013). They identified a unique seismic signature associated with landslides on glaciers. They found that one-third of the 29 landslides identified were previously unreported, including seven with a combined volume of c. 200 Mm^3 that descended onto Siachen Glacier in the Karakoram in September 2010. Otherwise, it seemed likely that these Siachen Glacier landslides would be misinterpreted as deposits of one or two extremely large slope failures or, perhaps, not recognized as landslides at all. Their technique is expected to facilitate identification of remote landslides.

Supraglacial rock avalanches that travel over moraines present an especially challenging interpretation problem (e.g., Deline and Kirkbride, 2009; Kirkbride and Winkler, 2012). Moreover, many recent supraglacial rock avalanches have been misinterpreted/identified as normal supraglacial cover, much as nonglacial examples have been interpreted as morainic complexes (Porter and Orombelli, 1981; Hewitt, 1999).

9.2.2 Supraglacial Motion

9.2.2.1 Flowing Processes

Rock avalanche movement has been described as "granular flow". Because of their high velocity, a collisional flow regime applies where energy is dissipated by contacts between individual grains. It has been suggested that frictional energy dissipation is reduced by dynamic, pore-pressure fluctuations due to grain rearrangement (Schneider et al., 2011b), but these are highly constrained, since clasts and original lithological units, although massively fractured and pulverized, maintain their relative positions in the mass. A slight dilation and intense local vibration or collisional chaos occurs. Rock fragmentation has been suggested to be a source of because it is the vibrational energy that reduces intergranular friction (Davies et al., 2010), because it is much more powerful than the purely vibrational energy due to grain rearrangement. The overall mass may spread, converge, split, or undergo large-scale shearing of superimposed debris sheets. However, in rock avalanche flow there is no turbulence and individual clasts cannot move through the mass. In other respects, the rock avalanche resembles a viscous fluid, and its deposits show well-defined flow features and preserve lithological relations in the original bedrock (Jibson et al., 2006). Distinctive properties of a glacier surface and its potential for erosion and ablation can modify rock avalanche behavior and also entrainment of ice and snow and their possible melting.

Depending on season, or descent over the glacier accumulation zone, the rock avalanche may encounter on-ice snowpack or firn. Particular rheological and movement features were identified with heavy snow covers on Alaskan glaciers in the post-1964 earthquake landslides (Post, 1967; Johnson and Ragle, 1968; McSaveney, 1978). Ice may be derived from failure zones, but is mostly entrained at the surface of the glacier, especially at icefalls (e.g., the Hochstetter Icefall during the 1991 Aoraki/Mount Cook rock avalanche; McSaveney, 2002). Mixed rock—ice deposits of rock avalanches have a wide range of ice contents (Table 9.1), from 15 percent at Sherman Glacier in 1964 (McSaveney, 1978) to 90 percent at Lyell Glacier (South Georgia) in 1975 (Gordon et al., 1978).

In 1997 at the Brenva Glacier (Mont Blanc massif, Italy), the collapse of ≥ 2 Mm^3 of granite rock mobilized a large amount of ice, firn, and snow along its 5.7-km-long path on the glacier. Although proportions of rock and ice/firn were roughly equal in the upper part, the distal part of the deposit (mean

FIGURE 9.2 Flowing processes of rock avalanches onto the surface of glaciers. (a) Trough carved by the 1997 rock avalanche on the Brenva Glacier surface, Mont Blanc massif. The 2 Mm³ scar is towering above the c. 15-m-deep trough *(Photo: P. Deline)*; (b) 2006 rock avalanche onto Jarvis Glacier, St Elias Mountains, British Columbia. The photograph, taken within a week after the event, shows melting of c. 2 m of firn as rock avalanche traveled over the glacier surface *(Photo: D. Capps.)*; (c) Gannish Chissh prehistoric rock avalanche, upper Barpu Glacier, Karakoram. It descended from Spantik Peak (SP), crossed the glacier to emplace crushed whitish crystalline limestone debris (RA) over moraines, and in >300 m run up of opposing slope (yellow arrow). Main lobe swung through 90° to travel 11 km downglacier to the left (dashed arrow). A surge-related lateral moraine (m) later covered the rock avalanche deposit *(Photo: K. Hewitt)*; (d) flow banding as the landslide changes direction, Black Rapids Glacier, Central Alaska Range; the arrow shows the flow path *(Photo: D. Shugar)*.

thickness >20 m) included 90 percent ice and firn. Some ice masses were as large as houses, even reported to be "as wide as the San Siro stadium in Milano, and as high as a 20-floor building". Total ice and firn volume was estimated at >4 Mm³ (Deline, 2009)—the wide trough excavated down the glacier corresponded to c. 3 Mm³ (Figure 9.2(a)). At that time (January), snow cover ranged from c. 150 cm at the glacier terminus to 300 cm at its head. Giani et al. (2001) modeled a mobilized snow volume of 5.4 Mm³, of which 45 percent would have been incorporated in the accompanying snow powder avalanche. With c. 1 Mm³ of packed snow included in the deposit, the total volume of ice/firn/snow mobilized by the rock avalanche was at least twice the collapsed rock volume.

When ice and snow are incorporated into a rock avalanche, thermal interactions occur: the frictionally heated debris may cause rapid melting of ice and snow and generate a water film. There is evidence of ice melting at the base and margins of some rock avalanches (Figure 9.2(b)), and of refrozen ice within the Sherman Glacier rock avalanche deposit (Sosio et al., 2012). At Bualtar Glacier (Karakoram Himalaya), rapid ablation and, possibly, impact erosion by the 1986 rock avalanches changed initially rough, heavily crevassed surfaces to relatively smooth ones (Hewitt, 1988).

As in nonglaciated environments, rock avalanches on glaciers generally produce very large dust clouds resulting from the rock fragmentation, sometimes with air blast that reflects the high energies involved. Examples include Aoraki/Mount Cook in 1991, Punta Thurwieser (Ortles-Cevedale massif, Italy) in 2004, Monte Rosa (Italy) in 2007, Mount Steele (Yukon) in 2007, and John Hopkins Glacier (Alaska) in 2012. The cloud can contain snow or ice, and may act like a dry snow avalanche. The air blast triggered by the 1997 Brenva rock avalanche destroyed a >200-year-old forest along the opposite valley side, with a run-up exceeding 500 m. Skiers present described being plunged into semidarkness, and tree branches were pulled up and thrown about. The mixed icy-rocky dust pitted the windscreens of cars, and a 5-cm-thick layer of ice powder was deposited downvalley (Deline, 2001). Porter and Orombelli (1980) reported that birds were killed during the 1717 Triolet rock avalanche on the Mont Blanc massif, presumably by the air blast. At Jarvis Glacier (BC), the air blast from the 2006 rock avalanche killed a bird and spattered it with mud.

The highly mobile rock avalanche mass is sensitive to the geometry of the surface, tending to extend or compress over convex or concave surfaces, respectively. Where it encounters larger obstacles, the sheet may split into separate debris lobes. Remarkable heights of run-up on opposing slopes may occur. Debris from the prehistoric Gannissh Chissh (Spantik Peak, Karakoram Himalaya, 7,027 m a.s.l.) landslide was emplaced on the opposing valley wall more than 300 m above the glacier surface (Figure 9.2(c)).

Dufresne and Davies (2009) argue that certain surface features such as longitudinal flow bands are fundamental characteristics of granular flows on glaciers, and differ from the typical elongate ridges found on nonglacial landslides though both may be similar in terms of the ratio ridge spacing/flow depth. Flow bands can separate zones of different lithologies (e.g., Shreve, 1968) or grain size (e.g., Shugar and Clague, 2011), or simply represent shear between debris moving at different velocities. At Black Rapids Glacier, flow bands on the three 2002 rock avalanches indicate a variety of flow paths (Figure 9.2(d)). The westernmost rock avalanche spread relatively uniformly across the glacier, with little deflection downglacier. Flow bands on the central rock avalanche indicate that it turned abruptly downglacier as it encountered the lateral moraine on the distal side. Flow bands on the easternmost rock avalanche indicate that some of the debris traveled north across the glacier, before it turned to the east, downglacier. Much of the debris however, traveled directly to the northeast. At Sherman Glacier, 1964 flow bands indicate that the landslide spread as it traveled downglacier, but with no abrupt turns as at Black Rapids Glacier.

9.2.2.2 Higher Mobility of Rock Avalanches on Glaciers

Whereas horizontal travel distance for most landslides approximately equals vertical travel, for rock avalanches the ratio is much greater, up to 10:1 (Friedmann, 1997) and, on glaciers the ratio can be even higher. Various mechanisms have been proposed to explain this, irrespective of whether they travel over ice, and include water or air lubrication (Kent, 1966; Shreve, 1968), acoustic and mechanical fluidization (Hsü, 1975; Melosh, 1979; Davies, 1982; Collins and Melosh, 2003), and fragmentation spreading (Davies and McSaveney, 2002, 2009). These theories can be broadly categorized as those that reduce basal friction, and those that reduce internal friction (Davies et al., 1999; for a recent summary see Davies and McSaveney (2012)).

In their seminal study of 17 cases, Evans and Clague (1988) found interaction with glacier ice significantly enhances rock avalanche mobility, and by an average of 24 percent. Other studies confirm this, including by Dutto and Mortara (1991) comparing 15 rock avalanches onto glaciers with 16 in nonglacial environments (Figure 9.3).

Evans and Clague (1988, 1994) offer several hypotheses to explain the greater mobility of rock avalanches on glaciers versus those in nonglacial environments:

1. the debris travels on low-friction (i.e., ice and snow) surfaces; the apparent coefficient of friction (μ, corresponding to height-over-length, H/L, ratio) of the interface between rock debris and ice is low (Table 9.1), but a preexisting supraglacial debris cover can markedly increase friction;

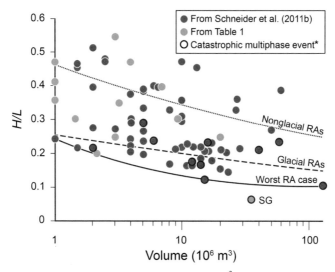

FIGURE 9.3 Mobility of rock avalanches (volume >1 Mm^3) on glaciers shown by the relationship between volume and ratio between vertical (H) and horizontal (L) travel distances; regression lines from Schneider et al. (2011b) and Evans and Clague (1988); SG, Steinsholtsjokull Glacier, Iceland; *, *sensu* Petrakov et al. (2008).

2. pore pressures are generated at the base of the debris by frictional melting. Sosio et al. (2012) model a reduction of the basal friction angle of >20 percent at Sherman Glacier through pore-pressure generation;
3. the debris is fluidized by melting ice and snow. A laboratory model of Schneider et al. (2011a) showed a reduction of the bulk friction angle is 13–50 percent for a mass of dry granular ice compared to a granular rock mass; and
4. the debris may be channelized or air-launched by moraines.

Evans and Clague (1999) later proposed that a significant volume of fine particles due to intense fragmentation of the source rock mass may enhance the mobility of a rock avalanche. It is not clear how this applies specifically to those descending onto glaciers. However, a significant difference between glacial and nonglacial rock avalanches relates to spreading. Most cases spread and thin, but Geertsema et al. (2006) suggested this is greater for travel over glaciers, a view supported by the fragmentation theory of Davies and McSaveney (1999). Compared to ice-free valleys in rugged terrain, glaciers tend to have more open surfaces allowing greater spread of the debris lobe and, consequently, producing much thinner debris sheets. Strom (2006) argued that, for unconfined rock avalanches, normalized debris deposit area is a better metric of mobility than linear travel distance, since friction acts over the entire basal surface. Recently, this has been demonstrated for a number of glacial and nonglacial rock avalanches (Shugar and Clague, 2011; Shugar et al., 2013a).

The high mobility of rock avalanches on glaciers causes enhanced velocities, generally >60–70 m/s (Sosio et al., 2012; Table 9.2).

9.2.3 Rock Avalanche Deposits and Sedimentary Properties

The character and fate of rock avalanche materials depend on original bedrock properties, the travel path, and where they come to rest (Table 9.3). On ice, subsequent modification occurs as part of supraglacial processes, including thermokarst, and final dispersal from, and deposition around the glacier. A critical distinction is between landslides that are wholly confined to a glacier surface, and those that travel over and beyond it. A suite of transitional situations may be involved. Where landslide material is encountered on ice, partial or intermediate rock avalanche properties may prevail. Deposits that are thick enough to prevent ablation may be passively transported. They can retain typical rock avalanche properties until finally released through thermokarst processes as the ice stagnates. However, dispersal to the margins increases moraine-like properties. Eventually some of the geomorphic and sedimentological diagnostics of rock avalanches may be lost.

If part or all the mass travels beyond the glacier terminus, deposits have the potential to survive in the landscape for millennia or longer. They may have some distinctive properties due to enhanced mobility and entrainment of snow

TABLE 9.2 Velocity of Rock Avalanches onto Glaciers

Rock Avalanche	U_{mean} (m/s)	U_{max} (m/s)	U at n km from Source (m/s)	References
1964 Sherman Glacier	26	67		Sosio et al. (2012)
1991 Aoraki/Mount Cook	55–58	100[1]		McSaveney (2002), Sosio et al. (2012)
1992 Mount Fletcher	80	120		McSaveney (2002)
2000 Tsar Mountain	22–45			Jiskoot (2011a)
2002 McGinnis Peak Glacier N			>54 at 4 km, >40 at 5 km	Jibson et al. (2006)
2002 Black Rapids Glacier E			>35 at 2 km	Jibson et al. (2006)
2002 Punta Thurwieser		63		Sosio et al. (2012)
2003 Iliamna Red Glacier		>80		Sosio et al. (2012)
2005 Mount Steller		100		Huggel et al. (2008)
2007 Mount Steele		65	>73 at 5 km	Lipovski et al. (2008)

[1]Modeled U_{max}.

and ice from the glacier. Materials emplaced along the ice margins may be incorporated into, or easily mistaken for, latero-terminal moraines. In general, materials emplaced beyond Neoglacial and, especially, Late-Glacial ice margins are disproportionately represented in the geological record. The incidence and full role of purely on-ice events can only be inferred.

For some time after initial emplacement, much or all of the material, even on the glacier, may retain distinctive rock avalanche properties in composition, facies, and assemblage morphology. Travel over the ice or entrainment of snow, ice, meltwater, and supraglacial debris may create some differences from nonglacial examples. In time, the materials undergo progressive modifications by supra- and englacial processes. On heavily crevassed surfaces, debris is irregularly emplaced and may create an envelope that is thicker over crevasses or depressions, and thinner over raised, convex, and smoother areas. McSaveney (1978) showed how such "wave crests and troughs" from the interaction of crevassed ice and rock avalanche trajectories, are amplified by subsequent differential ablation and ice movement (cf. Chillinji Glacier example in Section 9.3.2).

TABLE 9.3 Diagram of Four Emplacement Classes of Rock Avalanches Affecting Glaciers

	Origin		
Emplacement	On Slopes over Main Glacier	On Slopes over Tributary Glacier/s	On Slopes of Ice-Free Valleys
Over main glacier	Ia	Ib	—
Partially over main glacier or/and over its termini	Ib	Ib	III
In ice-free valleys with effect on main glacier	IIa	IIb	IV

Source: Modified after Hewitt (2009).

It is important to recognize the changing spatial dimensions and patterns of where, how quickly and in which parts of the on-ice deposit the modifications occur, including the role of past and contemporary changes in glacier mass balance. Large areas of pure, unmodified rock avalanche material may survive for decades, even as equally large volumes have been reworked and dispersed from other areas. What is found depends on the stage encountered. Eventually, on an active glacier, all the material will be transported to the margins and acquire the distinctive properties of ice-margin and proglacial deposition (Hewitt, 2009).

Where large quantities of moisture or wet sediment are absorbed, the dynamics and sediment properties of a rock avalanche may be transformed to those of debris avalanche, debris flow or more complex mass flows, whether deposited on or beyond the ice. Careful examination of these materials can, however, reveal sedimentary properties diagnostic of a rock avalanche origin, despite considerable alteration (Fauqué et al., 2009; Reznichenko et al., 2012).

9.2.3.1 Deposition onto Glacier Surface

The surface of rock avalanches is generally an openwork carapace of boulder-sized clasts (Figure 9.4(a)). This is, however, deceptive. Where the deposit is exposed below the carapace, it contains large quantities of finer-grained matrix materials (Figure 9.4(b)). The fragments consist of fractured (compressional) and shattered (impact) clasts, and pulverized fines, which form a matrix filling all the space between larger clasts (see below).

FIGURE 9.4   Deposit of the 1986 rock avalanches onto Bualtar Glacier, Karakoram. (a) Open-work carapace of boulder-sized clasts *(Photo: K. Hewitt)*; (b) cross section showing the large quantities of fine-grained matrix materials *(Photo: K. Hewitt)*.

Many researchers have qualitatively described geomorphic features of rock avalanche deposits, including raised rims, flow bands, and lithologic zonation. Little quantitative work, however, has been done to link debris sheet sedimentology to observed large-scale geormorphic features. Further, relatively little work has compared the sedimentary architecture of glacial and nonglacial rock avalanches, insofar as they relate to paleoclimatic reconstructions (e.g., Reznichenko et al., 2012; Shugar et al., 2013a). At Sherman Glacier, flow bands and subparallel grooves indicate the flow direction. At Black Rapids Glacier, the previously described longitudinal flow bands are alternating stripes of narrower fine debris and bands of coarse blocks on the rock avalanche surfaces. Flow bands at Black Rapids Glacier are characterized by differences in block (>1 m^2) size, with clusters of very large blocks (up to tens of meters long), especially in the distal reaches, separated by regions of much smaller blocks (Shugar and Clague, 2011). In comparison, the nonglacial 1903 Frank Slide (Canadian Rocky Mountains) exhibits no such flow bands, and the coarsest blocks are not concentrated in the distal reaches but in a band in the middle of the debris sheet (Shugar et al., 2013a).

9.2.3.1.1 Thickness of Rock Avalanche Deposits onto Glaciers

As a result of greater spreading across and down the glaciers, thicknesses of rock avalanche deposits are generally much less than those in nonglacial environments. In the Mont Blanc massif they are typically $1-5$ m, an order of magnitude thinner than most off-ice cases. In the Chugach Mountains, Uhlmann et al. (2013) mapped 123 supraglacial landslide deposits >0.1 km^2, and found thicknesses ranging from less than a meter to several meters. Deposits of the seven larger rock avalanches triggered by the 2002 earthquake in Alaska, whose volumes are in the range $4-20$ Mm^3, have an average thickness of $2-3$ m (Jibson et al., 2006; Shugar and Clague, 2011), whereas the deposits of the small Vampire rock avalanches of 2003 and 2008 onto Mueller Glacier ($0.12-0.15$ Mm^3) in New Zealand have an estimated average thickness of 0.5–1 m (Cox et al., 2008). At Jarvis Glacier (NW British Columbia), Evans and Clague (1999) observed average debris thickness of <0.5 m on a landslide triggered by the 1979 St Elias earthquake. At Bualtar Glacier, the debris thickness in 1987 was found to be quite variable, generally >2.5 m, in a few places over 10 m; exposures averaged between 3 and 5 m. After the 1991 Chillinji Glacier event, a few scattered vertical sections indicated similar thicknesses, averaging 3 m or more.

Thickness measurements are inherently difficult, and estimates are generally based on a few sampling pits or vertical sections revealed by erosion or ice collapse. For instance, McSaveney (2002) estimated that the 1991 Aoraki/ Mount Cook deposit was 1.2 m average, whereas 250-m-long ground-penetrating radar (GPR) profiling from one edge of the deposit in 2009 demonstrated a thickness of 5–10 m (Reznichenko et al., 2011). GPR profiling on the West Fork Robertson Glacier (Alaska) deposit shows a variable thickness of 1.5–5 m (S. Herreid, written communication). Reduced ablation below the debris sheet causes the debris surface to become elevated above the adjacent ice, giving the appearance of greater thickness; while, conversely, erosion of the ice surface by the avalanche may cause it to incise into the glacier surface, giving the appearance of lower thickness.

9.2.3.1.2 Morphology, Sedimentology, and Macrofabric of Rock Avalanche Carapace

The large clasts of a surface rock avalanche carapace can acquire distinctive patterns or fabric in at least four ways:

1. remnant lithological bands: in these bands, boulder sizes and shapes stand out due to strength, structural units, or partings in different bedrock. The largest boulders will derive from the more massive or resistant rocks. It is of some interest to identify the largest clasts where they are salient features, and record the survival and shape of large units. Their size, shape, and emergence above the surface could serve as indicators of the transporting competence of these events, hitherto little investigated (Hewitt, 2002);
2. preferential alignment of largest clasts subparallel or at right angles to the direction of movement: this suggests that the largest boulders are preferentially oriented with movement vectors (e.g., Shugar and Clague, 2011). Smaller, blade, or rod-shaped blocks appear jostled by interaction with the larger blocks and lack preferential orientation;
3. "final moment" developments: imbrication, collisional fractures, and torque are observed in the surface boulders suggesting their movement stops irregularly if quite suddenly (Figure 9.5). There may be detachment of individual boulders or parts of the surface boulder sheet, which continue to move after the main body has stopped, especially if the main body matrix materials become saturated with meltwater. This was observed in the Bualtar Glacier rock avalanches in 1986 where some surface areas were

FIGURE 9.5   Imbrication in the surface boulders of Bualtar Glacier rock avalanche deposit of 1986, suggesting their movement stops irregularly if quite suddenly *(Photo: K. Hewitt).*

left with few or no boulders, while concentrations of imbricated and jumbled boulders occurred downslide;

4. distal rim concentrations of largest boulders: this may be due to final moment detachment just mentioned. Possibly, blocks at the outer surface and lower, outer margins of the original failed rock slope escape the heaviest crushing forces in the initial descent, and maintain their outermost positions in the rock avalanche.

9.2.3.1.3 Matrix Particle-Size Distribution

The main body of a rock avalanche deposit generally contains a full spectrum of grain sizes from megaclasts (boulders exceeding 10 m diameter) to fine silts and clays. Grain-size distributions for matrix materials approximate a linear plot on a log-normal graph (Figure 9.6). Observations in the Karakoram indicate more than half the volume of rock avalanche deposits on glaciers usually lie in the range granule to fine-silt size grades. The Wentworth "sand-sized" fraction is the larger part of the matrix (Hewitt, 2002). Absence of clay-sized material may not only be due to little or no production. The moving mass may be sufficiently dilated for it to be dispersed and expelled with the compressed air to help create the great dust clouds observed in these events. Grain-size analyses of the 1991 Aoraki/Mount Cook rock avalanche sediment the day after the event showed 99.5 percent of the sampled fragments (by number) were <10 μm in diameter (McSaveney and Davies, 2007). A significant presence of fine particles was found in repeat studies nearly two decades later (Reznichenko et al., 2012). Grain-size analyses for the rock avalanche matrix on Black Rapids and Sherman Glaciers (Shugar and Clague, 2011) indicate predominant muddy sandy

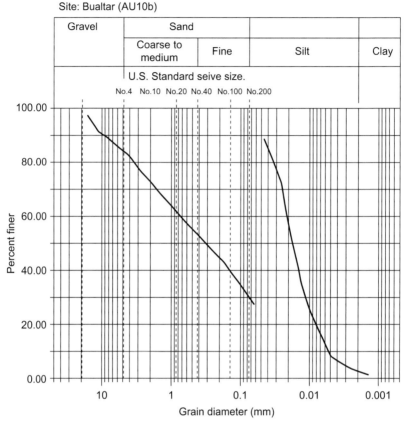

FIGURE 9.6   Grain-size curves for matrix material from Bualtar Glacier rock avalanche deposit of 1986. This sample is representative of a hundred samples, all very similar.

gravels, and generally less than 5 percent by mass of sediment <10 μm. The rock avalanche matrix samples from Black Rapids Glacier are significantly more poorly sorted than those from Sherman Glacier. One sample from a medial moraine on Black Rapids Glacier had c. 20 percent sediment <10 μm. This reported inconsistency of clay content in rock avalanche material may result from differences in the proportion of silt/clay size clasts to other size fractions varying between rock avalanches depending on the lithology of the parent rock, the volume of rock avalanche and its runout distance. Because in conventional grain-size measurements, it is rare to recognize grain sizes as small as <10 μm (Crosta et al., 2007), in many cases very fine particles adhere to the surfaces of coarser clasts or bond together into agglomerates, and thus may not be recognized as individual grains although formed during the event.

Reznichenko et al. (2012) developed a method to differentiate rock avalanche material from glacial sediment, using the presence of characteristic micron-scale features, missed by traditional grain-size analyses by sieve but seen under a scanning electron microscope (SEM). Silt- and clay-sized grains from rock avalanche sediment are typically agglomerates of numerous finer particles that survive the washing and dispersion of conventional grain-size procedures such as sieve analysis. The clasts appear angular and subangular, identical to parent material lithology, and clumped together by finer matrix of submicron size. This results from highly confined and nondispersed fragmentation of grains in the deposit during rock avalanche emplacement. The main point is that these agglomerates are entirely absent from glacial sediments: under SEM observations fine grains from glacial environments are solid and do not form clusters.

9.2.3.2 Postdepositional Modifications of Rock Avalanche Deposits on Glaciers

Rock avalanche deposits on large glaciers can survive supraglacially for decades and with little or no discernible change in the core areas. Where not otherwise disturbed, the whole deposit will gradually come to sit on a platform of ice 5–50 m thick, beneath whatever depths the landslide materials have. The apparent thickness of on-ice rock avalanche deposits observed after one or more ablation seasons, or in satellite images, must be treated with caution. The rate at which it can be raised is governed by absent or much lower ablation below the deposit compared to surrounding glacier surfaces (e.g., Reznichenko et al., 2010). However, thickening due to ablation protection generally seems to reach a limit at around 15–20 m suggesting that it is then compensated by increased ice flow at depth from beneath the raised areas (Hewitt, 2009). Differential ablation can also result in a hummocky topography and supraglacial thermokarst lakes with debris continuously reworked by relief inversion and backwasting.

9.2.3.2.1 Reworking of Rock Avalanche Debris

Rock avalanche deposits on a glacier are subject to a range of processes that can modify their composition, facies properties, and coherence. Ice movement, compression, divergence, and crevassing will disturb the debris sheet. Ice ablation can become the major modifier, especially at deposit margins and wherever ice becomes exposed or the debris thinned so that increased ablation causes local disturbance. Wind as well as meltwater winnows out the finer materials. Surface drainage can partially wash finer materials away, leading to coarsening and sorting of the residual deposits. However, these modifications are not necessarily rapid or complete. Transport on Tasman Glacier, and exposure to rain and snowmelt had not significantly winnowed the finer matrix noted in the 1991 Aoraki/Mount Cook deposit by Reznichenko et al. (2012; cf. 1.3.1.3).

Coarser fragments can be redistributed, comminuted, and crudely rounded by sliding and fallsorting, especially due to ablation-driven relief inversions. Freeze-thaw cycles, crushing between boulders, and splitting of large fractured boulders during supraglacial displacement, cause debris comminution, and lead to greater compactness. Material washed into on-ice ponds will become crudely sorted and stratified, but coarse blocks also tumble in to disturb and complicate grain-size distribution. On Black Rapids Glacier, Shugar and Clague (2011) noted dozens of large quartz diorite boulders that had been turned into sands and gravels in only 5 years of freeze-thaw activity.

Meanwhile, "normal" glacier processes continue and glacially derived debris becomes mixed with rock avalanche materials. They become mixed or interspersed with supraglacial moraines derived from vertical emergence of englacial debris septa (Eyles and Rogerson, 1978). Glacier tables can develop where rock avalanche boulders are or become dispersed away from the continuous deposit, but are more common in blocks fall-sorted along the margins of medial moraines, or from individual rockfalls.

9.2.3.2.2 Modification of the Pattern of Supraglacial Deposits

Eventually, rock avalanche deposits may be so modified as to resemble or be difficult to distinguish from other heavy debris covers. However, debris from other sources is rarely either as thick or monolithologic. Rock avalanche deposits also depart from the tendency of supraglacial moraines to show longitudinal and transverse gradients of debris thickness.

At Sherman Glacier for example, five decades of supraglacial transport have altered the morphology of the deposit. The flow bands created by shear during emplacement of the debris in 1964 now serve as passive indicators of differential ice velocity, and form arcuate lines, which are convex downglacier (Shugar and Clague, 2011). Fabric measurements in 2008 showed that the debris has been reoriented to reflect this post-depositional shear.

It can happen that several rock avalanche deposits are merged in a continuous supraglacial debris cover. The debris cover of the Miage Glacier (Mont Blanc massif) consists of several dozen morpholithological units resulting from rockfalls or rock avalanches. Units defined by lithology, fabric, and particle size, roundness and distribution, have areas in the range 5,000–300,000 m^2. Two types can be distinguished (Figure 9.7). Flow-parallel units (e.g., UD7, UD3) are stripes derived from rock avalanches and rockfalls onto the glacier accumulation area transported englacially, with some emerging at the glacier surface as debris septa. A straight stripe is associated with a short englacial transport path from emplacement close to the equilibrium line altitude (e.g., UD7). A curved stripe results from a longer englacial transport path (e.g., UD3a–c). The second type comprises irregular units (e.g., UD1) from rock avalanche and rockfall deposited in the

FIGURE 9.7 Morpholithological units of the debris cover of the Miage Glacier, Mont Blanc massif, in 1997, resulting from rockfalls and rock avalanches. 1, Moraine crest; 2, 3, megaboulders; 4, debris septum; 5, area not mapped.

ablation zone. They have been transported entirely supraglacially, and have irregular shapes independent of the flow, but have sharp limits, homogeneity, and a hummocky topography due to numerous megablocks. At Miage, three irregular units correspond to dated rock avalanches that occurred during the twentieth century (Figure 9.7). Large, flow-parallel units (e.g., UA1 or UD2) record earlier rock avalanches that affected the accumulation zone.

9.2.3.3 Deposition of Rock Avalanches outside the Glacier

Because of long runout, many rock avalanches may travel over and beyond smaller valley glaciers and sometimes even quite large ones. Dozens of examples are known in the Andes, the Caucasus, Pamir, Karakoram, Hindu Kush, and Greater Himalayan ranges, that spread far down river valleys below the ice; the longest so far identified has 28 km runout, at least half of which is beyond the glacier (Robinson et al., 2014). Those crossing glacier margins or valley junctions below, form some of the thickest deposits (>100 m) due to blocking and stalling of rock avalanche lobes against opposing slopes (Hewitt, 2006). Since these developments are more likely to pose serious hazards for human communities and infrastructure, the ability to reconstruct their role in the Holocene, at least, is a key to predicting future risk.

Deposits outside glaciers tend to have much less regular and thicker masses of debris, which may alternate with debris veneers. In humid mountains they can become obscured by vegetation after a few years. Ridges are common in mixed rock/ice deposits, but generally absent in the final debris deposit whose margins are usually raised and marked by a small talus. Finally, melting of ice and snow can result in a chaotic topography and chaotic sorting of debris. Clasts are mostly angular and subangular, but glacially rounded boulders entrained from moraines or from the proglacial margin are common. Smaller clasts are left piled and plastered over the larger boulders by ice melting. The silty-sand matrix may be reworked and flushed through the deposit.

Rock avalanche deposits outside a glacier may be mixed with those from other processes, and their morphology has led to confusing them especially with glacigenic or debris flow landforms and sediments. A critical example is in Val Ferret (Mont Blanc massif), whose morphology and composition led to prolonged discussion about the origin. There is an assemblage of morphologically distinct sectors with granite boulders covering a 2-km-long and 500 m-wide plain downstream of the moraine complex of the Triolet Glacier. In a topography dominated by hummocks and hollows are regions with chaotic blocks up to 1,000 m^3 in size, and with an openwork structure. There are 1–5 m-high concentric ridges consisting of matrix-supported diamicton, with megablocks on the crest. The deposits were reworked or partly buried by alluvial fill, late-LIA moraines, and polygenic debris cones supplied by runoff, debris flows, snow avalanches, and rockfall (Deline and Kirkbride, 2009).

Although de Saussure (1786) initially described a rock avalanche deposit on Triolet Glacier, many authors have since interpreted the entire deposit, or parts of it, variously as sets of moraines (Agassiz, 1845; Sacco, 1918; Zienert, 1965; Mayr, 1969; Aeschlimann, 1983), or glacial outburst flood deposits (Virgilio, 1883). Porter and Orombelli (1980) and Orombelli and Porter (1988) were the first to ascribe the entire Val Ferret deposit to the CE 1717 rock avalanche. This was discussed by Deline and Kirkbride (2009) but confirmed by Akçar et al. (2012, 2014) using geomorphic mapping and radiocarbon dating, and cosmogenic ^{10}Be dating, respectively.

Glaciers may also readvance over rock avalanche deposits that act as barriers early in the readvance. If not fully reworked and dispersed, a complex, "trans-glacial" geomorphology and sedimentology result (Iturrizaga, 2006; Hewitt, 2013b). An example is the White Horse complex at the terminus of Mueller Glacier, Southern Alps (NZ). The deposit has a large debris nucleus most probably of rock avalanche origin that has locally affected glacier readvances during the last couple of thousand years. As a result, the glacier has partly overridden it and been deflected around its moraines. Similarly, Cook et al. (2013) describe an advance of Fee Glacier, Switzerland, over a rock avalanche deposit. This small glacier, with limited ice flow and erosion potential, was not able to rework the rock avalanche deposit, which resulted in partial preservation of the deposit's characteristic features, including brecciation, a coarse carapace, angular clasts, and hummocky topography. These deposits in the proglacial valley affect glacier readvance, slow its progress and may help form another heterogeneous, or trans-glacial deposit.

9.3 CONSEQUENCES

9.3.1 Rock Avalanche Contribution to Supraglacial Debris Covers

Rock avalanches are just one, instantaneously large but relatively infrequent, source of material to mountain glacier surfaces. Other more common and widespread processes include debris delivered in snow and ice avalanches; rockfall and debris flows; and erosion along proximal sides of lateral moraines by gullying, wind action, boulder fall, and slide. The latter tend to contribute more where the glacier surface is lowered. Debris incorporated in accumulation zones emerges at the ice surface downglacier from sub- and englacial transport paths either along shear zones or from ablation and in vertical debris septa, or by exposure of englacial channels. The proportion delivered by each source depends upon terrain steepness, geologic, geomorphic, glaciological, and climatic conditions.

Diagnostics of morphology and sedimentology of supraglacial rock avalanche deposits developed in recent years help make these sources more readily discriminated. The debris sheet is usually some meters thick, lies on a raised platform of ice up to tens of meters thick, and is composed of coarse material capping a matrix of finely pulverized material. It has a relatively high bulk density compared with other supraglacial debris. Samples reveal the monolithological and unsorted angular clasts in all size fractions (Dunning, 2004; McSaveney and Davies, 2007; Hewitt et al., 2008; Reznichenko et al., 2012). Non-rock-avalanche sourced, supraglacial sediment has not undergone such intense fracturing and pulverizing so is generally coarser (Reznichenko et al., 2012), but also lacks the very large boulders characteristically associated with rock avalanches.

The sheer volume and physical properties of rock avalanche sediments help to discriminate them from more commonly observed supraglacial debris and materials from other subaerial sources. A rock avalanche immediately contributes an enormous mass of sediment (up to several km^3). Although relatively thin compared to rock avalanche deposits off-ice, the on-ice component is usually much thicker than even the heaviest, "normal" supraglacial debris. It can remain an area of conspicuously thicker debris during transport toward the glacial terminus. By comparison the more or less continuous exhumation of "normal" meltout debris annually or even over some decades adds very little and the cover may change little (Hewitt, 2009). Snow avalanche debris, boulder falls and most rockfalls are several orders of magnitude smaller than rock avalanches, and their individual clasts or blocks move largely in isolation from each other (Fort et al., 2009). In high mountain basins, the quasi-constant supply of material that results in "debris-covered glaciers" rarely exceeds an average thickness greater than 0.5–1 m, increasing gradually but systematically toward the glacier terminus (Nakawo et al., 2000). The debris contribution to the supraglacial cover originating from glacial erosion is estimated to be less than 1 percent of the volume of the ice melted in the ablation zone.

The 1991 Aoraki/Mount Cook rock avalanche deposit was on the order of 5–10 m thick (Reznichenko et al., 2011). Sediment delivery onto the Tasman Glacier from this one event exceeded 100 years of "normal" glacial sediment budget from all other processes.

The rock avalanche deposited on the Lyell Glacier, South Georgia, in 1975 is estimated to contribute at least 93 years' worth of sediment from subaerial erosion (Gordon et al., 1978). At Bualtar Glacier in the Karakoram, transport of the 1986 rock avalanche debris to the terminus in 30 years, including two surges, was equivalent to 500 years of supraglacial transport based on the prelandslide supraglacial cover and movement (Hewitt, 2009). At Miage Glacier, over 75 percent of debris derives from rockfall and rock avalanche. The lower Brenva Glacier was more than 90 percent covered by the 1997 rock avalanche (Deline, 2009).

Even with a low frequency, and certainly with one large event per century as indicated for many of the glacier basins identified here, rock avalanche inputs can still hugely increase the total sediment delivery. Also, for mountain glaciers, the substantial amounts of fines delivered almost instantaneously challenge the view that fine glacial sediment comes only from basal crushing and grinding by ice (Boulton, 1978; Davies, 2013).

9.3.2 Glacier Dynamics in Relation to Rock Avalanche Deposits

On-ice debris sheets can reduce or prevent ablation for years or decades and bring glacier thickening and advances out-of-phase with climatic conditions and surrounding glaciers (Shulmeister et al., 2009; Reznichenko et al., 2010, 2012). Unless recognized, these compromise glacial chronologies presumed to record only climatic fluctuations, especially in high mountains where massive rock slope failures are relatively frequent (Deline, 2009; Kirkbride and Winkler, 2012; Hewitt, 2013a).

9.3.2.1 Glacier Advance and Velocity Change due to Rock Avalanches

During times of climate warming, termini of debris-covered glaciers retreat less than those of clean glaciers. In the Mont Blanc massif, Mer de Glace (mainly clean) and Miage Glacier (debris-covered) have retreated 2400 and 300 m, respectively, since the 1820s LIA maximum (Figure 9.8). Conversely, advances of debris-covered glaciers last longer or reach farther downvalley. Triolet Glacier experienced a long-term advance after the CE 1717 rock avalanche: several decades later, de Saussure (1786) reported that it had been advancing for at least 8 years whereas the neighboring Pré-de-Bard Glacier was retreating. Triolet supraglacial debris cover, ascribed by Saussure to the rock avalanche, prevented ablation over much of the glacier, causing it to advance beyond the climatically controlled LIA limit. Although both glaciers currently have a similar area (c. 3 km^2) at a comparable elevation, the Triolet—which has not been debris-covered for about one century—has its current terminus standing c. 500 m higher than that of Pré-de-Bard although their LIA termini were at a similar elevation. Since then, the former retreated by 3 km but the latter by only 1.5 km (Figure 9.9: inset).

   In 1913, Brenva Glacier began advancing, and by 1919−1920, was creeping forward at a rate of 20−25 m/a. In 1920, a rock avalanche of >2.4 Mm^3 of rock and >7.5 Mm^3 of ice covered the whole lower glacier and part of the upper glacier (Deline, 2009). Glacier advance increased to 43 m/a in 1920−1924, peaking at 55 m/a between April 1922 and June 1923 (Valbusa, 1924). The glacier advanced a total of 490 m between 1920 and 1941, whereas neighboring glaciers in the Mont Blanc massif retreated from the mid-1920s (Figure 9.8). Brenva Glacier has been retreating since 1989. From an average of 24 m/a between 1993 and 1996, the retreat rate decreased after the 1997 rock avalanche to 12 m/a in 1999−2001, with a possible readvance in 2008−2009 (Imhof, 2010). Thus, twentieth century rock avalanche deposits sped up the 1920s Brenva Glacier advance, and slowed retreat since the 1990s.

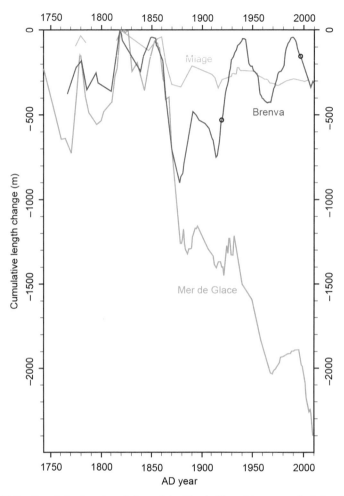

FIGURE 9.8   Cumulative length variations of the Mer de Glace (clean-type glacier), and Miage and Brenva Glaciers (debris-covered glaciers) since the end of the eighteenth century. Black circles: Brenva rock avalanches in 1920 and 1997. *After Imhof (2010), Nussbaumer et al. (2007), Vincent et al. (2012).*

Within 15 years of the 1964 rock avalanche at Sherman Glacier, Alaska, the debris sheet's distal edge had moved downglacier to the terminus. The landslide also caused the glacier to begin advancing (Figure 9.10). Between 1964 and 2011, the glacier advanced c. 0.5 km (Shugar et al., 2013b). At Black Rapids Glacier, Shugar et al. (2012) used satellite radar and ground surveying to measure surface ice velocity in the vicinity of the three 2002 rock avalanche deposits. It increased 44 percent within 2 years of the landslides, and then decreased to about the pre-2002 velocity. Interestingly, the velocity gradient, expressed as the difference between the velocity at the upstream and

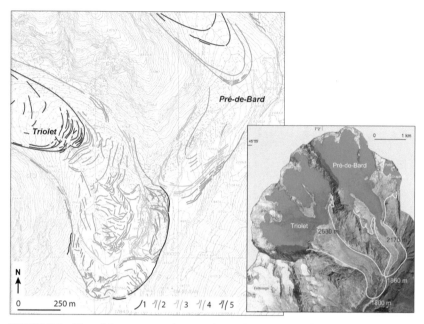

FIGURE 9.9 Map of the recent moraine complexes of Triolet and Pré-de-Bard Glaciers, Mont Blanc massif. Moraine ages: 1, possibly ante-CE 1717; 2, Triolet: post-CE 1717; Pré-de-Bard: seventeenth to eighteenth century; 3, CE 1820–1850; 4, CE 1860–1920; 5, 1920s. Inset: 2005 (blue area) and Little Ice Age (yellow line) extents of Triolet and Pré-de-Bard, with corresponding terminus elevation (Deline and Kirkbride, 2009 modified).

downstream ends of the debris sheet, dropped nearly to zero: in other words, in the vicinity of the landslide debris, the glacier was moving nearly uniformly. To confirm whether this significant change was triggered by the landslide, Shugar et al. (2012) used a Full Stokes 2D finite element model to elucidate the ice dynamics response to the debris sheet. The model reproduced the observed velocity gradient reduction, suggesting that landslide-induced mass balance changes were the cause. Changes in ice surface slope, resulting from reduced ablation under the debris and greater emergence velocities at the downglacier than at the upglacier end, pivoted the glacier surface toward the horizontal, resulting in a uniform longitudinal velocity profile. Black Rapids Glacier thickness is on the order of 600 m (Heinrichs et al., 1995), with the landslide debris just 2–3 m thick.

Chillinji Glacier in the Karakoram has had a somewhat similar response to Black Rapids since the 1991 rock avalanche (see Section 9.3.2). However, within a few months of the rock avalanche emplacements on their surfaces, Lokpar–Aling and Bualtar Glaciers went into surge-type accelerations 10 times faster than prelandslide velocities and massively disturbed the glacier tongues and ice-margin conditions (Hewitt, 2009).

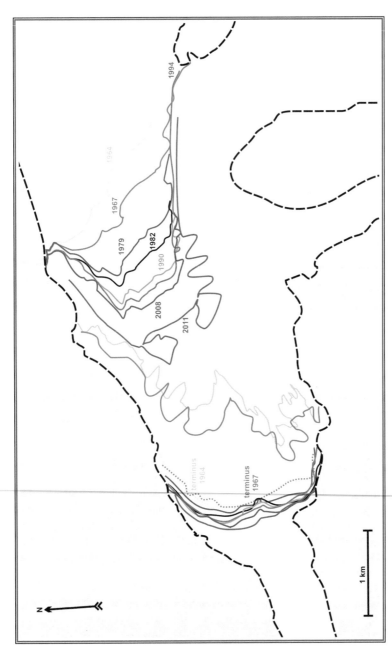

FIGURE 9.10 Changes of the rock avalanche deposit transportation since its 1964 supraglacial deposition onto the Sherman Glacier, Alaska, obtained from satellite and aerial images. During first 15 years the deposit distal edge reached glacier terminus and since caused glacier to start its readvance, while the upper edge moved >2 km downvalley, modified by the differential surface ice flow.

FIGURE 9.11   A 2-m-thick sheet of the 2002 rock avalanche debris overlie a 15-m-high ice platform in the ablation zone of Black Rapids Glacier, Alaska *(Photo: J.J. Clague)*.

9.3.2.2 Reduced Ablation due to Rock Avalanche Deposits

Debris cover on a glacier is an effective insulator that can substantially influence the ablation component of mass balance. Debris thickness and thermal conductivity together determine the thermal gradient, and impact on ablation rates. Debris thickness has the largest effect, hence the exceptional potential for rock avalanches to influence this factor. A 20 cm debris cover may reduce daily ablation by six times compared to clean ice and a meter or two shut it down altogether (Nakawo and Rana, 1999; Reznichenko et al., 2010). Debris-draped ice platforms may increase in thickness by several meters per year and reach heights of many tens of meters (Figure 9.11). Undisturbed rock avalanche sheets are near-perfect insulators under cyclic thermal inputs (Reznichenko et al., 2010). Thermal conductivity is important in normal debris covers and degraded landslide covers; in addition to the effect of thickness, thermal conductivity can vary with the mineralogy, porosity, and moisture content of the debris, its albedo, and water and air circulation (Mihalcea et al., 2007).

9.3.2.3 Effect of Load Increase and Subglacial Drainage Change

It seems unlikely that simple loading will significantly affect glacier dynamics, as the thickness of rock avalanche deposits on glaciers is limited by the increased mobility imparted by travel over ice—deposits >5-m-thick on average are generally associated with very rough, resistant substrates or stalling against an opposite slope. In the Karakoram for example, the thickness of ice protected from ablation in 4−6 weeks of summer by rock avalanche debris exceeds the landslide mass and thickness (Hewitt, 2009b). No known rock avalanche has been as thick or as great in mass as the brittle upper zone of ice that loads the underlying part where creep deformation occurs, so it is a small addition to the loading of the latter. On relatively thin glaciers, shock loading from rock avalanches may be significant—perhaps causing catastrophic collapse and rock—snow—ice avalanches (cf. Chapter 7 in this volume, and thereafter)—but, to date, no studies have examined this. The frictionally heated rock mass may generate a sudden increase of meltwater in the cold season, at least, but such an effect also awaits research.

Some have thought that the 2002 Kolka—Karmadon event, which claimed 125 lives in the Genaldon valley in North Ossetia, Caucasus, was a catastrophic effect of mass load increase. On September 20th the glacier detached catastrophically from its bed, and over 130 Mm^3 of ice and rock swept downvalley 19 km, then transforming into a mudflow (or debris flow) that traveled a further 15 km (Huggel et al., 2005; Evans et al., 2009b; Sosio, in this volume). The event is not unique in Kolka Glacier history. In July 1902, the glacier underwent a two-stage detachment, the resulting mass flow traveling 11 km and killing 36 people. Three hypotheses have been proposed to explain these events:

1. The glacier is surge-type. Kotlyakov et al. (2004a) proposed a 2002 surge may have resulted from high water pressures within and beneath the glacier. However, when it surged in 1969—1970, the glacier remained connected to its bed. Further, there was no indication of a surge during the summer of 2002 (Huggel et al., 2005).
2. A detachment triggered by the impact of a collapsing mass of 20 Mm^3 of ice and rock onto the glacier (Huggel et al., 2005). Kolka Glacier was relatively thin in 2002—ranging from 85 to 175 m. If a large rock avalanche, on the order of 10 Mm^3 as reported by Huggel et al. (2005) descended onto the glacier, conceivably large glaciological changes would occur. However, Evans et al. (2009b) could find no firm evidence for such a rock avalanche.
3. Load increase by ice and rock deposits due to repeated collapses (15 Mm^3). Sporadic but significant snow avalanches, rock and icefalls went onto the surface of Kolka Glacier through the preceding summer months. It has been suggested this was enough to disrupt drainage within the glacier, and generate high subglacial water pressures. Enhanced geothermal activity was another possibility. Decreased effective stress and frictional resistance may have detached the glacier at the bed (Evans et al., 2009b).

Among the hypotheses that have been proposed to explain the almost complete erosion/entrainment of Kolka Glacier, some favor a model where factors affecting basal and englacial water seem critical. The highest levels of subglacier water pressure and destabilized mass may have involved a combination of pre-event conditions, such as higher rate of precipitation and summer ablation 1.5 to 2 times higher than average (Kotlyakov et al., 2004b), and the presence of hot geothermal springs close to the glacier (Muravyev, 2004). Documented seismic and volcanic gas activity prior to the event, and eyewitness reports of explosive activity under the Kolka Glacier during the event, support the view that geothermal activity contributed. Loading by ice-rock debris may have been a factor further generating excess water pressure within and beneath the glacier (Fountain et al., 2005), perhaps causing a dramatic reduction of the frictional resistance that triggered the catastrophic failure (Evans et al., 2009b).

Bualtar was a known surge-type glacier before the 1986 landslides occurred but the landslide debris mass was less than annual ablation losses making it unlikely that loading was a factor. Massive injection of meltwater caused by the frictionally warmed debris is a possible factor—especially if the glacier was close to the surge threshold (Gardner and Hewitt, 1990). In all, these relations and mechanisms are poorly understood.

9.3.3 Moraine Complexes

High mountain glaciers, especially debris-covered ones and those that surge, often construct atypical moraine complexes, i.e., complexes with a disorderly pattern and/or at least partly disconnected from the regional climatic signal (Benn et al., 2003; Larsen et al., 2005; Hewitt, 2013b). When debris is supplied by rock avalanches even more complex developments occur, at odds with commonly inferred climate-driven glaciation and sedimentology.

9.3.3.1 Atypical Moraine Complexes and Implications for Paleo-Glacial Sequences/Reconstruction

Some 60 rock avalanches descended onto glaciers in the 1964 Alaskan earthquake. Dozens of rock avalanches are known to have affected the glaciers in the Karakoram Himalaya, possibly several thousand events in the Holocene alone (Hewitt et al., 2011b). At a more local scale, more than a dozen Holocene rock avalanches have occurred onto the Brenva Glacier below the Mont Blanc summit. These challenge the assumption that any major thickening, advance or moraine positions generally reflects regional climate cooling events, and related mass balance changes (Kirkbride and Winkler, 2012). They suggest the rock avalanches are recurrent, geologically "normal" processes in these, and perhaps most, high mountains. Major questions arise for glacial chronologies that fail to recognize and differentiate landslide-related episodes and deposits. This is compounded where surge-type instabilities are common, as in the Southern Andean, Alaska—Yukon, Caucasus, Pamirs, and Karakoram Glaciers. Moraine evidence for the major glaciations and longer-term climate-related glacial stages should be reliable but not, perhaps, their details.

Such problems have been identified with the Late-glacial Waiho Loop moraine in New Zealand, and debate about whether glaciers in the Southern Alps advanced during the last glacial—interglacial transition about 12—13 kyr BP, or "Younger Dryas" (e.g., Denton and Hendy, 1994; Barrows et al., 2008;

Applegate et al., 2008). A recently reinterpreted Antarctic Cold Reversal about
13 kyr BP is also involved (Putnam et al., 2010). A detailed sedimentological
and morphological examination of the moraine shows that the Waiho Loop
records the impact of a large, early Holocene rock avalanche on glacial
mass balance and dynamics (Tovar et al., 2008; Shulmeister et al., 2009;
Reznichenko et al., 2012). Given the typical arcuate terminal moraine shape of
the Waiho Loop, Shulmeister et al. (2009) suggest it was emplaced by a Franz
Josef Glacier advance, in turn due to a supraglacial rock avalanche. Using a
one-dimensional numerical ice-flow model, Vacco et al. (2010) confirmed that
this is plausible. Deposition of an extensive blanket of ablation moraine to its
rear was subsequently confirmed at the base of the moraine and under the
Tatare River gravels (Shulmeister et al., 2010). Geological and geochronological evidence (Wardle, 1978; Denton and Hendy, 1994; Tovar et al., 2008)
suggests the glacier terminus was located between Canavan's Knob and the
Waiho Loop when the rock avalanche occurred, requiring an advance of 3 km
or less. The reduced ablation rate caused by the ≥ 100 Mm^3 of estimated debris
is considered sufficient to affect mass balance enough to form the Waiho
moraine. More recent work (Alexander et al., 2014) suggests that in order to
form the uniquely high and steep moraine, the glacier terminus must have been
at the Loop position when the rock avalanche occurred; thus no advance would
have been involved, but instead a prolonged terminus still stand allowed the
advection of a large quantity of supraglacial debris to form a single arcuate
location.

Most of the Triolet moraine complex comprises a discontinuous and
disorderly pattern of short ridges, unlike the few subconcentric moraines of the
neighboring Pré-de-Bard foreland (Figure 9.9). The former relate to a heavy
supraglacial debris cover on the Triolet from the 1717 rock avalanche. The
post-1717 advance thus triggered was followed by a slow retreat over two
centuries. A chaotic disintegration of the stagnating, debris-laden terminus was
interspersed by minor glacial advances until the 1920s (Figure 9.9). Finally,
debris-covered dead ice persisted upstream of the 1920s moraines from 1935
to the 1980s after the separation of the valley tongue from the active glacier
(Deline and Kirkbride, 2009).

In these and other cases, the misinterpretation of moraine patterns resulting
from mass balance alteration by rock avalanches as due to climatic signals
raises doubts about glaciological, and hence, palaeoclimate reconstructions, at
least those based on valley glaciers. Many rock avalanche deposits have been
misidentified as glacial due to the morphological similarity and proximal
position. Some sediment characteristics of rock avalanche deposits (fragmented mass of angular to very angular clasts and usually monolithology) are
readily confused with supraglacial moraine in lithologically unvarying
catchments. Diagnostic techniques are required to reliably distinguish rock
avalanche affected moraines, such as that developed by Reznichenko et al.
(2012) using the SEM (cf. 1.3.1.3).

9.3.4 Postlandslide Developments and Hazards

Rock avalanches onto glaciers can trigger or eventually lead to other dangerous consequences. These include glacier advances (sometimes sudden) and large dambreak and outburst floods from lateral-margin, englacial, or proglacial lakes. In a few cases they may trigger or accelerate the incidence of glacier surges. One surge immediately followed a rock avalanche onto the Lokpar Glacier tributary (Karakoram) in 1990. It disturbed and released an ice-margin lake that had existed for a century or more; the glacial lake outburst floods (GLOF) destroyed a summer village, forest and pasturelands down the valley (Hewitt, 2009). The main Aling glacier advanced up to 3 km in a decade. Many old rock avalanche deposits in the NW Himalaya and Karakoram have also become sources of repeated debris flows in rainstorms and heavy snowmelt, and include places where the highways are most commonly blocked by them.

The most extreme chain of high magnitude processes has been called a "catastrophic glacier multiphase feature" by Petrakov et al. (2008). It can involve a large ice-rock or rock avalanche traveling onto a glacier that transforms into a catastrophic mass flow (debris avalanche or flow) whose momentum and rheology allow it to continue far beyond the source slope and glacier margins. It may also generate a rapid glacier response due to insulating supraglacial debris deposit and glacier drainage changes. The best-known examples with exceptional runout and megaslide dimensions have continued far downriver systems into inhabited areas where they can cause great devastation. They exhibit greater velocities and longer runout than pure rock avalanches, commonly with massive entrainment of materials and progressive fluidization along the path. Some of the best-known cases originated in combined massive rock slope failures and large ice collapses (e.g., Nevados Huascaran, 1971) or impact and travel over glaciers typically destabilizing them (e.g., Kolka-Karmadon, 2002), and/or incorporation of vast quantities of snow, ice, and moraine (Horcones, below).

In the upper Indus basin more than 15 prehistoric supraglacial rock avalanches have been identified that traveled far into fluvial valleys downstream. They include particularly deceptive ones that emplaced tens of meters of erosion-resistant hummocky, mixed debris in the lower part, damming valleys in most cases (Hewitt, 2006). These landslide dams can affect a much greater area than the landslide itself through inundation, effects on fluvial sedimentation, or through a subsequent dambreak flood. The outburst floods may be much larger, and duration of geomorphic control can be much longer than in ice or moraine-ice dams caused by glaciers.

9.4 CASE STUDIES

The systematic overview of the topic given above needs, finally, to be filled out by actual events, and a sense of the details and complexities that enter into them. We choose three here; two recent ones from the Southern Alps of

New Zealand and the Karakoram Himalaya, and a prehistoric case in the Argentinian Andes, reconstructed entirely from deposits and the landform legacy.

9.4.1 Recent Rock Avalanches onto Glaciers in Aoraki/Mount Cook Area, New Zealand

Numerous small and several large rock avalanches have recently descended onto glaciers in Aoraki/Mount Cook area, Southern Alps, New Zealand. Their frequency has varied from one in every ten to hundred years (Cox et al., 2008; Allen et al., 2011). In the past 25 years the largest originated from the High Peak of Mount Cook (3,754 m a.s.l.) on December 14, 1991. It traveled across the Grand Plateau, down the Hochstetter icefall and spread over the Tasman Glacier. The $11.8 \pm 2.4 \, \mathrm{Mm}^3$ mass of rock (96 percent) and ice/snow (4 percent) fell 2,720 m and traveled about 7.5 km in 2 min; an average speed of 60 m/s. During the travel over the Grand Plateau and the Hochstetter icefall the deposit incorporated a large amount of ice, increasing the ice content more than 10-fold (50 percent by mass) (McSaveney, 2002). The landslide generated a magnitude M 3.9 earthquake (McSaveney, 2002). By 2011 the Hochstetter ice stream had deformed and separated the deposit into two main parts. An almost stagnant northern-eastern part traveled supraglacially about 500 m. The south part was rapidly modified and carried downvalley by the Hochstetter ice stream (Figure 9.12(a) and (b)). The deposit, up to 10 m thick, almost eliminated underlying ice ablation and caused the ice to thicken more than 30 m (Reznichenko et al., 2011). The debris volume was roughly equivalent to 180 years of average prelandslide debris flux through the icefall (Kirkbride and Sugden, 1992). However, the rock avalanche deposit covered less than 4 percent of the glacier ablation zone, too little to substantially influence flow velocities or mass balance (Reznichenko et al., 2011). The Tasman Glacier downvalley of the deposit remains in a state of rapid downwasting and retreat through proglacial lake calving and thermokarst development (Hochstein et al., 1995).

In the last 150 years at least four other massive rock slope failures have descended from the Aoraki/Mount Cook range. In 1873, a rock avalanche from Aoraki/Mount Cook (Barff, 1873), similar to or larger than the 1991 event, descended onto the Hooker Glacier, on the west side of the range. Smaller events occurred at Mount Vancouver (c. 3,300 m a.s.l.) in 1974 and Anzac Peak (c. 2,530 m a.s.l.) in 1991 (McSaveney, 2002). In 2013 c. 1.5 Mm³ rock avalanche fell 500 m from Mount Dixon (c. 3,000 m a.s.l.) and covered about 1 km² of the Grand Plateau (Figure 9.12(a) and (c)). Satellite images and eyewitness video show that the rockslide partially eroded the glacier.

In November 2004 a rock avalanche from Mount Beatrice (2,528 m a.s.l.) buried a debris-free part of Hooker Glacier's ablation zone. It fell 440 m from an elevation between 1,620 and 1,700 m. Small rockfalls had been observed previously in the same area (Cox et al., 2008). GPR surveying in 2009 revealed

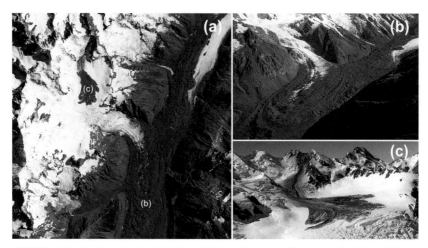

FIGURE 9.12   Rock avalanches in the Aoraki/Mount Cook area, New Zealand. (a) Aoraki/Mount Cook, Tasman Glacier and 1991 and 2013 rock avalanche deposits *(Image: NASA Earth Observatory, J. Allen and R. Simmon, February 5, 2013.)*; (b) transportation of the 1991 Aoraki/Mount Cook rock avalanche deposit through the Hochstetter icefall *(Photo: N. Reznichenko)*; (c) Rock avalanche deposit on Grand Plateau that fell from Mount Dixon in January 2013 *(Photo: Alpine Guides).*

an average debris thickness of 3–7 m, giving a volume of 0.14–0.20 Mm^3. The debris sat on a pedestal of ice 30–40 m thick (Reznichenko et al., 2011). Thus, in 5 years the rock avalanche protected 1.7 Mm^3 of ice and rock over an area of about 5 ha, or 34 m of ice (Reznichenko et al., 2011). Reported rock avalanches on Hooker Glacier suggest that they fall onto this glacier with about a decadal frequency.

The frequency of these events raises concern about of the hazard for tourist activities around Aoraki/Mount Cook National Park. To date, although there are many eyewitness accounts of rock avalanches (e.g., in 1991, 2004, and 2013) no fatal consequences are reported. However Plateau Hut, which holds tens of climbers most of the summer, was within a few hundred meters of being affected by both the 1991 and 2013 events.

9.4.2 The 1991 Chillinji Glacier Rock Avalanche (Western Karakoram)

A rock avalanche deposit was first observed in the upper ablation zone of Chillinji Glacier, Karambar valley, in June 1992. The landslide material had intact raised rims and distributary lobes, was buried in winter snowfall and showed no modifications by ablation (Figure 9.13(a)). This suggests emplacement in late 1991 or early 1992, possibly traveling over deep winter snow. About a third was buried under avalanche and wind deposited snow, and only revealed by melting in later years. Chillinji is an almost wholly

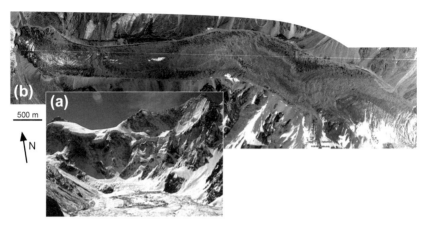

FIGURE 9.13   Chillinji Glacier, Karambar valley, Karakoram. (a) Upper ablation zone of the glacier in June 1992 *(Photo: K. Hewitt)*. (b) The rock avalanche material was still buried in winter snowfall. *Image: Google Earth, 25/07/2009.*

avalanche-nourished glacier or "Turkestan-type" and there are indications it is surge-type (Hewitt, 1998, 2013a).

When revisited in 1994, the rock avalanche area was raised above the surrounding glacier. A steep ice slope 10−20-m-high marked its downglacier edge. Material continually spilled down this slope while most of the rock avalanche deposit above seemed intact and unmodified, suggesting steady transport on the glacier surface. Below the cliff, a broad depression had developed across the glacier. Some 1−2 km further down, radial thickening was observed and, where this reached the glacier margins, ice was overriding the well-defined, tree-covered lateral moraines, indicating accelerated ice movement.

By summer 2009 the front of the rock avalanche debris had been carried 3 km down the glacier (176 m/a). The area of landslide debris had increased from an initial 1.3 km^2 to almost 2 km^2, as upper areas buried in snowpack and snow avalanches had been exhumed (Figure 9.13(b)). An extensive part of the debris had developed what McSaveney (1978) called "transverse folds", ridges and troughs extending radially across the glacier and associated with crevasses of a small icefall. A kilometer further downglacier, there was another, smaller transverse area of heavy debris on the ice, with many large, angular boulders. It seemed to record an older rockslide, possibly a rock avalanche. Rates of movement in this section suggested it was 8−12 years older than the 1991 event.

For some decades before the rock avalanche, and for 14 years afterward, the Chillinji terminus was stationary and a "moraine-dammed" ice tongue with heavy debris cover (Benn et al., 2003). The ice sat on a high ramp of moraine built out into the Karambar valley (Hewitt, 1998). An example of the

Karakoram "Ghulkin-type" of Owen (1994), it resembled the better-known Hatunraju Glacier in the Peruvian Cordillera Blanca (Iturrizaga, 2013), and the Shaigiri Glacier on Nanga Parbat (Kuhle, 1990; Benn et al., 2003). Heavy supraglacial sediment transport was occurring, as was build-up or inefficient removal at the margins.

However, quite suddenly in 2005 the Chillinji tongue broke through the terminal moraine-dam, as a breach-lobe. It advanced about a kilometer diagonally into the Karambar River, moving upvalley to rest against the far wall. Although the rock avalanche debris was 2.5 km from the location of the former, stable terminus and about 3 km from the extended terminus, it seems reasonable to suggest that accelerated ice, observed at the front of the rock avalanche in 1994, involved a kinematic wave that traveled ahead and faster than the landslide debris to reach the terminus first.

The Chillinji has dammed Karambar River in the past, but in this instance and so far, channels have been maintained under the ice (Hewitt, 1998; Hewitt and Liu, 2010). The terminal lobe remains heavily crevassed and debris tumbles down from a cliffed ice-edge and steep moraines into the river. It seems likely this condition will persist over the next several decades and further advance is likely as the thickened area, protected by rock avalanche debris, approaches the terminus. An ice dam could be sealed at some point, threatening GLOFs.

Benn et al. (2003) considered their moraine-dammed, "uncoupled" type to be typical of debris-covered glaciers. Their "outwash-head", "coupled" type is relatively free of surface debris. Change from one to the other type was seen to depend upon climate change. However, Chillinji changed from one to the other while the glacier remained debris-covered, and unrelated to climate. The change was associated with accelerated ice movement reaching the terminus from, but ahead of, the rock avalanche area.

Conditions at Chillinji illustrate how landslides can alter the usual picture of glacial fluctuations and sedimentation regulated by climate. For the past 23 years, and possibly for some decades into the future, the landslide influences glacier dynamics, mass balance, ice-margin conditions, and sedimentation. Its debris alters rates and patterns of glacier ablation, surface elevation, and movement geometry. The patterns, pace, and composition of sediment delivery are altered as landslide material is dispersed and modified by glacial activity. Over the half-century or more required to carry the landslide materials to the ice margin and beyond, they strongly buffer glacier conditions. Behavior is the reverse of what might be expected under present climatic warming—thickening not thinning, advance not retreat, a more vigorous tongue with increased sediment transport and release, not less. Positive mass balance and advance were observed at Chillinji when most glaciers in the western Karakoram and Hindu Raj were retreating or stable (Hewitt, 1998, 2005).

Of course, any broader significance of this event depends on whether the landslide is just an isolated occurrence, or if such events recur with sufficient frequency to be a perennial influence. Too few glaciers have been investigated to say how typical this case is. However, recently, reports show that rock avalanches descend on glaciers at least once in 2 years across the Greater Karakoram region. If at all representative, that suggests as many as 5,000 such events in the Holocene. If, as seems likely, major concentrations of events occur during and after large earthquakes, there would be many more. The potential of rare megaearthquakes in the NW Himalaya and Hindu Kush may mean recent events underestimate long-term incidence (Feldl and Bilham, 2006; Hough et al., 2009; Hewitt et al., 2011a). Repeated rock avalanches may have greater significance to long-term glacier behavior than has hitherto been considered.

Of course, in Holocene time, Chillinji and other glaciers in the region have not only been responding to rock avalanches. The LIA and other Neoglacial climatic advances were pervasive influences. However, it cannot be taken for granted that all glacial fluctuations are climate markers. And in Karambar Valley, as in much of the region, other conditions affect the sediment assemblages and glacier behavior (Figure 9.14).

First, several of the glaciers involved are known to be of surge-type. These advance, retreat and have mass balance cycles peculiar to the basin concerned, and largely independent of climate (Jiskoot, 2011b). An exceptional concentration of surge-type glaciers occurs in the Karakoram including several with identified rock avalanches (Gardner and Hewitt, 1990; Hewitt, 1998, 2009b).

Second, 10 glaciers, including Chillinji, have tongues that enter the channel of the Karambar River and have dammed it in the past 200 years (Hewitt and Liu, 2010). Even without dams, lateral moraines and other glacial deposits partly close the valley. They change the river's course and sediment transport. Thus, the whole river system is disturbed and fragmented by glacier interference. The main interest here is how rock avalanches from walls in glaciers basins have repeatedly helped control glacial processes (Evans, 2003; Hewitt, 2006, 2013b).

9.4.3 Holocene Horcones Mass Flow, Cerro Aconcagua (6961 m a.s.l.), Argentina

The Horcones deposit has long puzzled, and been disputed by, scientists. This applies, especially, to the great hummocky mass around the confluence of Horcones and Las Cuevas valley in Mendoza Province (Figure 9.15(a)). For some time the deposit had been identified and mapped as Late-Glacial moraine, but others suggested it could be a postglacial landslide deposit from collapsed glaciofluvial deposits. Matters are complicated by the Horcones Glacier being surge-type. It last surged in the 1990s (Figure 9.15(b)).

Resolving the dispute also happens to be of critical importance for hazard assessment of the nearby community of Puente del Inca, on a vital, trans-Andean transportation corridor. Also, the deposit is at the entrance of a

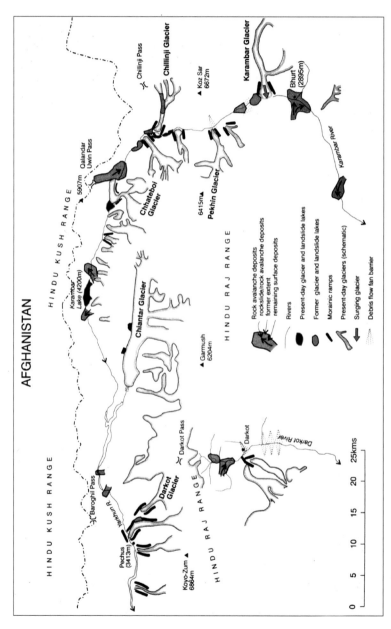

FIGURE 9.14   Glaciers and glacier dams in the Karambar Valley, Karakoram. *After Hewitt, 1998.*

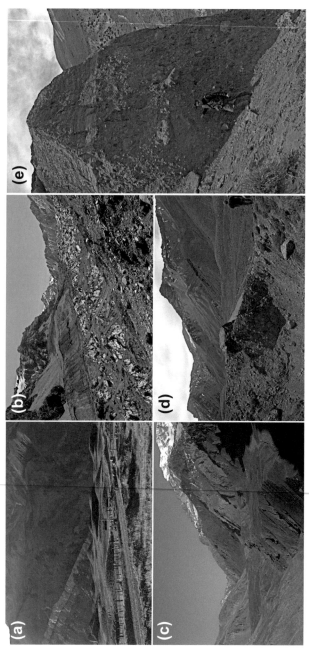

FIGURE 9.15 The Holocene Horcones deposit, Cerro Aconcagua, Argentina. (a) Hummocky surface of the lower Horcones lobe around the junction with Las Cuevas Valley (*Photo: K. Hewitt*); (b) lower Horcones Glacier shortly after its last surge. Initially the mass flow traveled over the glacier, entraining debris from the conspicuous reddish on-ice and lateral moraines (*Photo: K. Hewitt*); (c) the Horcones deposit at and below the junction of the Horcones Glacier (enters from the right middle ground, above small park camp site and below Aconcagua Peak, top right) (*Photo: K. Hewitt*); (d) lower Horcones deposit showing relations of gray and red material, and large boulders in a sink hole, evidently left where an incorporated ice block melted out (*Photo: K. Hewitt*); (e) terminal area of Horcones deposit in the Las Cuevas valley, showing typical, segregated gray and red materials, and lacustrine beds from the original impoundment by the mass flow (*Photo: K. Hewitt*).

national park where trekking and mountaineering are expanding. If it is strictly glacial, long periods would be required for the glaciers to grow enough to advance to the critical areas, even with surging. If it is related to a postglacial catastrophic landslide the question is whether conditions could bring a recurrence at any time, here or elsewhere.

A reconstruction, seen as more convincing here, does suggest that the deposit originated in a massive rock slope failure on Cerro Aconcagua. It is thought to have descended onto Horcones Glacier from Cerro Aconcagua as a rockfall—rock avalanche. Like the calamitous Nevados Huascaran landslide in Peru, 1970 (Plafker and Ericksen, 1978), large quantities of ice and snow that were part of the Horcones event may have come with the original mass failure on the peak. However some or most may have been picked up during its runout, along with additional debris, including moraine deposits scooped up from the Horcones Glacier margins and forefield. All of this contributed to a megaslide of more than 3 km^3 (Fauqué et al., 2009). During runout, it was transformed into a saturated mass flow and traveled another 8 km beyond the glacier terminus, to dam the main Las Cuevas valley, where the Trans-Andean Highway is located today.

Beyond the Horcones snout, at the junction of three valleys, the deposit separated into lobes entering each one, but the largest turned through 90° to descend the main valley. Morphology here is also hummocky; there were run-ups of over 100 m on slopes round the bend in the valley (Figure 9.15(c)). In the upper end of the valley below Horcones Glacier the deposit is composed of a single lithology from Cerro Aconcagua south face; a gray material mineralogically distinct from lateral moraine deposits of the last glaciation, and whose clasts are very angular and internally fractured. Reddish material is conspicuous below the junction and on the left flank of the valley.

There is a gap in the narrow, steeper gorge between the upper and lower main deposits, with no landslide material evident. Possibly this is due to subsequent erosion in the gorge, possibly to an acceleration of the flow here that carried most or all of the debris through the section. Rheological differences in the largest, lower mass may also be a factor; this is a hummocky deposit emplaced up to 20 km from Cerro Aconcagua, filling the lower 4 km valley floor. The lowest section has most intrigued and puzzled scientists. It is tens of meters thick, higher at midvalley than the margins and consists largely of thoroughly crushed and broken, gray rock of the Cerro Aconcagua lithology. However, many inclusions of red material occur as well, with subangular to subrounded clasts of various lithologies that outcrop along the Horcones valley and are found as the main components of lateral moraines there. In the deposits, boundaries between the two materials are sharp. They do not mix, although complexly contorted and intertwined in the mass (Figure 9.15(d)). Lacustrine beds over the landslide materials record how it dammed the Cuevas River for some decades or centuries (Figure 9.15(e)).

Age determinations for the mass flow by ^{36}Cl surface exposure dating vary between 8,800 and 11,100 years (Fauqué et al., 2009). The overlapping variations are possibly due to provenance and remnant ages in prelandslide materials, possibly sampling and technical uncertainties and conceivably more than one event. Some ^{14}C ages of underlying fluvial deposits are around 12,640 year BP. The same interval is identified by the lake sediments in the Las Cuevas valley (13,670−9,180 cal year BP). At least, these ages bracket events since the last major glaciation not, as previously assumed, Late-Glacial ones. Other rock avalanche deposits have been identified in adjacent valleys, and at least one has similar characteristics and age to the Horcones deposit. Postglacial climatic conditions, paraglacial adjustment of rock slopes, and earthquake activity may have been preparatory or triggering factors. Fauqué et al. (2009) suspected that climate warming could lead to similar events in future.

9.5 CONCLUSIONS

Rock avalanches can bring some of the most drastic and rapid changes in mountain landscapes, especially when they travel onto glaciers, where their velocity decreases more slowly than it would over ice free terrain, and their volume grows through incorporation of ice and supraglacial debris. Various preparatory factors, from glacier erosion, thinning and retreat, to earthquakes and permafrost degradation can act to destabilize rockwalls in glaciated, or formerly glaciated, basins. The forces unleashed in the rock avalanche, and the danger that some can continue past the glacier terminus into inhabited areas, make them highly threatening processes.

In most mountains landslide frequency is poorly documented, although new teleseismic techniques are facilitating their identification in remote areas. Further, rock avalanche deposits have commonly been confused with other, similar ones, especially in glacier environments, but new techniques promise to unequivocally identify rock avalanche deposits in the future.

Finally, the risks from these events are growing in many of mountain areas, due to greater populations, infrastructure, facilities, and activities. The problem is often not even recognized or, in cases where landslides have been misinterpreted as due to other processes, especially pre-Holocene glaciations, are assumed to represent other, less significant geohazards.

REFERENCES

Aeschlimann, H., 1983. Zur Gletschergeschichte des italienischen Mont Blanc Gebietes: Val Veni − Val Ferret − Ruitor (Ph.D. thesis). Universität Zürich, 105 pp.

Agassiz, L., 1845. Les glaciers et le terrain erratique du revers méridional du Mont-Blanc. In: Desor, E. (Ed.), Nouvelles excursions et séjours dans les glaciers et les hautes régions des Alpes de M. Agassiz et de ses compagnons de voyage. Kissling, Neuchâtel, 266 pp.

Akçar, N., Deline, P., Ivy-Ochs, S., Alfimov, V., Hajdas, I., Kubik, P.W., Christl, M., Schlüchter, C., 2012. The AD 1717 rock avalanche deposits in the upper Ferret Valley (Italy): a dating approach with cosmogenic [10]Be. J. Quat. Sci. 27, 383–392. http://dx.doi.org/10.1002/jqs.1558.

Akçar, N., Deline, P., Ivy-Ochs, S., Alfimov, V., Kubik, P.W., Christl, M., Schlüchter, C., 2014. Minor inheritance inhibits the calibration of the [10]Be production rate from the AD 1717 Val Ferret rock avalanche, European Alps. J. Quat. Sci. 29, 318–328. http://dx.doi.org/10.1002/jqs.2706.

Alexander, Davies, D.J., T.R.H., Shulmeister, J., 2014. Formation and evolution of the Waiho Loop terminal moraine New Zealand. J. Quat. Sci. 29 (4), 361–369. http://dx.doi.org/10.1002/jqs.2707.

Allen, S.K., Cox, S.C., Owens, I.F., 2011. Rock avalanches and other landslides in the central Southern Alps of New Zealand: a regional study considering possible climate change impacts. Landslides 8, 33–48.

Applegate, P.J., Lowell, T.V., Alley, R.B., 2008. Comment on "Absence of cooling in New Zealand and the adjacent ocean during the Younger Dryas chronozone". Science 320, 746.

Barff, E., 1873. A letter respecting the recent change in the apex of Mount Cook communicated by J. Hector. Trans. Proc. N. Z. Inst. 6, 379–380.

Barrows, T.T., Lehman, S.J., Fifield, L.K., De Deckker, P., 2008. Absence of cooling in New Zealand and the adjacent ocean during the Younger Dryas chronozone. Science 318, 86–89.

Benn, D., Kirkbride, M.P., Owen, L.A., Brazier, V., 2003. Glaciated valley landsystems. In: Evans, D.J.A. (Ed.), Glacial Landsystems. Hodder Arnold, London, pp. 372–406.

Boulton, G.S., 1978. Boulder shapes and grain-size distributions of debris as indicators of transport paths through a glacier and till genesis. Sedimentology 25, 773–799.

Collins, G.S., Melosh, H.J., 2003. Acoustic fluidization and the extraordinary mobility of sturzstroms. J. Geophys. Res. Solid Earth 108.

Cook, S.J., Porter, P.R., Bendall, C.A., 2013. Geomorphological consequences of a glacier advance across a paraglacial rock avalanche deposit. Geomorphology 189, 109–120.

Cossart, E., Braucher, R., Fort, M., Bourlés, D.L., Carcaillet, J., 2008. Slope instability in relation to glacial debuttressing in alpine areas (Upper Durance catchment, southeastern France): evidence from field data and [10]Be cosmic ray exposure ages. Geomorphology 95, 3–26.

Cox, S.C., Ferris, B.G., Allen, S., 2008. Vampire rock avalanches, Aoraki/Mount Cook National Park, New Zealand. In: GNS Science Report 2008.10, 34 pp.

Crandell, D.R., Fahnestock, R.K., 1965. Rockfalls and Avalanches from Little Tahoma Peak on Mount Rainier. U.S. Geological Survey Bulletin, 1221-A, Washington, 30 pp.

Crosta, G.B., Frattini, P., Fusion, N., 2007. Fragmentation in the Val Pola rock avalanche, Italian Alps. J. Geophys. Res. 112, F01006. http://dx.doi.org/10.1029/2005JF000455.

Davies, T.R.H., 1982. Spreading of rock avalanche debris by mechanical fluidization. Rock Mech. 15, 9–24.

Davies, T.R.H., 2013. Fluvial processes in proglacial environments. In: Shroder, J.F. (Ed.), Treatise on Geomorphology, vol. 8. Academic Press, San Diego, pp. 141–150.

Davies, M.C.R., Hamza, O., Harris, C., 2001. The effect of rise in mean annual temperature on the stability of rock slopes containing ice-filled discontinuities. Permafrost Periglac. 12, 137–144.

Davies, T.R.H., McSaveney, M.J., 1999. Runout of dry granular avalanches. Can. Geotech. J. 36, 313–320.

Davies, T.R.H., McSaveney, M.J., 2002. Dynamic simulation of the motion of fragmenting rock avalanches. Can. Geotech. J. 39, 789–798.

Davies, T.R.H., McSaveney, M.J., 2009. The role of rock fragmentation in the motion of large landslides. Eng. Geol. 109, 67–79.

Davies, T.R.H., McSaveney, M.J., 2012. Mobility of long-runout rock avalanches. In: Clague, J.J., Stead, D. (Eds.), Landslides: Types, Mechanisms and Modeling. Cambridge University Press, pp. 50–58.

Davies, T.R.H., McSaveney, M.J., Hodgson, K.A., 1999. A fragmentation-spreading model for long-runout rock avalanches. Can. Geotech. J. 36, 1096–1110.

Davies, T.R.H., McSaveney, M.J., Kelfoun, K., 2010. Runout of the Socompa volcanic debris avalanche, Chile: a mechanical explanation for low basal shear resistance. Bull. Volcanol. 72, 933–944. http://dx.doi.org/10.1007/s00445-010-0372-9.

Decaulne, A., Sæmundsson, þ., Pétursson, H.G., Jónsson, H.P., Sigurðsson, I.A., 2010. A large rock avalanche onto Morsarjökull Glacier, South-East Iceland. Its implications, or Ice-surface evolution and glacier dynamic. In: Iceland in the Central Northern Atlantic: Hotspot, Sea Currents and Climate Change, Plouzané, France hal-00482107.

Deline, P., 2001. Recent Brenva rock avalanches (Valley of Aosta): new chapter in an old story? Geogr. Fis. Din. Quat. 5 (Suppl.), 55–63.

Deline, P., 2009. Interactions between rock avalanches and glaciers in the Mont Blanc massif during the late Holocene. Quat. Sci. Rev. 28, 1070–1083. http://dx.doi.org/10.1016/j.quascirev.2008.09.025.

Deline, P., Kirkbride, M.P., 2009. Rock avalanches on a glacier and morainic complex in Haut Val Ferret (Mont Blanc massif, Italy). Geomorphology 103, 80–92. http://dx.doi.org/10.1016/j.geomorph.2007.10.020.

Denton, G.H., Hendy, C.H., 1994. Younger Dryas age advance of Franz Josef Glacier in the Southern Alps of New Zealand. Science 264, 1434–1437.

Dunning, S.A., 2004. Rock Avalanches in High Mountains (Ph.D. thesis). University of Luton, 309 pp.

Dutto, F., Mortara, G., 1991. Grandi frane storiche con percorso su ghiacciao in Valle d'Aosta. Rev. Valdôtaine Hist. Nat. 45, 21–35.

Dufresne, A., Davies, T.R.H., 2009. Longitudinal ridges in mass movement deposits. Geomorphology 105, 171–181.

Eberhardt, E., Stead, D., Coggan, J.S., 2004. Numerical analysis of initiation and progressive failure in natural rock slopes – the 1991 Randa rockslide. Int. J. Rock Mech. Min. Sci. 41, 69–87.

Eisbacher, G.H., Clague, J.J., 1984. Destructive mass movements in high mountains: hazard and management. Geol. Surv. Can. Pap. 84-16, 230 pp.

Ekstrom, G., Stark, C.P., 2013. Simple scaling of catastrophic landslide dynamics. Science 339, 1416–1419.

Evans, D.J.A. (Ed.), 2003. Glacial Landsystems. Hodder Arnold, London, 532 pp.

Evans, D.J.A., Rea, H.R., 2003. Surging glacier landsystem. In: Evans, D.J.A. (Ed.), Glacial Landsystems. Hodder Arnold, London, pp. 259–288.

Evans, S.G., Clague, J.J., 1994. Recent climatic change and catastrophic geomorphic processes in mountain environments. Geomorphology 10, 107–128.

Evans, S.G., Clague, J.J., 1988. Catastrophic rock avalanches in glacial environments. In: Bonnard, C. (Ed.), Proceedings of the 5th International Symposium on Landslides, vol. 2. Balkema, Rotterdam, pp. 1153–1158.

Evans, S.G., Clague, J.J., 1998. Rock avalanche from Mount Munday, Waddington Range, British Columbia, Canada. Landslide News 11, 23–25.

Evans, S.G., Clague, J.J., 1999. Rock avalanches on glaciers in the Coast and St. Elias Mountains, British Columbia. In: Proceedings of the 13th Annual Vancouver Geotechnical Society Symposium, Vancouver, pp. 115−123.

Evans, S.G., Bishop, N.F., Smoll, L.F., Valderrama Murillo, P., Delaney, K.B., Oliver-Smith, A., 2009a. A Re-Examination of the Mechanism and Human Impact of Catastrophic Mass Flows Originating on Nevado Huascarán, Cordillera Blanca, Peru in 1962 and 1970. Eng. Geology 108, 96−118.

Evans, S.G., Tutubalina, O.V., Drobyshev, V.N., Chernomorets, S.S., McDougall, S., Petrakov, D.A., Hungr, O., 2009b. Catastrophic detachment and high-velocity long-runout flow of Kolka Glacier, Caucasus Mountains, Russia in 2002. Geomorphology 105, 314−321.

Eyles, N., Rogerson, R.J., 1978. A framework for the investigation of medial moraine formation: Austerdalsbreen, Norway, and Berendon Glacier, British Columbia, Canada. J. Glaciol. 20, 99−113.

Fahnestock, R.K., 1978. Little Tahoma Peak rockfalls and avalanches. Mount Rainier, Washington, U.S.A.. In: Voight, B. (Ed.), Rockslides and avalanches, 1 Natural phenomena. Elsevier, Amsterdam, pp. 181−196.

Fauqué, L., Hermanns, R., Hewitt, K., Rosas, M., Wilson, C., Baumann, V., Lagorio, S., Di Tommaso, I., 2009. Mega-deslizamientos de la pared sur del Cerro Aconcagua y su relación con depósitos asignados a la glaciación pleistocena. Rev. Asoc. Geol. Argent. 65, 691−712.

Feldl, N., Bilham, R., 2006. Great Himalayan earthquakes and the Tibetan Plateau. Nature 444, 165−170. http://dx.doi.org/10.1038/nature05199.

Fischer, L., Kääb, A., Huggel, C., Noetzli, J., 2006. Geology, glacier retreat and permafrost degradation as controlling factors of slope instabilities in a high-mountain rock wall: Monte Rosa east face. Nat. Hazards Earth Syst. Sci. 6, 761−772.

Fountain, A.G., Jacobel, R.W., Schlichting, R., Jansson, P., 2005. Fractures as the main pathways of water flow in temperate glaciers. Nature 433, 618−621.

Friedmann, S.J., 1997. Rock avalanches of the Miocene Shadow Valley basin, eastern Mojave Desert, California: processes and problems. J. Sediment. Res. 67a, 792−804.

Fort, M., Cossart, E., Deline, P., Dzikowski, M., Nicoud, G., Ravanel, L., Schoeneich, P., Wassmer, P., 2009. Geomorphic impacts of large and rapid mass movements; a review. Géomorphol. Relief Processus Environ. 1, 47−64.

Gardner, J.S., Hewitt, K., 1990. A surge of Bualtar Glacier, Karakoram Range, Pakistan − a possible landslide trigger. J. Glaciol. 36, 159−162.

Geertsema, M., 2012. Initial observations of the 11 June 2012 Rock/Ice Avalanche, Lituya Mountain, Alaska. In: Wei, S., Ying, G., Chengcheng, Z. (Eds.), Proceedings of the First Meeting of Cold Region Landslides Network (International Consortium on Landslides). Harbin, pp. 49−53.

Geertsema, M., Clague, J.J., Schwab, J.W., Evans, S.G., 2006. An overview of recent large catastrophic landslides in northern British Columbia, Canada. Eng. Geol. 83, 120−143.

Giani, G.P., Silvano, S., Zanon, G., 2001. Avalanche of 18 January 1997 on Brenva Glacier, Mont Blanc Group, Western Italian Alps: an unusual process of formation. Ann. Glaciol. 32, 333−338.

Gordon, J.E., Birnie, R.V., Timmis, R., 1978. A major rockfall and debris slide on the Lyell Glacier, South Georgia. Arct. Alp. Res. 10, 49−60.

Gruber, S., Haeberli, W., 2007. Permafrost in steep bedrock slopes and its temperature-related destabilization following climate change. J. Geophys. Res. 112, F02S18. http://dx.doi.org/10.1029/2006JF000547.

Gruber, S., Hoelzle, M., Haeberli, W., 2004. Permafrost thaw and destabilization of Alpine rock walls in the hot summer of 2003. Geophys. Res. Lett. 31, L13504. http://dx.doi.org/10.1029/2004GL020051.

Guthrie, R.H., Friele, P., Allstadt, K., Roberts, N., Evans, S.G., Delaney, K.B., Roche, D., Clague, J.J., Jakob, M., 2012. The 6 August 2010 Mount Meager rock slide-debris flow, Coast Mountains, British Columbia: characteristics, dynamics, and implications for hazard and risk assessment. Nat. Hazards Earth Syst. Sci. 12, 1−18. http://dx.doi.org/10.5194/nhess-12-1-2012.

Haeberli, W., Wegmann, M., Vonder Mühll, D., 1997. Slope stability problems related to glacier shrinkage and permafrost degradation in the Alps. Eclogae Geol. Helv. 90, 407−414.

Haeberli, W., Kääb, A., Paul, F., Chiarle, M., Mortara, G., Mazza, A., Deline, P., Richardson, S., 2002. A surge-type movement at Ghiacciaio del Belvedere and a developing slope instability in the east face of Monte Rosa, Macugnaga, Italian Alps. Norw. J. Geogr. 56, 104−111.

Hasler, A., Gruber, S., Font, M., Dubois, A., 2011. Advective heat transport in frozen rock clefts − conceptual model, laboratory experiments and numerical simulation. Permafrost Periglac. 22, 378−389.

Heim, A., 1932. Bergsturz und Menschenleben. Frets und Wasmuth, Zürich, 218 pp.

Heinrichs, T.A., Mayo, L.R., Trabant, D., March, R., 1995. Observations of the surge-type Black Rapids Glacier, Alaska, during a quiescent period, 1970−92. U. S. Geol. Surv. Open File 94-512, 131 pp.

Herreid, S.J., Arendt, A.A., Hock, R., Kienholz, C., 2010. A New Inventory of Glaciers and Supraglacial Debris for the Alaska Range with a Case Study of Rock Avalanche Loading. AGU Fall Meeting, San Francisco.

Hewitt, K., 1988. Catastrophic landslide deposits in the Karakoram Himalaya. Science 242, 64−67.

Hewitt, K., 1998. Himalayan Indus streams in the Holocene: glacier-, and Landslide-'Interrupted' fluvial systems. In: Stellrecht, I. (Ed.), Karakorum-Hindu Kush-Himalaya: Dynamics of Change Part I. Rudiger Koppe Verlag, Koln, pp. 1−28.

Hewitt, K., 1999. Quaternary moraines vs catastrophic rock avalanches in the Karakoram Himalaya, Northern Pakistan. Quat. Res. 51, 220−237.

Hewitt, K., 2002. Styles of rock avalanche depositional complex in very rugged terrain, Karakoram Himalaya, Pakistan. In: Evans, S.G., DeGraff, J.V. (Eds.), Catastrophic Landslides: Effects, Occurrence, and Mechanisms, Reviews in Engineering Geology. Geological Society of America, Boulder, pp. 345−378.

Hewitt, K., 2005. The Karakoram Anomaly? Glacier expansion and the 'elevation effect', Karakoram Himalaya, Inner Asia. Mt. Res. Dev. 25, 332−348.

Hewitt, K., 2006. Disturbance regime landscapes: mountain drainage systems interrupted by large rockslides. Prog. Phys. Geogr. 30, 365−393.

Hewitt, K., 2009. Rock avalanches that travel onto glaciers: disturbance regime landscapes, Karakoram Himalaya, Inner Asia. Geomorphology 103, 66−79.

Hewitt, K., 2013a. Glaciers of the Karakoram Himalaya: Glacial Environments, Processes, Hazards and Resources. Springer, Heidelberg, 363 pp.

Hewitt, K., 2013b. The Great Lateral Moraine, Karakoram Himalaya, Inner Asia. Geogr. Fis. Din. Quat. 36, 1−14. http://dx.doi.org/10.4461/Gfdq.2013.36.0.

Hewitt, K., Clague, J.J., Orwin, J., 2008. Legacies of catastrophic rock slope failures in mountain landscapes. Earth Sci. Rev. 87, 1–38.

Hewitt, K., Liu, J., 2010. Ice-dammed lakes and outburst floods, Karakoram Himalaya: historical perspectives and emerging threats. Phys. Geogr. 31, 528–551.

Hewitt, K., Clague, J.J., Gosse, J., 2011a. Rock avalanches and the pace of late Quaternary development of river valleys in the Karakoram Himalaya. Geol. Soc. Am. Bull. 123, 1836–1850.

Hewitt, K., Clague, J.J., Deline, P., 2011b. Catastrophic rock slope failures and mountain glaciers. In: Singh, V.P., Singh, P., Haritashaya, U.K. (Eds.), Encyclopaedia of Snow, Ice and Glaciers. Springer, Dordrecht, pp. 112–126.

Hochstein, M.P., Claridge, D., Henrys, S.A., Pyne, A., Nobes, D.C., Leary, S.F., 1995. Downwasting of the Tasman Glacier, South Island, New Zealand: changes in the terminus region between 1971 and 1993. N. Z. J. Geol. Geophys. 38, 1–16.

Hough, S., Bilham, R., Bhat, I., 2009. Kashmir Valley megaearthquakes. Am. Sci. 97, 42–49. http://dx.doi.org/10.1511/2009.76.1.

Hsü, K.J., 1975. Catastrophic debris streams (sturzstroms) generated by rockfalls. Geol. Soci. Am. Bull. 86, 129–140.

Huggel, C., 2008. Recent extreme slope failures in glacial environments: effects of thermal perturbation. Quat. Sci. Rev. 28, 1119–1130.

Huggel, C., Zgraggen-Oswald, S., Haeberli, W., Kaab, A., Polkvoi, A., Galushkin, I., Evans, S.G., 2005. The 2002 rock/ice avalanche at Kolka/Karmadon, Russian Caucasus: assessment of extraordinary avalanche formation and mobility, and application of QuickBird satellite imagery. Nat. Hazards Earth Sys. Sci. 5, 173–187.

Huggel, C., Caplan-Auerbach, J., Waythomas, C.F., Wessels, R.L., 2007. Monitoring and modeling ice-rock avalanches from ice-capped volcanoes: a case study of frequent large avalanches on Iliamna Volcano, Alaska. J. Volcanol. Geotherm. Res. 168, 114–136.

Huggel, C., Caplan-Auerbach, J., Gruber, S., Molnia, B., Wessels, R., 2008. The 2005 Mt. Steller, Alaska, rock–ice avalanche: a large slope failure in cold permafrost. In: Kane, D.L., Hinkel, K.M. (Eds.), Proceedings of the 9th International Conference on Permafrost. Alaska, Fairbanks, pp. 747–752.

Imhof, P., 2010. Glacier Fluctuations in the Italian Mont Blanc Massif from the Little Ice Age until the Present. Historical Reconstructions for the Miage, Brenva and Pré-de-Bard Glaciers (MSc thesis). Universität Bern, 132 pp.

Iturrizaga, L., 2006. Transglacial landforms in the Karakoram (Pakistan): a case study from Shimshal Valley. In: Kreutzmann, H. (Ed.), Karakoram in Transition: Culture, Development and Ecology in the Hunza Valley. Oxford University Press, Karachi, 419 pp.

Iturrizaga, L., 2013. Bent glacier tongues: a new look at Lliboutry's model of the evolution of the crooked Jatunraju Glacier (Parón Valley, Cordillera Blanca, Perú). Geomorphology 198, 147–162.

Jibson, R.W., Harp, E.L., Schulz, W., Keefer, D.K., 2006. Large rock avalanches triggered by the M 7.9 Denali Fault, Alaska, earthquake of 3 November 2002. Eng. Geol. 83, 144–160.

Jiskoot, H., 2011a. Long-runout rockslide on glacier at Tsar Mountain, Canadian Rocky Mountains: potential triggers, seismic and glaciological implications. Earth Surf. Processes Landforms 36, 203–216.

Jiskoot, H., 2011b. Glacier surging. In: Singh, V.P., Singh, P., Haritashaya, U.K. (Eds.), Encyclopaedia of Snow, Ice and Glaciers. Springer, Dordrecht, pp. 415–428.

Johnson, N.M., Ragle, R.H., 1968. Analysis of Flow Characteristics of Allen II Slide from Aerial Photographs, the Great Alaska Earthquake of 1964-Hydrology, Part A. National Academy of Sciences, Washington, D.C, 369–373.

Kjartansson, G., 1967. The Steinsholtshlaup, central-south Iceland on January 15th, 1967. Jökull 17, 249−262.

Keefer, D.K., 1984. Rock avalanches caused by earthquakes: source characteristics. Science 223, 1288−1290.

Keefer, D.K., 1999. Earthquake-induced landslides and their effects on alluvial fans. J. Sediment. Res. 69, 84−104.

Keefer, D.K., 2002. Investigating landslides caused by earthquakes − a historical review. Surv. Geophys. 23, 473−510.

Kent, P.E., 1966. The transport mechanism in catastrophic rock falls. J. Geol. 74, 79−83.

Kirkbride, M.P., Sugden, D., 1992. New Zealand loses its top. Geogr. Mag. 64, 30−34.

Kirkbride, M.P., Winkler, S., 2012. Correlation of Late Quaternary moraines: impact of climate variability, glacier response, and chronological resolution. Quat. Sci. Rev. 46, 1−29.

Kotlyakov, V.M., Rototaeva, O.V., Nosenko, G.A., 2004a. The September 2002 Kolka Glacier catastrophe in North Ossetia, Russian Federation: evidence and analysis. Mt. Res. Dev. 24, 78−83.

Kotlyakov, V.M., Rototaeva, O.V., Desinov, L.V., Zotikov, I.A., Osokin, N.I., 2004b. Causes and effect of a catastrophic surge of Kolka glacier in the Central Caucasus. Z. Gletscherkd. Glazialgeol. 38, 117−128.

Kuhle, M., 1990. Ice marginal ramps and alluvial fans in semiarid mountains: convergence and difference. In: Rachocki, A.H., Church, M. (Eds.), Alluvial Fans: A Field Approach. Wiley, Chichester, pp. 55−68.

Larsen, S., Davies, T.R.H., McSaveney, M.J., 2005. A possible coseismic landslide origin of late Holocene moraines of the Southern Alps, New Zealand. N. Z. J. Geol. Geophys. 48, 311−314.

Lee, E.M., Jones, D.K.C., 2004. Landslide Risk Management. Thomas Telford Publishing, London, 464 pp.

Lipovsky, P.S., Evans, S.G., Clague, J.J., Hopkinson, C., Couture, R., Bobrowsky, P., Ekström, G., Demuth, M.N., Delaney, K.B., Roberts, N.J., Clarke, G., Schaeffer, A., 2008. The July 2007 rock and ice avalanches at Mount Steele, St. Elias Mountains, Yukon, Canada. Landslides 5, 445−455. http://dx.doi.org/10.1007/s10346-008-0133-4.

Malamud, B.D., Turcotte, D.L., Guzzetti, F., Reichenbach, P., 2004. Landslides, earthquakes, and erosion. Earth Planet. Sci. Lett. 229, 45−59.

Mauthner, T.E., 1996. Kshwan Glacier rock avalanche, Southeast of Stewart, British Columbia. Geol. Surv. Can. Curr. Res. 1996-A, 37−44.

Matsuoka, N., Murton, J., 2008. Frost weathering: recent advances and future directions. Permafrost Periglac. 19, 195−210. http://dx.doi.org/10.1002/ppp.620.

Mayr, F., 1969. Die postglazialen Gletscherschwankungen des Mont Blanc-Gebietes. Z. Geomorphol. Suppl. Band 8, 31−57.

McColl, S.T., 2012. Paraglacial rock-slope stability. Geomorphology 153−154, 1−16.

McColl, S.T., Davies, T.R.H., 2011. Evidence for a rock-avalanche origin for the Hillocks moraine, Otago, New Zealand. Geomorphology 127 (3−4), 216−224.

McColl, S.T., Davies, T.R.H., 2013. Large ice-contact slope movements; glacial buttressing, deformation and erosion. Earth Surf. Processes Landforms 38, 1102−1115. http://dx.doi.org/10.1002/esp.3346.

McColl, S.T., Davies, T.R.H., McSaveney, M.J., 2010. Glacier retreat and rock-slope stability: debunking debuttressing. In: Williams, A.L., Pinches, G.M., Chin, C.Y., McMorran, T.J., Massey, C.I. (Eds.), Geologically Active. Taylor and Francis, London, pp. 467−474.

McColl, S.T., Davies, T.R.H., McSaveney, M.J., 2012. The effect of glaciation on the intensity of seismic ground motion. Earth Surf. Processes Landforms 37, 1290—1301. http://dx.doi.org/10. 1002/esp.3251/full.

McSaveney, M.J., 1978. Sherman Glacier rock avalanche, Alaska, U.S.A. In: Voight, B. (Ed.), Rockslides and Avalanches, 1. Natural Phenomena. Elsevier, Amsterdam, pp. 197—258.

McSaveney, M.J., 1993. Rock avalanches of 2 May and 16 September 1992, Mount Fletcher, New Zealand. Landslides News 7, 32—34.

McSaveney, M., 2002. Recent rock falls and rock avalanches in Mount Cook National Park, New Zealand. In: Evans, S.G., DeGraff, J.V. (Eds.), Catastrophic Landslides: Effects, Occurrence, and Mechanisms, Reviews in Engineering Geology, vol. 15, pp. 35—70.

McSaveney, M.J., Davies, T.R., 2007. Rockslides and their motion. In: Sassa, K., Fukuoka, H., Wang, F., Wang, G. (Eds.), Progress in Landslide Science. Springer, Berlin, pp. 113—133.

Melosh, H.J., 1979. Acoustic fluidization: a new geologic process? J. Geophys. Res. 84, 7513—7520.

Mihalcea, C., Brock, B.W., Diolaiuti, G., D'Agata, C., Citterio, M., Kirkbride, M.P., Cutler, M.E.J., Smiraglia, C., 2007. Supraglacial surface temperature from ASTER and ground-based measurements analysed to investigate debris-covered pattern on Miage Glacier (Mont Blanc, Italy). Cold Reg. Sci. Technol. 52, 341—354.

Muravyev, Y.D., 2004. Subglacial geothermal eruption — the possible reason of catastrophic surge of Kolka Glacier in Kazbek volcanic massif (Caucasus), bulletin of Kamchatka regional association "Educational-Scientific Center". Earth Sci. 4, 6—20.

Nagai, H., Fujita, K., Nuimura, T., Sakai, A., 2013. Southwest-facing slopes control the formation of debris-covered glaciers in the Bhutan Himalaya. The Cryosphere 7, 1303—1314. http://dx. doi.org/10.5194/tc-7-1303-2013.

Nakawo, M., Rana, B., 1999. Estimate of ablation rate of glacier ice under a supraglacial debris layer. Geogr. Ann. 81A, 695—701.

Nakawo, M., Raymond, C.F., Fountain, A., 2000. Debris-Covered Glaciers. International Association of Hydrological Sciences Publication 264, Wallingford, 288 pp.

Nussbaumer, S.U., Zumbühl, H.J., Steiner, D., 2007. Fluctuations of the Mer de Glace (Mont Blanc area, France) AD 1500—2050: an interdisciplinary approach using new historical data and neural network simulations. Z. Gletscherkd. Glazialgeol. 40 (2005/2006), 1—183.

Orombelli, G., Porter, S.C., 1988. Boulder deposit of upper Val Ferret (Courmayeur, Aosta valley): deposit of a historic giant rockfall and debris avalanche or a late-glacial moraine? Eclogae Geol. Helv. 81, 365—371.

Owen, L., 1994. Glacial and non-glacial diamictons in the Karakoram Mountains and Western Himalaya. In: Warren, W.P., Croots, D. (Eds.), The Formation and Deformation of Glacial Deposits. Balkema, Rotterdam, pp. 9—24.

Pacione, M., 1999. Applied Geography: Principles and Practice. Routledge, London, 664 pp.

Petrakov, D.A., Chernomorets, S.S., Evans, S.G., Tutubalina, O.V., 2008. Catastrophic glacial multi-phase mass movements: a special type of glacial hazard. Adv. Geosci. 14, 211—218.

Pirulli, M., 2009. The Thurwieser rock avalanche (Italian Alps): description and dynamic analysis. Eng. Geol. 109, 80—92. http://dx.doi.org/10.1016/j.enggeo.2008.10.007.

Plafker, G., Ericksen, F.E., 1978. Nevados huascaran avalanches, Peru. In: Voight, B. (Ed.), Rockslides and Avalanches. 1 Natural Phenomena. Elsevier, Amsterdam, pp. 277—314.

Porter, S.C., Orombelli, G., 1980. Catastrophic rockfall of September 12, 1717 on the Italian flank of the Mont Blanc massif. Z. Geomorphol. 24, 200—218.

Porter, S.C., Orombelli, G., 1981. Alpine rockfall hazards. Recognition and dating of rockfall deposits in the western Italian Alps lead to an understanding of the potential hazards of giant rockfalls in mountainous regions. Am. Sci. 69, 67−75.

Post, A., 1967. Effects of the March 1964 Alaska earthquake on glaciers. U.S. Geol. Surv. Prof. Pap. 544-D, 42.

Prager, C., Ivy-Ochs, S., Ostermann, M., Synal, H.-A., Patzelt, G., 2009. Geology and radiometric ^{14}C-, ^{36}Cl- and Th-/U-dating of the Fernpass rockslide (Tyrol, Austria). Geomorphology 103, 93−103.

Putnam, A.E., Denton, G.H., Schaefer, J.M., Barrell, D.J.A., Andersen, B.G., Finkel, R.C., Schwartz, R., Doughty, A.M., Kaplan, M.R., Schlüchter, C., 2010. Glacier advance in southern middle-latitudes during the Antarctic Cold Reversal. Nat. Geosci. 3, 700−704.

Reid, J.R., 1969. Effects of a debris-slide on "Sioux Glacier", South-central Alaska. J. Glaciol. 8, 353−367.

Reznichenko, N., Davies, T.R.H., Shulmeister, J., McSaveney, M.J., 2010. Effects of debris on ice-surface melting rates: an experimental study. J. Glaciol. 56, 384−394.

Reznichenko, N., Davies, T.R.H., Alexander, D.J., 2011. Effects of rock avalanches on glaciers behaviour and moraines formation. Geomorphology 132, 327−338.

Reznichenko, N.V., Davies, T.R.H., Shulmeister, J.P., Larsen, S.H., 2012. A new technique for identifying rock avalanche-sourced sediment in moraines and some palaeoclimatic implications. Geology 40, 319−322.

Robinson, T.R., Davies, T.R.H., Reznichenko, N.V., De Pascale, G.P., 2014. The extremely long-runout Komansu rock avalanche in the Trans Alai range. Pamir Mountains, southern Kyrgyzstan. Landslides. http://dx.doi.org/10.1007/s10346-014-0492-y.

Sacco, F., 1918. I ghiacciai italiani del gruppo del Monte Bianco. Boll. Com. Glaciol. Ital. 3, 21−102.

de Saussure, H.-B., 1786. Voyages dans les Alpes, précédés d'un essai sur l'histoire naturelle des environs de Genève, vol. 2. Barde, Manguet et Compagnie, Genève. XVI−641 pp.

Schneider, D., Kaitna, R., Dietrich, W.E., Hsu, L., Huggel, C., McArdell, B.W., 2011a. Frictional behavior of granular gravel-ice mixtures in vertically rotating drum experiments and implications for rock-ice avalanches. Cold Reg. Sci. Technol. 69, 70−90.

Schneider, D., Huggel, C., Haeberli, W., Kaitna, R., 2011b. Unravelling driving factors for large rock-ice avalanche mobility. Earth Surf. Processes Landforms 36, 1948−1966.

Schwab, J.W., 2002. In: Catastrophic Rock Avalanche: Howson Range, Telkwa Pass. Forest Sciences, Prince Rupert Forest Region, Extension Note, vol. 46, 5 pp.

Scott, K.M., Vallance, J.W., 1995. Debris Flow, Debris Avalanche, and Flood Hazards at and Downstream from Mount Rainier. Hydrologic Investigations Atlas (USGS), HA-729, Washington, 9 pp.

Selby, M.J., 1993. Hillslope Materials and Processes. Oxford University Press, Oxford, 451 pp.

Sharp, M., 1988. Surging glaciers; geomorphic effects. Prog. Phys. Geogr. 12, 9015−9022.

Shreve, R.L., 1968. Sherman landslide. In: The Great Alaska Earthquake of 1964 (Hydrology, Part A), Publication 1603. National Academy of Sciences, Washington, D.C, pp. 395−401.

Shugar, D.H., Clague, J.J., 2011. The sedimentology and geomorphology of rock avalanche deposits on glaciers. Sedimentology 58, 1762−1783.

Shugar, D.H., Rabus, B.T., Clague, J.J., 2010. Elevation changes (1949−1995) of Black Rapids Glacier, Alaska, derived from a multi-baseline InSAR DEM and historical maps. J. Glaciol. 56, 625−634.

Shugar, D.H., Clague, J.J., Giardino, M., 2013a. A quantitative assessment of the sedimentology and geomorphology of rock avalanche deposits. In: Margottini, C., Catani, F., Trigila, A.,

Iadanza, C. (Eds.), Landslide Science and Practice, Global Environmental Change, vol. 4. Springer-Verlag, pp. 321–326.

Shugar, D.H., Clague, J.J., McSaveney, M.J., August 2013b. Late Holocene Behavior of Sheridan and Sherman Glaciers, Chugach Mountains, Alaska. Canadian Quaternary Association. Edmonton, Canada.

Shugar, D.H., Rabus, B.T., Clague, J.J., Capps, D.M., 2012. The response of Black Rapids Glacier, Alaska, to the Denali earthquake rock avalanches. J. Geophys. Res. 117. http://dx.doi.org/10.1029/2011jf002011.

Shulmeister, J., Davies, T.R., Evans, D.J.A., Hyatt, O.M., Tovar, D.S., 2009. Catastrophic landslides, glacier behaviour and moraine formation – a view from an active plate margin. Quat. Sci. Rev. 28, 1085–1096.

Shulmeister, J., Davies, T.R.H., Reznichenko, N., Alexander, D.J., 2010. Comment on "Glacial advance and stagnation caused by rock avalanches" by Vacco, D.A., Alley, R.B. and Pollard, D. Earth Planet. Sci. Lett. 297, 700–701.

Sosio, R., Crosta, G.B., Hungr, O., 2008. Complete dynamic calibration for the Thurwieser rock avalanche (Italian Central Alps). Eng. Geol. 100, 11–26.

Sosio, R., Crosta, G.B., Chen, J.H., Hungr, O., 2012. Modelling rock avalanche propagation onto glaciers. Quat. Sci. Rev. 47, 23–40.

Strom, A., 2006. Morphology and internal structure of rockslides and rock avalanches: grounds and constraints for their modeling. In: Evans, S.G., Scarascia Mugnozza, G., Strom, A., Hermanns, R.L. (Eds.), Landslides from Massive Rock Slope Failure, NATO Science Series IV, Earth and Environmental Sciences, 49. Springer, Dordrecht, pp. 305–326.

Tovar, D.S., Shulmeister, J., Davies, T.R., 2008. Evidence for a landslide origin of New Zealand's Waiho Loop moraine. Nat. Geosci. 1, 524–526.

Uhlmann, M., Korup, O., Huggel, C., Fischer, L., Kargel, J.S., 2013. Supra-glacial deposition and flux of catastrophic rock-slope failure debris, south-central Alaska. Earth Surf. Processes Landforms 38, 675–682.

Vacco, D.A., Alley, R.B., Pollard, D., 2010. Glacial advance and stagnation caused by rock avalanches. Earth Planet. Sci. Lett. 294, 123–130.

Valbusa, U., 1924. Il ghiacciaio della Brenva (M. Bianco) dal 20 Aprile 1923 al 15 Giugnio 1924. Riv. Club Alp. Ital. 43, 270–281.

Vincent, C., Six, D., Berthier, E., Le Meur, E., 2012. Ecoulement et fluctuations de la Mer de Glace. In: Deline, P., Nussbaumer, S., Vincent, C., Zumbuhl, H. (Eds.), La Mer de Glace, art et science. Esope, Chamonix, 192 pp.

Virgilio, F., 1883. Sui recenti studi circa le variazione periodiche dei ghiacciai. Boll. Club Alp. Ital. 50, 50–70.

Voight, B. (Ed.), 1978. Rockslides and Avalanches, 1. Natural Phenomena. Elsevier, Amsterdam, 843 pp.

Wardle, P., 1978. Further radiocarbon dates from Westland National Park and the Omoeroa River mouth, New Zealand. N. Z. J. Bot. 11, 349–388.

Wegmann, M., Gudmundsson, G.H., Haeberli, W., 1998. Permafrost changes in rock walls and the retreat of alpine glaciers: a thermal modelling approach. Permafrost Periglac. 9, 23–33.

Whalley, W.B., 1984. Rockfalls. In: Brunsden, D., Prior, D.B. (Eds.), Slope Instability. Wiley, Chichester, pp. 217–256.

Zienert, A., 1965. Gran Paradiso-Mont Blanc. Prähistorische und historische Gletscherstände. Eiszeitalter Ggw. 16, 202–225.

Chapter 10

Paleolandslides

John J. Clague

Centre for Natural Hazard Research, Simon Fraser University, Burnaby, BC, Canada

ABSTRACT

Paleolandslides are mass movements that predate the historical period and are documented using geologic and geomorphologic evidence. Catalogs of paleolandslides allow scientists to extend a brief, documentary record of mass movements, an exercise that is commonly required to evaluate hazard and risk to existing or proposed development. Paleolandslides also provide the context for determining the contribution to long-term erosion of mass movements. Finally, extremely large paleolandslides have no historical analogs; their study yields insights into unusual effects that have not been witnessed, including large-scale liquefaction, undrained loading of the substrate over which a large landslide travels, and the streaming behavior of rock avalanches. A serious problem in using paleolandslides in quantitative risk assessments is the temporal bias in landslide catalogs. The record of landslides becomes increasingly incomplete with elapsed time and biased toward progressively larger events. Corrections to frequency–magnitude (F–M) curves must thus be made to evaluate risk at long return periods. Reliable F–M curves also require accurate age dating of events, which can be achieved with the careful use of dendrochronological, radiocarbon, and surface exposure dating techniques.

10.1 INTRODUCTION

Landslides depend only on gravitational forces and thus occur on all celestial bodies with surface relief. Aside from the Earth, they have been identified in our solar system on the terrestrial planets Mercury; Venus; and Mars; on the Martian moons of Phobos and Diemos; the Moon; the moons of Io, Ganymede, Callisto, and Europa orbiting Jupiter; Iapetus orbiting Saturn; and the asteroids of 433 Eros, 253 Mathilde, 951 Gaspra, 243 Ida, and 25,143 Itokawa (Bulmer 2012). However, landslides are by far most common on active terrestrial (rocky) planets, of which there is only one in our solar system—Earth. Convection in the Earth's mantle moves continents and deforms the crust, elevating and deforming crustal rocks that fail under the influence of gravity. Volcanic activity, earthquakes, weathering, and erosion are ancillary processes

Landslide Hazards, Risks, and Disasters. http://dx.doi.org/10.1016/B978-0-12-396452-6.00010-0
321

that promote instability on our planet. Relief also exists on the Earth's ocean floor, notably at the drowned edges of continents and in ocean trenches; thus, mass movements are also common at these sites.

The subject of this chapter is "paleolandslides," and it is sensible to define the term at the outset. Landslides have happened on Earth for at least 3 billion years and perhaps for more than 4 billion years, ever since the first continents formed and drifted on the planet's surface. Evidence for the transport of sedimentary particles by water can be gleaned from some of the oldest rocks on Earth's cratons, indicating that rivers and seas, and, by extension, mass movements date back to the first billion years of the planet's history. By any definition, such landslides are "paleo." Plate tectonic processes and erosion have destroyed all direct evidence of these ancient features. Aside from the fragmentary deposits of landslides preserved in sedimentary rocks, we rely on landforms to inform us about ancient landslides. Our planet is sufficiently active that the oldest preserved land surfaces, located on the continental cratons of South America, Africa, and Australia, are not more than 250 million years old. We have a rudimentary understanding of continental landscapes since that time from fits of continental margins, patterns of ocean-floor magnetic anomalies, and positions of paleomagnetic poles. For these earlier periods, paleomagnetic poles are supplemented by geologic evidence such as orogenic belts that mark the edges of ancient plates and past distributions of flora and fauna. However, as one moves further and further back in time, the data become scarcer and harder to interpret.

Any attempt to derive useful insight from landslides that are divorced from their geomorphic context seems fruitless. Thus, for the purpose of this paper, I focus on "paleolandslides" that retain at least some expression at the surface, perhaps only their scarps or their deposits, but more commonly both. In arid environments that were not glaciated in the Pleistocene, landslides with remnant surface expression may date back millions of years. In contrast, in areas that were covered by alpine glaciers and continental ice sheets during the Pleistocene Epoch, most paleolandslides are Holocene in age (<11,600 years old).

Clear value exists in studying landslides that have happened before the period of historical observations. The largest subaerial landslides are prehistoric, and an understanding of the kinematics and dynamics of these events relies on geologic and geomorphic studies. Classic examples include landslides >10 km^3 in volume, such as Flims in Switzerland (Poschinger et al., 2006), Saidmarreh (Seymareh) in Iran (Roberts and Evans, 2013), and Green Lake in New Zealand (Hancox and Perrin, 2009). Even these terrestrial "megalandslides" are dwarfed by huge submarine failures including the Agulhas Slide off South Africa, which is Pleistocene in age (Dingle, 1977); the Storegga Slide in the Norwegian Sea in the early Holocene (Jansen et al., 1987); and the landslide triggered by the 1929 Grand Banks earthquake in the North Atlantic off Newfoundland (Clifton, 1988).

In this chapter, I discuss the methods that earth scientists use to identify, map, and understand paleolandslides. I then briefly review the common dating methods used to establish the ages of paleolandslides. Next, I consider the role that landslides play in landscape evolution and discuss the complication that results from temporal biasing in the preserved record of paleolandslides. Finally, I discuss the use of paleolandslide data in risk assessments related to development in potentially hazardous areas.

10.2 SIGNIFICANCE OF PALEOLANDSLIDES

Local and regional studies of paleolandslides are important for several reasons. First, in most parts of the world, the period during which landslides have been witnessed and recorded is short, typically <200 years. Even in long-settled alpine areas, such as the European Alps, details on contemporary large landslides are lacking prior to the eighteenth century (Eisbacher and Clague, 1984); in western North America, South America, Asia, and several other areas, the record is shorter, <100 years. In this context, well described and accurately dated paleolandslides substantially augment and extend the archived record of landslide events. The largest paleolandslides provide an insight into rare events that have not happened in the historic period. An example is the huge Flims landslide in Switzerland (Poschinger et al., 2006). About 9400 years ago, about 10 km^3 of limestone crashed into the valley of the Rhine River, liquefying nearly 1 km^3 of valley fill (Poschinger and Kippel, 2009). The liquefied sediment flowed down the Rhine Valley and 16 km up its major tributary, rafting hill-shaped masses of rockslide material, known as "tomas," several hundreds of meters across and many tens of meters high.

Second, a local catalog of paleolandslides is a requirement for determining the risk when land is being considered for development in potentially hazardous areas. Assessments of landslide risk are based on reliable frequency–magnitude (F–M) curves, which are plots of the size of landslides against their average return period. On an F–M curve, smaller events with shorter average return periods are commonly anchored with historical data, but larger, more rare events are based on paleolandslide data. When performing such risk assessments, different types of landslides with different possible causes and triggers must be treated separately. An example of risk assessment for debris flows is provided below in this paper.

Third, many scientists are interested in the role that landslides play in shaping landscapes on a regional or global scale. A fundamental question that arises in this context is, "How important is landsliding compared to other erosional agents, notably fluvial erosion and glaciation?" This question can only be answered after thorough documentations of both historic and paleolandslides and of other erosion processes. The task is daunting, although such databases are now being produced and regularly updated (e.g., Hermanns et al., 2012). The single greatest difficulty in assembling landslide databases to evaluate the regional or global erosional efficacy of landslides is the temporal bias inherent in cataloging paleoevents. This problem, which is discussed later in this chapter, can also compromise F–M curves.

10.3 RECOGNITION AND MAPPING

Central to all paleolandslide studies are identification and mapping of landforms produced by mass movements. Geomorphologic study based on remotely sensed imagery commonly is complemented by field characterization of the source and debris. The three-dimensional morphology of landslide scarps and deposits and the physical characteristics of the debris allow researchers to infer mass movement processes, causes, and triggers. In this section, I discuss the roles of geomorphology and stratigraphy in identifying, mapping, and characterizing paleolandslides. I do not discuss geological details of landslide scarps, for example, stratification, fractures, and joints, because they are dealt with exhaustively in other chapters of this book and relate to failure causes, which is beyond the scope of this chapter.

10.3.1 Role of Geomorphology

The morphology of landslide scarps and deposits is governed by the characteristics of the terrain in which the landslide occurs, the geology of the source materials, the size of the failure, and the type and velocity of movement. In addition, water is important because it commonly causes or triggers slope failure and also strongly influences movement. Landslides are classified according to these factors (e.g., Varnes, 1978; Cruden and Varnes, 1996).

The main landslide types are falls, topples, slides (including slumps), and flows. A "fall" involves the rolling and bouncing of rock and, less commonly, sediment from a cliff. The characteristic geomorphic expression of a repeated rock fall is a talus cone sloping away from the base of the cliff (Figure 10.1);

FIGURE 10.1   Talus aprons, such as this one in British Columbia, are landforms indicative of repeated rock falls. *Photograph - John J. Clague.*

coalescent talus cones are referred to as aprons. Talus cones and aprons comprise angular blocks with little matrix; they are resistant to erosion and therefore may persist in the unglaciated landscapes for hundreds of thousands of years or more. In contrast, falls of weakly consolidated sediment such as silt or till have little or no preservation potential.

A "slide" involves the downslope movement of an intact or disintegrating mass of rock or sediment along a well-defined basal surface. Slides are sub-divided into translational and rotational types. A translational slide moves on a planar or undulating surface. The slide mass commonly disintegrates as it moves downslope, but the fragments tend to retain their positions with respect to one another. A rotational slide, or slump, involves the translation of rock or sediment along a curved failure surface that is concave upward. The preser-vation potential of slides depends on its size and the type of earth material involved. Large rockslides leave distinctive inset source scarps in slopes that remain in the landscape long after failure (Figure 10.2); geomorphologists use these slope insets to identify paleolandslides. If the failure involves a rock slab hundreds of meters thick, the scarp will be deeply inset into the slope and may resemble a depression created by glacier erosion (Turnbull and Davies, 2006). Smaller rockslides involving rock slabs less than a few tens of meters thick leave much less distinctive recessed scarps and have correspondingly less preservation potential. Many rockslides fall onto glaciers and cover the snow and ice with blocky debris. The debris is carried to the glacier terminus where it forms moraines and may not be recognized as being of landslide origin (Reznichenko et al., 2012).

Paleorockslide deposits are readily identifiable, both on remotely sensed images and on the ground. In some cases, the failed slab does not greatly fragment as it moves downslope from the source scarp. These slabs are

FIGURE 10.2 Amphitheater marking the source of the Tamins landslide in southeast Switzerland. *Photograph—John J. Clague.*

FIGURE 10.3 The partly
forested, blocky debris sheet of the
Holocene Mystery Creek rock
avalanche in southwest British
Columbia. The maximum dimension of the circled block is 18 m.
Google Earth image.

geomorphic anomalies in that they are elevated above the slope on which they
came to rest and have a tabular shape with steep margins. The spatial association of such a slab with a tabular-shaped hollow on the source slope above
is apparent to a geomorphologist. More commonly, however, rockslides
fragment as they move downslope. Again, the recess in the source slope is
generally obvious, as is the blocky character of the debris sheet (Figure 10.3).
The debris sheet has an irregular or undulating morphology that differs from
the form of adjacent parts of the slope that have not been overrun by the
landslide. The deposit of a large rockslide also will appear as a convex bulge
on the slope on which it has come to rest.

Large rockslides can transform into rock avalanches (also known as
sturzströms), which are large, fast-moving flows of fragmenting rock (Hewitt
et al., 2008). Their rapid motion appears to result, at least in part, from the
release of large amounts of energy due to particle interactions and comminution. Rock avalanches are the fastest of all landslides, and in some cases,
they achieve speeds of ≥ 100 m/s. They also travel long distances and, where
unimpeded by topography, spread out and come to rest as broad thin sheets of
debris (Figure 10.4). A continuum exists between rockslides and rock avalanches; the former do not achieve the mobility or travel as far as the latter.
Surface features produced by rock avalanches include ridges and other linear
features parallel and perpendicular to flow. These features are produced by
internal shear, extension, and compression of the flowing debris. Where rocks
of different lithologies crop out in the source zone of a rockslide or rock
avalanches, the debris sheet may be lithologically zoned, with the lowest rock
type in the scarp moved to the front of the deposit.

"Flows" are a large and diverse group of landslides, but they have one
similarity—the failed material moves as fluid, commonly a non-Newtonian
fluid. In wet flows, rock fragments are partly supported by interstitial water.
Debris flows are the most common type of wet flow; they consist of a mixture
of water, rock fragments, and plant material that moves down a steep stream

FIGURE 10.4   Blocky deposit of the 1903 Frank rock avalanche viewed from the top of Turtle Mountain, the source of the landslide. Where unimpeded by topography, rock avalanches spread out and deposit relatively thin debris sheets over large areas. The Frank Slide Interpretive Center is circled. *Photograph—John J. Clague.*

course or a ravine as a slurry. Mudflows are similar to debris flows, but the solid fraction consists mainly of sand, silt, and clay, with little or no coarser material. Sediment flows also happen in oceans and lakes, especially off deltas and at the heads of submarine canyons. Those that travel down submarine canyons into deep ocean waters are termed "turbidity currents." Deposits of turbidity currents and other subaqueous flows cannot be imaged using traditional remote-sensing tools. However, detailed three-dimensional models of the floors of lakes or the sea can be acquired with multibeam echo sounders. Deep-sea submarine fans have a distinctive geomorphology—a low-gradient, radially sloping surface extending from the mouth of a submarine canyon to abyssal depths; levees border channels within the canyon and on the fan itself.

Many landslides that are termed "flows" are misnomers in that they do not involve a thorough mixing of the failed materials. An example is earth flows, which are large masses of rock debris that move slowly downslope as a viscous mass, deforming in a laminar fashion (Figure 10.5).

The preservation potential of flows, like slides, depends strongly on their size. Most individual debris flows are small ($<100,000$ m^3) and thus are not preserved as landforms. Debris flow fans, however, are the product of a large number of debris flows and, like talus cones, can persist in the landscape for long periods. It may be difficult to distinguish a debris flow fan from an alluvial fan that consists entirely or largely of fluvial deposits. Debris flow fans, however, are steeper and more hummocky than are alluvial fans. Boulders on the fan surface and levees bordering channels argue for deposition by debris flows rather than stream flow.

FIGURE 10.5   The Pavilion earthflow in southwest British Columbia. This large landslide has been active episodically throughout the Holocene and has a bulging irregular surface indicative of slow downslope movement of a viscous body of debris. The inset geomorphologic map shows areas affected by four phases of movement (1 is oldest; 4 is the youngest). *Photograph—John J. Clague, map—Bovis (1985), Figure 10.7.*

Many landslides do not fit comfortably into existing classification schemes. A notable example, and one with high preservation potential, is sackung, a German word meaning "sag." A sackung is produced by the slow downslope movement of a large, internally broken rock mass, commonly without a single, well defined basal failure surface. Movement is manifested at the land surface by cracks, trenches, and scarps at midslope and upper slope positions, and by bulging lower on the slope (Figure 10.6). Sackungs are notable in many ways. First, they are some of the largest landslides on Earth; many have volumes >1 km^3. Second, they can move for thousands or tens of thousands of years,

FIGURE 10.6   Sackung at Fels Glacier, Alaska. Note the bulging of the toe of the slope and extensive surface cracking. The field of view in the foreground is about 400 m wide. *Photograph—John J. Clague.*

although typically at very low velocities, on the order of millimeters per year (Ambrosi and Crosta, 2006). Third, sackungs occur in nearly all rock types. They are most common in foliated metamorphic rocks, especially schist, with foliation dipping parallel to the slope, and in fine-grained sedimentary rocks with beds dipping parallel to the slope. Sackungs, however, have been documented in highly jointed granitic and volcanic rocks. Fourth, some sagging slopes may deteriorate until they fail catastrophically as rockslides or rock avalanches. Whether or not a sackung will fail catastrophically depends on the topography, the orientation of discontinuities (foliation, bedding, and joints) within the failing rock mass, and the absence or presence of favorably oriented, low-strength layers (e.g., clay layers at Vajont, Italy; Hendron and Patton, 1985; a felsite sheet at Hope, British Columbia; Mathews and McTaggart, 1978). Small displacements or catastrophic failure may occur during earthquakes (e.g., 2002 Denali earthquake). Many sackungs lack pronounced head scarps, but are recognizable from their commonly downstepped irregular surfaces, uphill- and downhill-facing scarps, and bulging toes.

Finally, many landslides, including most large ones, are "complex," that is, they involve more than one type of movement. A rockslide, for example, may evolve into a debris flow as it entrains water or saturated sediments along its path. Complex landslides may have geomorphic signatures of two or more types of movement. For example, the 1965 Hope Slide in British Columbia began as a rockslide with the collapse in two phases of 47 million cubic meters of rock from a steep valley wall (Figure 10.7). The fragmented rock mass struck a lake on the valley floor, and its leading edge incorporated water and valley-floor sediment, transforming into debris flows that ran up and down the valley, past the limits of the rockslide (Mathews and McTaggart, 1978).

Scientists have misinterpreted some landforms produced by large paleolandslides as evidence of other processes. For example, steep linear, uphill-facing scarps below mountain ridges (Figure 10.8) have been interpreted as products of recent (Holocene) earthquakes. They are, however, gravitational scarps and part of the suite of sackung features occurring on slowly deforming slopes (e.g., Clague and Evans, 1994). Although not necessarily earthquake generated, many long and straight uphill-facing scarps mark the traces of ancient crustal faults and weak gouge zones (Thompson et al., 1997; Hensold, 2011). Another example is the misinterpretation of some paleolandslide deposits as being glacial in origin. Huge rock avalanches in the Karakoram Himalaya, for example, have left deposits many kilometers from their sources. In some cases, these deposits have been interpreted to be moraines left by Pleistocene valley glaciers (Hewitt et al., 2008).

10.3.2 Role of Stratigraphy and Sedimentology

A second tool used by paleolandslide researchers is stratigraphy. It supplants geomorphology when mass movement landforms are destroyed by erosion.

FIGURE 10.7   The 1965 Hope Slide initiated as a rockslide, but it transformed into debris flows when the rock debris struck a lake on the valley floor at the center of this photograph. The debris flowed up and down the valley.

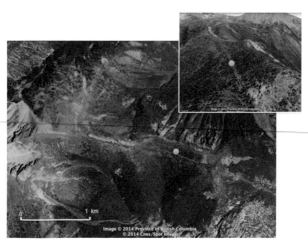

FIGURE 10.8   The Hell Creek "fault" in southwest British Columbia. The filled blue circle shows the same location on each image. Although this linear uphill-facing scarp has developed in the gouge zone of a Neogene fault, it is a product of Holocene gravitational movements, not earthquakes. Many scarps associated with sackung have been misinterpreted as having formed during earthquakes. *Satellite image − Google Earth, photograph—John J. Clague.*

The likelihood that geomorphic evidence of landslides is preserved decreases with time; in the case of Pleistocene events, all that may be left are the scarp and remnants of the deposit.

Natural exposures and excavated sections provide information on the deposit characteristics, source material, degree of fragmentation or mixing of substrate materials, and transport mechanism. In particular, the often-important role of water in failure and transport can be inferred. Deformation of the substrate over which the landslide traveled provides insights into shear and extensional stresses. It may also allow scientists to test proposals to explain the excess mobility of some large landslides, for example, by undrained loading. The position of a landslide deposit within a sequence of sediments may help constrain the age of the event (next section). Unfortunately, a remnant deposit, without geomorphic context, provides little or no information on the size and source of the landslide.

10.4 DATING PALEOLANDSLIDES

Establishing the age of individual landslides is an important element of paleolandslide research. Accurate ages of landslides are required to establish F−M relations that can be used to estimate risk. They are also needed to assess possible causative factors, such as a lengthy period of wetter-than-normal climate, or triggers such as large earthquakes.

Geoscientists use one of three methods to provide absolute (calendric) ages of paleolandslides: dendrochronology, radiocarbon dating, and terrestrial cosmogenic nuclide (TCN or surface exposure) dating. Although an exhaustive review of these methods is beyond the scope of this chapter, I discuss their specific applications to dating landslides, as well as their limitations. Relative dating methods, including lichenometry (Bull and Brandon, 1998) and weathering (Whitehouse et al., 1986), are not discussed here.

10.4.1 Dendrochronology

The use of tree rings to date past events is referred to as dendrochronology. Three general approaches are used in landslide applications (Clague, 2010). First, the age of the oldest tree living on a landslide deposit is determined. Trees obviously colonized the surface of the deposit after the landslide; thus, the oldest tree provides a minimum age for the event. This strategy has been successfully used, for example, to date debris flows triggered by exceptional regional rainstorms in British Columbia (Schwab, 1998). The assumption underpinning this approach is that the tree seeded soon after the landslide. However, trees do not immediately seed on new landslide surfaces—there is a period, termed the ecesis interval, between the time the new surface becomes available and the time that trees become established on it. Ecesis intervals in northwest North America range from a year or two to almost 100 years,

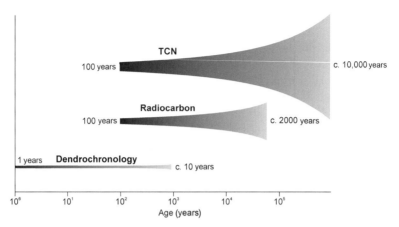

FIGURE 10.9   Time intervals over which the common dating techniques can be applied to landslides. Also shown are approximately uncertainties (1 sigma) associated with each of the techniques.

depending on the tree species, the nature of the substrate, and climate (e.g., Sigafoos and Hendricks, 1969; Wiles et al., 1999; Luckman, 2000). A correction may be applied for a specific area if the ecesis time is known in the modern environment. A more serious problem, which commonly restricts the strategy to landslides <1,000 years old (Figure 10.9), is that the living trees on a landslide deposit may not be first cohort. In other words, the original cohort may have died and been replaced by second-growth or younger forest. In such a case, the living trees provide, at best, a gross minimum age for the landslide.

A second strategy is to date trees that were damaged, but not killed, by a landslide. Catastrophic landslides, including rockslides and rock avalanches, remove trees along their paths, but some trees at the lateral margins and fronts of these landslides survive the trauma. Damage to the crown or root systems may produce a sudden reduction in the thickness of the annual rings, known as suppression. In other cases, landslides produce a marked increase in ring width (release) by eliminating competition from nearby trees (Paolini et al., 2005). In both cases, counting back from the outermost ring to the ring where tree growth changed easily dates the landslide. Trees also are scarred or tilted by landslides (Clague, 2010). After the landslide, the trees repair the damage—new tissue is progressively emplaced over scars—and the tilted trees gradually right their stems to a vertical position. In both instances, the age of the landslide can be easily determined by counting back from the cambium to the first ring emplaced after the landslide damaged the tree. Dating is not so straightforward in the case of slow-moving earth flows and sackung. Tilted trees may record the initiation of movement, but if the movement is continuous, the tree may never right itself. Episodic movement may leave a complex pattern of stem rings, comprising asymmetric rings formed during times of movement and symmetric ones when no movement occurred (Shroder, 1978; Braam et al., 1987). Such patterns are difficult to interpret.

A third approach is used to date trees that were killed by the landslide. In some cases, scarred or tilted trees at the margin of the debris, although dead, still stand. Floating ring-width chronologies derived from these stems might be crossdated with local or regional master chronologies derived from living trees. In other cases, where trees were overridden and destroyed by the landslide, growth-position roots and boles of trees may remain beneath the debris or exposed along the path of the landslide. Again, ring-width chronologies from the boles possibly can be crossdated with local or regional master chronologies (Bégin and Filion, 1988). In still other instances, logs and stems, entrained by the landslide, occur within the debris (Filion et al., 1991). The strategy of crossdating the subfossil series into living tree-ring chronologies can be problematic. The outer-ring age of an entrained stem or in situ root bole will coincide with the age of the landslide only if the tree was killed by the event. This assumption probably is valid in the case of overridden in situ boles with preserved bark. However, the assumption becomes more tenuous when dealing with incorporated stems and other woody debris lacking bark. In such cases, the researcher must date many stems and also consider the possibility that the landslide entrained deadfall, in which case the age of the outermost ring is greater than that of the landslide.

10.4.2 Radiocarbon Dating

The radiocarbon dating technique can be used to date landslides over a much longer period than dendrochronology can, but at a price—accuracy (Figure 10.9). Dendrochronology offers annual, or in rare cases, subannual resolution, whereas radiocarbon dating provides ages with uncertainties of not less than 10 years and more commonly 50−100 years. There are two sources of uncertainty in radiocarbon ages. First, there is unavoidable imprecision in the laboratory-calculated age. In ultraclean laboratories, analytical uncertainty can be as low as ±10 years (1 sigma), but in most laboratories, it is of the order of ±50 years (1 sigma). Second, radiocarbon years are not the same as sidereal (calendar) years; thus, radiocarbon ages must be "calibrated" to bring them into a standard time frame. With well-established calibration data sets (e.g., Stuiver et al., 2010), the procedure is straightforward, but calculated 1-sigma and 2-sigma calibrated age ranges commonly are larger than the corresponding radiocarbon age ranges. Radiocarbon ages on marine and freshwater shells, foraminifera, and aquatic mosses carry additional uncertainty that must be considered when using these organisms to date subaqueous landslides or lakes impounded by landslides. Mollusks and freshwater mosses commonly incorporate carbon dissolved in water, rather than the atmosphere, and the isotopic ratio of ^{14}C to ^{12}C in these organisms may be lower than that in the atmosphere. As a result, dated shells and aquatic mosses may be anomalously old. Corrections can be made for "old carbon" effects on marine shells, although these introduce additional uncertainties in calculated ages.

Care is also required in interpreting radiocarbon ages. Plant fossils in growth position on a surface over which a landslide travels are ideal material for dating because the landslide killed the plants; in other words, the ages of the fossils and the landslide are the same, given the aforementioned un-certainties. Similarly, radiocarbon ages on growth-position fossils associated with a surface directly below lake sediments deposited in a landslide-dammed lake can be reliably linked to the age of the event. Unfortunately, such occurrences are extremely rare. All fossil plant material and bones entombed in landslide deposits are "detrital," that is, they have been transported to their depositional positions. As such, there is no certainty that the organism was killed by the landslide, which is the conceptual assumption many researchers make when interpreting radiocarbon ages on such fossils. More correctly, the derived ages should be considered "maxima" for the time of the event. It is possible, for example, that a dated piece of wood recovered from landslide debris came from a dead tree, rather than from a living one. Or the wood may have been recycled from older sediments that were eroded by the landslide and incorporated into its deposit.

Another important consideration is the position of the dated rings within the tree. Old trees can be of an age of many hundreds of years. The outer rings of trees are those that are nearest the time of death of the tree and should be preferentially selected for dating. It is unlikely, however, that the outer rings of a tree will be preserved in a fragment of wood recovered from a landslide deposit; if they are not, the derived age will be some unknown number of years older than the landslide.

Uncertainties related to detrital wood ages can be minimized by dating several fragments of wood. Assuming that each radiocarbon age is reliable, the youngest age will be nearest the age of the landslide. The outer-ring issue can be circumvented by dating twigs, leaves, seeds, and other delicate plant fossils, although such materials commonly are not present in landslide deposits.

Some landslides can be dated indirectly, using plant material contained within sediments deposited in an upstream lake impounded behind the debris dam or within outburst flood sediments deposited when the dam is breached by overflow. The same limitations discussed above apply when dating plant material recovered from lacustrine and flood sediments.

Radiocarbon dating is the most widely used tool for dating landslides, but like dendrochronology, it has a temporal limitation. Because the half-life of ^{14}C is about 5700 years, the technique provides reliable ages only back to about 40,000 radiocarbon years (Figure 10.9). Some scientists claim reliable ages as old as 50,000 years, but the amount of modern carbon contamination that will produce an age of 50,000 years from a sample that contains only

^{12}C and thus is much older is miniscule. It is thus dangerous to assume that a 50,000-year age is real and not merely an unknown age >50,000 years.

10.4.3 Terrestrial Cosmogenic Nuclide Dating

Some landslides older than 40,000 years can be dated using TCN methods (Goss and Phillips, 2001; Dunai, 2010). The Earth is constantly bombarded with cosmic rays that interact with atoms in atmospheric gases to produce a cascade of secondary particles, including neutrons. When a neutron strikes an atom within the crystal lattice of a mineral at or within about 10 m of the Earth's surface, it can dislodge one or more protons or neutrons, producing a different element or a different isotope of the original element, called a cosmogenic nuclide. By measuring the concentration of certain cosmogenic nuclides, a scientist can determine how long a particular surface has been exposed. The two most frequently measured cosmogenic nuclides are ^{10}Be and ^{26}Al; they are produced when cosmic rays strike oxygen and silicon, respectively (Nishiizumi et al., 1993). ^{36}Cl is also used to measure the age of surface rocks; it is produced from calcium and potassium (Stone et al., 1996).

TCN dating can be used to date moraines and landslide deposits as young as about 100 years (Schaefer et al., 2009) and as old as about 1 million years (Figure 10.9). However, the technique, like those discussed above, is not without problems. An assumption is that the dated surface, whether the scarp of a landslide or a block deposited at the surface of a debris sheet, has not changed since it first formed. A dated block cannot have changed position since emplacement, nor can it have been exhumed due to erosion of overlying material or weathered significantly (or at least if it has weathered, the rate of weathering must be independently known). A scarp must be the original exposed slip surface of the landslide and cannot have degraded through weathering and erosion since it formed. These are severe limitations, and it becomes increasingly likely that they will be violated with increasing duration of exposure. Another issue is that the cumulative flux of cosmic rays at the landslide site must be known, but it is affected by many factors, including elevation, latitude, the varying intensity of Earth's magnetic field over time, solar winds, and shielding due to topography and snow cover. A consequence is that calculated TCN ages, like radiocarbon ages, carry considerable uncertainty, and the uncertainties increase with age. Uncertainties can be reduced by measuring two radiogenic isotopes in tandem and by acquiring a suite of cosmogenic exposure measurements from the surface of a sedimentary deposit to depths of several meters.

10.5 TEMPORAL BIAS

An issue that relates to both identification and dating of paleolandslides is the lack of completeness of event catalogs. Even a historic catalog is biased

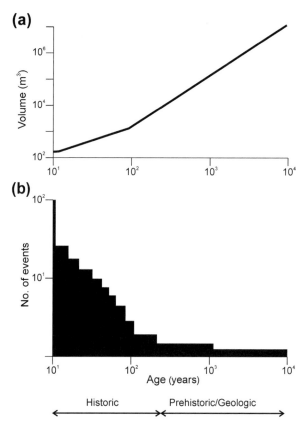

FIGURE 10.10 Temporal of biasing of landslide records. (a) Conceptual graph showing the threshold minimum size of landslides still preserved in a temperate landscape as a function of age. For example, only 100-year-old landslides that are larger than about 10^3 m^3 remain in the landscape. (b) Hypothetical histogram of the number of landslides recorded in a region over the past 10,000 years. Fewer and fewer events are recorded back in time, even in the historic period. The record of small landslides is much more biased than that of large landslides.

toward more recent events (Figure 10.10). During early settlement periods, and even extending to recent decades in many areas, many events go unrecorded or are poorly documented. Any event catalog is therefore biased toward more recent and larger events.

The problem is compounded when the historic record of landslides is extended into prehistory using geologic and geomorphologic data. The record becomes progressively less complete a farther back in time and more biased toward larger events that are preserved longer in the landscape (Figure 10.10). Small paleolandslides generally are poorly preserved and thus are not recorded. Only the largest pre-Holocene landslides (typically >100 million cubic meters) are likely to be detected in glaciated areas, and, as mentioned above, dating these events is itself an issue.

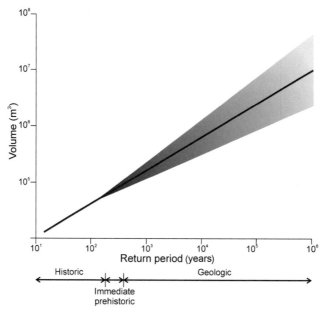

FIGURE 10.11   Hypothetical F−M curve for debris flows within a watershed. The curve is a composite of (1) precisely dated historic events, (2) events in the immediate prehistoric period that have scarred trees and dated using dendrochronology, and (3) large older events documented stratigraphically and radiocarbon dated. The envelope of uncertainty is the grey toned area. Temporal biasing and uncertainties in dating affect the older part of the record.

This issue becomes important when the paleolandslide record is used to estimate long-term landscape erosion rates or to assess landslide hazard and risk. The bias toward younger events could lead to underestimates of erosion rates by landslides or to incorrect F−M relations. F−M curves are at the heart of landslide risk assessments (see Section 10.6); thus, risk potentially could be underestimated or at the least subject to considerable uncertainty. Temporal bias is unavoidable, but the experienced landslide researcher can, to some extent, correct for it. Corrections are based on "expert opinion" and are subjective. For example, historic events should be used to anchor the lower end of the F−M curve, with emphasis on events for which the landslide catalog is deemed to be reasonably complete (Figure 10.11). The geologic record must be used to define the upper end of the F−M curve, but expert judgment is required to define this part of the curve based on the limited subset of those older events, bearing in mind not only the temporal bias in the data but also uncertainties in magnitude and age. Uncertainties, which expand from the lower end to upper end of the F−M curve, should be captured and clearly articulated when using the curve for risk assessment (Figure 10.11).

10.6 ROLE IN LANDSCAPE EVOLUTION

Mass movements are intimately connected with other processes that shape the landscape. They are among the denudational processes that counter constructional tectonic processes that elevate landmasses. They also are closely linked to fluvial and glacial processes—streams and glaciers remove the materials produced by landslides and also condition slopes for further failure. These competing forces are most strongly at work in mountains, and over geologic time are responsible for shaping mountain landscapes. Mass movement processes ultimately limit the maximum elevation of orogens—it is unlikely, given the pace of lithospheric plate interactions and with the Earth's present atmosphere, that mountains higher than the Himalayas are possible on our planet. Most mountain ranges have morphologies controlled as much by mass movements as by glacial erosion. The term "glacial buzzsaw," which has been used as a metaphor for the role that climate and glaciers play in countering and balancing uplift of tectonically active orogens (Montgomery et al., 2001; Spotilla et al., 2004; Mitchell and Montgomery, 2006) does not adequately acknowledge the important role that landslides play.

10.7 RISK ASSESSMENT

Here, I illustrate the use and value of paleolandslide research for evaluating risk to people and property with a current example from British Columbia. The site is "Cheekye Fan," a large debris flow fan directly north of the rapidly growing community of Squamish, southwest British Columbia (Figure 10.12). The Cheekye River, which formed the fan, flows from the steep west face of the Mount Garibaldi massif, a stratovolcano that last erupted at the end of the Pleistocene.

Studies relevant to the hazard on Cheekye Fan date back by >60 years. Mathews (1952, 1958) showed that Mount Garibaldi erupted during the waning stage of the last glaciation and that part of the volcano was built out onto the glaciers in adjacent Squamish and Cheakamus valleys. As these glaciers thinned and receded, the west flank of the volcano collapsed to form ice-contact terraces in the Cheekye River watershed just east of Cheekye Fan. Cheekye River subsequently incised the collapsed this volcanic debris and redeposited it in the confluence area of Cheakamus and Squamish rivers as the Cheekye Fan (Figure 10.12). Consultancy studies, starting in the mid-1970s and continuing to the present, have been completed in response to pressure to develop the fan and in light of concern that large debris flows might threaten that development. These studies, and related studies by academic researchers, have produced an impressive body of data on the frequency and magnitude of debris flows that have contributed to the growth of the fan over the past 10,000 years.

Based on geologic investigations in the Cheekye basin, inspection of available exposures on the fan, and consideration of soil development, Crippen

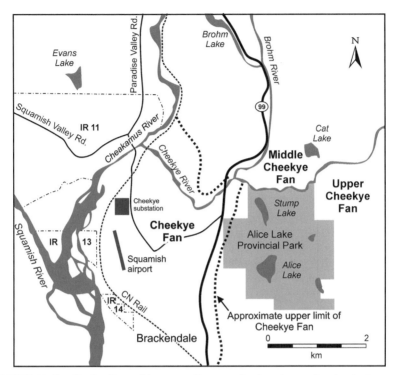

FIGURE 10.12   Cheekye Fan and its surrounding area, showing places referred to in the text.

Engineering (1974) concluded that the Cheekye Fan is a relic of the geologic past, with little possibility of a damaging debris flow today. This conclusion was challenged following further studies in 1991 and 1992, which included excavation and logging of about 50 test pits, review of water well logs, and excavations at a local landfill on the fan. Based on these studies, Thurber Engineering—Golder Associates (1993) produced a probability-of-exceedence plot for debris flows based on the knowledge of sediment deposited in the past 7,500 years and the estimated magnitude of the two largest events known at the time.

Clague et al. (2003) used cores recovered from Stump Lake (Figure 10.12), just upstream of the fan apex, to constrain the maximum size of the debris flows that had affected the fan during the Holocene. Stump Lake is fed by ground water and has no inlet streams; thus, the sediment that normally accumulates within it is gyttja. Its outlet sill, however, is <10 m above the base of the modern Cheekye River channel and only 50 m from the edge of the lake. Thus, the lake could receive sediment only from very large debris flows traveling down the Cheekye River. Based on the core stratigraphy and diatom analysis, Clague et al. (2003) concluded that an early debris flow deepened the lake basin before 11,600 years ago, and that a second debris flow entered the lake between about 7,500 and 7,100 years ago; no other debris flows affected the lake during the Holocene. They used two surveyed cross-sections extending north from the lake across Cheekye River and a Newtonian flow model (Hungr et al., 1984) to estimate velocity and peak discharge of the debris that happened 7,100–7,500 years ago.

Kerr Wood Leidal (2003) reviewed the results of Clague et al. (2003) and estimated that the largest debris flow to affect the fan during the Holocene was no larger than 5.5 million cubic meters. They also modeled scenarios for 3, 5.5, and 7 million cubic meter events and concluded that only the two larger of the three events would impact residential areas of Brackendale on the southern sector of the fan.

Renewed development pressure over the past decade led to a comprehensive review of existing data and a new test-pitting program on the fan. Ninety-five new test pits, ranging in depth from 2 to 6 m, were dug and 16 new radiocarbon ages were obtained on wood fragments and charcoal from paleosols between debris flow units. These new data, together with data on scarring of living trees along the banks of the Cheekye River by debris flows <200 years ago, allowed BGC Engineering (2008) to produce a new F–M curve for debris flows affecting the Cheekye Fan (Figure 10.13) and to conclude that debris flows exceeding 3 million cubic meters are unlikely in the future. This conclusion is significant in that modeling has shown that debris flows of this size, although not larger, could be contained if a large debris flow retention basin were to be built above the fan.

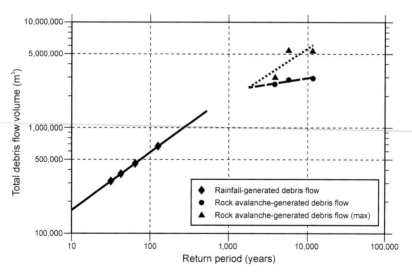

FIGURE 10.13 Frequency–magnitude (F–M) curve for debris flows affecting Cheekye Fan produced by BGC Engineering (2008).

The Cheekye Fan is one of best, if not the best, studied debris flow fans in the world, yet considerable uncertainties still remain, and these uncertainties are pivotal in future decisions about new development on the fan. Risk in such situations can never be reduced to zero, so any decision on land use on the Cheekye Fan must consider what is an acceptable risk. As in most jurisdictions, "acceptable risk" from natural hazards is not clearly quantified in British Columbia. The trend, however, is to disallow new development at sites where there is a 1 in 10,000 chance of a single fatality in a single year. A problem in defining the size of a debris flow with a return period of 10,000 years is that the uncertainty in the size of that debris flow is large, perhaps 50 percent or more. In the case of the Cheekye Fan, geologists have only a 10,000-year record of events to work with, and it is doubtful that the largest debris flow of the past 10,000 years is the same as the largest debris flow that will occur over the next 10,000 years. Further, with uncertainties of say 50 percent, the largest debris flow of the past 10,000 years might be as large as 5 million cubic meters, and if such an event were to occur, it would overwhelm any foreseeable engineered protective works and cover much of the fan.

10.8 CONCLUSION

Study of prehistoric, or "paleo," landslides provides insights into mass movement processes, notably those that are rarely observed and difficult or impossible to model realistically. Examples include large-scale liquefaction, undrained loading at real-world scales, and the dynamic behavior of rock avalanches. Landslide hazard and risk studies also benefit from study of paleolandslides. Reliable landslide F—M curves, which commonly underpin decisions as to whether an area of concern can be developed, require identification, characterization, and dating of larger, rare landslides that have not happened historically but could happen within the period used to benchmark "acceptable" risk. Caution is to be exercised, however, in the use of paleolandslides in hazard and risk assessments. Events must be accurately dated, and inevitable uncertainties considered and explicitly captured in risk assessments. The largest source of uncertainty lies in the temporal bias that exists in all landslide catalogs. The record of landslides becomes increasingly incomplete with time and biased toward progressively larger events.

REFERENCES

Ambrosi, C., Crosta, G.B., 2006. Large sackung along major tectonic features in the Central Italian Alps. Eng. Geol. 83, 183—200.

Bégin, C., Filion, L., 1988. Age of landslides along Grande Rivière de la Baleine estuary, eastern coast of Hudson Bay, Quebec (Canada). Boreas 17, 289—299.

BGC Engineering, 2008. Cheekye River Debris Flow Frequency and Magnitude. Report prepared for Kerr Wood Leidal Associates by BGC Engineering Ltd., Vancouver, BC.

Bovis, M.J., 1985. Earthflows in the interior plateau, southwest British Columbia. Can. Geotech. J 22, 313−334.

Braam, R.R., Weiss, E.E.J., Burrough, A., 1987. Spatial and temporal analysis of mass movement using dendrochronology. Catena 14, 573−584.

Bull, W.B., Brandon, M.T., 1998. Lichen dating of earthquake-generated regional rockfall events, Southern Alps, New Zealand. Geol. Soc. Am. Bull. 110, 60−84.

Bulmer, M.H.K., 2012. Landslides on other planets. In: Clague, J.J., Stead, D. (Eds.), Landslides; Types, Mechanisms and Modeling. Cambridge University Press, Cambridge, pp. 393−408.

Clifton, H.E. (Ed.), 1988. The 1929 "Grand Banks" Earthquake, Slump, and Turbidity Current. Geological Society of America Special Paper 229.

Clague, J.J., 2010. Dating landslides with trees. In: Stoffel, M., Bollschweiler, M., Butler, D.R., Luckman, B.H. (Eds.), Tree Rings and Natural Hazards: A State-of-the-Art. Advances in Global Change Research. Springer, Dordrecht, pp. 81−89.

Clague, J.J., Evans, S.G., 1994. A gravitational origin for the 'Hell Creek fault', British Columbia. In: Current Research 1994-A, pp. 193−200. Geological Survey of Canada Paper 1994-A.

Clague, J.J., Friele, P.A., Hutchinson, I., 2003. Chronology and hazards of large debris flows in the Cheekye River basin, British Columbia, Canada. Environ. Eng. Geosci. 8, 75−91.

Crippen Engineering, 1974. Report on Geotechnical−Hydrological Investigation of the Cheekye Fan. Report prepared for B.C. Department of Housing by Crippen Engineering Ltd., Vancouver, BC.

Cruden, D.M., Varnes, D.J., 1996. Landslide types and processes. In: Turner, A.K., Schuster, R.L. (Eds.), Landslides—Investigation and Mitigation. National Research Council, pp. 36−75. Transportation Research Board Special Report 247.

Dingle, R.V., 1977. The anatomy of a large submarine slump on a sheared continental margin (SE Africa). J. Geol. Soc. London. 134, 293−310.

Dunai, T.J., 2010. Cosmogenic Nuclides: Principles, Concepts and Applications in the Earth Surface Sciences. Cambridge University Press, Cambridge, MA.

Eisbacher, G.H., Clague, J.J., 1984. Destructive Mass Movements in High Mountains: Hazard and Management. Geological Survey of Canada Paper 84-16, 230 pp.

Filion, L., Quinty, F., Bégin, C., 1991. A chronology of landslide activity in the valley of Rivière du Gouffre, Charlevoix, Quebec. Can. J. Earth Sci. 28, 250−256.

Goss, J.C., Phillips, F.M., 2001. Terrestrial in situ cosmogenic nuclides: theory and application. Quat. Sci. Rev. 20, 1475−1560.

Hancox, G.T., Perrin, N., 2009. Green Lake landslide and other giant and very large landslides in Fiordland, New Zealand. Quat. Sci. Rev. 28, 1020−1036.

Hendron, A.J., Patton, F.D., 1985. The Vaiont Slide, a Geotechnical Analysis Based on New Geologic Observations of the Failure Surface. U.S. Army Corps of Engineers Technical Report GL-85-5, 324 pp.

Hensold, G., 2011. An Integrated Study of Deep-Seated Gravitational Slope Deformations at Handcar Peak, Southwestern British Columbia. (M.Sc. thesis). Simon Fraser University, Burnaby, BC, 136 pp.

Hermanns, R.L., Hansen, L., Sletten, K., Böhme, M., Bunkholt, H., Dehls, J.F., Eilertsen, R., Fischer, L., ĹHeureux, J.-S., Høgaas, F., Nordahl, B., Oppikofer, T., Rubensdotter, L., Solberg, I.-L., Stalsberg, K., Yugsi Molina, F.X., 2012. Systematic geological mapping for landslide understanding in the Norwegian context. In: Eberhardt, E., Froese, C., Turner, A.K., Leroueil, S. (Eds.), Landslide and Engineered Slopes: Protecting Society through Improved Understanding. Taylor & Francis Group, London, pp. 265−271.

Hewitt, K., Clague, J.J., Orwin, J.F., 2008. Legacies of catastrophic rock slope failures in mountains. Earth Sci. Rev. 87, 1–38.

Hungr, O., Morgan, G.C., Kellerhals, R., 1984. Quantitative analysis of debris torrent hazards for design of remedial measures. Can. Geotech. J. 21, 664–677.

Jansen, E., Befring, S., Bugge, T., Eidvin, T., Holtedahl, H., Sejrup, H.P., 1987. Large submarine slides on the Norwegian continental margin: sediments, transport and timing. Mar. Geol. 78, 77–107.

Kerr Wood Leidal, 2003. Preliminary Design Report for Cheekye Fan Deflection Berms. Report prepared for District of Squamish, Squamish by Kerr Wood Leidal Associates Ltd., Burnaby, BC.

Luckman, B.H., 2000. The Little Ice Age in the Canadian Rockies. Geomorphology 32, 357–384.

Mathews, W.H., 1952. Mount Gariabldi, a supraglacial Pleistocene volcano in southwestern British Columbia. Am. J. Sci. 250, 553–565.

Mathews, W.H., 1958. Geology of the Mount Garibaldi map area, southwestern British Columbia, Canada. Part II: geomorphology and Quaternary volcanic rocks. Geol. Soc. Am. Bull. 69, 179–198.

Mathews, W.H., McTaggart, K.C., 1978. Hope rock slides, British Columbia. In: Voight, B. (Ed.), Rockslides and Avalanches, Natural Phenomena, vol. 1. Elsevier, New York, NY, pp. 259–275.

Mitchell, S.G., Montgomery, D.R., 2006. Influence of a glacial buzzsaw on the height and morphology of the cascade range in central Washington State. Quat. Res. 65, 96–107.

Montgomery, D.R., Balco, G., Willett, S.D., 2001. Climate, tectonics, and the morphology of the Andes. Geology 30, 1047–1050.

Nishiizumi, K., Kohl, C.P., Arnold, J.R., Dorn, R., Klein, I., Fink, D., Middleton, R., Lal, D., 1993. Role of in situ cosmogenic nuclides [10]Be and [26]Al in the study of diverse geomorphic processes. Earth Surf. Processes Landforms 18, 407–425.

Paolini, L., Villalba, R., Grau, H.R., 2005. Precipitation variability and landslide occurrence in a subtropical mountain ecosystem of NW Argentina. Dendrochronologia 22, 175–180.

Poschinger, A., Kippel, Th, 2009. Alluvial deposits liquefied by the Flims rock slide. Geomorphology 103, 50–56.

Poschinger, A. v., Wassmer, P., Maisch, M., 2006. The Flims rock slide; history of interpretation and new insights. In: Evans, S.G., Scarascia-Mugnozza, G., Strom, A., Hermanns, R.L. (Eds.), Massive Rock Slope Failure. Kluwer Academic Publishers, Dordrecht, pp. 341–369.

Reznichenko, N.V., Davies, T.R.H., Shulmeister, J., Larsen, S.H., 2012. A new technique for identifying rock avalanche–sourced sediment in moraines and some paleoclimatic implications. Geology 40, 319–322.

Roberts, N.J., Evans, S.G., 2013. The gigantic Seymareh (Saidmarreh) rock avalanche, Zagros fold-thrust belt, Iran. J. Geol. Soc. London 170, 685–700. http://dx.doi.org/10.1144/jgs2012-090.

Schaefer, J.M., Denton, G.H., Kaplan, M., Putnam, A., Finkel, R.C., Barrell, D.J.A., Andersen, B.G., Schwartz, R., Mackintosh, A., Chinn, T., Schluechter, C., 2009. High-frequency Holocene glacier fluctuations in New Zealand differ from the northern signature. Science 324, 622–625.

Schwab, J.W., 1998. Landslides on the Queen Charlotte Islands: processes, rates, and climatic events. In: Hogan, D.J., Tschaplinksi, P.J., Chatwin, S. (Eds.), Carnation Creek and Queen Charlotte Islands Fish/Forestry Workshop: Applying 20 Years of Coast Research to Management Solutions. B.C., Land Management Handbook, vol. 41 Ministry of Forests, pp. 41–48.

Shroder, J.F., 1978. Dendrogeomorphological analysis of mass movements on Table Cliffs Plateau, Utah. Quat. Res. 9, 168—185.

Sigafoos, R.S., Hendricks, E.L., 1969. The time interval between stabilization of alpine glacial deposits and establishment of tree seedlings. U.S. Geological Survey Professional Paper 650-B pp. B89—B93.

Spotilla, J.A., Buscher, J.T., Meigs, A.J., Reiners, P.W., 2004. Long-term glacial erosion of active mountain belts: example of the Chugach-St. Elias Range, Alaska. Geology 32, 501—504.

Stone, J., Allan, G., Fifield, L., Cresswell, R., 1996. Cosmogenic chlorine-36 from calcium spallation. Geochim. Cosmochim. Acta 60, 679—692.

Stuiver, M., Reimer, P.J., Reimer, R.W., 2010. CALIB 6.0. http://calib.qub.ac.uk/calib/.

Thompson, S.C., Clague, J.J., Evans, S.G., 1997. Holocene activity of the Mt. Currie scarp, Coast mountains, British Columbia, and implications for its origin. Environ. Eng. Geosci. 3, 329—348.

Thurber Engineering — Golder Associates, 1993. The Cheekeye River Terrain Hazard and Land-Use Study, Final Report. Report prepared for British Columbia Ministry of Environment, Lands and Parks, by Thurber Engineering — Golder Associates, Vancouver, BC.

Turnbull, J.M., Davies, T.R.H., 2006. A mass movement origin for cirques. Earth Surf. Processes Landforms 31, 1129—1148.

Varnes, D.J., 1978. Slope movement types and processes. In: Schuster, R.L., Krizek, R.J. (Eds.), Landslides—Analysis and Control. National Research Council, pp. 11—23. Transportation Research Board Special Report 176.

Whitehouse, I.E., McSaveney, M.J., Knuepfer, P.L.K., Chinn, T.J., 1986. Growth of weathering rinds on Torlesse Sandstone, Southern Alps, New Zealand. In: Coleman, S.M., Dethier, D.P. (Eds.), Rates of Chemical Weathering of Rocks and Minerals. Academic Press, New York, NY, pp. 419—535.

Wiles, G.C., Barclay, D.J., Calkin, P.E., 1999. Tree-ring-dated little Ice Age histories of maritime glaciers from western Prince William Sound. Holocene 9, 163—173.

Chapter 11

Remote Sensing of Landslide Motion with Emphasis on Satellite Multitemporal Interferometry Applications: An Overview

Janusz Wasowski [1] and Fabio Bovenga [2]

[1] *National Research Council, CNR-IRPI, Italy,* [2] *National Research Council, CNR-ISSIA, Italy*

ABSTRACT

Landslide hazard reduction can benefit from increased exploitation of affordable remote sensing systems, with a focus on early detection of ground deformations, long-term monitoring, and possibly early warning of catastrophic failure. Among several innovative space-based remote-sensing techniques, synthetic aperture radar (SAR) and multitemporal interferometry (MTI) hold the most promise, because of its capacities and strengths: (1) wide-area coverage (tens of thousands of square kilometers) combined with a high spatial resolution (up to 1 m for the new generation of radar sensors) and hence the possibility of conducting multiscale investigations with the same data sets (from regional to slope-specific); (2) systematic, high-frequency (from a few days to weeks) measurements over long periods (years); (3) a high precision of surface displacement measurements (millimeters—centimeters) only marginally affected by poor weather conditions; (4) cost effectiveness, especially in the case of long-term, large-area investigations (catchment to regional scale); and (5) integration of landslide monitoring (based on new satellite imagery) with retrospective studies (archived imagery) to investigate slope failure history or landslide reactivation/acceleration processes. We illustrate the potential of MTI and explain how it can be used to detect and monitor landslide motion by considering applications in areas with a broad range of geomorphic, climatic, and vegetation conditions. The chosen examples of local-to-catchment-scale MTI case studies focus on unstable hill slopes and landslides in the Apennines (Italy), the European Alps, and on the island of Haiti. The potential of MTI is further assessed by also considering the strengths and limitations of other innovative applications of remote sensing in landslide monitoring, which rely on several recent or emerging techniques: Corner Reflector SAR interferometry, which

Landslide Hazards, Risks, and Disasters. http://dx.doi.org/10.1016/B978-0-12-396452-6.00011-2
345

exploits artificial targets installed on the ground and radar satellite imagery; ground-based InSAR; air- and ground-borne Light Detection And Ranging (LiDAR); and air/space-borne image matching. These applications, however, typically focus on single failed slopes and their use for regular, wide-area mapping of ground surface changes is at present economically prohibitive. We foresee that MTI will make landslide monitoring more effective and more affordable in more situations, and will become increasingly more important in cases where little or no conventional monitoring is feasible (e.g., remote locations and limited funds). We also expect that the role of prevention in slope hazard management can be enhanced by capitalizing more on the presently underexploited advantage of MTI, that is, its ability to regularly provide vast amounts of quantitative information on slope/ground stability conditions in large areas currently unaffected (or thought to be unaffected) by landslides, but where the terrain geomorphology and geology may indicate potential for future failures. Finally, we stress that high spatial and temporal resolution satellite remote sensing of ground deformations open new possibilities for landslide research and for more timely and detailed slope hazard assessment.

11.1 INTRODUCTION

Landslide hazards represent a worldwide problem in as much as failures occur not only in mountainous settings, but can also affect even modest-relief coastal zones, river banks, and artificial slopes. The population growth, with increasing impact of humans on the environment and urbanization of areas susceptible to slope failures, coupled with the ongoing change in climate patterns, will require a shift in the approaches to landslide hazard assessment and reduction. Indeed, there is evidence that landslide-related, socioeconomic losses have been increasing worldwide, especially in developing countries (Schuster, 1996; Petley, 2012). It follows that the protection of preexisting and newly developed areas only via traditional engineering stabilization works and in situ monitoring is no longer considered economically feasible. Further, even where available, ground control systems are typically installed only post factum and for short-term monitoring, and hence, their role in preventing disasters is limited.

Considering the global dimension of the problem of slope instability, a sustainable way towards landslide-hazard reduction seems to be via increased exploitation of affordable remote-sensing systems, with focus on early detection, long-term monitoring, and possibly early warning. In this context, satellite-based remote sensing has great potential thanks to wide-area coverage, regular schedule with increasing revisit frequency, improving resolution of space-borne sensors, as well as availability of free or low cost imagery through international organizations (e.g., European Space Agency—ESA, Bally, 2012).

New and already-planned satellite missions can not only provide global capacity for research-oriented and practical applications, such as mapping, characterizing and monitoring of landslides, but they can also offer a possibility to push the research frontier and prompt innovative studies on slope-movement dynamics and processes. We believe that among a number of emerging space-based remote sensing techniques, synthetic aperture radar (SAR), multitemporal interferometry (MTI) is the most promising for important innovation in landslide-hazard assessment and monitoring. The technique as applied to landslide research (e.g., Colesanti et al., 2003; Hilley et al., 2004; Colesanti and Wasowski, 2004, 2006), is appealing to scientists and end users concerned with slope hazards because it can provide very precise information on slow displacements of the ground surface over vast areas where vegetation cover is limited. The outstanding capacities and strengths of MTI applications include

1. Wide-area coverage (tens of thousands of square kilometers) combined with a high spatial resolution (up to 1 m for the new generation of radar sensors) and hence the possibility of conducting multiscale investigations with the same data sets (from regional to slope-specific).
2. Systematic high-frequency (from few days to weeks) measurements over long periods (years).
3. High precision of surface displacement measurements (millimeters– centimeters) only marginally affected by poor weather conditions.
4. Cost effectiveness, especially in the case of long-term, large-area investigations (from catchment to regional scale).
5. Integration of landslide monitoring (based on new satellite imagery) with retrospective studies (archived satellite) to investigate slope failure history or landslide reactivation/acceleration processes.
6. Regular, regional-scale update on the persistence (or not) of slope/ground stability conditions in inhabited areas or those intended to be urbanized (prevention and land-use planning).

Considering the strengths listed above, we foresee a possibility of future widespread exploitation of MTI technology for slope hazard assessment. It is also expected that MTI applications will become increasingly important in cases where little or no conventional monitoring is feasible (e.g., remote locations and limited funds). Further, we envision that the current approach to assessment of slope hazard can be transformed by capitalizing more on the presently underexploited advantage of the technique, that is, the unique capability to provide regularly, and in a cost-efficient way, enormous amounts of quantitative information on the slope/ground stability conditions in large areas currently unaffected by landslides, but where the terrain geomorphology and geology may indicate potential for future failures. Indeed, conventional geotechnical landslide monitoring efforts typically focus on single failed slopes (post factum investigations). This is also true for the relatively recent or emerging remote-sensing techniques including CRInSAR (Corner Reflector SAR interferometry) that exploits artificial targets installed on the ground and radar satellite imagery (e.g., Froese et al., 2008; Crosetto et al., 2013); GBInSAR (ground-based InSAR) (e.g., Tarchi et al., 2003; Monserrat et al.,

2013); air- and ground-borne such as LiDAR (Light Detection And Ranging) (e.g., Jabayedoff et al., 2010; Abellán et al., 2013), or air/space-borne image matching (e.g., Debella-Gilo and Kääb, 2012; Booth et al., 2013). The use of air-borne LiDAR for regular, wide-area mapping of ground surface changes, although technically possible, is prohibitively expensive at present. A considerable limitation of the air/space-borne image matching techniques is their relatively low measurement precision.

Although MTI techniques are considered to have already reached the operational level (Bally, 2012), it is apparent that in both research and practice we are at present only beginning to benefit from the high-resolution imagery that is currently acquired by the new generation of radar satellites (e.g., Notti et al., 2010; Bovenga et al., 2012a; Herrera et al., 2013). At the same time, some technical and data interpretation issues still limit practical applications to landslide investigations and require much attention. These issues have recently been discussed in detail by Wasowski et al. (2012a) and Wasowski and Bovenga (2014).

The main purpose of the present work is to illustrate the tremendous potential of MTI and explain what this remote-sensing technique can deliver in terms of detection and monitoring motion of landslides. This is done by considering different areas characterized by a wide range of geomorphic, climatic, and vegetation conditions, and presenting selected case study examples of local to catchment scale MTI applications comprising hill slopes in the Southern Apennines (Italy), the European Alps, and unstable slopes in the island of Haiti. The ultimate goal is to show that by taking full advantage of the strengths of new space technologies, remote sensing can help improve our methods of slope hazard assessment and make landslide monitoring more effective and more affordable in more situations than in the past.

We are aware that slope failure mechanisms are characteristically complex and that the determination of the depth and shape of the slip surface is needed for a detailed landslide assessment (Ch 13, in this volume). Furthermore, spatiotemporal patterns of postfailure landslide motions could be similarly complex so that one single measurement technique of ground surface displacements may not be applicable everywhere or sufficient to provide all the necessary information (e.g., about the lateral and vertical movements within a slide body, distribution and rates of these movements, and the areal extent/boundaries of a slide). Many different field-based and remote-sensing technologies can be employed for landslide investigation and monitoring (e.g., Turner and Schuster, 1996; Metternicht et al., 2005; Delacourt et al., 2007), depending, among others, on the objectives and scale of a project, vulnerability level and expected losses, and financial constraints. Although a comprehensive review of ground-, air-, and space-based techniques used in landslide monitoring is beyond the scope of this paper, these different methods are introduced in the summary discussion section, and their strengths and limitations are compared with those of MTI applications. In doing this, the emphasis remains on innovative approaches providing high spatial and temporal resolution data, which can open new possibilities for landslide research and for more timely and detailed hazard assessment.

Finally, it is also recognized that monitoring of surface displacement based on innovative satellite remote sensing, though very appealing to scientists, may not on its own be sufficient to advance our understanding of landslide causative factors. The remotely sensed displacements will have to be integrated with other observations regarding temporal variations of environmental factors, such as precipitation, as well as with information derived from in situ controls and subsurface investigations (e.g., material properties and hydrogeological data) to provide a sound assessment of landslide hazard and facilitate mitigation efforts (e.g., Fell et al., 2008).

11.2 BRIEF INTRODUCTION TO DIFFERENTIAL SAR INTERFEROMETRY AND MULTITEMPORAL INTERFEROMETRY

Theoretical and technical details on space-borne SAR and DInSAR can be easily found in the literature on remote sensing. Further, the principles of radar interferometry have been presented in several articles published in two special volumes of Engineering Geology dedicated to remote sensing of landslides and unstable slopes (e.g., Colesanti et al., 2003; Bovenga et al., 2006; Colesanti and Wasowski, 2006). For additional, updated information on the subject, as well as a detailed reassessment of MTI techniques applied to landslide investigations, the readers are referred to a recent review article by Wasowski and Bovenga (2014). Therefore, DInSAR is only briefly mentioned here, and the introduction to MTI focuses on the aspects most relevant to landslide motion detection and monitoring.

11.2.1 DInSAR and MTI

Space-borne SARs are active sensors that use microwave signals for imaging the Earth surface. A SAR image is made by pixels of complex value (amplitude and phase) representing the electromagnetic signal backscattered by a resolution cell on the Earth's surface and arranged along two dimensions: the sensor—target distance direction (line of sight—LOS or slant range) and the satellite flight direction (azimuth). The resolution cell size varies from few tens of meters to few tens of centimeters, depending on the microwave signal bandwidth and on the antenna length. The SAR amplitude is related to the orientation of the resolution cell with respect to the sensor, the roughness of the surface, and the dielectric properties of the target, whereas the phase is related to the ratio between the sensor—to-target distance (R) and the signal wavelength (λ).

To perform distance measurements, phase difference images (interferograms) are generated by coupling two or more SAR images acquired over the same area

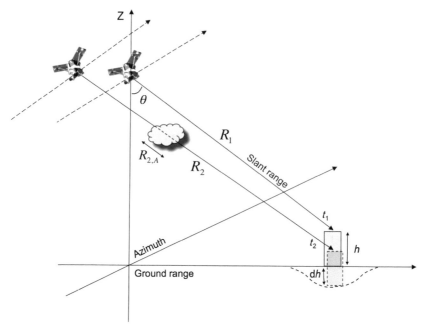

FIGURE 11.1   Sketch of two SAR acquisitions occurring at successive satellite passes.

during successive satellite passes. The interferometric phase is proportional to the wavelength and to the difference between the two sensor–target distances $dR = R_1 - R_2$ (Figure 11.1), which depends on (1) the target elevation (h in Figure 11.1) with respect to a reference surface (dR_T); (2) the target displacement (dh in Figure 11.1) that occurred between the acquisition times of the two satellite passes (dR_D); (3) the refractivity index changes due to the presence of the atmosphere (dR_A); and (4) decorrelation sources (dR_N):

$$\Phi = \frac{4\pi}{\lambda}\left(dR_T + dR_A + dR_D + dR_N\right). \tag{11.1}$$

According to Eqn (11.1), the interferometric phase can be used both to produce three-dimensional (3D) images of the Earth's surface and to measure possible movements of the ground surface occurred between the SAR image acquisitions along the LOS (DInSAR principle). DInSAR processing requires an independent elevation model (Hanssen, 2001) to remove the target elevation contribution to the interferometric phase, and is able to measure the movement with a sensitivity of $\lambda/2$ (Eqn (11.1)). Since the wavelength of the different SAR bands ranges from about 3 to 20 cm (Table 11.1), DInSAR displacement measurements can potentially reach millimeter precisions, provided that dR_N and dR_A in Eqn (11.1) are either negligible or can be reliably removed. These are the major limiting factors for the applicability of DInSAR to movement

TABLE 11.1 Selected Characteristics of Principal SAR Sensors

Satellite Mission	Wave Length (cm)	Life Status	Resolution Az./Range (m)	Repeat Cycle (Days)	Swath Width (km)	Maximum Velocity (cm/Year)	Incident Angle (Degree)
			C-Band				
ERS-1/2	5.6	1992–2001	≈ 6/24	35	100	14.6	23
ENVISAT	5.6	2003–2010	≈ 6/24	35	100	14.6	19–44
RADARSAT-1	5.5	1995–	≈ 8–30	24	45 (Fine) 100 (Strip) 200 (Scan)	20.4	20–50
RADARSAT-2	5.5	2007–	≈ 3/3 ≈ 8/8 ≈ 26/25	24	10 (Spot) 40 (Strip) 200 (Scan)	20.4	20–50
Sentinel-1	5.6	2014–2024	5–20	6, 12	250	85	30–46
RADARSAT constellation mission (3 Sat)	5.5	2018–2026	5–50	3, 12	30–350	163,2	20–55
			L-Band				
J-ERS	23.5	1992–1998	18	44	75	48.7	35
ALOS PALSAR	23.6	2006–2011	≈ 5/7–88	46	40–70	46.8	8–60

(Continued)

TABLE 11.1 Selected Characteristics of Principal SAR Sensors—cont'd

Satellite Mission	Wave Length (cm)	Life Status	Resolution Az./Range (m)	Repeat Cycle (Days)	Swath Width (km)	Maximum Velocity (cm/Year)	Incident Angle (Degree)
L-Band							
ALOS PALSAR-2	22.9	2014–2019	1/3 3–10/3–10 100/100	14	25 (Spot) 50–70 (Strip) 350 (Scan)	149.2	8–70
SAOCOM (2 Sat)	23.5	2014–2021	10–50	8, 16	20–150	268	20–50
X-Band							
COSMO-SkyMED (4 Sat)	3.1	2007–2014	≈2.5/2.5 1.0/1.0	2, 4, 8, 16	10 (Spot) 40 (Strip) 200 (Scan)	17.7 35.4 70.7 141.4	20–60
TerraSAR-X	3.1	2007–2018	≈3.3/2.8 1.0/1.0	11	10 (Spot) 30 (Strip) 100 (Scan)	25.7	20–55
COSMO-SkyMED-2 (2 Sat)	3.1	2015–2023	1–3		10–40		
TerraSAR-X-2	3.1	2015–2018	0.5–4		10–40		

detection and monitoring, in particular for landslides (e.g., Bovenga et al. 2006; Colesanti and Wasowski, 2006). These problems are mitigated by using the Persistent Scatterer (PS), Interferometry (PSI), and other MTI techniques that rely on processing of long temporal series of SAR data (usually >15 radar images, Hilley et al., 2004) to remove the atmospheric disturbances (dR_A), and on the identification of radar targets that provide a backscattered phase signal coherent in time ($dR_N \approx 0$).

Following Wasowski and Bovenga (2014), in the present work, we refer to all the measurable radar targets as coherent targets (CTs), regardless of the differences in terms of both scattering mechanism involved and strategies adopted for their identification. In this context, the coherence is not intended in the interferometric sense, but reflects the more general property of the target to remain measurable through time.

Since 2000, several MTI algorithms for precision detection of the deformation signal of the long-term ground surface have been developed and used to study different geophysical phenomena (e.g., Sansosti et al., 2010; Hooper et al., 2012). All these algorithms perform a joint nonlinear estimation of stochastic fields due to atmospheric signal, displacements, and errors in the reference topography. The atmospheric signal estimation is performed exploiting its stochastic behavior, which is known to be totally uncorrelated in time but correlated in the space domain with a power-law frequency spectrum. The assessment of the displacement field is much more complicated, because ground deformations are known to have certain correlation properties in both time and space. Further, a reliable discrimination of these two contributions requires careful consideration of both signals' characteristics, such as typical spatial−temporal scales.

The selection of CTs is performed according to the interimage coherence, $\gamma \in [0,1]$, which is related to the standard deviation, σ_Φ, of the phase residuals obtained after removing all the estimated signal components (dR_T, dR_D, dR_A in (1)): the higher the γ, the lower the σ_Φ. Assuming negligible processing errors (mainly due to an unreliable deformation model and inaccurate orbital records), γ values measure the decorrelation degree within the resolution cell (dR_N). For each CT, the MTI processing provides the LOS displacement time series, the mean LOS velocity (or equivalent LOS displacement), and the refined elevation estimates.

The MTI algorithms can be grouped into two general categories: (1) the PSI (Ferretti et al., 2001) and similar techniques (Werner et al., 2003; Duro et al., 2004; Hooper et al., 2004; Bovenga et al., 2005; Kampes, 2005; van der Kooij et al., 2006; Crosetto et al., 2008), relying on the phase information from single isolated objects characterized by a high temporal phase stability; and (2) the Small Baseline Subset (SBAS) and related techniques (Berrardino et al., 2002; Usai, 2003; Mora et al., 2003a), exploiting distributed scatterers (without any dominant element within the resolution cell), which are more sensitive to both temporal and volume decorrelation than PS. It follows that considerable differences in the number and spatial distribution of the measurable targets can occur depending on the scattering characteristics of the ground surface (e.g., Lauknes et al., 2010). Recently, new processing strategies have been proposed for efficiently combining PSI and SBAS approaches, thereby extending the analysis to both isolated and distributed targets (Hooper, 2008; Ferretti et al., 2011), and improving spatial sampling of the investigated deformation phenomenon.

11.2.2 Technical and Practical Aspects of MTI Applied to Landslide Motion Detection and Monitoring

Geologists, geomorphologists, and other users need to be aware of many different factors that influence the applicability of MTI to landslide investigations. This has recently been discussed in detail by Wasowski and Bovenga (2014), and here, only some of the most basic factors to be considered are noted:

- CT selection, coherence threshold (CT quality index), and number of images
- CT density and use of artificial reflectors (AR) where CTs are lacking
- CT geocoding accuracy (varies according to pixel size, i.e., from about 1 to 20 m)
- CT types (natural, e.g., rock outcrops and artificial objects, e.g., buildings and other artificial structures)
- Geometrical distortions related to the side-looking acquisition mode of radar sensors and slope visibility issues (cf. Figure 11.1)
- Historic, recent, and future satellite radar missions and data characteristics (long, medium, and short wavelengths, respectively, L-, C-, and X-bands)
- High-resolution (1−3 m) X-band data and improved MTI application capabilities

Additional technical and practical aspects of MTI directly related to landslide motion detection and monitoring are recalled and briefly discussed below.

11.2.2.1 Stable Reference Point Selection

The interferometric phase values are relative in time and space. Therefore, to derive absolute displacement values from the DInSAR phase values computed through MTI, a reference point is required both in time and in space. The temporal reference is usually the date of acquisition of the SAR image used as reference (master image) during the interferometric processing; this adds only an offset on the displacement time trend. The reference point in space is a CT selected within the radar scene area; its stability is either known, for example, from independent Global Navigation Satellite

System (GNSS) or from other in situ measurements, or can be inferred on the basis of the site's geomorphologic/geological characteristics. A proper reference point selection is most important, because it has a direct influence on the identification of stable and unstable areas. In some areas (e.g., those characterized by a high rate of tectonic deformations), motionless reference points may not be available. In those cases, the MTI analysis can be done by using a reference CT whose 3D motion is known.

11.2.2.2 Surface Displacement/Deformation Model

MTI algorithms follow different strategies in the multidimensional analysis of long temporal series of SAR data to suppress the atmospheric component, identify CT, and measure the displacement signal. According to the algorithm, the final results can depend on both the model adopted for the deformation and the selected reference CT.

MTI algorithms do not necessarily require a specific deformation model (e.g., Hooper et al., 2004), and in general, a parametric function can be adopted to describe the temporal trend of the ground displacement. Most often, linear kinematics are assumed for the displacements to speed up the processing, especially in the case of regional-scale investigations where ground information is lacking. However, where strong nonlinear displacements occur, the MTI products may lack CTs and/or significantly underestimate the movement velocity, because the linear model does not fit the actual phase trend. In particular, nonlinear components of deformation have to be taken into account when, for example, long time periods are considered, or high-resolution X-band data are used; this can be especially important in urban areas including artificial structures potentially affected by thermal dilation processes. In fact, some reference CTs may be affected by periodic displacements due to thermal dilation, and, if not properly taken into account, these oscillations may propagate to other CT time series (Notti et al., 2012).

11.2.2.3 Phase Aliasing Problem and Maximum Detectable Motion Velocity

Since interferometric phase samples are known only modulo 2π (wrapped phase), it is not possible to measure unambiguously phase differences $>\pi$, which correspond to LOS deformation $>\lambda/4$. This constrains the measurable displacement velocity.

MTI procedures compute phase differences in both space (between CT pairs) and time (between consecutive acquisitions—dt), and thereby the maximum detectable velocity depends on (1) the spatial gradient of the deformation, (2) the CT spatial sampling, and (3) the temporal sampling of the acquisitions (dt), which is the only a priori known parameter. In this work, we use the maximum velocity between neighboring CT as a reference value (cf. Table 11.1), which can be considered as a lower bound limit, and is computed as

$$v < \frac{\lambda}{4 \cdot dt} = v_{\max}. \tag{11.2}$$

In the case of aliasing ($v > v_{\max}$), the temporal coherence used to select CT decreases, leading to the loss of CT; this can take place even when the deformation model adopted by the MTI algorithm is correct. Therefore, very fast displacements can lead to a lack of CT despite the ground surface showing coherent backscattering.

Instead, a false CT detection can occur in the case of short-term acceleration (e.g., landslide movement triggered by rain) corresponding to a phase discontinuity of a few 2π phase shifts, which causes unwrapping errors. In such cases, the target temporal coherence is preserved and the actual movement underestimated leads to false (negative) detection (Notti et al., 2012).

It should be remembered that one of the products generated by MTI processing is the average yearly estimate of displacement velocity. This information is typically used to construct synoptic deformation maps to provide an overview of ground instability conditions. Obviously, on such maps, faster (or slower) nonlinear movement episodes tend to be averaged out (maximum velocity underestimated).

11.2.2.4 Three-Dimensional Surface Displacement versus LOS Measurement from MTI

DInSAR techniques measure only the component of the actual displacement along the sensor LOS defined by the satellite acquisition geometry. This is sketched in Figure 11.2 for both ascending and descending satellite passes and a slope facing east. The LOS and the slope inclination are defined, respectively, by the angles θ and β measured with respect to the vertical direction. The orbital vector and the slope aspect are defined, respectively, by the angles α and δ measured clockwise from the north. The LOS unit vector provides the sensitivity of the SAR LOS measurement to the displacement components along both vertical (Z) and horizontal (E and N) directions, and can be computed according to the symbols in Figure 11.2:

$$\text{LOS} = [\text{LOS}_Z, \text{LOS}_N, \text{LOS}_E] = [\cos(\theta), -\sin(\theta) \cdot \sin(\alpha), -\sin(\theta) \cdot \cos(\alpha)]. \tag{11.3}$$

The full 3D displacement vector can be recovered by using at least three independent InSAR data sets (corresponding to noncoplanar acquisition geometries), or by exploiting additional information or assumptions on ground motion. This limits the displacement vector measurement to the plane orthogonal to the approximately south—north direction, with sensitivity to vertical and horizontal components depending on the sensor look angle (Hanssen, 2001). SAR satellites operate on nearly polar orbits providing

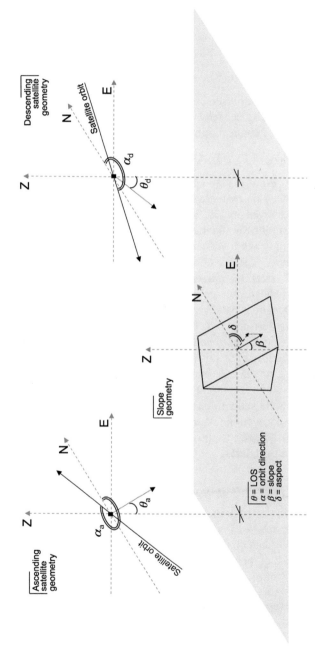

FIGURE 11.2 Satellite acquisition geometry for ascending and descending passes and scope facing east.

mainly two independent InSAR measurements on opposite pass directions (ascending or descending in Figure 11.2).

A reasonable assumption for a sliding body is that the movement occurs along the steepest slope. In this case, the movement direction is defined by the angles β and δ of the slope (Figure 11.2):

$$\overline{d} = [d_Z, d_N, d_E] = d \cdot [\sin(\beta), -\cos(\delta) \cdot \cos(\beta), -\sin(\delta) \cdot \cos(\beta)]. \qquad (11.4)$$

The modulus of the displacement along the LOS direction can be computed as the projection of displacement vector $\overline{d}$ along the LOS direction, being positive in the case of target motion toward the satellite:

$$d_{LOS} = [d_Z, d_N, d_E] \cdot [LOS_Z, LOS_N, LOS_E]^T. \qquad (11.5)$$

By combining Eqns (11.3)–(11.5), the displacement modulus can be computed. Depending on both slope angles β and δ, and LOS direction, the sensitivity to the movement of the SAR measurement can be very poor, leading to noise amplification in the inversion process. The availability of data from both ascending and descending orbits improves the accuracy in displacement retrieval (Cascini et al., 2010).

Further, some satellite missions, such as COSMO-SkyMED (CSK) constellation, are able to provide acquisitions with two incidence angles on ascending and descending passes with the radar both left- and right-looking. This can allow the derivation of the south–north deformation with an error twice as large as the error in radar range direction (Wright et al., 2004). However, the availability of acquisitions with these multiple geometries is very limited as such data are only occasionally collected following specific requests.

Finally, recent high-resolution satellites make it possible to measure displacement along the azimuth direction (approximately south–north), based on amplitude crosscorrelation (Bamler and Eineder, 2005) or spectral diversity (Erten et al., 2010). The results, however, are characterized by lower precision with respect to MTI products.

11.2.2.5 Precision and Quality Assessment of MTI Measurements

The assessment of MTI measurement precision was attempted in several studies. It is a complex issue due to the presence of critical, nonlinear steps in the MTI processing, which depend on the environmental setting of the study area and the temporal distribution of the available SAR acquisitions. Submillimeter accuracy was reported from controlled experiments conducted in ideal settings (e.g., Ferretti et al., 2007), while comparisons with independent in situ measurements (such as leveling, GNSS) demonstrate that the achievable accuracy is about 5 mm for a single displacement measurement and around 1 mm/year for the average displacement rate (e.g., Casu et al., 2006; Ferretti et al., 2007; Bovenga et al., 2013).

In general, the measurement precision is lower in the case of (1) significant vegetation coverage that reduces the CT spatial density and (2) high or steep topography that can induce major atmospheric signals. Further, the distance from the reference CT also affects the reliability of the MTI estimates, depending on the specific algorithm used for the atmospheric signal removal (Casu et al., 2006). Considering the ideal setting (absence of temporal and spatial aliasing), the loss in measurement precision results mainly from low spatial frequency signals due to atmospheric noise and orbital errors (Rucci et al., 2012); these factors, however, have a limited effect in the case of areally restricted phenomena such as landslides.

During the MTI processing, a number of implicit assumptions are made, some of which are study area- and method-specific. This, and the fact that the final output depends on a large variety of factors (Hanssen, 2005; Mahapatra et al., 2012), make the quality assessment of MTI results a complex task requiring different validation procedures depending on the type of deformation phenomena, available ground truth, and ground-surface characteristics (cf. Hanssen et al., 2008). In general, better quality MTI products (and CT) can be obtained with a higher number of images, homogenous temporal distribution of acquisitions, and longer temporal extent of image stack.

In conclusion, with reference to the availability of different SAR imagery and ground data, the following consistency checks of the MTI measurement results can be attempted (Wasowski and Bovenga, 2014):

1. Using both ascending and descending data from the same SAR sensor (acquired over the same area); with such independent data even simple qualitative comparisons of the results are very useful; with a sufficient number of CT targets detected from both ascending and descending acquisitions (and ideally corresponding to the same targets on the ground), a quantitative assessment of product quality can be attempted (Bovenga et al., 2013).
2. Comparisons of MTI result similarity using data from different sensors (e.g., Bovenga et al., 2012a, 2013; Duro et al., 2012).
3. Intercomparisons involving independent results based on different MTI-processing techniques (e.g., Wasowski et al., 2007; Lauknes et al., 2010; Bovenga et al., 2013; see also Figure 11.3).
4. Validation of MTI results using ground truth, for example, global positioning system (GPS), leveling data (e.g., Raucoules et al. (2009) and references therein).

11.3 EXAMPLES OF DIFFERENT SCALE MTI APPLICATIONS TO LANDSLIDE MOTION DETECTION AND MONITORING

Here, we mainly illustrate the potential of space-borne MTI for detecting and monitoring landslide motion by considering regions characterized by a wide range of geomorphic, climatic, and vegetation conditions, and presenting selected case study examples of local-to-catchment-scale applications. The examples address hill slope areas in the European Alps, the Southern

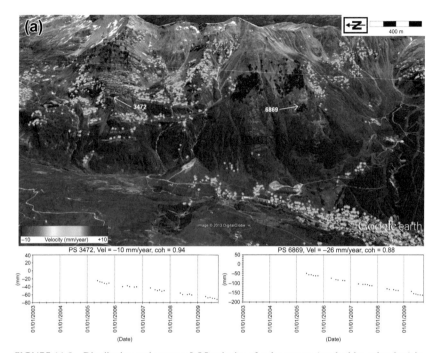

FIGURE 11.3   Distribution and average LOS velocity of radar targets (marked by color dots) in the Madesimo area, Italian Alps. Velocity saturated at ±10 mm/year for visualization purposes. Background optical images are from Google Earth™. (a) Velocity map derived by processing ENVISAT descending imagery through SPINUA algorithm (GAP srl); (b [next page]) Velocity map derived by processing RADARSAT-1 descending imagery through SqueeSAR™ algorithm (courtesy of TRE srl). For comparative reasons, also displacement time series of radar targets selected from the same unstable slope areas are shown.

Apennines (Italy), and moderate-relief mountains in the island of Haiti. Further, we also present one case study of GBInSAR applied in a slope scale investigation in the Southern Apennines, Italy.

11.3.1 Reliability of MTI Results

The case study examples presented below rely on the use of the SPINUA (Stable Point INterferometry over Unurbanized Areas) MTI algorithm. This PSI-like algorithm, originally developed for detection and monitoring of CT targets in nonurbanized or scarcely urbanized areas (Bovenga et al., 2005, 2006), has been updated to increase its flexibility also in the case of applications in densely urbanized areas, as well as to assure proper processing of new high-resolution X-band data from both CSK and TSX (Bovenga et al., 2012a).

SPINUA has already been applied to investigate different geophysical processes such as landslides, subsidence and postseismic ground deformations (e.g., Bovenga et al., 2006; Nitti et al., 2009; Reale et al., 2011). Moreover, the

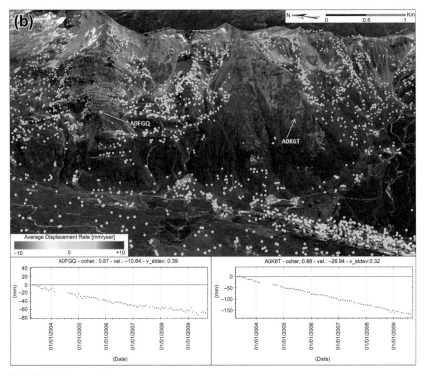

FIGURE 11.3   cont'd

results obtained through SPINUA have been cross-compared with those derived by exploiting other MTI techniques, as well as validated by using measurements from both GPS, GNSS, and leveling (e.g., Wasowski et al., 2007; Reale et al., 2011; Bovenga et al., 2013).

To provide an example of a simple check on the reliability of MTI results, we utilise here the Madesimo case (Fig. 11.3) study of an area of the Italian Alps afflicted by slope instabilities. For this purpose, our MTI results based on SPINUA processing of 32 ENVISAT images from the period June 2005–August 2010 (Wasowski et al., 2012a; Wasowski and Bovenga, 2014) are compared with the outcomes obtained from an independent SqueeSAR™-based processing by Tele-Rilevamento Europa (TRE srl) of 83 RADARSAT-1 images acquired between April 2003 and August 2009.

Figure 11.3 shows that the MTI results obtained from ENVISAT and RADARSAT-1 are closely comparable despite the differences in processing approaches and in radar data (acquired by different sensors). In particular, the same two major unstable slopes can be easily recognized on both surface displacement maps. Also, the same stable areas are consistently identified. Further, in spite of some differences in temporal coverage, the average LOS velocities and deformation time series of radar targets selected from the same slope areas result in being nearly identical (Figure 11.3).

Some differences occur, however. The much higher number of RADARSAT-1 images along with SqueeSAR™ processing resulted in a greater density and better (more homogeneous) coverage of the overall study area (Figure 11.3). Clearly, with fewer information gaps (e.g., in vegetated areas), the interpretation of MTI results and the assessment of slope conditions are more straightforward. Further, thanks to much higher average temporal frequency of RADARSAT-1 data (∼13/year as opposed to ∼6/year of ENVISAT), the displacement time series provide a more complete and more detailed picture of ground deformation trends (Figure 11.3).

11.3.2 Examples of MTI Application from the Italian *Alps*: Issues of Radar Visibility and Sensitivity to Down-Slope Movements

11.3.2.1 Issues of Radar Visibility and Sensitivity to Down-Slope Movements

The application of MTI in high mountains such as the European Alps is constrained by the geometrical distortions originating from the side-looking acquisition mode of satellite SAR sensors. Depending on the relations between the geometry (aspect and inclination) of local slopes and the sensor's LOS characteristics (ascending or descending look, incidence angle, cf. Figure 11.2), certain areas may or may not be suitable for MTI-based study. Importantly, no useful information can be obtained for the areas affected by radar layover or shadowing (Hanssen, 2001; Colesanti and Wasowski, 2006). This and other issues relevant for MTI feasibility assessment have recently been reviewed in detail by Wasowski and Bovenga (2014).

Where appropriate, especially in the case of local-scale investigations, a priori radar visibility maps can be prepared by combining the satellite orbit information and the DEM of the area to be studied (cf. Cascini et al., 2009; Notti et al., 2010; Plank et al., 2012). Apart from their reconnaissance value for MTI suitability evaluations, such maps can be very useful for the selection of the optimal acquisition geometry (ascending or descending look, incidence angle). In this context, a simple and generally applicable rule states that the descending and ascending geometries are favorable for west and east facing slopes, respectively (Colesanti and Wasowski, 2006). It follows that the visibility constraints linked to the mountain topography can be considerably reduced when both descending and ascending acquisitions are available. This, however, means doubling the radar data-processing effort.

Slope inclination and aspect also have direct impacts on the SAR sensor sensitivity to down-slope movements. In fact, as detailed in Section 2.2.4, satellite radars register the LOS projection of a 3D slope deformation, and only the movements that are parallel to LOS can be fully resolved.

In practice, with typical incidence angles of current satellite radar systems (ranging between 20 and 50°), the measurement sensitivity is 6–13%, 34–75%, and 64–94%, respectively, for the north, east, and vertical components of motion (Wasowski and Bovenga, 2014). This means that in the case of slope-parallel translational movements along directions approximating north–south, the real displacements (or rates) can be greatly underestimated. The same problem afflicts the MTI-derived estimates of lateral movements on shallow slopes.

As long as the displacements are slope parallel and largely unidirectional, the LOS (underestimated) velocities can be readily converted to the so-called slope velocity, by inverting Eqn (11.5). The direction of the movement vector $\bar{d}$ can be inferred from ground data or, if unknown, assumed to be parallel to direction of the maximum slope inclination (defined by the angles β and δ in Eqn (11.4)).

11.3.2.2 MTI Application Example from the Italian Alps

To demonstrate the use of radar visibility and sensitivity to down-slope movement, we present a case study example from the Italian Alps. The area of interest includes the northern part of the Lombardy Region, bordering with the Switzerland, and is characterized by a high summit (>3,000 m above sea level) and deep valley topography. The presence of high relief and intensely sheared rock masses makes Alpine slopes susceptible to different types of failures, including very large, slow deep-seated landslides that are commonly suitable for MTI applications.

Figure 11.4 (upper part) shows two maps of radar visibility and sensitivity to down-slope movement constructed using DEM-derived information (slope and aspect) and considering the ENVISAT ascending and descending acquisitions with average incident angles of 22.8°. Although radar shadowing is not an issue, layover effects are locally present in the entire area. Significantly, these effects are less pronounced in the case of ascending geometry. Further, a simple visual comparison of the two maps indicates that the use of both ascending and descending data would lead to a considerable improvement in radar visibility. Some areas are also identified where radar has low sensitivity to down-slope movements. These are more evident in the northeastern portion of both maps, which includes many slopes with directions facing close to north or south (Figure 11.4, upper part).

The issues of MTI suitability become more obvious when examining the mountain terrain on a local scale. This is illustrated in Figure 11.4 (Lower part; small zoomed portion of Figure 11.4, Upper part), which shows three radar

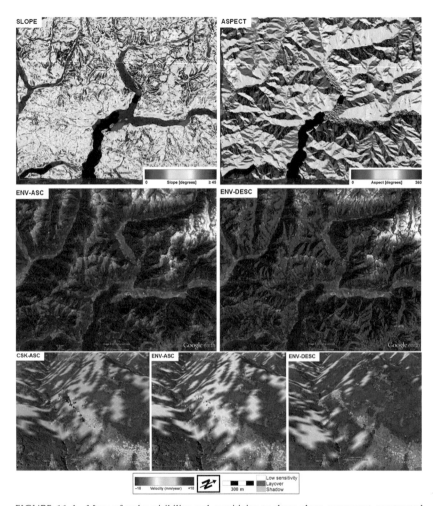

FIGURE 11.4 Maps of radar visibility and sensitivity to down-slope movement constructed using DEM-derived information (slope and aspect) and considering different satellite acquisition geometries: (Upper) two wide-area maps obtained for the ENVISAT ascending and descending acquisitions with average incident angles of 22.8°; (Lower) map of the Albano River valley and the Garzeno–Catasco landslide (in the center) obtained for the Cosmo–SkyMed ascending acquisition (Stripmap mode) with an average incident angle of 26.6°, compared with the zoomed portions of the same ENVISAT-based maps shown in the Upper figure. Color dots indicate the actual distribution of radar targets obtained from MTI processing; for visibility reasons, the LOS velocity scale is saturated to ±10 mm/year.

visibility maps of the Albano River valley area. For practical reasons, it is convenient to focus on the facing, about 30°-inclined valley slope affected by a large (∼2 km long and wide), deep landslide involving the towns of Garzeno and Catasco (see also Figure 11.5). Again, layover effects are more

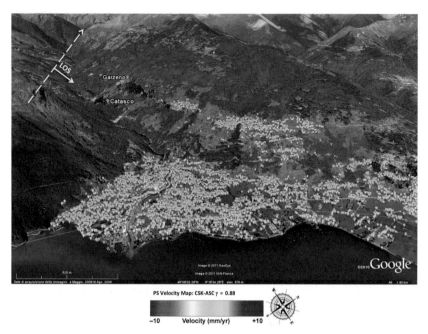

FIGURE 11.5 Distribution and average LOS velocity of radar targets in the Albano River valley area including the large, deep-seated Garzeno—Catasco landslide (left of center); results derived from MTI processing of Cosmo—SkyMed data. For visibility reasons the LOS velocity scale is saturated to ±10 mm/year. The instability of the distal portion of the Albano River fan delta (western shore of the Lake Como) is likely due to subsidence. The background optical image is from Google Earth™.

pronounced for the descending acquisition, while low susceptibility to downslope movement is somewhat more evident in the case of ascending geometry; this is consistent because the latter has an orbit direction closer to the slope-facing direction, that is, SSE.

The visibility and sensitivity to down-slope movement maps in Figure 11.4 (Lower part) also show the actual distribution of radar targets obtained from MTI processing of 34 ascending and 32 descending ENVISAT images with a 22.8° mean incident angle (covering the period 2004—2010), as well as of 26 CSK ascending images with a 26.6° mean incident angle and 3-m resolution (period August 2008—August 2010). The greater density of the targets retrieved from the ascending data is consistent with the results of the radar visibility maps and confirms their predictive capability. The use of CSK ascending data leads to the highest density of radar targets, mainly because of the better spatial resolution of the imagery.

Bovenga et al. (2012a) examined in detail the influence of radar imagery spatial resolution and other factors on radar target density achievable from MTI processing. Their comparative study indicated that from 3 to 11 times greater target densities can be obtained from high- (3 m) resolution CSK imagery with respect to medium-resolution (20 m) ENVISAT data. One of the sites examined was the Garzeno–Catasco landslide. In particular, by considering the overall failed slope, the calculated spatial densities (number of targets per square kilometer) ranged from 97 to 329 for ENVISAT descending and ascending data, respectively, to 846 for CSK ascending data.

The distribution of radar targets derived from MTI processing of CSK data is shown in Figure 11.5 using Google Earth 3D visualization tools and high-resolution optical imagery. This shows the general lack of targets in heavily vegetated areas (mainly tree cover), and high densities in urbanized areas. It is also apparent that the gently sloping urbanized areas have higher target densities; this is evident for the Albano River fan delta located on the western shore of Lake Como (Figure 11.5).

Further details regarding the distribution of CSK radar targets are illustrated in Figure 11.6, which also shows two typical deformation time series,

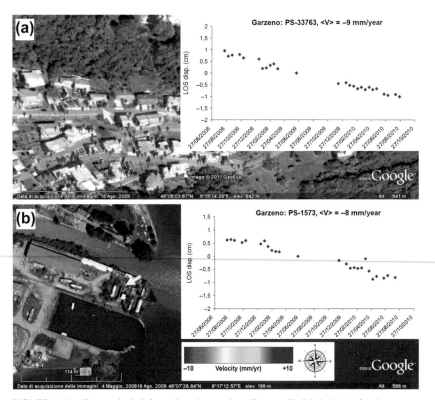

FIGURE 11.6   Two typical deformation time series (Cosmo–SkyMed data) of radar targets located in (a) the eastern part of the town of Garzeno (on the landslide); and (b) in the distal portion of the Albano river fan delta; see Figure 11.5 for more context information.

one located in the town of Garzeno (on the landslide) and the other showing the subsidence affecting the distal (the youngest), urbanized portion of the Albano river fan delta. The two targets correspond to artificial structures and exhibit very similar displacement trends and LOS velocities (8 and 9 mm/year). It is apparent that these data alone do not allow the distinction between the two different deformation phenomena.

However, in the case of a landslide or sloping terrain, the registered LOS velocities are often much influenced by local variations in slope aspect and inclination (Colesanti and Wasowski, 2006; Bovenga et al., 2012a). This can be well demonstrated through the example of the Garzeno–Catasco landslide by comparing the LOS velocity maps with the slope velocity maps (Figure 11.7); the latter represent the radar target velocity values calculated by assuming displacements parallel to ground surface and along the slope-maximum direction. By considering both the ascending and descending ENVISAT acquisition geometries, it is shown that in both cases the radar target velocities become spatially less variable after the transformation to slope velocity values. Further, the velocities of targets obtained from ascending and descending data become more similar. Significantly, the identical maximum slope velocities ($\sim$26 mm/year) in both cases are more than twice the maximum LOS velocities (Figure 11.7). All this lends credibility to the interpretation of the MTI results as reflecting mainly the movements of a deep translational landslide (Bovenga et al., 2012a).

Interestingly, the same slope failure has recently been considered in the work of Frattini et al. (2013), who called it the Catasco rockslide. The authors reported that despite the long history of damage in the town of Catasco, the landslide has only recently been identified and mapped. Further, in situ monitoring that started in 2010 revealed via inclinometer measurements the presence of up to 30-m-deep movements. Frattini et al. (2013) also documented the presence of moderate-to-heavy damage to the buildings in Catasco and related it to the ground surface displacements obtained from PSInSAR™ processing of ascending and descending RADARSAT data sets covering the period 2003–2007. The displacement results were comparable to those reported here and by Bovenga et al. (2012a). Such comparisons of the results obtained using different satellite sensor imagery and independent MTI processing techniques can clearly provide a useful consistency check for the MTI measurements, especially where ground-monitoring data are not available.

11.3.3 Examples of MTI Application from the Apennine Mountains: Instability of Hilltop Towns

11.3.3.1 Characteristics of the Study Area and Previous MTI Investigations

For comparative reasons, we present some examples of MTI applications in the southern portion of the Apennines (Daunia region, Southern Italy), where

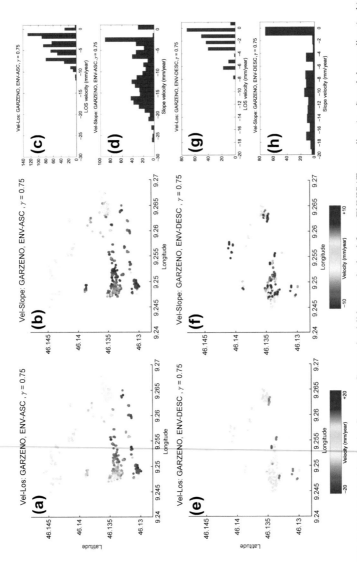

FIGURE 11.7 Displacement velocities in the Garzeno–Catasco landslide area obtained from ENVISAT ascending (a–d) and descending (e–h) data: maps of LOS velocity (a and e) and corresponding maps of slope velocity (b and f); histograms of velocity values computed along LOS (c and g) and along the maximum slope direction (d and h).

the environmental characteristics are distinctly different from those of the
Italian Alps. In particular, unlike the Alps, the Daunia region is characterized
by moderate-relief mountains occasionally exceeding 1000 m above sea level
(Figure 11.8). The slopes have inclinations typically around 10—15°. The
steepest, highest elevation areas include a significant percentage of forested
land. Elsewhere, the vegetation cover is mainly cultivated land (cereals) and
locally grassland. The barren land areas are very limited and correspond to
clay-rich units, which represent the predominant lithology. Unlike in the Alps,
rocky outcrops (sandstone and limestone) are scarce. The proportion of
developed land (urban and rural settlements, roads, and other infrastructure) is
low.

Clay-rich flysch formations with poor geotechnical properties are widespread in Daunia and this, along with the presence of poorly drained slopes
and intense ground disturbance by agriculture, results in high susceptibility to
landsliding (Wasowski et al., 2010, 2012b). Further, periurban areas and road
networks are often affected by landslide problems.

The Daunia Mountains include >20 small hilltop towns (Figure 11.8), and
in the 1990s, an apparent increase in landslide occurrence was noticed in
several urban and periurban areas (Wasowski et al., 2008, 2010). It is

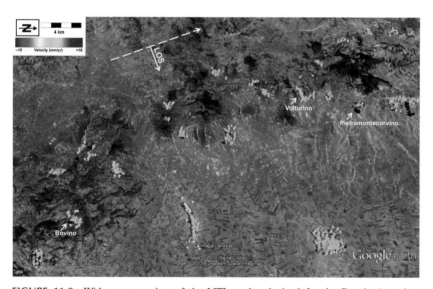

FIGURE 11.8   Wide-area overview of the MTI results obtained for the Daunia Apennines
(Southern Italy) by exploiting ENVISAT ascending imagery; for visibility reasons, the LOS
velocity scale is saturated to ±10 mm/year. Patchwise processing resulted in >20 clusters of radar
targets, which correspond to small hilltop towns distributed along the NW—SE trending mountain
front. The MTI results pertaining to the towns of Bovino, Volturing, and Pietra Montecorvino are
shown in more detail in Figures 11.9—11.11. The background optical image is from Google
Earth™.

suspected that the stability of slopes bordering the hilltop towns has gradually worsened, because of residential development over recent decades.

The Daunia region has already been investigated using MTI and the ESA medium resolution C-band ERS-1/2 SAR data covering the period 1995–1999 (Bovenga et al., 2006; Wasowski et al., 2008). The results from four urban areas (Casalnuovo Monterotaro, PietraMontecorvino, Motta Montecorvino and Volturino) showed that extremely slow displacements, apparently related to slope failure processes or simply to ground instability, occur mainly in the peripheral parts of the Daunia towns. There the urban expansion is more recent and the artificial structures acting as radar targets are situated on or close to the marginally stable slopes bordering the hill tops.

Subsequently, Wasowski et al. (2012a) and Nutricato et al. (2013) presented the MTI results obtained using the radar imagery acquired by the ENVISAT satellite (C-band, medium spatial resolution sensor) and the images acquired by the X-band high-resolution (~ 3 m) sensor aboard the German TerraSAR-X satellite. Thanks to its finer spatial resolution, the X-band MTI proved to be very useful for monitoring single artificial structures and slope/ground instability, especially in areas where the density of targets derived from C-band data was low.

11.3.3.2 Instability of Hilltop Towns in the Daunia Apennines

Figure 11.8 presents a wide-area overview of the MTI results obtained for the Daunia region by exploiting ENVISAT ascending imagery acquired between 2002 and 2010. Radar targets are overlaid on a Google Earth optical image to provide environmental context. One can note >20 clusters of radar targets, which correspond to small hilltop towns distributed along the NW–SE trending front of the Daunia Apennines. Significant tree cover also occurs along the higher portions of the mountain front, whereas agricultural land predominates on the lower slopes and in the foreland area (in the ENE part). These two types of land cover and land use result in decorrelation of the SAR signal, which makes MTI processing impractical. Therefore, in this case, the MTI results were obtained by using the SPINUA algorithm and adopting a patchwise processing scheme. This consists of processing small image patches (usually a few square kilometers) selected according to the density and the distribution of potential radar targets. Such a solution provides fast results on small areas by processing even scarcely populated stacks of SAR images (for more details, see Bovenga et al., 2012a).

The MTI results shown in Figure 11.8 indicate that the majority of the hilltop towns are affected by extremely slow ground deformations. Significantly, the instability affects the peripheral areas of the towns that slope away from the hill tops. Further, along the mountain front, the moving radar targets often concentrate on the east sides of the towns, that is, the slopes facing the foreland.

In the following section, we present a more detailed discussion of these MTI results, with reference to three selected areas including the towns of Bovino, Volturino, and Pietra Montecorvino (Figure 11.8).

11.3.3.2.1 The Hilltop Town of Bovino

The first MTI results based on the ENVISAT ascending data were presented by Nutricato et al. (2013). They reported a cluster of extremely slow (6–7 mm/year) radar targets in the southern periphery of the town of Bovino, on a slope affected by an old landslide. This outcome is reproduced in Figure 11.9, which, in addition, includes the displacement results obtained from the ENVISAT descending imagery.

The moving targets fall on a gently inclined ($\sim 5°$) part of a southeast-facing slope, within the middle part of an old, deep earth slide (Mossa et al., 2005). The failed slope has been subjected to stabilization works and currently does not present any obvious signs of instability. However, inclinometer borehole monitoring conducted since 2008 indicated the presence of extremely slow displacements (up to 15 mm/year) occurring in the central portion of the slide at a depth of a few tens of meters (F. Santaloia, CNR-IRPI, unpublished data).

Although the MTI and inclinometer measurements have millimetric precision, the comparison of the results may not be straightforward because of the differences between their temporal and spatial reference systems. Also, while an inclinometer measures lateral deformation along a slip zone at a depth, the MTI provides surface displacements in the satellite LOS. Nevertheless, the SAR sensor's ability to detect the horizontal component of the landslide movement can be approximated by taking into account ENVISAT's LOS geometry and the local slope aspect and inclination. Assuming that the shear displacement is parallel to the ground surface at the midportion of the landslides (i.e., the one with the moving targets), the radar sensitivity is between 20 and 25%. This implies about 30-mm/year slope velocities, if one considers only lateral movements. However, these slope velocities are twice the displacement rates registered through the inclinometer monitoring.

It is apparent that without the inclinometer data indicative of significant lateral displacement, the 1D MTI deformation results could have been differently interpreted. For example, in such a gently sloping area, one could expect some ground settlements in response to loading by houses and other engineering structures. In the Bovino case, this is plausible, taking into account the presence of disturbed, clay-rich (compressible) landslide materials and the relatively recent urbanization history of the middle part of the slide (which started in the late 1970s).

If ground truth is not available, the uncertainties regarding the significance of 1D MTI measurements can be resolved by using both ascending and descending data. As shown in Figure 11.9, the MTI results based on the latter

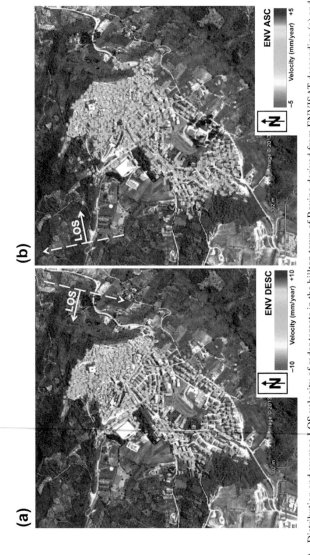

FIGURE 11.9 Distribution and average LOS velocity of radar targets in the hilltop town of Bovino derived from ENVISAT descending (a) and ascending (b) data sets; the LOS velocity scales saturated to ±10 and ±5 mm/year; see Figure 11.8 for more context information. Note spatial coincidence of the bluish and reddish cluster of radar targets, which correspond to the central part of a deep landslide. Background optical images are from Google Earth™.

data set also indicate the presence of extremely slow displacements in the midpart of the slope. Thus, the occurrence of lateral movements is confirmed. Finally, although the inclinometer and MTI results at Bovino may not be expected to match each other, the local setting and recent slope history suggest that the radar target displacements capture the cumulative effect of landslide and ground settlement processes.

11.3.3.2.2 The Hilltop Town of Volturino

This case is of interest, because thanks to the new high-resolution (3-m) TerraSAR-X imagery, we can now update the earlier MTI results obtained using medium-resolution (20-m) C-band data sets. The first MTI-based investigation of the Volturino area was presented in Bovenga et al. (2006), who exploited 40 ERS ascending radar images from the period 1992–1999. The results indicated the presence of slope/ground instabilities in the southern periphery of the town (especially in the Fontana di Monte area), which was known to have experienced landslide events in the past (1980s and 1990s). Recently, Nutricato et al. (2013) presented MTI results based on the 47 ENVISAT images acquired between November 2002 and January 2010. These new results confirmed the persistence of the instabilities detected earlier. Furthermore, in both periods, very similar, extremely slow deformation rates (generally up to 5 mm/year in LOS) were measured.

As in Bovino, there were some data available from borehole inclinometer monitoring. In particular, the monitoring conducted since late 2008 in the SW part of Volturino (Fontana di Monte area) registered displacements averaging about 15 mm/year (Lollino et al., 2010). The much slower movements obtained from the MTI analysis can be linked in part to the low sensitivity of the satellite sensor to horizontal displacements.

The results based on the ENVISAT data are compared in Figure 11.10 with the MTI displacement map obtained using 13 high-resolution (3-m) ascending TSX images acquired in the period January 2010–March 2011. The TSX results confirm again the persistence of instability in the SW part of Volturino. Importantly, thanks to the nearly 13 times greater density of radar targets (Table 11.2) and hence much better coverage, a few additional slope/ground instabilities were detected in the area of Volturino. These included not only the urban/periurban zones but also different parts of the local road network (Figure 11.10). Landslide damage to roads occurs in different zones of the Daunia Mountains (e.g., Wasowski et al., 2010, 2012b), and the applicability of MTI to road stability monitoring in landslide-prone areas has recently been discussed by Wasowski and Bovenga (2014).

11.3.3.2.3 The Hillside Town of Pietra Montecorvino

The MTI displacement maps obtained from processing ENVISAT (2002–2010) and TSX (2010–2011) data are shown in Figure 11.11. The

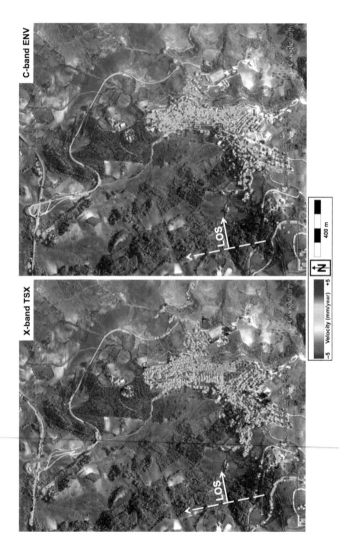

FIGURE 11.10 Comparison of the MTI displacement maps of the hilltop town of Volturino obtained from high-resolution TerraSAR-X (left) and medium-resolution ENVISAT ascending data sets (right); for visibility reasons. the LOS velocity scale is saturated to ±5 mm/year. Background optical images are from Google Earth™.

TABLE 11.2 Comparison Between the Number of Coherent radar Targets (PS) obtained From C-band (ENVISAT) and X-band (TerraSAR-X) Imagery for Hilltop Town Sites in the Daunia Apennines (see Figure 11.8 for Location)

Site	#PS (C Band)	#PS (X Band)	#PS (X Band)/ #PS (C Band)
Alberona	573	6233	10.9
Carlantino	529	6996	13.2
Casalnuovo Monterotaro	1062	14,829	14.0
Casalvecchio di Puglia	1154	14,767	12,8
Castelnuovo della Daunia	1187	15,364	12.9
Celenza Valfortore	504	7942	15.8
Motta Montecorvino	542	8074	14.9
Pietra Montecorvino	1140	18,398	16.1
San Marco la Catola	182	7605	41.8
Volturara Appula	335	4959	14.8
Volturino	1269	16,103	12.7

results are consistent and indicate the presence of extremely slow movements, which are especially evident in the southern part of the town. Significantly, very small ground deformations in the southern periphery of Pietra Montecorvino have already been reported by Wasowski et al. (2008), who used ERS C-band data from the period 1992–1999.

Figure 11.11 also demonstrates that the MTI map based on high-resolution X-band data offers more complete information thanks to the 16 times higher density of radar targets (Table 11.2). Clearly, the presence of X-band radar targets is particularly valuable where no usable C-band targets are available. This situation is especially evident in the peripheral parts of the town.

The presence of many extremely slow displacements (typically ~5 mm/ year) at Pietra Montecorvino does not come as a surprise, given that a good portion of the town is built on an old large landslide (Zezza et al., 1994) and that surrounding slopes have been affected by active landsliding in recent years (Wasowski et al., 2008). The persistence of ground and/or structural instability conditions is consistent also with the presence of many damaged buildings.

However, in the absence of in situ monitoring data, much caution is warranted when interpreting extremely small surface displacements measured in

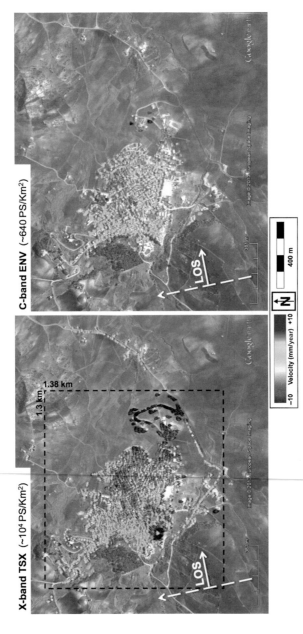

FIGURE 11.11 Comparison of the MTI displacement maps of the hilltop town of Pietra Montecorvino area obtained from high-resolution TerraSAR-X (left) and medium-resolution ENVISAT ascending data sets (right); for visibility reasons, the LOS velocity scale is saturated to ±10 mm/year. Black dashed rectangle shows the area for which the densities of radar targets (PS/km²) were calculated. Note over 10 times greater density of PS obtained from high-resolution data. Background optical images are from Google Earth™.

LOS. Unless both ascending and descending radar data sets are available, the MTI information alone may not be sufficient to ascertain the exact origin of the detected deformations, that is, whether, for example, they represent predominantly down-slope creeping landslide movements, or vertical movements (settlements), or a combination of both. The presence of clay-rich deformable lithologies susceptible to volumetric changes makes the interpretation of MTI results even more difficult. This and additional MTI interpretation issues are discussed in detail in the final part of this chapter.

11.3.3.3 Utility of MTI for Monitoring Slope/Ground Instability Hazards in Urban/PeriUrban Areas

The examples from the Daunia Apennines show that, regardless of the exact origin, any ground instability detected by MTI in an urbanized area will be of much interest to local administrators responsible for slope or ground deformation risk reduction. Although the average deformation rates are very low and may be typical of long-term postfailure creep or settlement processes, such persistent instabilities can produce significant damage to artificial structures in the peripheries of the hilltop towns. Therefore, in these cases, long-term monitoring would be ideal. MTI based on high-resolution imagery seems to be best suited for the task because of (1) its performance is excellent in urbanized zones, typically characterized by very high density of radar targets (Table 11.2); (2) in situ measurements are commonly impractical or difficult in towns because of the visibility problems (e.g., GPS or other methods of topographic surveying) and the issues of benchmark installation and security; (3) the difficulties in recognizing (and interpreting) the evidence of extremely slow deformations through noninstrumental observations; and (4) the relatively high cost and typically limited spatiotemporal coverage of ground monitoring.

The millimeter precision and high spatial density of MTI measurements thus offer the unique possibility to detect and accurately map the areas affected by extremely slow deformations. This is important in hilltop town environments, where susceptibility to slope failure is poorly known (e.g., lack of historical evidence) or where the local conditions may have changed in time (e.g., through recent urban expansion involving slope loading or undercutting).

Finally, the benefits of MTI applications to monitoring urban landslides or towns built on large, ancient slope failures have also been discussed by Colesanti and Wasowski (2006) and Bovenga et al. (2012a), with reference extremely slow, deep-seated movements in the European Alps (Italy, Liechtenstein, and Switzerland). Similar conclusions can be drawn from the works of Bovenga et al. (2013) and Calò et al. (2014), who presented results of detailed MTI investigations of an urban landslide affecting the town of Assisi (Central Apennines, Italy).

11.3.4 Example of MTI Application From the Mountains of Haiti

The main reason for presenting an example from Haiti is to show that MTI can furnish very useful data on slope instability even in the case of a tropical region characterized by dense and rapidly growing vegetation, as well as by significant climatic variability (two rainy seasons) with intense precipitation events. Regions with such environmental characteristics can be difficult for MTI applications, because of temporal decorrelation problems. Significant difficulties can be expected especially when using shorter wavelength (C- or X-band) data.

In the Haiti case, nearly 100 high-resolution (3 m) CSK images (June 2011–August 2013) were available for MTI processing. Despite the unfavorable setting, the identification of a suitable density of radar targets was possible even in some rural (inhabited) areas thanks to the high resolution of CSK radar imagery, the adoption of a patchwise processing SPINUA approach, and the presence of many artificial structures dispersed in heavily vegetated terrain.

Figure 11.12 shows an example of a landslide and associated instabilities detected on the basis of MTI results in the Montagnes du Trou D'Eau. These moderate elevation mountains (up to ∼1,350 m above the sea level) are located few tens of kilometers northeast of the Haiti's capital Port-au-Prince. The landslide affects a slope in stratified sedimentary rocks (mainly sandstone and secondary limestones), and is approximately 300 m long and >200 m wide (with the toe eroded by a stream). Many small artificial structures, mainly poor housing belonging to the village of Triano and the adjacent National Route N°3, are the sources of radar targets identified over the slide and in its surroundings. The LOS velocities of the radar targets on the slide are rather uniform (typically ∼15 mm/year); this and the overall geomorphic expression of the Triano slope are indicative of a deep-seated movement. Notably, the displacement rates of targets within the slide limits are on the average greater than those of other moving targets (typically much less than 15 mm/year); the latter denote localized instabilities affecting the road or isolated artificial structures (Figure 11.12). Although no information is available on the origin of the main failure, the examination of "historical" optical images in Google Earth shows that the Tiano landslide predates the 2010 7 Mw Haiti earthquake.

Figure 11.12 also shows three examples of different types of deformation time series from the radar targets in the area of the Traino. Some targets located on the deep slide do not exhibit linear displacement trends, but show a stepwise pattern with accelerations alternated with periods of near standstill. Two other examples of displacement trends are from the targets on different portions of the Route N°3. One target shows an approximately linear trend and the other indicates a significant decrease in the deformation rate (down to about 0 mm/year) in the second part of the monitoring period. Upon site

FIGURE 11.12 The slope area affected by the Triano landslide and traversed by the National Route N°3, Montagnes du Trou D'Eau, Haiti: MTI displacement map based on high-resolution Cosmo−SkyMed ascending data; for visibility reasons, the LOS velocity scale is saturated to ±10 mm/year. Three representative displacement time series of radar targets pointed out by black numbers 1, 2, 3, and arrows are also shown. The movement of targets on the slide is away from the SAR sensor, that is, presumably toward the east assuming the lateral component of motion is predominant. The background optical image is from Google Earth™.

inspection, we found out that the latter target corresponds to a road embankment support structure (gabions). Since the road has recently been improved, it is tempting to speculate that the target's displacement time history reflects the initial settlement period (after the gabion emplacement), followed by more stable conditions.

11.3.5 Example of GBInSAR Application from the Southern Apennines, Italy

GBInSAR is based on a terrestrial remote sensing system consisting of a SAR sensor moving along a rail track and covering an area up to few square kilometers. GBInSAR has recently gained much interest among landslide researchers and practitioners, and ample information on the technique and its applications is available in the international literature (e.g., Tarchi et al., 2003; Leva et al., 2003; Gischig et al., 2011; Lowry et al., 2013; Monserrat et al., 2013).

In comparison to space-borne MTI techniques considered above, the unique features of GBInSAR include the capability to provide precise measurements for a wide range of slope movement velocities (from extremely slow to moderate motions, Table 11.3) and suitability for emergency, near real-time monitoring (Casagli et al., 2010). The technique is appropriate for slope-scale monitoring, and we present here one example of such application adapted from the work of Di Pasquale et al. (2013).

The area of interest is located in the Apennine Mountains of the Basilicata Region, Southern Italy. Figure 11.13 shows the investigated slope, where an apparently active landslide threatens the stability of the bridge and the viability of the road to the town of Aliano. The slope is characterized by moderate dips and sandy—silty—clayey lithologies.

TABLE 11.3 Applicability of Selected Satellite, Air-borne and Terrestrial Remote Sensing Techniques to Detection and Monitoring of Different velocity Landslides; velocity Classification After WP/WLI (1995)

Velocity Class	Description	Velocity	MTI/ DInSAR/ CRInSAR	GBInSAR	LIDAR	Image Cross-correlation
1	Extremely slow	<16 mm/year	Yes	Yes	Yes	Yes
2	Very slow	16 mm/year–1.6 m/year	Yes(1)	Yes	Yes	Yes
3	Slow	1.6 m/year–13 m/month	No	Yes	Yes(2)	Yes(3)
4	Moderate	13 m/month–1.8 m/hr	No	Yes	No	No
5	Rapid	1.8 m/hr–3 m/min	No	No	No	No
6	Very rapid	3 m/min–5 m/s	No	No	No	No
7	Extremely rapid	>5 m/s	No	No	No	No

Notes: (1) Rates exceeding few/several tens of centimeters per year are not measurable at present; (2) Only terrestrial LIDAR; (3) Only using high-frequency image acquisitions (typically via terrestrial remote sensing); (4) Rates >0.1 m/hr (or few tens of meters per month) are not measurable.

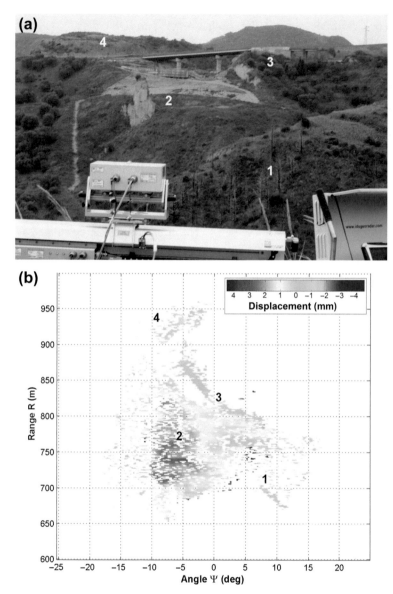

FIGURE 11.13 Example of ground-based radar interferometry GBInSAR application to monitoring slope instability and associated infrastructure: (a) Photograph showing the unstable slope as viewed by the radar; (b) Displacement map of the slope based on 24-h-long monitoring. The numbers 1–4 facilitate the recognition of the same features on both figures (e.g., "2" marks the central portion of the landslide affecting the slope).

The GBInSAR system (IBIS-FL hardware from IDS company, processing algorithm from DIAN srl) was installed at about 900-m distance from the bridge to guarantee adequate coverage (Figure 11.13). Although the system was used only for 24-h-long monitoring, the data acquired provided valuable information on the deformations affecting the slope. In particular, a ground surface displacement map was derived by processing two SAR images, one acquired at the beginning and the other at the end of the monitoring period. The map clearly showed the limits of the unstable slope area (Figure 11.13). The GBInSAR results also indicated that the bridge structure was not affected by any appreciable displacements during the monitoring period.

Further, to investigate the temporal trend of deformations, the entire set of images acquired by GBInSAR was interferometrically processed. An example of time series with mean displacement values and associated standard deviations is shown in Figure 11.14. The time series was calculated for a 10 x 10 pixel area ($\sim 40 \, m^2$) in the central part of the unstable slope (cf. Figure 11.13). In spite of only using a 24-h monitoring period, the results revealed the highly nonlinear nature of the instability process. In particular, GBInSAR detected a small but significant acceleration of movements (velocity approximating 0.2 mm/h), which started at about 5 AM (Figure 11.14). This was tentatively linked to the occurrence of rain on the site during the night (Di Pasquale et al., 2013).

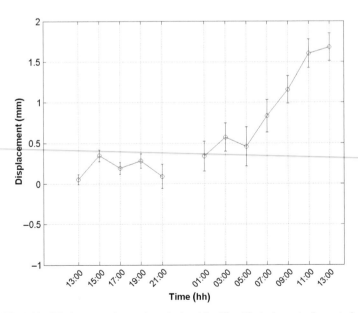

FIGURE 11.14 Displacement time series calculated for 10 × 10 pixel area in the central part of the unstable slope (cf. point 2 in Figure 11.13).

Finally, we use the above example also to highlight the fact that currently GBInSAR exploitation is generally limited to cases where people and/or important infrastructure are at risk. Indeed, monitoring relies on expensive and bulky instrumentation, and requires careful setup and manpower in the field (cf. Table 11.4). Importantly, GBInSAR represents one of the best remote sensing methods for detailed surveying of higher velocity and/or episodic nonlinear movements. Further, in comparison with space-borne MTI, ground-based radar interferometry is more suitable for conducting measurements of instabilities on steeper slopes.

11.4 SUMMARY DISCUSSION

Despite some persisting limitations (see the review by Wasowski and Bovenga (2014)), MTI techniques, particularly those based on high-spatial−temporal resolution data, today offer unmatched capability for combining regular long-term (years-decades) with precision (millimeters-centimeters) measurements of very slow slope movement over large areas. Thanks to this, MTI represents an excellent reconnaissance tool for different-scale slope hazard assessments based on the direct indicator of instability, that is, the presence and rate of ground surface deformation. Long-term displacement data can be vital in unraveling the mechanics of very slow landslides and predicting their future movements. Such quantitative information is difficult to acquire through other remote sensing techniques, because of the high costs involved, especially in the case of multitemporal surveys (e.g., LiDAR). Also, ground-based slope instability investigations are notoriously expensive, and in situ monitoring is typically reserved for special cases (e.g., high risk areas) and maintained for limited periods of time (often not more than a seasonal cycle of weather).

Below we provide a summary discussion on remote sensing of landslide motion, with reference to the strengths and limitations of MTI. In particular, we focus on (1) landslide detection and monitoring using MTI and other remote sensing techniques; and (2) presently underexploited and future MTI application opportunities.

11.4.1 Landslide Motion Detection and Monitoring Using MTI and Other Remote Sensing Techniques

Numerous remote space-, air-, and ground-based remote sensing techniques can be usefully applied to measure landslide motion (e.g., Turner and Schuster, 1996; Metternicht et al., 2005; Delacourt et al., 2007; Mansour et al., 2011). Clearly, each technique has its merits depending on the purpose (e.g., research or conventional engineering practice) and scale of the investigation (e.g., local, regional), environmental setting of the area of interest (e.g., degree of vegetation cover, slope steepness), and landslide characteristics (e.g., velocity and type or mechanics of movement).

In this context, among the most important strengths of MTI are

- the wide-area coverage (up to tens of thousands of square kilometers) combined with a high spatial resolution (meters) and high precision of measurement (millimeters;
- regular, long-term (years) acquisition of data with short revisit times (from weeks to days) and the availability of long time series of historical data (since 1992) suitable also for retrospective studies.

These strengths afford a certain flexibility to the MTI applications such as, for example, a possibility of using the same series of images for variable scale assessments (from regional to local). Furthermore, long time series of radar satellite records can be exploited both to unravel the past and to monitor the current trends in landslide activity. Importantly, patterns of change in landslide motion can now be investigated at high spatiotemporal resolutions over large areas.

However, the possibility of obtaining useful information from MTI is critically dependent on the availability of a sufficient number of radar targets in an area of interest. Since the exact spatial distribution of radar targets cannot be predicted in advance, for some areas no MTI information will be available, or, for some slopes and landslides, there will be significant spatial inconsistencies in data density (Wasowski and Bovenga, 2014). The gaps in spatial coverage together with other limitations of the technique imply that MTI could best be exploited in an exploratory stage of a slope hazard assessment project, to be followed by a more detailed investigation and monitoring of sites at risk. As in many other satellite remote sensing applications, the larger the area investigated (and the higher spatial frequency and areal extent of active landslides or marginally stable slopes), the more attractive will the MTI be in terms of costs and benefits. It will also be very advantageous to employ MTI for extensive landslide areas that require longer term (several years or more) monitoring (e.g., regions affected by large magnitude earthquakes).

The applicability of MTI to motion detection and monitoring also depends on the rates of ground surface change, in this case caused by landsliding. The technical parameters of the available radar systems (Table 11.1) imply that displacements with average annual velocities exceeding several tens of centimeters per year may no longer be resolved by MTI. Also, the lifetime of slope instabilities matters. For example, rockfalls or rock avalanches are too fast and too brief for MTI, while large, very slow deep-seated landslides that can persists for centuries or millennia represent very good investigation targets. Many case studies reported in the literature, and our direct experience (Colesanti and Wasowski, 2006; Bovenga et al., 2012a; Wasowski and Bovenga, 2014 and references therein), show that MTI can be most usefully exploited in the case of long-lived, slowly evolving mass movements. Further, the technique is most effective in detecting gradual, continuous ground surface displacements, whereas it can be essentially "blind" to abrupt ground surface changes produced by strongly nonlinear deformation mechanisms. In those cases, and for landslides belonging to slower and higher velocity classes (WP-WLI, 1995), one will have to resort to other remote sensing monitoring techniques (Table 11.3) or to in situ instrumentation.

In Table 11.4, the main strengths and limitations of MTI are compared with common and emerging remote sensing techniques that are applicable to detecting and monitoring landslide surface motion. In relation to each technique, comments also focus on the best or unique application task or scenario. Although the detailed description of these techniques is outside the scope of this work, in Table 11.4, we provide also a series of representative references (including the most recent ones), where the interested reader can find relevant information.

In particular, MTI is first compared with two other methods relying on space-borne radar imagery. These include standard or conventional DInSAR and interferometry based on the use of corner reflectors that are installed in situ (CRInSAR). This is followed by GBInSAR, which, similarly to CRInSAR, is suitable for slope-scale monitoring. In comparison to other remote sensing techniques considered here, the unique features of GBInSAR are the capability to provide precision measurement for a wide range of slope movement velocities (from extremely slow to moderate motions, Table 11.3) and the suitability for emergency, near real-time monitoring. Then, LiDAR techniques are considered, which are based on a laser beam scanning the ground surface. A distinction is made between Airborne Laser Scanner (ALS) and Terrestrial Laser Scanner (TLS) applications, because this implies differences in scale (regional or local) of investigation and in data resolution. Subsequently, optical image crosscorrelation (or image matching) techniques are presented. These can rely on images acquired by space-borne or air-borne platforms (including the new Unmanned Aerial Vehicles—UAV technology), and also on the imagery obtained from ground-based digital cameras (another approach now being tested). The emerging interest in the applications involving UAV and ground-based cameras can be linked to their low cost, high spatial–temporal resolutions, and flexibility in scheduling data acquisitions. The final method considered is photogrammetry-based air-borne imagery. The major application pros and cons are, respectively, the viability of long-term (often several tens of years) historical data and low acquisition frequency (typically years). Indeed, photogrammetry represents one of the first remote sensing techniques applied in research and practice to landslide motion detection and monitoring (e.g., Chandler and Moore, 1989). However, at present, it does not seem to be used routinely in landslide investigations. This could be in part due to the fact that, unlike some of the above-discussed more recent or emerging techniques, photogrammetry requires commonly complex, expert processing, and sophisticated software.

Finally, it is also useful to recall that in engineering geology practice, conventional slope/landslide monitoring systems (e.g., involving borehole

TABLE 11.4 Comparison of MTI with Common and Emerging Remote Sensing Methods of Detecting and Monitoring Landslide Surface Motion

Method	Main Strengths	Main Limitations	Special Remarks	Main References
MTI	Wide-area coverage (tens of thousands of km^2) combined with a high spatial resolution (up to 1 m for the new-generation satellite radar sensors) and high measurement precision (mm-cm) only marginally affected by poor weather; day/night capability; systematic, high frequency (from few days to weeks) measurements over long periods (years); very high target density (>1000 per km^2) in urbanized or rocky (bare ground) areas	No data in areas with few natural radar targets (<5 per km^2, e.g., heavily vegetated or very steep); unable to detect strong nonlinear deformations and displacements over few tens of cm/year; LOS (1D) measurements (commonly not aligned in the slide motion direction); fixed measurement frequency; need few months to build up new imagery stack required for processing; very sophisticated software and expert processing; high-resolution radar satellite data are currently very expensive	Multiscale investigations with the same data set (from regional to slope-specific); cost effective, especially in the case of long-term (years-decades), large area monitoring; best for reconnaissance/routine surveying of extremely slow–very slow slides; retrospective studies (using archived imagery) to investigate landslide reactivations or slope failure history; can retrieve 2D and 3D motion if ascending and descending data are available	Colesanti et al. (2003); Hilley et al. (2004); Colesanti and Wasowski (2006); Farina et al. (2006); Cascini et al. (2010); Lauknes et al. (2010); Bovenga et al. (2012a); Herrera et al. (2013); Calò et al. (2014); Wasowski and Bovenga (2014)
Standard DInSAR	Under suitable conditions (e.g., scarce vegetation) can provide spatially	No information over vegetated or very steep slopes; very susceptible to	Appropriate to study very large, slow landslides in high coherence areas	Colesanti and Wasowski (2006); Rott et al. (1999); Berardino et al. (2003);

	Capabilities	Limitations	Applications	References
	"semicontinuous," high-precision (cm) results over large areas (tens of thousands of km²); day/night measurement capability; limited number of images required to obtain results; simple processing	coherence loss and to atmospheric artifacts; very low system sensitivity with respect to translational displacements along the north–south direction; unfeasible if differential deformations are significant	(e.g., mountains above the tree line), if only few images available; L-band data are most useful; best for qualitative assessment of deformation; not suitable for monitoring	Squarzoni et al. (2003); Strozzi et al. (2005, 2010); Roering et al. (2009); Calabro et al. (2010); Garcia-Davillo et al. (2013)
Corner reflector interferometry-CRInSAR	High precision (mm-cm) day/night measurement capability; high frequency (from few days to weeks) monitoring (based on satellite revisits); preselected, controlled target positions	Requires much field work and experience to install custom designed (satellite-specific) corner reflectors; logistic, network setup and maintenance issues; vulnerable to vandalism; low target density	Suitable for slope scale, midterm (few years) monitoring when people or key infrastructure (e.g., lifelines) are at risk; can well complement MTI and GNSS monitoring	Froese et al. (2008); Wenxue et al. (2010); Bovenga et al. (2012b); Crosetto et al. (2013)
GBInSAR	Very high target density ("semicontinuous"); high measurement precision (mm); wide range of detectable landslide velocities (from extremely slow to moderate); day/night, all weather capability; very high-frequency (minutes) measurements; can detect strongly nonlinear signal;	Very expensive and bulky instrumentation; requires careful setup and assistance in the field; loss of coherence in the case of fast surface change; slope coverage limited by LOS geometry (visibility depends on local topography); often impractical to retrieve accurate displacements in 2D and 3D (need	Appropriate for emergency, near real-time monitoring (prefailure or postfailure situations); early warning potential; cost effective for short-term (daily/seasonal) monitoring, when people or key infrastructure (e.g., lifelines) are at risk; one of the best methods for detailed surveying of higher velocity and/or episodic	Tarchi et al. (2003); Leva et al. (2003); Casagli et al. (2010); Gischig et al. (2011); Lowry et al. (2013); Monserrat et al. (2013)

(Continued)

TABLE 11.4 Comparison of MTI with Common and Emerging Remote Sensing Methods of Detecting and Monitoring Landslide Surface Motion—cont'd

Method	Main Strengths	Main Limitations	Special Remarks	Main References
	possibility of very rapid processing and delivery of measurement results (hours)	measurements from two or more locations or detailed DEM reflecting the slope/landslide conditions during the monitoring period)	movements; good for measurements of instabilities on steep slopes	
LiDAR ALS and TLS	Up to regional scale coverage (ALS); provides spatially "continuous" information (clouds of points); cm (TLS) to dcm (ALS) spatial resolution and dcm-sub-cm measurement precision; can obtain useful results even in the case of densely vegetated slopes; relatively easy set up (TLS); straightforward processing (commercial software)	Very expensive and relatively large instruments; significant costs of airborne surveys preclude frequent systematic repetition of ALS measurements; sensitive to ambient conditions; in general, ground control points and DEM needed for high quality 3D motion results; TLS requires man assistance in the field	Can retrieve landslide motion from repeat surveys, but more feasible for single slopes using TLS; appropriate for detecting mass movements ranging from extremely slow to slow; good for documenting very small to large instabilities on steep slopes (e.g., rockfalls, rockslides)	Roering et al. (2009); Gischig et al. (2011); Rosser et al. (2007); Teza et al. (2007); Dewitte et al. (2008); Monserrat and Crosetto (2008); Oppikofer et al. (2008); Jaboyedoff et al. (2010); Mackey and Roering (2011); DeLong et al. (2012); Abellan et al. (2013); Corsini et al. (2013)
Optical/radar image crosscorrelation/pixel matching/pixel offset (space-	Spatially "continuous" 2D background information (local to regional scale); good measurement precision (subpixel, i.e., from dcm to meters for	Low frequency (typically years) of airborne data (high acquisition costs); weekly–monthly resolution of satellite data, but currently significant costs;	Appropriate for 2D monitoring of larger displacements (from several dcm to several meters), which cannot be resolved by MTI or DInSAR; good	Kääb (2002); Casson et al. (2003); Delacourt et al. (2004, 2007); Chadwick et al. (2005); Debella-Gilo and Kääb (2012); Travelletti et al. (2012); Niethammer

air-borne and terrestrial data)	images with resolutions ranging from submeters to few meters); imagery from UAVs and ground-based cameras offer, at a lower cost, high spatial-temporal resolutions and flexibility in survey scheduling	strongly affected by weather conditions (cloud cover) and sensitive to vegetation growth (decorrelation); need ground control points and detailed DEM to obtain reliable results and 3D motion vectors; often complex and/or long (post) processing	precision 3D motion from ascending and descending radar images; retrospective studies using historical imagery (e.g., kinematics and displacement history of long-lived, preferably large landslides)	et al. (2011); Raucoules et al. (2013); Singleton et al. (2014)
Photogrammetry (from air-borne imagery)	Spatially "continuous" 3D background information (local to regional scale); historical data archives (often cover >50 years); spatial resolution and measurement precision (dcm-meters); UAV can offer at a lower cost very high spatial-temporal resolutions and much flexibility in data acquisition programing	Low-frequency (typically years) of stereo air-borne data (high acquisition costs); a consistent number of ground control points necessary for reliable results; strongly affected by weather conditions and sensitive to vegetation growth; often complex and/or long (post) processing; requires sophisticated software and expert processing	Photogrammetry is now used for volume estimates rather than for 3D motion monitoring; pixel offset method, and UAV photogrammetry based on high-resolution dual look radar images may become viable alternatives for local to subregional scale 3D monitoring of slow to moderate velocity slides	Strozzi et al. (2005); Dewitte et al. (2008); Chandler and Moore (1989); Mora et al. (2003b); Brückl et al. (2006); Walstra et al. (2007)

inclinometers, extensometers) are typically selected and employed on the basis of individual site/instability process characteristics. In situ monitoring typically serves to protect people and property and possibly provide adequate warning of movement so that the immediate or long-term hazard can be reduced (Schuster and Kockelman, 1996). On the other hand, apart from retrospective studies relying on archived data, the application of MTI techniques requires a significant time period (currently at least a few months) for the initial acquisition and build-up of temporal series of imagery. Thus, a short-term (days) response to impending slope hazard is not possible via MTI at present. However, once the initial time series is built and the first MTI results are available, the subsequent updates can be made on an approximately monthly basis considering the acquisition frequency of new-generation sensors. All this indicates that currently MTI can be best exploited for monitoring aimed at midterm (years) and long-term (decades) landslide risk reduction. Importantly, the wide-area coverage of MTI implies also a good possibility of detecting previously unknown movements (e.g., involving landslides assumed stable or, possibly, slopes near incipient failure stage) and directing appropriate attention to these.

11.4.2 Underexploited and Future MTI Application Opportunities

MTI is able to furnish an unprecedented quantity of precise information on ground surface displacements occurring on slopes. The potential to advance our knowledge of slope and landslide processes seems great, but the MTI-derived results are yet to be fully explored, in particular those based on high spatiotemporal resolution data. Some of the research and application areas that may benefit more from MTI are discussed below.

11.4.2.1 Long-Term Behavior and Climatic Controls of Very Slow Deeper Landslides

The temporally and spatially extensive observation and monitoring available by MTI presents an underexploited opportunity to study and develop a better understanding of the dynamics of very slow landslides, as well as of the longer-term climatic controls of their activity and variations in mobility. For example, Hilley et al. (2004) demonstrated that historical data acquired by satellites may in some cases be suitable for identifying interannual or pluriannual variability in slope movements or landslide activity.

However, despite extensive research on the topic, it is still extremely difficult to predict the motion pattern of very slow landslides (e.g., Crosta and Agliardi, 2003; van Asch et al., 2006). Further, only a limited number of very slow deeper landslides may be monitored for the long periods of time necessary to derive specific rainfall/groundwater thresholds (Coe, 2012). However, the cumulative activity of such very slow landslides can result in significant infrastructure damage (e.g., Mansour et al., 2011), and these phenomena deserve appropriate attention in the context of slope hazard assessment and mitigation (cf. Glastonbury and Fell, 2008).

Obtaining MTI-based precise, high-resolution ground surface deformation measurements at the same time over a large area offers a unique possibility of capturing the natural variability of slope movement patterns and inferring spatial—temporal variations in cause—effect relations at a catchment or regional scale (e.g., taking into account locally variable response to a common destabilizing factor such as rainstorm). In particular, thanks to their high sampling frequency, the new generation of sensors such as CSK or TSX are best suited for exploring short-term linkages between slope instability and variation of ambient factors. Landslide behavior or slope performance evaluation can now be conducted at a seasonal or even event-specific resolution by combining regular MTI-based monitoring with, for example, rainfall data series. It follows that the relative susceptibility of slopes to failure (or the susceptibility of landslides to being reactivated) can be better assessed by linking the short- and longer-term response of many slopes/landslides (presence or absence of movements, accelerations) to specific preparatory and triggering factors (e.g., seasonal precipitation and rainstorms). This is an important opportunity because historical landslide records have generally poor temporal resolution, and it is generally difficult to establish clear links between triggering factors and failure episodes (or reactivation events).

Further, with the availability of decade-long historical radar imagery and the commitment of international space agencies to maintain the existing repeat coverage of large areas and to guarantee continuous data acquisition in the future, MTI offers the monitoring capability of suitable duration to investigate longer-term climate change (or variability) effects on activity and movement mechanics of very slow landslides.

Indeed, when considering the potential impact of changing climate conditions on the huge Triesenberg slide (Liechtenstein Alps), Tacher and Bonnard (2007) noted that very few investigations cover sufficiently extensive periods of time to ascertain the long-term behavior of very large landslides. It is apparent that thanks to the temporally and areally extensive coverage, MTI-derived data could contribute to efforts aimed at predicting future deep-seated landslide mobility (e.g., the forecast modeling based on regional scale moisture balance recently proposed by Coe, 2012).

11.4.2.2 Long-Term Dynamics of Very Large, Deep-Seated Landslides and Deep-Seated Gravitational Slope Deformations

The works of, for example, Colesanti and Wasowski (2006), Dehls et al. (2012), and Frattini et al. (2013), indicated that MTI could be also better exploited for monitoring spatial—temporal changes in the dynamics of very large, deep-seated landslides, as well as deep-seated gravitational slope deformations. These phenomena often occur in high mountain environments and can be characterized by long-term, intermittent or continuous extremely slow to very slow displacements. Due to the sheer scale involved, and commonly difficult access, ground surveying data, if available, are typically limited. Therefore, MTI may often represent the only or the most viable monitoring option. Some very large, long-lived (centuries or more) landslides are known to exhibit continuous very slow movements (that can last for tens of years) with short periodic accelerations. Evidence from investigations of very large Alpine landslides in Switzerland (e.g., Tacher et al., 2005; Tacher and Bonnard, 2007) shows that accelerated movements can take place several times per century, and that short-term (months) velocities during "crisis" periods can be, in some cases, up to 100 times higher than pluriannual or decadal average velocities; the latter may typically range from a few centimeters per year to a few decimeters per year. Even outside of the "crisis" period, such very slow landslides need to be monitored, because their prolonged activity can lead to a significant damage of buildings and infrastructure (e.g., Mansour et al., 2011). Long-term MTI monitoring may indeed represent one of the most cost-effective options of landslide hazard management in these cases. Where the density of measurable radar targets is high, MTI will offer also a unique possibility to detail spatial variations in deformation patterns and rates within large landslide bodies (Colesanti and Wasowski, 2006; Bovenga et al., 2012a). This is important because structural damage can often be more directly linked to very local gradients in movement rate (differential deformations) than to the average displacement velocity.

11.4.2.3 Numerical Modeling of Very Slow Persistent Landslides and Long-Term Evolution of Slopes

We expect that in the future, MTI-derived surface displacement results integrated with data on material properties will be used more commonly to assist in landslide numerical/geotechnical modeling efforts. For example, very precise detection of ground surface changes would be needed to test some mathematical models of slow landslide motion regulation through dilatancy and pore-pressure feedback mechanisms (Iverson, 2005). Similarly, in addition to typically shorter term in situ monitoring data, MTI displacement results covering longer periods could provide useful input or cross-checks on geotechnical modeling of very slow, large, deep landslides that can be characterized by complex deformation patterns (e.g., Puzrin and Schmid, 2012; Kalenchuk et al., 2013). This opportunity was first recognized by Hilley et al. (2004), whose model of very slowly deforming landslides in the San Francisco Bay area was based on several years' worth of MTI surface displacement data. In another case, Francois et al. (2007) examined the MTI displacement results presented by Colesanti and Wasowski (2006) and found them to be consistent with the outcomes of their numerical modeling of the hydrogeological and geomechanical behavior of the large, very slow moving Triesenberg landslide in the Liechtenstein Alps. Most recently, Calò et al. (2014) presented an interesting example of numerical modeling of a creeping landslide in Italy. The modeling took advantage of the integration of long-term (~ 20 years) MTI displacement results and subsurface geotechnical information that included data from borehole inclinometer monitoring.

11.4.2.4 Postearthquake Landslide Activity and Evolution of Slopes

MTI potential is also underexploited in the case of postseismic activity of earthquake-induced mass movements, which, to the best of our knowledge, have not yet been investigated with the aid of radar remote sensing (Wasowski et al., 2011). Obviously, given the episodic and instantaneous nature of earthquakes, MTI will be able to document only very slow, postevent evolution of slopes and landslides. However, a possibility also exists that delayed reactivations of preexisting deep-seated landslides could be captured. The literature provides evidence that such postseismic phenomena can occur from hours to days or even weeks after the mainshock (e.g., Keefer, 2002; Wasowski et al., 2002).

The use of MTI will be especially beneficial in the case of large magnitude events when slope failures occur in great numbers over very wide areas (e.g., >60,000 landslides triggered by the 7.9 Mw 2008 Wenchuan earthquake in an area exceeding 50,000 km^2, Gorum et al., 2011). Large earthquakes are known to have a long-lasting legacy of producing instability. This particularly affects mountainous environments where, long after the primary seismic triggering event has occurred, slopes remain for years or perhaps decades more susceptible to generate postseismic failures or landslide reactivations. While the former are mostly fast movements (e.g., debris flows), the latter could be slow (e.g., reactivations of deep-seated slides) and thus suitable for MTI-based monitoring. The example of the 2008 Wenchuan earthquake shows that widespread landsliding continues to have significant impacts on lives and livelihoods. In high mountain settings like the one affected by the Wenchuan event, in situ monitoring of landslides is costly and impractical in most situations, because of limited accessibility of slopes and the very large numbers of potentially unstable sites that require surveying.

MTI applications in postseismic investigations could be very useful especially where event-based slope hazard maps or landslide inventory maps are incomplete or may not be prepared due to time and cost constraints. The same can be said considering landslide susceptibility maps (Lee et al., 2008; Guzzetti et al., 2012), which, if available, necessitate periodic updating, especially in the case of environments characterized by high rates of geomorphic change (e.g., high seismicity areas) and pronounced land-use variations.

11.4.2.5 Nonlinear Kinematics of Landslides, Maximum Velocities, and Accumulated Displacements

MTI could be better exploited in detailed studies of nonlinear landslide kinematics by paying more attention to the information content in displacement time series. Although Colesanti and Wasowski (2006) highlighted the possibility of recognizing nonlinear trends in MTI displacement time series, it should be recalled that in the absence of a priori information on the deformation pattern, many MTI algorithms adopt a linear kinematic model to measure the displacements. This implies that nonlinear components of motion can be significantly underestimated, especially where the actual deformation process is strongly nonlinear. Thus, the maximum displacement rates tend to be overlooked or missed. In the case of very slow landslides, the difficulty in the recognition of nonlinear displacement trends can be related in part to insufficient temporal frequency of sampling. Indeed, MTI times series often reveal linear patterns of ground deformation, while in reality, the movement rates of most landslides are not constant. For example, many slow landslides are characterized by unsteady, nonuniform motions in relation to wet and dry season cycles (e.g., Iverson and Major, 1987; Iverson, 2005). Nevertheless, the Haiti landslide case study presented above demonstrates that this problem can be in part mitigated by exploiting the higher sampling frequency typical of the new generation of radar satellites.

The issues linked to the more effective interpretation and use of displacement time series have recently received desirable attention in the landslide community (e.g., Notti et al., 2012; Cigna et al., 2012; Berti et al., 2013), and further advances can be expected in the future. In this context, we foresee more efforts aimed at the exploitation of MTI results for the recognition of nonlinear kinematics of slow landslides and investigation of their causes. These efforts can be facilitated by the higher temporal resolution of data available from the new and upcoming sensors. The higher frequency information implies also that ground movements related to a single causative event (e.g., a few days of rainfall) should be more easily differentiated from the longer-term deformation trend. Further, the retrieval of maximum displacement rates from MTI is more feasible now. This is recommended, because maximum velocity information bears more significance for slope hazard assessment (cf. Cruden and Varnes, 1996) than time-averaged information.

Finally, it seems that in addition to the most common visualization and analysis of the MTI results in terms of the average annual deformation velocities, more attention should be directed to the significance of long-term cumulative displacements recorded on radar targets. There are two practical reasons for this. First, different materials can accommodate different amounts of strain before failure, but large cumulative displacements measured by MTI in engineering soils can be used to indicate near-residual-strength conditions. Second, the performance of engineering structures on slopes or the degree of expected damage could be related to the slide velocity and accumulated displacement (Mansour et al., 2011; Frattini et al., 2013).

ACKNOWLEDGMENTS

ENVISAT and CSK data for Garzeno, and Madesimo (Italy) were provided, respectively, by the European Space Agency (ESA) and by ASI in the framework of the MORFEO project funded by ASI (Contract n. I/045/07/0). ENVISAT data for Volturino site were provided by ESA under CAT-1 project ID 2653 in the framework of Puglia Region project "FRANE PUGLIA—Rilevamento di deformazioni al suolo con tecniche satellitari avanzate." We thank Raffaele Nutricato and Davide Oscar Nitti of GAP srl for providing MTI processing through SPINUA algorithm. We are grateful to Alessandro Ferretti and his colleagues at TRE srl (Milan, Italy) for sharing with us the results of SqueeSAR™ processing for the Madesimo study area. We also thank Giovanni Nico (CNR-IAC, Bari) and his colleagues at DIAN srl (Matera, Italy) for providing the example of GBInSAR application. Finally, we are indebted to this Volume Editor Tim Davies for his help and suggestions.

REFERENCES

Abellán, A., Oppikofer, T., Jaboyedoff, M., Rosser, N.J., Lim, M., Lato, M.J., 2013. Terrestrial laser scanning of rock slope instabilities. Earth Surf. Processes Landforms 39, 80−97.

Bally, P., 2012. In: Scientific and Technical Memorandum of the International Forum on Satellite EO and Geohazards, May 21−23, 2012, Santorini, Greece. http://dx.doi.org/10.5270/esa-geo-hzrd-2012.

Bamler, R., Eineder, M., 2005. Accuracy of differential shift estimation by correlation and split-bandwidth interferometry for wideband and delta-k SAR systems. IEEE Geosci. Remote Sens. Lett. 2 (2), 151−155.

Berardino, P., Fornaro, G., Lanari, R., Sansosti, E., 2002. A new algorithm for surface deformation monitoring based on small baseline differential SAR interferograms. IEEE Trans. Geosci. Remote Sens. 40, 2375−2383.

Berrardino, P., Costantini, M., Franceschetti, G., Iodice, A., Pietranera, L., Rizzo, V., 2003. Use of differential SAR interferometry in monitoring and modelling large slope instability at Maratea (Basilicata, Italy). Eng. Geol. 68 (1−2), 31−51.

Berti, M., Corsini, A., Franceschini, S., Iannacone, J.P., 2013. Automated classification of persistent scatterers interferometry time-series. Nat. Hazards Earth Syst. Sci. 13, 1945−1958. http://dx.doi.org/10.5194/nhess-13-1945-2013.

Booth, Adam M., Roering, Josh J., Rempel, Alan W., 2013. Topographic signatures and a general transport law for deep-seated landslides in a landscape evolution model. Geophys. Res. Earth Surf. 118, 1−22. http://dx.doi.org/10.1002/jgrf.20051.

Bovenga, F., Refice, A., Nutricato, R., Guerriero, L., Chiaradia, M.T., 2005. SPINUA: a flexible processing chain for ERS/ENVISAT long term interferometry. In: Proceedings of ESA-ENVISAT Symposium, September 6−10, 2004, Salzburg, Austria. ESA Special Publication SP-572. April 2005, CD. ISBN 92-9092-883-2, ISSN 1609−042X.

Bovenga, F., Nutricato, R., Refice, A., Wasowski, J., 2006. Application of Multi-temporal Differential Interferometry to Slope Instability Detection in Urban/Peri-urban Areas. Engineering Geology 88 (3–4), 218–239.

Bovenga, F., Wasowski, J., Nitti, D.O., Nutricato, R., Chiaradia, M.T., 2012a. Using Cosmo/SkyMed X-band and ENVISAT C-band SAR Interferometry for landslide analysis. Remote Sens. Environ. 119, 272–285. http://dx.doi.org/10.1016/j.rse.2011.12.013.

Bovenga, F., Refice, A., Pasquariello, G., 2012b. Using corner reflectors and x-band SAR interferometry for slope instability monitoring. In: Proceedings of Tyrrhenian Workshop on Advances in Radar and Remote Sensing (TyWRRS), September 12–14, 2012, Naples, Italy, pp. 114–120. http://dx.doi.org/10.1109/TyWRRS, 2012.6381114.

Bovenga, F., Nitti, D.O., Fornaro, G., Radicioni, F., Stoppini, A., Brigante, R., 2013. Using C/X-band SAR interferometry and GNSS measurements for the Assisi landslide analysis. Int. J. Remote Sens. 34 (11), 4083–4104. http://dx.doi.org/10.1080/01431161.2013.772310.

Brückl, E., Brunner, F.K., Kraus, K., 2006. Kinematics of a deep-seated landslide derived from photogrammetric, GPS and geophysical data. Eng. Geol. 88 (3–4), 149–159.

Calabro, M.D., Schmidt, D.A., Roering, J.J., 2010. An examination of seasonal deformation at the Portuguese Bend landslide, southern California, using radar interferometry. J. Geophys. Res. 115, F00A20. http://dx.doi.org/10.1029/2009JF001314.

Calò, F., Ardizzone, F., Castaldo, R., Lollino, P., Tizzani, P., Guzzetti, F., Lanari, R., Angeli, M.-G., Pontoni, F., Manunta, M., 2014. Enhanced landslide investigations through advanced DInSAR techniques: the Ivancich case study, Assisi, Italy. Remote Sens. Environ. 142, 69–82.

Casagli, N., Catani, F., Del Ventisette, C., Luzi, G., 2010. Monitoring, prediction, and early warning using ground-based radar interferometry. Landslides 7 (3), 291–301.

Cascini, L., Fornaro, G., Peduto, D., 2009. Analysis at medium scale of low-resolution DInSAR data in slow-moving landslide affected areas. ISPRS J. Photogramm. Remote Sens. 64 (6), 598–611. http://dx.doi.org/10.1016/j.isprsjprs.2009.05.003.

Cascini, L., Fornaro, G., Peduto, D., 2010. Advanced low- and full-resolution DInSAR map generation for slow-moving landslide analysis at different scales. Eng. Geol. 112 (1–4), 29–42.

Casson, B., Delacourt, C., Baratoux, D., Allemand, P., 2003. Seventeen years of the "La Clapière" landslide evolution analysed from ortho-rectified aerial photographs. Eng. Geol. 68 (1–2), 123–139.

Casu, F., Manzo, M., Lanari, R., 2006. A quantitative assessment of the SBAS algorithm performance for surface deformation retrieval from DInSAR data. Remote Sens. Environ. 102, 95–210.

Chadwick, D.J., Dorsch, S., Glenn, N., Thackray, G., Shilling, K., 2005. Application of multi-temporal high-resolution imagery and GPS in a study of the motion of a canyon rim landslide. J. Int. Soc. Photogramm. Remote Sens. (ISPRS) 59, 212–221.

Chandler, J.H., Moore, R., 1989. Analytical photogrammetry: a method for monitoring slope instability. Q. J. Eng. Geol. Hydrogeol. 22, 97–110.

Cigna, F., Tapete, D., Casagli, N., 2012. Semi-automated extraction of deviation indexes (DI) from satellite persistent scatterers time series: tests on sedimentary volcanism and tectonically induced motions. Nonlin. Processes Geophys. 19, 643–655. http://dx.doi.org/10.5194/npg-19-643-2012.

Coe, J.A., 2012. Regional moisture balance control of landslide motion: Implications for landslide forecasting in a changing climate. Geology 40, 323–326.

Colesanti, C., Wasowski, J., 2004. Satellite SAR interferometry for wide-area slope hazard detection and site-specific monitoring of slow landslides. In: Proceedings of the Ninth International Symposium on Landslides, Rio de Janeiro (Brazil), June 28—July 2, 2004, vol. 1, pp. 795—802.

Colesanti, C., Wasowski, J., 2006. Investigating landslides with space-borne synthetic aperture radar (SAR) interferometry. Eng. Geol. 88 (3—4), 173—199.

Colesanti, C., Ferretti, A., Prati, C., Rocca, F., 2003. Monitoring landslides and tectonic motion with the permanent scatterers technique. Eng. Geol. 68, 3—14.

Corsini, A., Castagnetti, C., Bertacchini, E., Rivola, R., Ronchetti, F., Capra, A., 2013. Integrating airborne and multi-temporal long-range terrestrial laser scanning with total station measurements for mapping and monitoring a compound slow moving rock slide. Earth Surf. Processes Landforms 38, 1330—1338.

Crosetto, M., Biescas, E., Duro, J., 2008. Generation of advanced ERS and Envisat interferometric SAR products using the stable point network technique. Photogramm. Eng. Remote Sens. 4, 443—450.

Crosetto, M., Gili, J., Monserrat, A.O., Cuevas-González, M., Corominas, J., Serral, D., 2013. Interferometric SAR monitoring of the Vallcebre landslide (Spain) using corner reflectors. Nat. Hazards Earth System Sci. 13, 923—933.

Crosta, G., Agliardi, F., 2003. Failure forecast for large rock slides by surface displacement measurements. Can. Geotech. J. 40 (1), 176—191.

Cruden, D.M., Varnes, D.J., 1996. Landslide types and processes. In: Turner, A.K., Schuster, R.L. (Eds.), Landslides. Investigation and mitigation. Transportation Research Board Spec. Rep., vol. 247. Nat. Academy Press, p. 673.

Debella-Gilo, M., Kääb, A., 2012. Measurement of surface displacement and deformation of mass movements using least squares matching of repeat high resolution satellite and aerial images. Remote Sens. 4 (12), 43—67. http://dx.doi.org/10.3390/rs4010043.

Dehls, J.F., Fischer, L., Böhme, M., Saintot, A., Hermanns, R.H., Oppikofer, T., Lauknes, T.R., Larsen, Y., Blikra, L.H., 2012. Landslide monitoring in western Norway using high resolution TerraSAR-X and Radarsat-2 InSAR. In: Eberhardt, E., Froese, C., Turner, A.K., Leroueil, S. (Eds.), Landslides and Engineered Slopes, Proceedings of the 11th International and 2nd North American Symposium on Landslides, Banff (Canada), June 3—8, 2012, vol. 2. CRC Press/Balkema, Leiden (The Netherlands), pp. 1321—1325.

Delacourt, C., Allemand, P., Casson, B., Vadon, H., 2004. Velocity field of the "La Clapière" landslide measured by the correlation of aerial and QuickBird satellite images. Geophys. Res. Lett. 31, L15619. http://dx.doi.org/10.1029/2004gl020193.

Delacourt, C., Allemand, P., Berthier, E., Raucoules, D., Casson, B., Grandjean, P., Pambrun, C., Varel, E., 2007. Remote-sensing techniques for analyzing landslide kinematics: a review. Bull. Soc. Geol. Fr. 178 (2), 89—100.

DeLong, S.B., Prentice, C.S., Hilley, G.E., Ebert, Y., 2012. Multitemporal ALSM change detection, sediment delivery, and process mapping at an active earthflow. Earth Surf. Processes Landforms 37 (3), 262—272. http://dx.doi.org/10.1002/esp.2234.

Dewitte, O., Jasselette, J.C., Cornet, Y., Van Den Eeckhaut, M., Collignon, A., Poesen, J., Demoulin, A., 2008. Tracking landslide displacements by multi-temporal DTMs: a combined aerial stereophotogrammetric and LIDAR approach in western Belgium. Eng. Geol. 99 (1—2), 11—22.

Di Pasquale, A., Corsetti, M., Guccione, P., Lugli, A., Nicoletti, M., Nico, G., Zonno, M., June 36, 2013. Proc. 33rd EARSel Symp., Matera, Italy [Online]. Available. http://www.earsel.org/symposia/2013-symposium-Matera/pdf_proceedings/EARSeL-Symposium-2013_4_3_Dipasquale.pdf.

Duro, J., Inglada, J., Closa, J., Adam, N., Arnaud, A., 2004. High resolution differential inter-ferometry using time series of ERS and ENVISAT SAR data. In: Proceedings of Fringe 2003 Workshop, December 1-5, 2003. ESA Special Publication SP-550, Frascati (Italy). August 2004, CD.

Duro, J., Gaset, M., Koudogbo, F.N., Arnaud, A., 2012. Combination of x-band high resolution SAR data from different sensors to produce ground deformation maps. In: Proceedings of Fringe 2011 Workshop, September 19–23, 2011. ESA Special Publication SP-697, Frascati (Italy). January 2012, CD. ISBN 978-92-9092-261-2, ISSN 1609-042X.

Erten, E., Reigber, A., Hellwich, O., 2010. Generation of three-dimensional deformation maps from InSAR data using spectral diversity techniques. ISPRS J. Photogramm. Remote Sens. 65 (4), 388–394.

Farina, P., Colombo, D., Fumagalli, A., Marks, F., Moretti, S., 2006. Permanent scatterers for landslide investigations: outcomes from ESA-SLAM project. Eng. Geol. 88, 200–217.

Fell, R., Corominas, J., Bonnard, C., Cascini, L., Leroi, E., Savage, W.Z., 2008. Guidelines for landslide susceptibility, hazard and risk zoning for land-use planning. Eng. Geol. 102 (3–4), 99–111.

Ferretti, A., Prati, C., Rocca, F., 2001. Permanent scatterers in SAR interferometry. IEEE Trans. Geosci. Remote Sens. 39 (1), 8–20. http://dx.doi.org/10.1109/36.898661.

Ferretti, A., Savio, G., Barzaghi, R., Borghi, A., Musazzi, S., Novali, F., Prati, C., Rocca, F., 2007. Submillimeter accuracy of InSAR time series: experimental validation. IEEE Trans. Geosci. Remote Sens. 45, 1142–1153. http://dx.doi.org/10.1109/TGRS, 894440.

Ferretti, A., Fumagalli, A., Novali, F., Prati, C., Rocca, F., Rucci, A., 2011. A new algorithm for processing interferometric data-stacks: SqueeSAR. IEEE Trans. Geosci. Remote Sens. 49 (9), 3460–3470. http://dx.doi.org/10.1109/TGRS, 2124465.

François, B., Tacher, L., Bonnar, Ch, Laloui, L., Trigero, V., 2007. Numerical modeling of the hydrological and geomechanical behavior of a large slope movement: the Triesenberg land-slide (Liechtenstein). Can. Geotech. J. 44, 840–857.

Frattini, P., Crosta, G., Allievi, J., 2013. Damage to buildings in large slope rock instabilities monitored with the PSInSAR^{TM} technique. Remote Sens. 5, 4753–4773. http://dx.doi.org/10.3390/rs5104753.

Frosee, C.R., Poncos, V., Skittow, R., Mansour, M., Martin, D, 2008. Characterizing complex deep seated landslide deformation using corner reflection InSAR: Little Smoky Landslide, Alberta. In: Proceedings of the 4th Canadian Conference on Geohazards: From Causes to Management, May 20–24, 2008, Quebec City. Presse de l'Universitè Laval, Quebec.

García-Davalillo, J., Herrera, G., Notti, D., Strozzi, T., Álvarez-Fernández, I., 2013. DInSAR analysis of ALOS PALSAR images for the assessment of very slow landslides: the Tena Valley case study. Landslides, 1–22.

Gischig, V., Amann, F., Moore, J.R., Loew, S., Eisenbeiss, H., Stempfhuber, W., 2011. Composite rock slope kinematics at the current Randa instability, Switzerland, based on remote sensing and numerical modeling. Eng. Geol. 118, 37–53.

Glastonbury, J.P., Fell, R., 2008. Geotechnical characteristics of large slow, very slow, and extremely slow landslides. Can. Geotech. J. 45, 984–1005.

Gorum, T., Fan, X., van Westen, C.J., Huang, R., Xu, Q., Tang, C., 2011. Distribution pattern of earthquake-induced landslides triggered by the May 12, 2008 Wenchuan earthquake. Geomorphology 133, 152–167.

Guzzetti, F., Mondini, A.C., Cardinali, M., Fiorucci, F., Santangelo, M., Chang, K.-T., 2012. Landslide inventory maps: new tools for an old problem. Earth-Sci. Rev. 112, 1–25.

Hanssen, R., 2001. Radar Interferometry: Data Interpretation and Error Analysis. Kluwer Academic Publishers, Dordrecht (The Netherlands).

Hanssen, R., 2005. Satellite radar interferometry for deformation monitoring: a priori assessment of feasibility and accuracy. Int. J. Appl. Earth Obs. Geoinfo. 6, 253–260.

Hanssen, R., van Leijen, F.J., van Zwieten, G.J., Dortland, S., Bremmer, C.N., Kleuskens, M., February 2008. Validation of PSI results of Alkmaar and Amsterdam within the Terrafirma validation project. In: Proceedings of Fringe 2007 Workshop, November 26–30, 2007, Frascati, Italy. ESA Special Publication SP-649. CD. ISBN 92-9291-213-3, ISSN 1609–042X.

Herrera, G., Gutiérrez, F., García-Davalillo, J.C., Guerrero, J., Notti, D., Galve, J.P., Fernández-Merodo, J.A., Cooksley, G., 2013. Multi-sensor advanced DInSAR monitoring of very slow landslides: the Tena Valley case study (Central Spanish Pyrenees). Remote Sens. Environ. 128, 31–43.

Hilley, G.E., Burgmann, R., Ferretti, A., Novali, F., Rocca, F., 2004. Dynamics of slow-moving landslides from permanent scatterer analysis. Science 304, 1952–1955.

Hooper, A., Zebker, H., Segall, P., Kampes, B., 2004. A new method for measuring deformation on volcanoes and other natural terrains using InSAR persistent scatterers. Geophys. Res. Lett. 31, L23611.

Hooper, A., 2008. A multi-temporal InSAR method incorporating both persistent scatterer and small baseline approaches. Geophys. Res. Lett. 35, L16302.

Hooper, A., Bekaert, D., Spaans, K., Arıkan, M.T., 2012. Recent advances in SAR interferometry time series analysis for measuring crustal deformation. Tectonophysics 514–517, 1–13. http://dx.doi.org/10.1016/j.tecto.2011.10.013.

Iverson, R.M., 2005. Regulation of landslide motion by dilatancy and pore pressure feedback. J. Geophys. Res. 110, F02015. http://dx.doi.org/10.1029/2004JF000268.

Iverson, R.M., Major, J.J., 1987. Rainfall, groundwater flow, and seasonal movement at Minor Creek landslide, north western California—physical interpretation of empirical relations. Geol. Soc. Am. Bull. 99, 579–594.

Jaboyedoff, M., Oppikofer, T., Abellán, A., Derron, M.-H., Loye, A., Metzger, R., Pedrazzini, A., 2010. Use of LIDAR in landslide investigations: a review. Nat. Hazards, 1–24.

Kääb, A., 2002. Monitoring high-mountain terrain deformation from repeated air- and spaceborne optical data: examples using digital aerial imagery and ASTER data, ISPRS. J. Photogramm. Remote Sens. 57, 39–52.

Kalenchuk, K., Hutchinson, D., Diederichs, M., 2013. Geomechanical interpretation of the Downie Slide considering field data and three-dimensional numerical modeling. Landslides 10, 737–756.

Kampes, B.M., 2005. Deformation Parameter Estimation Using Permanent Scatterer Interferometry (Ph.D. thesis). Delft University of Technology, Delft.

Keefer, D.K., 2002. Investigating landslides caused by earthquakes—a historical review. Surv. Geophys. 23, 473–510.

Lauknes, T.R., Piyush Shanker, A., Dehls, J.F., Zebker, H.A., Henderson, H.C., Larsen, Y., 2010. Detailed rockslide mapping in northern Norway with small baseline and persistent scatterer interferometric SAR time series methods. Remote Sens. Environ. 114, 2097–2109.

Lee, C.T., Huang, C.C., Lee, J.F., Pan, K.L., Lin, M.L., Dong, J.J., 2008. Statistical approach to earthquake-induced landslide susceptibility. Eng. Geol. 100, 43–58.

Leva, D., Nico, G., Tarchi, D., Fortuny-Guasch, J., Sieber, A.J., 2003. Temporal analysis of a landslide by means of a ground-based SAR interferometer. IEEE Trans. Geosci. Remote Sens. 41, 745–752.

Lollino, P., Elia, G., Cotecchia, F., Mitaritonna, G., February 20–24, 2010. Analysis of landslide reactivation in Daunia clay slopes by means of limit equilibrium and FEM methods. In: Proceedings of Geo-Florida 2010. West Palm Beach, pp. 3155–3164.

Lowry, B., Gomez, F., Zhou, W., Mooney, M.A., Held, B., Grasmick, J., 2013. High resolution displacement monitoring of a slow velocity landslide using ground based radar interferometry. Eng. Geol. 166, 160–169.

Mackey, B.H., Roering, J.J., 2011. Sediment yield, spatial characteristics, and the long-term evolution of active earthflows determined from airborne LiDAR and historical aerial photographs, Eel River, California. Geol. Soc. Am. Bull. 123 (7–8), 1560–1576. http://dx.doi.org/10.1130/b30306.1.

Mahapatra, P., Samiei-Esfahany, S., Hansen, R., January 2012. Towards repeatibility, reliability and robustness in time-series InSAR. In: Proceedings of Fringe 2011 Workshop, September 19–23, 2011, Frascati, Italy. ESA Special Publication SP-697. CD. ISBN 978-92-9092-261-2, ISSN 1609-042X.

Mansour, M., Morgenstern, N., Martin, C., 2011. Expected damage from displacement of slow-moving slides. Landslides 8, 117–131.

Metternicht, G., Hurni, L., Gogu, R., 2005. Remote sensing of landslides: an analysis of the potential contribution to geo-spatial systems for hazard assessment in mountainous environments. Remote Sens. Environ. 98, 284–303.

Monserrat, O., Crosetto, M., 2008. Deformation measurement using terrestrial laser scanning data and least squares 3D surface matching. ISPRS J. Photogramm. Remote Sens. 63, 142–154.

Monserrat, O., Moya, J., Luzi, G., Crosetto, M., Gili, J.A., Corominas, J., 2013. Non-interferometric GB-SAR measurement: application to the Vallcebre landslide (eastern Pyrenees, Spain). Nat. Hazards Earth Syst. Sci. 13, 1873–1887.

Mora, O., Mallorqui, J.J., Broquetas, A., 2003a. Linear and nonlinear terrain deformation maps from a reduced set of interferometric SAR images. IEEE Trans. Geosci. Remote Sens. 41, 2243–2253.

Mora, P., Baldi, Casula, G., Fabris, M., Ghiotti, M., Mazzini, E., Pesci, A., 2003b. Global positioning systems and digital photogrammetry for the monitoring of mass movements: application to the Ca'di Malta landslide (northern Apennines, Italy). Eng. Geol. 68, 103–121.

Mossa, S., Capolongo, D., Pennetta, L., Wasowski, J., 2005. A GIS-based assessment of landsliding in the Daunia Apennines, southern Italy. In: Proceedings of the International Conference "Mass Movement Hazard in Various Environments" 20–21 October 2005, Cracow, Poland, pp. 86–91.

Niethammer, U., James, M.R., Rothmund, S., Travelletti, J., Joswig, M., 2011. UAV-based remote sensing of the Super-Sauze landslide: evaluation and results. Eng. Geol. 128 (9), 2–11.

Nitti, D.O., Bovenga, F., Refice, A., Wasowski, J., Conte, D., Nutricato, R., 2009. L- and C-band SAR interferometry analysis of the Wieliczka salt mine area (UNESCO heritage site, Poland). In: Proceedings ALOS PI 2008 Symposium, November 3–7, 2008, Rhodes, Greece. ESA SP-664, January 2009.

Notti, D., Davalillo, J.C., Herrera, G., Mora, O., 2010. Assessment of the performance of X-band satellite radar data for landslide mapping and monitoring: upper Tena valley case study. Nat. Hazards Earth Syst. Sci. 10, 1865–1875.

Notti, D., Meisina, C., Zucca, F., Crosetto, M., Montserrat, O., 2012. Factors that have an influence on time series. In: Proceedings of Fringe 2011 Workshop, September 19–23, 2011. ESA Special Publication SP-697, Frascati (Italy). January 2012, CD. ISBN 978-92-9092-261-2, ISSN 1609-042X.

Nutricato, R., Wasowski, J., Bovenga, F., Refice, A., Pasquariello, G., Nitti, D.O., Chiaradia, M.T., 2013. C/X-Band SAR interferometry used to monitor slope instability in Daunia, Italy. In: Margottini, C., et al. (Eds.), Landslide Science and Practice, vol. 2. © Springer-Verlag Berlin Heidelberg, pp. 423−430. http://dx.doi.org/10.1007/978-3-642-31445-2_55.

Oppikofer, T., Jaboyedoff, M., Keusen, H.-R., 2008. Collapse at the eastern Eiger flank in the Swiss Alps. Nat. Geosci. 1 (8), 531−535.

Petley, D., 2012. Global patterns of loss of life from landslides. Geology 40 (10), 927−930.

Plank, S., Singer, J., Minet, C., Thuro, K., 2012. Pre-survey evaluation of the differential synthetic aperture radar interferometry method for landslide monitoring. Int. J. Remote Sens. 33 (20), 6623−6637.

Puzrin, A.M., Schmid, A., Schwager, M.V., 2012. Case studies of constrained creeping landslides in Switzerland. In: Eberhardt, E., Froese, C., Turner, A.K., Leroueil, S. (Eds.), Landslides and Engineered Slopes, Proceedings of the 11th International and 2nd North American Symposium on Landslides, Banff (Canada), June 3−8, 2012, vol. 2. CRC Press/Balkema, Leiden (The Netherlands), pp. 1795−1800.

Raucoules, D., Bourgine, B., De Michele, M., Le Cozanet, G., Closset, L., Bremmer, C., Veldkamp, H., Tragheim, D., Bateson, L., Crosetto, M., Agudo, M., Engdahl, M., 2009. Validation and intercomparison of persistent scatterers interferometry: PSIC4 project results. J. Appl. Geophys. 68 (3), 335−347.

Raucoules, D., de Michele, M., Malet, J.P., Ulrich, P., 2013. Time-variable 3D ground displacements from high-resolution synthetic aperture radar (SAR). Application to La Valette landslide (South French Alps). Remote Sens. Environ. 139, 198−204.

Reale, D., Nitti, D.O., Peduto, D., Nutricato, R., Bovenga, F., Fornaro, G., 2011. Post-seismic deformation monitoring with the COSMO/SKYMED constellation. IEEE Geosci. Remote Sens. Lett. 8 (4), 696−700. http://dx.doi.org/10.1109/LGRS, 2010.2100364.

Roering, J.J., Stimely, L.L., Mackey, B.H., Schmidt, D.A., 2009. Using DInSAR, airborne LiDAR, and archival air photos to quantify landsliding and sediment transport. Geophys. Res. Lett. 36 (19), L19402.

Rosser, N., Lim, M., Petley, D., Dunning, S., Allison, R., 2007. Patterns of precursory rockfall prior to slope failure. J. Geophys. Res. 112 (F4), F04014.

Rott, H., Scheuchel, B., Siegel, A., 1999. Monitoring very slow slope movements by means of SAR interferometry: a case study from mass waste above a reservoir in the Otztal Alps, Austria. Geophys. Res. Lett. 26, 1629−1632.

Rucci, A., Ferretti, A., Monti Guarnieri, A., Rocca, F., 2012. Sentinel 1 SAR interferometry applications: the outlook for sub millimeter measurements. Remote Sens. Environ. 120, 156−163. http://dx.doi.org/10.1016/j.rse.2011.09.030.

Sansosti, E., Casu, F., Manzo, M., Lanari, R., 2010. Space-borne radar interferometry techniques for the generation of deformation time series: an advanced tool for Earth's surface displacement analysis. Geophys. Res. Lett. 37 (20), L20305.

Schuster, R.L., 1996. Socioeconomic significance of landslides. In: Turner, K., Schuster, R.L. (Eds.), Landslides Investigation and Mitigation Transportation Research Board, Special Report 247. Washington, DC, pp. 91−105.

Schuster, R.L., Kockelman, W.J., 1996. Principles of landslide hazard zonation. In: Turner, K., Schuster, R.L. (Eds.), Landslides Investigation and Mitigation Transportation Research Board, Special Report 247. Washington, DC, pp. 12−35.

Singleton, A., Li, Z., Hoey, T., Muller, J.-P., 2014. Evaluating sub-pixel offset techniques as an alternative to D-InSAR for monitoring episodic landslide movements in vegetated terrain. Remote Sens. Environ. 147, 133−144.

Squarzoni, C., Delacourt, C., Allemand, P., 2003. Nine years of spatial and temporal evolution of the La Valette landslide observed by SAR interferometry. Eng. Geol. 68, 53–66.

Strozzi, T., Farina, P., Corsini, A., Ambrosi, C., Thüring, M., Zilger, J., Wiesmann, A., Wegmüller, U., Werner, C., 2005. Survey and monitoring of landslide displacements by means of L-band satellite SAR interferometry. Landslides 2 (3), 193–201.

Strozzi, T., Delaloye, R., Kääb, A., Ambrosi, C., Perruchoud, E., Wegmüller, U., 2010. Combined observations of rock mass movements using satellite SAR interferometry, differential GPS, airborne digital photogrammetry, and airborne photography interpretation. J. Geophys. Res. 115 (F1). http://dx.doi.org/10.1029/2009JF001311.

Tacher, L., Bonnard, Ch, Laloui, L., Parriaux, A., 2005. Modelling the behaviour of a large landslide with respect to hydrogeological and geomechanical parameter heterogeneity. Landslides 2, 3–14. http://dx.doi.org/10.1007/s10346-004-0038-9.

Tacher, L., Bonnard, C., 2007. Hydromechanical modelling of a large landslide considering climate change conditions. In: Proc. Int. Conf. On Landslides and Climate Change, Ventnor, Isle of Wight, 21–24 May 2007, pp. 131–141.

Tarchi, D., Casagli, N., Fanti, R., Leva, D.D., Luzi, G., Pasuto, A., Pieraccini, M., Silvano, S., 2003. Landslide monitoring by using ground-based SAR interferometry: an example of application to the Tessina landslide in Italy. Eng. Geol. 68, 15–30.

Teza, G., Galgaro, A., Zaltron, N., Genevois, R., 2007. Terrestrial laser scanner to detect landslide displacement fields: a new approach. Int. J. Remote Sens. 28 (16), 3425–3446.

Travelletti, J., Delacourt, C., Allemand, P., Malet, J.-P., Schmittbuhl, J., Toussaint, R., Bastard, M., 2012. Correlation of multi-temporal ground-based optical images for landslide monitoring: application, potential and limitations. ISPRS J. Photogramm. Remote Sens. 70, 39–55.

Turner, A.K., Schuster, R.L., 1996. In: Landslide Investigation and Mitigation. National Research Council, Transportation Research Board Special Report 247. Washington, 673 pp.

Usai, S., 2003. A least squares database approach for SAR interferometric data. IEEE Trans. Geosci. Remote Sens. 41 (4), 753–760 (part 1).

Van Asch, T.W.J., Van Beek, L.P.H., Bogaard, T.A., 2006. Problems in predicting the mobility of slow-moving landslides. Eng. Geol. 91, 46–55.

Van der Kooij, M., Hughes, W., Sato, S., Poncos, V., February 2006. Coherent target monitoring at high spatial density, examples of validation results. In: Proceedings of Fringe 2005 Workshop, November 28–December 2, 2005, Frascati, Italy. ESA Special Publication SP-610. CD. ISBN 92-9092-261-9, ISSN 1609–042X.

Walstra, J., Chandler, J.H., Dixon, N., Dijkstra, T.A., 2007. Aerial Photography and Digital Photogrammetry for Landslide Monitoring. Geol. Soc. London Spec. Publ. 283 (1), 53–63.

Wasowski, J., Bovenga, F., 2014. Investigating landslides and unstable slopes with satellite multi temporal interferometry: current issues and future perspectives. Eng. Geol. 174, 103–138. http://dx.doi.org/10.1016/j.enggeo.2014.03.003.

Wasowski, J., Del Gaudio, V., Pierri, P., Capolongo, D., 2002. Factors controlling seismic susceptibility of the Sele valley slopes: the case of the 1980 Irpinia earthquake reexamined. Surv. Geophys. 23, 563–593.

Wasowski, J., Lamanna, C., Casarano, D., 2010. Influence of land-use change and precipitation patterns on landslide activity in the Daunia Apennines, Italy. Q. J. Eng. Geol. Hydrogeol. 43, 1–17.

Wasowski, J., Casarano, D., Bovenga, F., Conte, D., Nutricato, R., Refice, A., Berardino, P., Manzo, M., Pepe, A., Zeni, G., Lanari, R., 2007. A comparative analysis of the DInSAR results achieved by the SBAS and SPINUA techniques: a case study of the Maratea valley,

Italy. In: Proceedings of the Envisat Symposium, April 23–27, 2007. ESA Special Publication SP-636, Montreux (Switzerland). July 2007, CD.

Wasowski, J., Casarano, D., Bovenga, F., Refice, A., Conte, D., Nutricato, R., Nitti, D.O., 2008. Landslide-prone towns in Daunia (Italy): PS-interferometry based investigation. In: Chen, Z., et al. (Eds.), Proceedings of the 10th International Symposium on Landslides and Engineered Slopes, Xi'an, pp. 513–518.

Wasowski, J., Lee, C., Keefer, D., 2011. Toward the next generation of research on earthquake-induced landslides: current issues and future challenges. Eng. Geol. 122 (1–2), 1–8. http://dx.doi.org/10.1016/j.enggeo.2011.06.001.

Wasowski, J., Bovenga, F., Nitti, D.O., Nutricato, R., 2012a. Investigating landslides with persistent scatterers interferometry (PSI): current issues and challenges. In: Eberhardt, E., Froese, C., Turner, A.K., Leroueil, S. (Eds.), Landslides and Engineered Slopes, Proceedings of the 11th International and 2nd North American Symposium on Landslides, Banff (Canada), 3–8 June, 2012, vol. 2. CRC Press/Balkema, Leiden, The Netherlands, pp. 1295–1301.

Wasowski, J., Lamanna, C., Gigante, G., Casarano, D., 2012b. High resolution satellite imagery analysis for inferring surface-subsurface water relationships in unstable slopes. Remote Sens. Environ. 124, 135–148. http://dx.doi.org/10.1016/j.rse.2012.05.007.

Wenxue, F., Huadong, G., Qingjiu, T., Xiaofang, G., 2010. Landslide monitoring by corner reflectors differential interferometry SAR. Int. J. Remote Sens. 31 (24), 6387–6400.

Werner, C., Wegmuller, U., Strozz1, T., Wiesmaenn, A., 2003. Interferometric point target analysis for deformation mapping. In: IEEE Proceedings of Geoscience and Remote Sensing Symposium (IGARSS), 21–25 July 2003, Toulouse, France, 7, pp. 4362–4364. http://dx.doi.org/10.1109/IGARSS, 2003.1295516.

International Geotechnical Societies' UNESCO Working Party on World Landslide Inventory WP/WLI, 1995. A suggested method for describing the rate of movement of a landslide. Int. Assoc. Eng. Geol. Bull. 52, 75–78.

Wright, T.J., Parsons, B.E., Lu, Z., 2004. Toward mapping surface deformation in three dimensions using InSAR. Geophys. Res. Lett. 31, L01607. http://dx.doi.org/10.1029/2003GL018827.

Zezza, F., Merenda, L., Bruno, G., Crescenzi, E., Iovine, G., 1994. Condizioni di instabilità e rischio da frana nei comuni dell'Appennino Dauno Pugliese. Geol. Appl. Idrogeol. 29, 77–141.

Chapter 12

Small Landslides—Frequent, Costly, and Manageable

Elisabeth T. Bowman

Department of Civil and Structural Engineering, University of Sheffield, South Yorkshire, UK

ABSTRACT

Small landslides make up the vast majority of landslide incidents that affect communities and individuals globally. However, this reality is not well reflected in the landslide literature, which tends to focus on larger, high-individual-impact events. Globally, the large number of small to medium landslides that occur each year in hill slope and mountainous areas results in large economic, social, and environmental costs. These costs are mostly associated with human habitation and transport linkages. The decision whether to actively manage small landslides—either by event reduction using slope stabilization methods or by mitigation of their impacts—is a function of cost-benefit analyses that are undertaken either explicitly or implicitly. Although the direct costs and benefits are generally assessed to inform a decision about whether to mitigate the effects of a potential or actual landslide, relatively few analyses attempt to quantitatively determine the wider (economic, social, and political) implications of slope failure on communities, although evidence suggests that doing so would probably result in a greater number of interventions. This chapter considers a range of case studies from different countries that include larger and smaller landslide scenarios in order to examine and illustrate current management practice and likely future trends.

12.1 INTRODUCTION

This chapter concerns the management of small landslides and the costs that they pose to society. Four questions must be asked in this context. The first question is "What are the costs that landslides pose?" The second question: "How frequent is frequent?" The third: "What is a manageable landslide?" The fourth and final question is "Just how small is a 'small' landslide?" These questions are discussed in turn here, followed by the examination of some specific case studies that serve to illustrate some of the issues raised.

Landslide Hazards, Risks, and Disasters. http://dx.doi.org/10.1016/B978-0-12-396452-6.00012-4

12.2 COSTS OF SMALL-MEDIUM LANDSLIDES

Small to medium-sized slope instabilities and associated landslides result globally in large economic, financial, environmental, and social costs each year. However, in contrast to the impacts from very large landslides, their effects are commonly hidden within statistics associated with other hazards and events (e.g., landslides that occur during storms or earthquakes) and in the background noise of daily life, with smaller landslide events rarely producing national headlines, although they may figure in regional and local reports.

12.2.1 Financial and Economic Losses

The dominant impact of small slope instabilities and associated landslides is economic and financial. In this sense, small landslides are of particular interest where they directly affect civil infrastructure (usually transport corridors, e.g., roads and rail, less commonly power production, water, wastewater, and reticulation facilities) or the built environment (housing and business premises).

Several studies have attempted to determine the economic costs of landslides compared with other natural and artificial hazards on a region or country basis. Published figures for given countries are highly variable, and reflect the difficulty in pinning down the true costs of these hazards. In particular, published data for less developed regions are sparse to nonexistent. Much recent information on landslides costs is referred to Emergency Events Database: The Office of U.S. Foreign Disaster Assistance/Centre for Research on the Epidemiology of Disasters (EM-DAT:OFDA/CRED) International Disaster Database maintained by the Catholic University of Leuven Belgium. By their criteria, in order for a disaster to be entered into this database, at least one of the following must be fulfilled: 10+ people reported killed, 100+ people reported affected, a call for international assistance, or a declaration of a state of emergency. This removes many of the smaller landslides that this chapter is intended to cover, although the data are still extremely useful.

Some further recent published figures include the following:

USA: Schuster and Highland (2001) quoted the estimated direct and indirect losses due to landslides as US$ 1−2 billion at 1985 values. In 2014 terms, this is equivalent to US$ 2.2−4.4 billion, so on average around US$ 3.3 billion assuming little change in the distribution of populations and infrastructure during 1994−2014. However, according to the US census reports, between 2000 and 2010, the US population increased by 9.7% (the slowest rate since the 1950s), and more of this increase occurred in the south and west of the USA (U.S. Census Bureau, 2012). Extrapolating from this rate of change, the US population will have grown by approximately 20% during 1994−2014. It is likely that indirect losses will therefore have increased by this proportion; hence, a better midrange estimate of the total annual losses in 2014 may be US$ 3.5−4 billion.

Canada: Guthrie (2013) details 56 "notable" landslides between 1841 and 2012 that cost CAD\$ 10 billion (2009 value) over this period in direct and indirect costs. This equates to US\$ 60 million/year at 2014 values. However, the author noted also that these events were only a part of the picture in not including smaller and more frequent landslides. In contrast, Cruden et al. (1989) suggested that by extending the work of Fleming and Taylor (1980) on the costs of landslides in the USA to Canada, the minimum losses would be CAD\$ 200 million (1989 value), equating to around US\$ 300 million/year in 2014 terms. In addition, the authors suggested that by virtue of the low population density of Canada and the large transport network to be maintained, the actual costs could be as high as CAD\$ 1 billion (1989 value)—equating to around US\$ 1.5 billion (2014 value).

Italy: Trezzini et al. (2013) determined that, based on data from 1952 to 2009, landslides cost 52 billion Euros, or approximately one billion Euros/year (order of US\$ 1.3 billion/year). This was compared with approximately 40 billion Euros spent over the same period.

Switzerland: Hilker et al. (2009) reported on a study in which direct damages between 1972 and 2007 due to landslides amounted to nearly eight billion Euros at 2007 values, equating to around US\$ 340 million/year on average in 2014.

China: Zhang and Shan (2006) have suggested that the annual cost of all geological hazards between 1998 and 2006 was around 10 billion Yuan (~US\$ 1.2 billion at 2006 values or US\$ 1.4 billion at 2014 values). In a separate analysis of regional trends, they attributed 92% of the total geological hazards in vulnerable areas to those of slope hazards including landsliding, rock avalanche, and debris flow.

Japan: Estimates of the annual cost of Japanese landslides vary from US\$1 billion (Petley et al., 2005) to US\$ 4–6 billion (UNU News Release, 2006), giving a range of US\$ 1.2–7 billion in 2014. The disparity in these estimates is likely to reflect the exclusion or inclusion of the broader economic impacts of this hazard.

Clearly, a deficiency exists in the data available on economic losses due to smaller landslides on a per country basis. Some recent studies have attempted to address this by subdivision into types or scales of events or infrastructure affected, and developing specific methodologies for these in order that future estimates of costs can be more accurately compiled and compared. A number of studies have been published detailing different approaches for regional assessment of losses (e.g., Wang et al. (2002); Blöchl and Braun (2005); Vranken et al. (2013), whereas others have focused on particular elements of infrastructure, such as Klose et al. (2014) on transport corridors. The recently completed European project "Safeland," among other aims, attempts to compare and harmonize different methods for Quantitative Risk Assessment used across Europe (Safeland, 2012).

Finally, it is important to understand the global effects of climate change on the current and future incidences of small, costly, and potentially manageable landslides. DARA (2012) finds that, due to global climate change, by 2010, floods and related landslides (i.e., excluding those triggered by earthquakes) have already cost an additional US$ 10 billion PPP (purchasing power parity) globally compared to the hypothetical situation in which climate change had not occurred. According to their models, this will increase to an additional economic cost of around $ 95 billion USD PPP (nondiscounted) by 2030 according to their best estimates of climate and population trends. This is over and above "background" business-as-usual levels.

Accordingly, this chapter, where available, specifically seeks to include published costs for case studies given that management and mitigation options are inextricably linked to the financial cost of implementation weighed against the determined economic benefits.

12.2.2 Social Costs

While, fortunately, small landslides are responsible for relatively few human casualties annually, this does not mean that social effects are not felt. Instead, the majority of these effects are related to impacts on civil or urban infrastructure, whereby people's daily habits and businesses are disrupted by the temporary closure of transport links, or by damage to services (power and water) and urban infrastructure. In addition, public perceptions of the reliability of infrastructure affected by landslides can affect political relations within and external to a region—in particular where transport corridors run through multiple countries. Some specific examples of social costs are given in the case studies that follow.

12.2.3 Environmental Costs

Changes in land morphology can be brought about by slope recession due to landslides occurring over time, leading to accumulated losses of land and associated vegetation and soils. In addition, the tracks left by mobile landslides such as debris flows may result in a loss of forestry, other deep-rooted vegetation, grasslands, and continued erosion (Schuster and Highland, 2004). Probably, the largest environmental cost of small landslides is where they enter water courses. In this scenario, they are responsible for generating pulses of sediment loading that can affect the aquatic life of streams and rivers (Acharya et al., 2011). Landslide damming of rivers and streams, though more associated with large landslides, can also occur, causing problems to migrating fish while in place, and the failure of which can lead to devastation of downstream natural resources and anthropic infrastructure.

12.3 FREQUENCY OF LANDSLIDES

Small landslides are specifically omitted from many databases (e.g., EM-DAT); they often occur as clusters within periods of heavy rainfall so that determining the true number generated by a specific storm or typhoon event can be difficult. In addition, the geological/geophysics literature generally is focused on large landslides, given that they pose the most catastrophic of scenarios in terms of risk to life and because the mechanics of such landslides is of particular research interest. The order of the size of landslides that receive the most research attention is generally from 0.1 to 10^5 km^2. In a study on large landslide emplacement behavior, Dade and Huppert (1998), for example, report on landslides with an areal extent of between 0.6 km^2 and up to 45,000 km^2; in Legros (2002), which examines the mobility of landslides, 80 of 86 landslides of which areal extent is reported are larger than 0.1 km^2.

In contrast, an examination of frequency—magnitude graphs created from inventories of landslides caused by known events—such as incidences of high rainfall or snowmelt (Malamud et al., 2004; Guzzetti et al., 2003) generally indicate "upper" boundaries of the order of 0.1–1 km^2 with rollover (most frequent size of landslide) occurring at an order of 10^{-4}–10^{-3} km^2 (100–1,000 m^2) (Guzzetti et al., 2003). Given that the resolution of mapping is consistently below the rollover value determined, rollover values may be deemed as being "real" in reflecting a peak in the frequency of landslide size, and reasonably accurate as a result of the survey work (aerial and field) occurring within a week or two of the event. Historical inventories, in which smaller landslides tend to be absent due to geological weathering (Guzzetti et al., 2008) still show upper boundaries of 1–10 km^2 but with a rollover occurring between 10^{-3} and 10^{-2} km^2 (1,000–10,000 m^2) (Hurst et al., 2013), that is, the accordant determined rollover shifts to a higher value due to a reduction in smaller landslides in the record.

Comparing these two ranges (Figure 12.1), it is interesting to note that the maximum size of landslides in which mitigation and management are deemed even possible (as discussed later) occurs at the lowest magnitude at which the interests of the geological community commonly ends, while the most frequent landslide size is approximately two to three orders of magnitude smaller again.

12.4 MANAGEMENT OF LANDSLIDES

The third question posed earlier—"what is management?"—may be partially answered by addressing the fourth—"how small is a small landslide?." If by "management" we mean active management by an engineered physical solution, rather than say, by avoidance, then we need to understand what options are available to manage or mitigate slope instability while recognizing there is an upper limit to size of landslide that may be economically and physically dealt

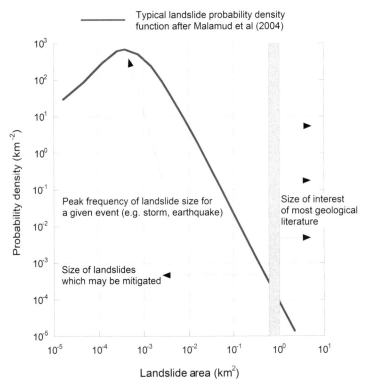

FIGURE 12.1 Frequency—magnitude probability distribution curve for a typical landslide inventory modified after Malamud et al. (2004), compared with the typical size of interest within the geological literature and size of landslides that may be mitigated.

with. In this section, the first point is examined by first looking at the analysis of landslide mechanisms and then the mitigation options that are available.

12.4.1 Analysis Methods—Understanding Mechanisms

Mitigation options must be based on an understanding of the landslide mechanism in question, which means that as well as useful ground-based investigation, an appropriate geotechnical analysis is undertaken. This section focuses on typical analysis approaches.

12.4.1.1 Limit Equilibrium Methods

In routine geotechnical analysis, the use of limit equilibrium (LE) methods with a Mohr—Coulomb failure criterion tends to be the preferred option to examine the failure initiation of slopes using a 2D approximation that corresponds as closely as possible to the worst credible scenario, and there are various commercial software packages available to do this. The advantage of the LE approach is twofold: First, the analysis is simple and parametric studies can easily be undertaken. Second, the Mohr—Coulomb $c' - \phi'$ strength parameters are relatively easy to obtain: by back analysis of a known failure (i.e., by varying the material strength and/or pore pressures and setting the factor of safety to unity); by material strength testing in a shear box, triaxial test, or ring shear; or by carrying out index testing (e.g., Atterberg plastic and liquid limits for clay) and using published correlations of these indices with shear strength (Mitchell and Soga, 2005; Muir Wood, 1990). With known parameters, the factor of safety, defined as the ratio of resisting forces to driving forces within the slope may be determined, and this is easily communicated as a concept to nonspecialists.

The disadvantages of LE methods are that prefailure deformations are not able to be modeled, that the user must specify the failure mode (rotational, planar, or wedge-type) and then which method to use to analyze the failure (e.g., Bishop's or Spencer's methods of slices for rotational failures; Janbu or Sarma's method for nonrotational failures), and that resisting forces provided by retaining structures or reinforcing elements may not be modeled well. Neither are three-dimensional situations easily dealt with.

12.4.1.2 Continuum Numerical Methods

Finite element (FE) and finite difference (FD) continuum numerical models, of which several commercial software options are available, have several advantages over LE methods, and a number of reasons exist as to why the investigator or designer may wish to use this approach. First, the difficulty with (albeit usefulness of) LE methods is that one must, a priori, have an idea of the potential failure or deformation mechanism for a given slope. FE and FD models remove this requirement. A 2D geometric model at a landslide section may be constructed with specified ground conditions and material parameters and the slope allowed to deform spontaneously. Regions of highest shear then become apparent without being specified by the user, and hence, the failure mechanism may be found to be some hybrid of rotational and translational slide or wedge.

The second reason is that even the simplest of constitutive models available within FE and FD require both deformation and failure parameters to be specified. This means that both failure and deformation states may be modeled, which may better reflect field observations, and constitutive models can be selected that improve model behavior. Of course, with increased complexity of a constitutive model comes the increased need for calibration of the model using element testing of the soil in question, a requirement that is too-often ignored (Muir Wood, 2003). It should be noted that the simplest model for soil used in the FE or FD methods generally involves a Mohr—Coulomb failure criterion coupled with linearly elastic parameters for prefailure deformation. Using this model, the investigator can check the results of a "strength reduction" study that is available in most commercial software against the factor of safety calculation provided using a simpler LE method with the same failure mode. The factor of safety results should be very similar, and this can be used as a useful "sanity check" before proceeding with more complex constitutive models or geometric scenarios.

The third reason for using FE or FD models is that complex soil–structure interactions involving structural elements may be incorporated as well as 3D effects. Soil–structure interaction can be particularly important when the element has a stiffness that is intermediate between "stiff" (in which case it can be modeled as fully rigid) and "soft" (in which case it can be modeled as having the same stiffness as the soil) in comparison to the soil. In that case, differential strain development can lead to local effects that may be important to how the soil (or the structure) behaves. In conducting these analyses, extreme care should be taken to examine the degree of mesh refinement around structural elements such as piles. These may be limited by the overall model size, leading to important local effects—such as strain-induced localized failure or slippage between the element and the soil—being ignored. Also, given that complex and large models may require days or weeks of computational time, it is common to model only segments of the slide that interact with the structural element in question; hence, boundary conditions need to be carefully checked.

12.4.1.3 Discrete Numerical Methods

An alternative approach to the modeling of landslides is through the use of the discrete element method (DEM), which inherently models soil and rock as discontinuous particles that interact with each other according to Newton's laws of motion (Cundall and Strack, 1979). Both commercial software and open-source codes exist, and different approaches are available for soil and rock problems (O'Sullivan, 2011; Jing and Stephansson, 2007). A detailed discussion of DEMs is beyond the scope of this chapter; however, it is noted that the method is becoming increasingly popular for modeling of slide-flow processes, jointed rock failure, rock fracture, and rockfall. The major limitations for modeling of soil-based landslides is that pore pressures and fluid flow can be difficult to incorporate numerically in a heavily shearing or flowing particle system, and that even without these effects, computational times can be long. Codes that do include fluid effects may be computationally very expensive, particularly those in which fluid flow and particle movement are fully coupled. As developments in this field progress, however, it is expected that the routine use and utility of such models will increase.

12.4.2 Mitigation Methods

Proposed remedial measures for slopes should directly relate to the potential causes of their instability—which is why having a good understanding of soil and rock mechanics is so important. Four standard approaches exist to mitigation: regrade the slope, reduce the pore pressures by incorporating drainage, reinforce the slope internally using structural or "soft" engineering approaches, or retain it externally.

12.4.2.1 External Retaining Solutions

External retaining solutions most often involve a structural solution based on reinforced concrete (e.g., piles or retaining walls), and although this was in common use in the twentieth century, it has become the option of last resort due to the high cost of implementation, high embedded energy (leading to higher CO_2 emissions), rigid-brittle behavior in the face of seismic loading, and often poor aesthetics. On the other hand, such solutions are robust and may allow much steeper slopes (as much as vertical), which may be necessary in cases of low land availability (e.g., road cuttings for transport corridors through cities) or high relief (e.g., transport corridors in hilly or mountainous terrain).

12.4.2.2 Regrading/Reprofiling Slopes

At the other extreme, cut and fill solutions can be very cost effective and have low maintenance. Grading to a flatter slope is most effective for preventing shallow forms of instability in which the ground movement is confined to soil layers that are close to the surface. For deep-seated instabilities, the overall height of the slope theoretically may be reduced; however, often practical limitations exist (such as the case with a road embankment); hence, more often, toe loading will be carried out using soil fill, rock, rock gabions, or construction of a berm. Finally, although regrading a slope to reduce the driving stresses may be the preferred option to enhance stability—often, it is not physically possible due to land-take issues. Hence, more often than otherwise, other options may need to be considered.

12.4.2.3 Drainage Implementation

Drainage, which can be either shallow or deep, can be a very effective method of stabilizing slopes, but installed drains must be maintained to be effective at all times. Unfortunately, however, proper maintenance is often not considered at the planning stage, and neither is it practised subsequently. The effectiveness of drains should be considered in the long term because the installation of drains will not stop slope movement immediately.

Shallow drainage is designed to control the movement of surface water and hence the hydraulic boundary conditions of the seepage regime in the slope. Shallow drains, which are usually limited to around 1 m in depth, are generally constructed along the top surface of the slope to divert surface runoff. They can be lined with concrete to prevent scour or be unlined ditches filled with gravel, although solutions can also include shallow closed perforated plastic pipes.

Deep drains are designed to modify the seepage pattern within the soil or rock mass. They are more expensive than shallow drains, but are generally very effective because they remove or decrease the pore water pressure directly at the seat of the problem. Typically, deep vertical drains may comprise bored sand- and gravel-filled drains that convey water from within the area of the slope to some permeable strata away from the slope. The increasing use of adits and large diameter drainage tunnels for very deep drainage in recent years is noted—these have the benefit of, in themselves, requiring relatively little maintenance due to their size, although elements that connect to them (e.g., vertical drains) may still require this. One of the earliest of such schemes was completed in 1993 to stabilize deep-seated landslides within the Cromwell Gorge in New Zealand during the construction of the Clyde Dam. This work was achieved at a cost of NZ$ 936 million (2005 value) (Crozier. 2012)—a figure described as "staggering" (Rush, 2009). Some more recent projects are discussed below.

12.4.2.4 Internal Reinforcement of Slopes

During recent decades, internal reinforcement of slopes has come into increasing use. This has followed advances in technology (e.g., in the production of durable plastics for reinforced soil, equipment to mix soil with stabilizers in situ for deep soil mixed (DSM) columns, and methods for the quick installation of soil nails).

Internally, reinforced soil has advantages over many other construction techniques as it is rather flexible in application, being constructed within or as part of the landscape as a design element, or as a remedial measure (e.g., DSM columns used for stabilization of road embankments (Tatarniuk and Bowman, 2012) and geosynthetic-reinforced soil used for embankment reconstruction, Figure 12.2). It also exhibits greater ductility than conventional retaining or restraint methods do, meaning that under seismic loading, gradual deformation rather than sudden failure tends to occur (Jackson et al., 2012). Finally, internal reinforcement can provide for improved aesthetics over conventional methods in allowing for gradual changes in profile (Figure 12.3) and coverage

FIGURE 12.2 Postearthquake reconstruction in Christchurch, New Zealand, using in situ reinforcement. (a) Fitzgerald Road embankment reinstated by the use of geogrid retaining wall; (b) close-up of geogrid rolls being laid. Photographs: Author.

FIGURE 12.3   Rock bolts and nets above small retaining walls with ornamental weep holes ensuring that slope stabilization solutions are not visually intrusive in the historic hillside village of Savoca, Sicily. Photograph: Author.

by vegetation (Choi and Cheung, 2013). The detailed design of internal reinforcement is beyond the scope of this chapter but may be found in the geosynthetics (e.g., (Holtz, 2009), bioengineering (Howell, 1999; Schuster and Highland, 2004), ground improvement, and remediation literatures.

12.5 SIZE OF MANAGEABLE LANDSLIDES

The physical constraint on what size of landslide can or should be mitigated can be relaxed by expanding the economic constraint. In other words, many slope stability (and other!) problems can be fixed if enough money is spent on them, although the only reason why one would implement a very expensive solution would be if the economic benefits in doing so were clear in comparison to the alternative options, or there was simply no other alternative.

When we are looking at slope stability scenarios, the most common large-scale and costly problems that are "chosen to be managed through engineering solutions" are associated with maintaining transport links. In contrast, for landslides that affect towns, the economic arguments for engineering mitigation can be far more difficult to make, meaning this is much less common (rather, planning rules may be changed or other management measures such as early warning systems implemented). This section begins by illustrating this point through an examination of several case studies involving larger landslides that have been successfully mitigated by use of geotechnical/structural engineering solutions (note that the term "larger" landslides is used in this chapter to denote those landslides that are at the upper end of the typical rollover curves as discussed in Section 12.3 but which are still considered "small" in geological terms).

The chapter then examines cases of larger landslides in which no engineering mitigation has been undertaken, situations that generally relate solely to towns—and some of the reasons why. We finally examine some cases which illustrate both common and innovative approaches to mitigating smaller landslides and finally, to some situations in which landslides have not been actively managed. From this, an understanding of what constitutes "manageable" may be derived, including what options are available to do this and when it is desirable to do so.

12.5.1 Larger Managed Landslides

12.5.1.1 Macesnik & Slano Blato Landslides, Slovenia

In Slovenia, several larger landslides have been identified as activating or reactivating in recent years (e.g., the Slano Blato, Stoze, and Strug landslides in Western Slovenia and the Macesnik landslide in Northern Slovenia) due to high rainfall. In terms of engineering management, the most interesting of these are possibly the Macesnik and Slano Blato landslides. Both may be considered to be larger landslides (one to two million cubic meters), yet both are being actively managed using structural and geotechnical engineering solutions. In this section, these are briefly examined to understand what constraints exist on the active management of larger landslides and therefore what is the true limit of landslide size that can be managed in this way.

The Macesnik landslide is located in Northern Slovenia near the Austrian border within the Savinja River basin, above the village of Solčava. The landslide, which was triggered in 1989, is >2,500 m long and from 50 m wide in the upper section to >100 m wide in the lower section, with an estimated volume of the sliding mass of about two million cubic meters (Mikos et al., 2005). On average, it is 10−15 m deep, ranging from approximately 5 m thick at the slide head to around 24 m toward the toe. The inclination of the basal sliding surface is found to vary from 9 to 18°, dependent on location within the slip. The active landslide is located within a much larger (8−10 million cubic meters) and deeper-seated (up to 50 m) fossil landslide, although it is not clear what effect this has on the present slope movement. The geological conditions are complex, consisting of Carboniferous, Triassic, and Oligocene rock formations with the landslide mass being described as consisting of heterogeneous stiff clay with layers of more permeable clayey gravels of different thicknesses at different depths (Mikos et al., 2005).

The landslide was initially triggered in 1989 in a forested area during a period of intense precipitation. However, because it initially did not affect infrastructure, no action was taken until 1994 when the slide began to enlarge and show headscarp retrogression. Between 1994 and 1998, remedial action consisted of the construction of ditches and small concrete canals in an attempt to intercept surface water—although this did not have any measurable effect on the rate of movement of the slide. By 1996, the Solčava-Sleme road, which traverses the slide at several points, had been destroyed in three places, being replaced by a pontoon-style bridge in one area. Downslope motion of the landslide was arrested in 1999 as a result of the slide reaching a natural rock outcrop, so that the bottom of the slide lies at 840 m asl while the upper portion lies at 1,360 m asl.

Monitoring of the slide was put in place in 2001. This initially consisted of conventional survey methods and boreholes equipped with piezometers and inclinometers—although some of these were found to be difficult to maintain (and were in fact destroyed over time) due to the relatively rapid rate of movement of the landslide (Mikoš et al., 2005). These methods were later supplemented by newer measurement methods such as light detection and ranging (Lidar) and those based on global positioning system, which are less reliant on the establishment of discrete monitoring points (Pulko et al., 2014). Between 2000 and 2004, the monitoring revealed that high movement rates were related to high rainfall and piezometric levels (Mikos et al., 2005), and an average rate of movement of 15−25 cm/day was found in the upper portion of the slide, with a maximum rate of movement of 50 cm/day.

This rate of movement was too rapid for the installation of structural mitigation works. Instead, surficial and deep drainage works were implemented from 2002, which slowed down the rate of movement dramatically to between 0.2 and 1 cm/day by the end of 2004 (Pulko et al., 2014). The maximum depth of this drainage was 8 m, and measures consisted of sand-filled perforated trenches. The reduction of movement then allowed two pairs of 5 m-diameter, hollow, reinforced concrete shafts or "wells" to be constructed on the upper and middle sections of the landslide in 2004 and 2006, respectively (Figure 12.4). The hollow shafts had the dual purpose of doweling the landslide into the stable bedrock and acting as deep drainage into which ground water could flow out from the sliding mass during periods of heavy rainfall, reducing the pore pressures that were seen to be chiefly responsible for the movement.

The analysis of the landslide and its mitigation measures were carried out using a commercial FE software package for geotechnical analysis. For the slide, Mikos et al. (2005) quote geotechnical strength parameters as the internal angle of shearing resistance $\phi' = 24.6°$ and apparent cohesion $c' = 1$ kPa, whereas Pulko et al. (2014) quote $\phi' = 32°$ and $c' = 0$ kPa. This difference is due to computational effort, whereupon different portions or reaches of the landslide were analyzed independently. With the variation in basal slip plane angle β, and a determination of strength parameters based on back-analysis with the water table assumed to be located at the ground surface (i.e., considering this as the worst case scenario for each section analyzed), this will lead to different strength parameters being determined for different

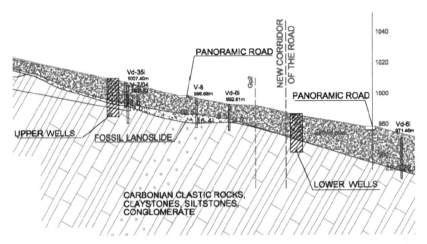

FIGURE 12.4 Central longitudinal profile of the Macesnik landslide in Slovenia, showing remediation solutions via a 5.0-m-diameter reinforced concrete wells (Mikos et al., 2005).

sections. Here, using a simple infinite slope geometry and an average basal slope angle that is assumed to reflect the overall dimensions of the slide, an average basal slope angle β of $12°$ may be determined as representative of the whole slide. Assuming slope-parallel seepage and full saturation, this would give marginal stability (i.e., a factor of safety of unity using LE resulting in a stable slope angle that is approximately $\phi'/2$) for a strength of $\phi' \sim 24°$ ($c' = 0$ kPa)—in close agreement with Mikos et al. (2005) and typical of clay.

By the same process, the internal angle of shearing resistance found as $32°$ in the case of Pulko et al. (2014) applies to an average slope β of approximately $16°$, which corresponds reasonably to the slope of the section directly below the pontoon bridge where the mitigation measures were installed. The ϕ' value is relatively high for clay, but reflects the contribution to resistance provided by the sidewalls in the narrow section of the slide in which the slope was the steepest (Pulko, pers. comm. June 12, 2014). It is also likely to reflect the connectedness of the different parts of the slide, with the true driving and resisting stresses being modified at the ends compared to the model, which is a simplification of the field reality—as pointed out in Pulko et al. (2014). Indeed as stated in Mikos et al. (2005), the part of the landslide that showed the highest rate of movement prior to stabilization was the area in which the basal slope was the steepest. This suggests that it was, in part, being supported by the shallower section below, and hence, a numerical model that considers equal driving and resisting stresses at the top and bottom of the slope will not reflect this. In any case, given the complexity of numerical models and the potential for errors to be induced at various stages of analysis, the quick type of "back of envelope" infinite slope LE calculation carried out here can be an extremely valuable check in assessing the reasonableness of numerical results and can be used for a simple parametric study on the effects of pore pressures and strength parameters.

Mikos et al. (2005) state that by 2005, 500,000 Euros of damage to pastures and forests were estimated to have occurred, while mitigation measures consisting of drainage and structural stabilization of the upper part of the landslide had cost five million Euros. As a result of the ongoing threat to water supplies and additional damage to infrastructure, Mikos et al. (2005) mooted further mitigation measures at an additional cost of 11 million Euro. Pulko et al. (2014) state that only the middle portion of works suggested by Mikos et al. (2005) was undertaken by 2014, and the lower section is yet to be mitigated. In spite of this, the landslide until 2014 has remained stable during periods of heavy rain, and monitoring has shown that installed drainage is working well.

The Slano Blato landslide is located in western Slovenia, in the Ajdovščina municipality, relatively close to the Italian border. The landslide, which is classed as an earth flow, is approximately 1,600 m long, 60–250 m wide, with a total sliding mass of about 900,000 m^3 (area $\sim$240,000 m^2, 4 m deep) located between 270 and 650 m asl. The Slano Blato landslide is approximately 1,300 m long, 60–200 m wide, and 3–11 m deep with a volume of about 700,000 m^3.

Following the success of mitigation measures applied to the Macesnik landslide, a similar approach was applied to the Slano Blato landslide whereby hollow reinforced concrete shafts were intended to be installed to both dowel the moving mass into the bedrock and to provide drainage. Geotechnical differences exist however: Slano Blato has had a history of landsliding encompassing at least 200 years (Logar et al., 2005); the soils are predominantly medium-plasticity silty clay with up to 45% sand and gravel; and most importantly, the underlying flysch bedrock has been found to contribute to very high influxes of water to the soil during periods of heavy rainfall. This led to problems in installing the five 5.4 -m diameter shafts initially (Pulko et al., 2014) and therefore to modifications to the design, which finally led to a total of 11 shafts being installed between 6.5 and 5.4 m in diameter at one section of the slide. During and after construction of the shafts, it was found that the drainage provided by the shafts is a key factor in stabilizing the landslide by reducing the pore pressures both in the landslide mass itself and in the surrounding ground.

These two case studies show that the size of a landslide that may be mitigated by structural solutions is not limited so much by volume but rather by depth and cost. For planar landslides, it is economically and technologically feasible to provide deep drainage via wells, stone or gravel columns, and porous plastic pipes to a limited depth. In the case of the Macesnik landslide, 8 m was given as the maximum that was technically feasible for trench provision by Mikos et al. (2005), but the reinforced concrete shafts provided deeper drainage to a depth of 22 m (including an 18-m landslide depth). For the Slano Blato slide, the drainage provided by the reinforced concrete shafts proved to be the key solution to reducing if not entirely stopping slope movements.

Landslide Hazards, Risks, and Disasters

12.5.1.2 The Manuwatu Gorge, New Zealand

From October 2011 until November 2012, the largest landslide ever to affect transport infrastructure in New Zealand closed a major arterial road, State Highway 3 (SH3) between Woodville and Ashhurst in the North Island. The Manawatu gorge through which SH3 passes has had a history of slope failures since the road was constructed in 1872. The slopes of the steeply sided gorge range from 40 to 60° and consist of weak argillite, sandstone, and colluvium (Hancox, 2011). Failures have been recorded along a 1-km section of the road through the 1930s, 1960s, 1970s and as specific events in 1995, 1998, and 2004 (Hancox, 2011; Anonymous, 2014), when the failure of approximately 100,000 m^3 of soil and rock caused the road to be shut for 10 weeks.

On August 18, 2011, heavy rainfall led to a failure of approximately 20,000 m^3 of soil, rock, and vegetation, which shut the highway for several weeks. This was followed by minor slips that extended the headscarp by 70-m upslope throughout the following two months, leading to 33 days of closure by mid-September due to concerns over safety of the traveling public. Another very large landslide, estimated at 60,000 m^3, then occurred on October 18, causing the highway to be shut once more due to blockage of the road by debris. Investigations during attempts to clear the road revealed that damage had also occurred to a key bridge, while concerns over the potential for further landsliding of unstable material (estimated by Hancox (2011) as constituting up to 100,000 m^3) led to remedial works being proposed to stabilize the slope.

An integrated approach was taken toward stabilization of the slope due to its difficulty of access (the roadway is located ~60 m above the Manawatu River), the range of geological materials involved (from soil to weak rock), and the criticality of the infrastructure. Methods included removal of loose debris, construction of an anchored whaling wall, the application of rock mesh, rockfall benches, and an active attenuator rockfall net system, and revegetation. The eventual volume of debris that was cleared was estimated at 370,000 m^3 (Anonymous, 2014), while the highway was finally fully reopened in November 2012, approximately 13 months after closure, at a total final cost of $21.4 million NZD (McKay, 2012)—of which $5.4 million was spent on upgrading and maintaining alternative routes. This cost can be compared with daily losses to the region of between $62,000 and $74,500 NZD, based on estimates from Vision Manawatu, an organization representing local businesses and engineering consultants (MWH), respectively (Anonymous, 2014). Such losses would total $25–30 million NZD, if taken over the full closure period. In their study, Vision Manawatu included lost productivity ($34,000/day due to extra travel time of 13–20 min, assuming 500 trucks and 6,300 other vehicles would normally travel via SH3 daily), extra fuel required to divert via alternative routes ($20,000/day), and the extra pay for drivers

($8,500/day). The study excluded lost business earnings and damage to the region's reputation as a business hub (Forbes, 2011).

This case study highlights the importance of cost-benefit analyses to engineering decision making. In the case of State Highway 3, which is one of New Zealand's main arterial routes, a "do nothing" approach was clearly unacceptable, and the intention would always have been to reopen the road. However, the decision to have construction carried out at an accelerated pace—with around-the-clock working where deemed safe—enabled the construction time to be reduced from an initially estimated 23 months to 13 months. This decision was based on the need to both reopen the road as quickly as possible and facilitate alternative travel during the construction process, and was weighed against the costs to the local (and national) economy of carrying out the repairs at a "normal" pace (Anonymous, 2014). Unfortunately, in New Zealand as elsewhere, this approach is relatively rare. That is, although closures of more minor roads due to small landslide and rockfall events is common (ODESC, 2007), rarely has there been a detailed attempt to determine the true costs associated with disruption, increased travel time, additional fuel due to rerouting, directly lost business, and the possibility of business going elsewhere in the long term. If this type of study were routinely conducted, based on published cost data, it is likely that greater attempts would be made to ensure that failures occurred less often than they do.

12.5.1.3 La Frasse, Switzerland

Matti et al. (2012) and Matti (2008) have reported on a slow moving landslide at La Frasse above the river "Grande-Eau" in the Canton of Vaud in Switzerland, which has been known to move for over 200 years. The currently active part of the landslide (located—as is disturbingly common—in a wider and deeper inactive landslide) is 2,000 m long, 500 m wide, and has a basal shear depth of between 60 and 100 m with a volume of 42 million cubic meters. The landslide is crossed by two main roads of high economic value leading to tourist resorts (Prina et al., 2004), as well as forests, agricultural land, and some individual dwellings. Between 1975 and 2000, the highest rate of movement of the slide was approximately 60 cm/year, although different segments of the slide move at considerably different rates. In addition, unlike many other slides, although hydraulic pressures are regarded as the main destabilizing feature of the slide, piezometric readings cannot be easily correlated to movement rates due to the complex flysch formations underlying the slide mass, which deliver water via discrete fractures.

Reducing the displacement rate of the landslide had previously been attempted in 1994 by using a system of 22 pumped boreholes, but with limited success as the boreholes increasingly silted up reducing the groundwater flows into them, so that only 15 boreholes were still being pumped by 2003 (Matti, 2008). Two test adits were installed in 2001 in part to assist with drainage and in part to examine the feasibility of constructing a larger drainage gallery—which was adopted as a scheme in 2005. Between 2007 and 2008, the landslide was partially remediated by the construction of a 725-m drainage gallery below the sliding surface, linked to a series of vertical boreholes spaced around 10 m apart, which was built at a cost of approximately 16 million CHF (Matti, 2008). Postconstruction, differential interferometric synthetic aperture radar has been used to monitor slope movements (Michoud et al., 2009). While results have not yet been published, numerical models reported in Tacher et al. (2005) and Matti et al. (2012) indicate that movements should reduce to between 5% and 10% of the original rate of movement.

Readers are also referred to a report (Eberhardt et al., 2007) on a very deep moving rockslide at Campo Villemaggia in Switzerland, which has a volume of 800,000 m^3 and a maximum depth of 300 m; this has been successfully mitigated (i.e., movement has halted) by the construction of deep drainage adits.

12.5.1.4 Hong Kong—Po Shan Managed Drainage Tunnels

On June 18, 1972, during heavy rain, a landslide of approximately 40,000 m^3 occurred above the residential area of Po Shan Road in Hong Kong, resulting in the deaths of 67 people. This event, and several other landslides that occurred during this period up to 1976, led to the formation of the Geotechnical Engineering Office (GEO) in 1977, with the remit to regulate all aspects of slope safety in Hong Kong (Choi and Cheung, 2013).

Geotechnical investigations at Po Shan Road revealed high natural water levels in the hillside and geology consisting of colluvium overlying highly to completely decomposed tuff, which then overlay bedrock (Chau et al., 2011). To reduce the risk of further landslides, drainage to the hillside was implemented in the form of subhorizontal drains in 1984—1985. These were found to perform well in mitigating larger and deeper-seated landslides; however, during periods of heavy rainfall, the ground water level was still found to rise significantly, leading to small, shallow landslides, and damage to the subhorizontal drains—putting these measures in jeopardy.

In 2005, a private engineering company, AECOM, was commissioned to design a more robust solution to prevent landslides, aimed at protecting the residential area of Po Shan Road (Chau et al., 2011). They devised a unique scheme, completed in 2009, consisting of two 3.0-m diameter tunnels or adits (259 and 188 m long, respectively) constructed deep inside the hillside bedrock with a total of 172 subvertical drains drilled from the adits outward and upward into the overlying soil.

The unique element here is that the drainage to the tunnel is actively managed so as to regulate the ground water level. Drainage valves to the tunnel are opened during periods of high precipitation so that the ground water level does not rise significantly, and are closed during quiescent periods so as not to draw the water table down further than historic levels, which could cause further destabilization due to shrink-swell deformation. This active management is enabled by extensive groundwater monitoring in real time using wireless technology. Monitoring of the outcomes has indicated the success of this approach (Figure 12.5) (Cheuk, 2011).

Although this chapter has noted that consideration of risk to residential areas alone is rarely sufficient to justify mitigation of large-scale slope instabilities, the case of Po Shan Road may be the exception to prove the rule. In this case, Hong Kong's high population density (it is the fourth most populous territory in the world) is a major factor, that is, the options to build elsewhere are few. This, and Hong Kong's high degree of technological development, allows for such a costly solution, which requires constant monitoring and response, to be adopted. Further to this, the innovative mitigation measures devised for Po Shan Road may inspire similar schemes to be implemented elsewhere in areas of high land value and vulnerability (Cheuk, 2011).

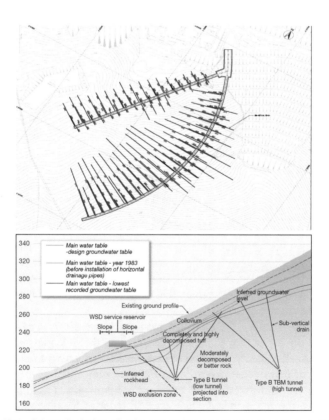

FIGURE 12.5 Schematic design of adits and drainage in the Po Shan Project, Hong King (Cheuk, 2011). Top: Plan view of tunnels and fanned drains in the Po Shan project; Bottom: Longitudinal section through drainage system and water table. *Figure courtesy of Civil Engineering and Development Department, HKSAR Government.*

12.5.2 Larger Unmanaged Landslides

In contrast to the large translational landslides discussed in the previous section, for some deep rotational landslides, available engineering solutions may not be adequate to arrest the movement of a landslide sufficiently to enable maintain existing infrastructure within budgetary constraints.

12.5.2.1 Mam Tor, United Kingdom

One notable historic case study is the approximately 3,000 -year-old Mam Tor landslide located in the North Derbyshire Peak District of the United Kingdom, which the busy A625 road between Sheffield and Manchester twice crossed (Skempton et al., 1989; Donnelly, 2006). The road was originally built in 1810, and although undoubtedly earlier works were required to maintain it, available documents from 1907 show that it had major repairs carried out in 1912, 1933, 1946, 1952, and 1977 (Donnelly, 2006). It was finally abandoned in 1979 due to the inexorable movement of what can be classed as an earth flow. The landslide is approximately 1,000 m long and 350 m wide (Skempton et al., 1989). Crucially, the depth extends to >32 m in the main shearing part of the slide, where a residual angle of shearing resistance of approximately $14°$ is thought to apply to this medium plasticity clay (this compared to a basal slope of $\sim 13°$ in the vicinity of the shear zone). Skempton et al. (1989) carried out detailed LE analyses of the slide under various alternative scenarios and found that increases in pore pressures above the usual background water levels during periods of high rainfall were enough to drive the seasonal movements recorded. The paper by Skempton et al. (1989) provides an exemplary analysis of the slide—including consideration of geological history, monitoring and historical data, borehole data, and stability analyses using the results of soil-index testing correlated to strength parameters (as is commonly possible in the United Kingdom via the large body of soil data published). Between 1990 and 2000, regular surveys were carried out on the Mam Tor landslide that indicated slip rates of approximately 0.25 m per year in the midportion in which the basal shear plane was the steepest and the rate of movement was the fastest (Waltham and Dixon, 2000). Comparing relatively wet and dry years, Waltham and Dixon (2000) found that slip rates increased approximately tenfold in the former.

Proposals to stabilize the landslide made after the 1977 slope movement included the driving of four 1-m diameter adits into the slide mass as far as the slip surface to alleviate the build-up of pore pressure at a deep level, as well as realignment of part of the road to avoid the most active part of the slide. The costs of the works was estimated to be £2 million ($\sim$£11 million in 2014 values). However, the road was abandoned in favor of building an alternative route (Figure 12.6). This route was never actually built, initially because budgetary constraints caused its delay and then because traffic adapted to the alternative routes available. The key comment here by Waltham and Dixon (2000)

FIGURE 12.6   Remains of the original A625 road near Mam Tor, United Kingdom, June 2014. *Photo: Author.*

is that the road over the Mam Tor slide was considered "major, but nonessential." Waltham and Dixon (2000) also stated that although they believe the adits would have been the most appropriate course of action to reduce movement of the slide, their subsequent monitoring showed that even when pore pressures were low, minor movements of the most active part would have continued albeit at a slow pace. Considering Skempton et al. (1989), this ongoing movement may be attributed to the fact that the gradient of basal slip surface (13°) nearly coincides with the residual angle of internal friction of the soil (14° taken from both index test correlation and back analysis). Under these circumstances, even minor elevation of pore pressures above zero along the slip plane would be enough to generate instability; hence, no amount of drainage would have been enough to halt the slide completely without additional structural measures.

12.5.2.2 Taihape, New Zealand; Hunter's Crossing, USA

Further examples of slow, potentially deep-seated failures in which infrastructure has been affected and intervention has not been considered in detail are at Taihape in New Zealand (Massey, 2010) and Hunter's Crossing in the USA (Latham et al., 2010), both of which involve small towns from which residents have had to move away.

The depth of the landslide base is thought to be 13−32 m at Taihape with an average thickness around 22m, with an overall landslide volume of approximately 1 million m^3 (Assets Group, 2009) and covering 140,000 m^2 in the currently active lower part (Massey, 2010). At Hunter's Crossing, the landslide depth is thought to be greater than 11m while the landslide volume is over 50,000 m^3 (Latham et al. 2010). In each of these cases, high pore pressures appear to be responsible for reactivating these creeping slides. The depth of the basal shear layer compared to the size of the overall slide has meant that mitigation has not been seriously considered.

12.5.3 Smaller Manageable Landslides

12.5.3.1 New Zealand—Northland Slips

Small slope instabilities are common in the Northland region of New Zealand as a result of periods of intense rainfall that may last a few hours to a few days during the passage of extratropical storms. The Northland region does not have particularly high relief in comparison to many other parts of New Zealand (Figure 12.7), but the residual clay and mudstone soils from the Northland Allocthon are particularly prone to landsliding in the upper layers due to a high degree of weathering, creating fissured layers within the upper soil, Figure 12.8 (Tatarniuk, 2014).

From July 10—11, 2007, heavy rainfall of >400 mm in 24 h caused approximately $NZ80 million damage to the region's infrastructure including $NZ20 million to the roading infrastructure alone (NIWA, 2007). Smart (2007) notes that much of the serious damage in the far north was caused primarily by unstable hill slopes, whereas many of the recent hill slope failures were

FIGURE 12.7 Northland in New Zealand—a region of relatively low relief but highly impacted by small landslides due to the fissured mudstones and residual overlying soil (Tatarniuk, 2014).

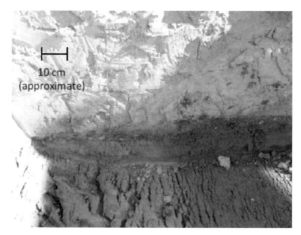

FIGURE 12.8   Residual soil overlying fissure transitional soil—rock at an approximately 2- to 9-m depth in Northland region of New Zealand (Tatarniuk, 2014). Many landslides slip surfaces occur at this interface.

reactivated old slips. At the time, hundreds of overslips (i.e., small landslides onto roads in cuttings) and scores of underslips (where roads embankments or sidings failed, so that road itself required reinstatement) were recorded. Although the 2007 event was mooted as a possible 1 in 150 year event, Northland was again struck by heavy rain (>280 mm) during periods in 2011—causing flood and landslide damage in the region of $7 million NZD and further flooding and rainfall-induced landsliding has occurred through July 2014, as this is written.

With respect to highway infrastructure, various different in situ mitigation options have been trialled to reinstate the roads at different times, taking the form of reinforced soil walls, gabions, H-piles, deep-soil mixed columns, and soil nails—and combinations of these. In general, these approaches appear to be successful and are reasonably economic in terms of both cost and timeliness (O'Sullivan et al., 2008; Terzaghi et al., 2004).

12.5.3.2 Hong Kong—Cut and Fill Slopes

Between the 1940s and mid-1970s, 470 people died due to landslide events in Hong Kong. In response to this, in 1977, the Hong Kong government implemented the Landslip Preventive Measure (LPM) program to systematically retrofit substandard, artificial slopes. According to Choi and Cheung (2013), by completion of the LPM program in 2010, around 4,500 slopes under government control had been upgraded and 5,100 slopes on private land had been safety screened. The total expenditure over the course of the program was about HK$ 14 billion or US$ 1.8 billion. In 2010, at the end of the LPM program, the government inaugurated the Landslip Prevention and Mitigation program to continue the works for medium risk slopes—work that is ongoing. A history and recent updates on the program are given in Chan (2011), while Choi and Cheung (2013) also documented the types of remedial measures used.

Slope stabilization measures primarily involved slope regrading during the 1980s, where possible the implementation of drainage measures, and the provision of internal soil reinforcement since the 1980s. Regarding the last element, given the versatility of soil nailing and its particular utility in the stabilization of shallow slopes, a notable effort has been made to improve the state of knowledge by the use of physical modeling via numerical and centrifuge testing (Choi and Cheung, 2013; Cheuk et al., 2013)—Figure 12.9.

Mitigation of debris flows has also been undertaken via the development of numerical codes and design methods, for example, GEO (2008); GEO-HKIE (2011), and implementation of structural measures, check dams, and most recently flexible barriers. Various relevant documents are accessible at http://www.cedd.gov.hk/eng/publications/geo/.

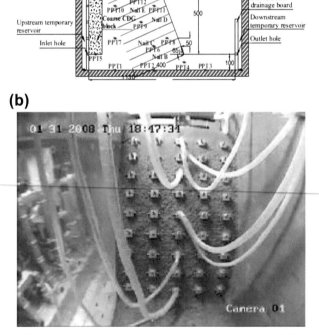

FIGURE 12.9 Geotechnical centrifuge test set up to examine the behavior of soil nails as a slope stabilization measure (Choi and Cheung, 2013). (a) Set-up of a nailed-slope model in centrifuge test. (b) Front view of the slope model in centrifuge test.

12.5.3.3 Swiss/Japanese/Austrian Debris Flows

Switzerland, Austria, and Japan, have high population densities and a propensity to landsliding due to high geological relief. In these countries, active engineering management of small landslides and debris flows near areas of greatest population density, such as towns and cities, is achieved through provision of engineering structures such as soil-nailed slopes, reinforced soil walls and gabions, surficial and deep drainage channels to reroute water, drainage basins, and check dams. These structural provisions are supported by regular maintenance and clearance to ensure that the channels do not become filled with sediment that would reduce their efficacy. A number of well-documented examples exist of these highly managed flows in the literature, including those cited in Jakob and Hungr (2005) and Takahashi (2007). There is little publicly available literature that explicitly examines costs and benefits of different engineering mitigation solutions for debris flows, but several papers exist that discuss management philosophy (e.g., Lateltin et al. (2005) on the Swiss approach) and more general costs of hazard mitigation (e.g., Pfurtscheller and Thieken (2013) on the management and mitigation costs of European Alpine hazards).

12.5.3.4 Laos Road Network

Laos is a landlocked country of medium to high topographic relief with a relatively low population density compared to surrounding countries, such as Vietnam, Thailand, Cambodia, and China. It has a subtropical climate and the incidences of landslides, which are predominantly shallow and small, are largely controlled by its summer rainy season (Hearn et al., 2008). In a study on the cost of small landslides on the road network of Laos, Hearn et al. (2008)— Figure 12.10—reported expenditure of around $US 3–5 million between 2004 and 2007 in remediation of landslides. This only includes direct costs.

FIGURE 12.10   Above road (overslip) and below road (underslip) on a road network in Laos (Hearn et al., 2008).

These authors then undertook an additional analysis including statistics on road usage, costs of vehicles, gross domestic product, working age population, etc. and hence, on the economic impact of road closures. From this, the study adjudged that the cost of road closure due to a landslide would increase exponentially with time so for a road carrying 300 vehicles per day, this would increase from around US$ 2,300 for a 3-h closure to US$ 150,000 for a 24-h closure. In addition Hearn et al. (2008) estimated the environmental costs (increased sedimentation of waterways, loss of trees, etc.) of landslides associated with roads to be US$ 8,150 per event.

Finally, the authors estimated the costs and benefits of using engineering works to prevent or mitigate slope instability, including measures such as bioremediation, provision of drainage, and retaining walls. Based on direct costs alone, they determined that such measures applied to the design and construction of new roads would lead to an increase in initial costs of 10% and a decrease in annual maintenance of 28%. This meant that, using net present value discounting, the investment would only become economical over a 50-year period. They also estimated the costs and benefits associated with improving slope stability for existing roads. Depending on the discount rate used with time, and the estimated storm period leading to significant landsliding, the improvements in terms of engineering costs were found to be marginal (i.e., slight negative or positive return). However, it was pointed out that both these approaches entirely ignore the wider economic and environmental consequences and associated costs as discussed above. For roads of higher traffic volume and where landslides are more common, it was considered that proactive slope management would indeed provide a beneficial return on the investment. Further, the national roads in Laos provide international links with China, Thailand, and Vietnam, and there is a growing expectation of accessibility and reliability of the network.

This study highlights the importance of assessing the broader issues around small slope failures and how these should include not only economic and environmental costs but also consideration of social and political values.

12.5.3.5 Mossaic—Slope Instabilities in Vulnerable Communities in the Caribbean

Mossaic ("Management of Slope Stabilities in Communities"; Anderson and Holcombe, 2013), funded by the World Bank and originally developed in the Caribbean, describes an approach aimed at reducing the incidence of frequent small- to medium-sized rainfall-induced landslides within settled areas. The specific focus is on socially, economically, and physically vulnerable members of society, noting that poorer communities are disproportionately affected by such landslides. This is because, due to population pressure, the land upon which these communities have developed is commonly the most marginal in terms of slope stability and because of the prevalence of poorly constructed

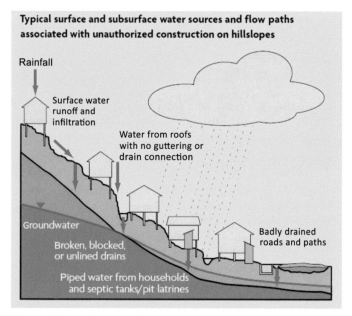

FIGURE 12.11   Influence of communities on water sources due to unauthorized construction on hill slopes (Anderson and Holcombe, 2013).

dwellings in such areas. It is particularly pointed out that slope instability can be exacerbated by human activities such as removal of vegetation and increased runoff due to water use and poor wastewater drainage design (Figure 12.11).

Noting that engineering design and implementation can lead to the reduction of risk to acceptable levels (Figure 12.12), Anderson and Holcombe (2013) espoused a community-focused approach to landslide mitigation using

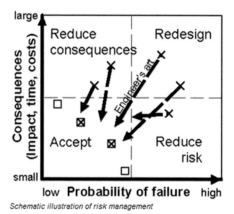

FIGURE 12.12   Concept of engineering risk management toward risk reduction, in Anderson and Holcombe (2013) after Solheim et al. (2005).

low cost but highly beneficial solutions, that is, the application of drainage measures and improved slope design. The handbook produced covers elements such as hazard recognition, geotechnical design, community engagement, project management, and project evaluation.

12.5.3.6 Scotland, A83 Rest and Be Thankful Pass

In the United Kingdom, the A83 trunk road in Scotland has been closed on numerous occasions since 2004 when 57 traveling members of the public were trapped between two debris flow streams and had to be airlifted out. On October 28, 2007, around 150 m^3 was deposited by a debris flow over the road at the Rest and Be Thankful Pass (Transport Scotland, 2014). According to Parkman et al. (2012), on this notable occasion, the road, classed as a low traffic rural road with average daily traffic of 2,250 vehicles, was closed for 12 days, creating an additional mean travel distance of 41 km at an estimated cost of UK £ 320,000 (2008 values) as a result of the increased travel time alone (i.e., excluding other costs, such as lost earnings and business). Further debris flow events involved volumes up to 550 m^3 resulting in closures of several days.

As a result of this activity, a series of engineering assessments of the options to mitigate was commissioned (Jacobs, 2012; Winter et al., 2008), resulting in a flexible barrier system derived from rockfall protection measures being selected as the most cost-effective and least intrusive option. Between 2010 and 2014, 18 debris flow barriers were constructed based on these flexible systems (NCE Editorial 2014). The overall cost of the measures implemented at Rest and Be Thankful is around UK £ 8 million (Smith, 2014), while additional funds of UK £ 6 million have been allocated to other parts of the A83 to further mitigate the effects of shallow landslides. This case study shows that, even for low volume roads, specific landslide mitigation solutions may be put into place if the costs and benefits warrant the intervention.

12.5.3.7 Matata, New Zealand

In New Zealand, management of debris flow hazards is generally carried out by issuing general safety weather warnings and removal of small bridges on walking tracks during winter months when debris flows are most common (i.e., by avoidance), whereas some small simple barriers, such as provided by steel posts, have been installed for at-risk individual dwellings (Tonkin and Taylor Ltd, 2013).

An exception to this was considered for the small town of Matata (population ∼700) in the North Island, which in May 2005 during heavy rainfall, experienced the worst debris flow event in living memory (McSaveney et al., 2005). This event resulted in the destruction some 100 houses (many in a new subdivision) within the small town but fortunately no fatalities. It was determined that during the event, two streams, the Waitepuru and the Awatarariki, contributed 100,000 and 300,000 m^3, respectively, to the high volume of material deposited on the fan $(250,000 \text{ m}^3)$ upon which the town was constructed (note that the discrepancy is due to comparing total volumes of sediment displaced in the streams versus that which arrived at Matata itself). It is significant that there are records of earlier events of a smaller volume occurring in the same locations, and evidence in the form of large boulders, which were not considered when the subdivision was approved.

After the event, some stream-training works were completed on the Waiterpuru stream, and the installation of a very large flexible catchment barrier along the Awatarariki stream was seriously considered (CPG Ltd, 2012). In this case, for both economic and cultural reasons (Domain Environmental Ltd, 2012), more conventional structural solutions were not considered acceptable, and the flexible mesh barrier was thought to provide a feasible alternative for what was considered to be a one in 150-year event. Initial estimated costs of this option grew by 2011 to be $5−7 million NZD. Upon review, it was felt that the provision of such a barrier, which was expected to reduce the hazard but would not eliminate it, would also encourage the construction of houses in inappropriate high risk areas, increasing potential insurance costs and council liability (CPG NZ Ltd, 2012; Domain Environmental Ltd, 2012). Alternative solutions including a chute out-to-sea and a deflection bund were also considered with construction costs at $10 million NZD for the chute option and $5.6 million NZD for the bund option, with annual maintenance of $0.2 and $0.3 million NZD, respectively. Such costs when compared to building property and road assets of approximately $2.6 million NZD within the area expected to be most affected were considered to be too high to enable these options to proceed (CPG NZ Ltd, 2012). Management, in this scenario, has instead taken the form of community engagement and consultation regarding modifications to the District Plan (WDC, 2013).

12.5.3.8 Low-Volume Roads

Exact specified ranges are not set globally, however in general, a low-volume road (LVR) may be defined as having average annual traffic of $<1,000$ vehicles per day, whereas a very low volume road (VLVR) has <400 vehicles per day. Worldwide, an estimated 34 million km of road exist, of which LVRs make up approximately 80−85%, while VLVRs make up 70−75%. Such roads are generally poorly designed to cope with erosion and slope instability.

Although obtaining exact figures is difficult, the economic and social importance of rural and low-volume roads has been recognized by various governments and organizations. Accordingly, a number of manuals and best practice guides for LVRs, aimed at improving their resilience have been produced in recent years. Several recent publications (Fay et al., 2012; Department of Roads − Nepal, 2003; Hearn, 2011), deal specifically with slope stability.

12.5.3.9 Small Unmanaged Landslides/Debris Flows

It is notable that countries that have relatively few, but potentially significant debris flow events each year (e.g., UK, Australia) have few, if any, actively managed catchments given that the economics of control may well outweigh the costs of the event itself. In countries that are sparsely populated, but with high relief (e.g., Canada, New Zealand), the reasons for a lack of active management is more due to the stochastic and unpredictable nature of the events, which, while numerous debris flows occur each year, do not necessarily constitute a high threat to life or economic cost. Instead active management is confined to specific areas of relatively high land value or high population.

12.6 CONCLUSIONS

Small landslides may be defined as those that affect an area per event less than around 1 km^2, although the vast majority of landslides that occur annually are around two to four orders of magnitude smaller than this. Such landslides, although featuring relatively little in the geological literature, impact on civil and urban infrastructure and pose significant problems worldwide. The losses generated by small landslides are mainly economic, although social and environmental costs are also important. The direct costs of landslides that affect transport linkages include designing, implementing, and maintaining solutions to produce stable slopes and mitigating failed slopes. Indirect costs associated with disruption of transport routes include those due to rerouting (e.g., additional fuel, lost time), lost business, and lost economic confidence. Direct costs for disruption of habitations include reinstatement of housing, roads, and services (including drainage, the loss of which can exacerbate slope instability). Indirect costs include increased insurance premiums, loss of business, and disruption to or even destruction of communities.

An examination of cases when small landslides are actively managed shows that the size of a landslide is generally not, in its own right, a limiting factor in whether it will be mitigated. That is, for structural solutions, rather, the landslide depth is important, with maximum economically constructable depths for piles usually not exceeding 20 m. For deeper-seated landslides, deep drainage using tunnels and adits is possible, should the cost warrant such a solution. Monitoring in real time with active structural management is also possible, once a commitment is made to doing so—although this is only likely to be undertaken for areas of very high land value.

Overall, the decision whether or not to mitigate a landslide is made on the basis of economics. From this, we see that costly interventions of larger landslides occur more often in the maintenance of key transportation routes, particularly roads, rather than habitations alone, unless the local value of land is extremely high. This is an understandable distinction from the perspective that transportation, that is, connecting people and goods, affects a much larger proportion of society and the economy. At the other end of the scale, for low volume and rural roads, as well as for rural communities sited in hill slope areas, good geometric design of slopes and slope drainage is increasingly being promulgated to most cost effectively reduce the economic and social impacts of small landslides. Such solutions do not require a high degree of technical expertise, but rather require that a good understanding of soil mechanical behavior is put into practice, coupled with community engagement—particularly since anthropic influences (drain maintenance and clearance, removal or collection of piped water, sewerage, and runoff) are so important to the success of drainage measures and hence slope stability in these communities.

Finally, it is seen that although cost-benefit analyses are generally undertaken to inform decision making on whether to mitigate the effects of landslides, they rarely include a quantitative analysis of the wider (economic, social, political, and environmental) implications of slope failure on communities, Evidence here suggests that including such costs would probably result in a greater number of active engineering interventions.

ACKNOWLEDGMENTS

The author would like to thank Dr CY Johnny Cheuk of AECOM for providing helpful insights on Hong Kong landslide management practices and information on the Po Shan mitigation project. She would also like to thank Dr Liz Holcombe of Bristol University for useful discussions on Mossaic.

REFERENCES

Acharya, G., Cochrane, T., Bowman, E., Davies, T., 2011. Quantifying and modeling post-failure sediment yields from laboratory-scale soil erosion and shallow landslide experiments with silty loess. Geomorphology 129 (1−2), 49−58.

Anderson, M.G., Holcombe, E., 2013. Community-Based Landslide Risk Reduction: Managing Disasters in Small Steps. World Bank, Washington, DC. http://dx.doi.org/10.1596/978-0-8213-9456-4.

Anonymous, 2014. How to Think Fast, Act Smart, Play Safe and Keep Everyone Happy. Infra-structure, NZ ed., vol. 3. Media Solutions Ltd, Takapuna.

Assets Group, 2009. The West Taihape Area: Management Strategy. Rangitikei District Council.

Blöchl, A., Braun, B., 2005. Economic assessment of land slide risks in the Swabian Alb, Germany—research framework and first results of homeowners' and experts' surveys. Nat. Hazards Earth Syst. Sci. 5, 389−396.

Chan, R.K.S., 2011. Evolution of LPM policy in the past thirty five years. In: Paper Presented at the 31st Annual Seminar of HKIE—Landslide Risk Reduction through Works: Thirty-five Years of Landslip Preventive Measures Programme and beyond, Hong Kong.

Chau, S.F., Cheuk, J.C.Y., Lo, J.Y.C., 2011. Innovative approach for landslide prevention—a tunnel and sub-vertical drain system. In: Paper Presented at the 31st Annual Seminar of HKIE—Landslide Risk Reduction through Works: Thirty-five Years of Landslip Preventive Measures Programme and beyond, Hong Kong.

Cheuk, C.Y., 2011. An innovative approach to landslide prevention. Tunnels Tunnelling Int. (vol September), 29–31.

Cheuk, C.Y., Ho, K.K.S., Lam, A.Y.T., 2013. Influence of soil nail orientations on stabilizing mechanisms of loose fill slopes. Can. Geotech. J. 50 (12), 1236–1249. http://dx.doi.org/10.1139/cgj-2012-0144.

Choi, K.Y., Cheung, R.W.M., 2013. Landslide disaster prevention and mitigation through works in Hong Kong. J. Rock Mech. Geotech. Eng. 5, 354–365. http://dx.doi.org/10.1016/j.jrmge.2013.07.007.

CPG NZ Ltd, 2012. Matata Debris Flows Mitigation Structure—Project Review. Whakatane District Council,.

Crozier, M.J., 2012. Landslides. Te Ara—the Encyclopedia of New Zealand.

Cruden, D.M., Thomson, S., Bornhold, B.D., Chagnon, J.Y., Locat, J., Evans, S.G., Heginbottom, J.A., Moran, K., Piper, D.J.W., Powel, R., Prior, D., Quigley, R.M., 1989. Landslides: extent and the economic significance in Canada. In: Paper Presented at the 28th International Geological Congress, Symposium on Landslides, Washington, DC.

Cundall, P.S., Strack, O.D.L., 1979. Discrete numerical model for granular assemblies. Géotechnique 29 (1), 47–65.

Dade, W.B., Huppert, H.E., 1998. Long-runout rockfalls. Geology 26 (9), 803–806.

DARA, 2012. Floods and Landslides. Climate Vulnerability Monitor, second ed. (A Guide to the Cold Calculus of a Hot Planet).

Department of Roads – Nepal, 2003. Guide to Road Slope Protection Works Department of Roads. Ministry of Physical Planning and Works, His Majesty's Government of Nepal, Kathmandu.

Domain Environmental Ltd, 2012. Matata Debris Flow Management—the Way Forward. Whakatane District Council.

Donnelly, L., 2006. The Mam Tor landslide, geology and mining legacy around Castleton, peak district National Park, Derbyshire, UK. In: IAEG2006 Field Trip Guide: Mam Tor & Castleton. Geological Society of London, London.

Eberhardt, E., Bonzanigo, L., Loew, S., 2007. Long-term investigation of a deep-seated creeping landslide in crystalline rock. Part II. Mitigation measures and numerical modelling of deep drainage at Campo Vallemaggia. Can. Geotech. J. 44, 1181–1199.

Editorial, N.C.E., 2014. Geotechnical Road Protection: The New Rock Stars. New Civil Engineer.

Fay, L., Akin, M., Shi, X., 2012. Cost-effective and sustainable road slope stabilization and erosion control. NCHRP 430 (Transportation Research Board).

Fleming, R.W., Taylor, F.A., 1980. Estimating the Cost of Landslide Damage in the United States. United States Geological Survey Circular 832.

Forbes, M., September 22, 2011. Manawatu Gorge Closure Costs about $62,000 a Day. Dominion Post.

GEO-HKIE, 2011. Design Illustrations on the Use of Soil Nails to Upgrade Loose Fill Slopes.

GEO, 2008. Guide to Soil Nail Design and Construction.

Guthrie, R., 2013. Canadian Technical Guidelines and Best Practices Related to Landslides: A National Initiative for Loss Reduction. Geological Survey of Canada. http://dx.doi.org/10.4095/292241.

Guzzetti, F., Aleotti, P., Malamud, B.D., Turcotte, D.L., 2003. Comparison of three landslide event inventories in central and northern Italy using frequency-area statistics. In: Paper Presented at the 4th Plinius Conference on Mediterranean Storms, Mallorca, Spain, October 2002.

Guzzetti, F., Ardizzone, F., Cardinali, M., Galli, M., Reichenbach, P., Rossi, M., 2008. Distribution of landslides in the Upper Tiber River basin, central Italy. Geomorphology 96, 105–122.

Geovert, 2012. Manawatu Gorge Slip Remediation. Ground Engineering Project. Geovert.

Hancox, G.T., 2011. Manawatu Gorge Landslide. New Zealand Transport Agency. http://www.nzta. govt.nz/traffic/current-conditions/highway-info/docs/manawatu-gorge-landslide-description-20111026.pdf.

Hearn, G., Hunt, T., Aubert, J., Howell, J., 2008. Landslide Impacts on the Road Network of Lao PDR and the Feasibility of Implementing a Slope Management Programme. South East Asia Community Access Programme (SEACAP). Department for International Development, United Kingdom.

Hearn, G.J., 2011. Engineering Geology Special Publications. Slope Engineering for Mountain Roads, vol. 24. Geological Society London.

Hilker, N., Badoux, A., Hegg, C., 2009. The Swiss flood and landslide damage database 1972—2007. Nat. Hazards Earth Syst. Sci. 9, 913—925.

Holtz, R.D., 2009. Geosynthetics for soil reinforcement. In: Ang, A.H.S., Chen, S.-S. (Eds.), Frontier Technologies for Infrastructures Engineering. CRC Press, pp. 143—163.

Howell, J., 1999. Roadside Bio-engineering—Reference Manual. Department of Roads, Government of Nepal.

Hurst, M.D., Ellis, M.A., Royse, K.R., Lee, K.A., Freeborough, K., 2013. Controls on the magnitude-frequency scaling of an inventory of secular landslides. Earth Surf. Dyn. 1, 67—78. http://dx.doi.org/10.5194/esurf-1-67-2013.

Jacobs, 2012. A83 Trunk Road Route Study Part A—A83 Rest and Be Thankful Summary—Draft. Transport Scotland.

Jakob, M., Hungr, O., 2005. Debris-flow Hazards and Related Phenomena. Springer, Chichester.

Jing, L., Stephansson, O., 2007. Developments in Geotechnical Engineering. Fundamentals of Discrete Element Methods for Rock Engineering: Theory and Applications, vol. 85. Elsevier.

Jackson, P., Bowman, E.T., Cubrinovski, M., 2012. Seismic testing of model-scale geosynthetic reinforced soil walls. Bulletin of the New Zealand Society for Earthquake Engineering, 45(4).

Klose, M., Damm, B., Terhorst, B., 2014. Landslide cost modeling for transportation infrastructures: a methodological approach. Landslides. http://dx.doi.org/10.1007/s10346-014-0481-1.

Lateltin, O., Haemmig, C., Raetzo, H., Bonnard, C., 2005. Landslide risk management in Switzerland. Landslides 2, 313—320. http://dx.doi.org/10.1007/s10346-005-0018-8.

Latham, R.S., Wooten, R.M., Billington, E.D., Gillon, K.A., Witt, A.C., Bauer, J.B., Fuemmeler, S.J., Douglas, T.J., Kinner, D., Waters-Tormey, C., 2010. The hunters crossing weathered rock slide, Haywood County, North Carolina, USA. In: Weathering as a Predisposing Factor to Slope Movements. Geological Society Engineering Geology Special Publications, London, pp. 149—166. http://dx.doi.org/10.1144/EGSP23.9.

Legros, F., 2002. The mobility of long-runout landslides. Eng. Geol. 63, 301—331.

Logar, J., Fifer, B.K., Kočevar, M., Mikoš, M., Ribičič, M., Majes, B., 2005. History and present state of the Slano blato landslide. Nat. Hazards Earth Syst. Sci. 5, 447—457.

Malamud, B., Donald, D., Turcotte, L., Guzzetti, F., Reichenbach, P., 2004. Landslide inventories and their statistical properties. Earth Surf. Processes Landforms 29 (6), 687—711. http://dx. doi.org/10.1002/esp.1064.

Massey, C.I., 2010. The Dynamics of Reactivated Landslides: Utiku and Taihape, North Island, New Zealand. Durham University, UK.

Matti, B., 2008. Geological Heterogeneity in Landslides: Characterization and Flow Modelling. École Polytechnique Fédérale de Lausanne.

Matti, B., Tacher, L., Commend, S., 2012. Modelling the efficiency of a drainage gallery work for a large landslide with respect to hydrogeological heterogeneity. Can. Geotech. J. 49, 968—985. http://dx.doi.org/10.1139/T2012-061.

McKay, C., October 19, 2012. Gorge Road Back after 13 Months, $21m. Hawkes Bay Today.

McSaveney, M.J., Beetham, R.D., Leonard, G.S., 2005. The 18 May 2005 Debris Flow Distaster at Matata: Causes and Mitigation Suggestions. Institute of Geological and Nuclear Sciences.

Michoud, C., Rune Lauknes, T., Pedrazzini, A., Jaboyedoff, M., Tapia, R., Steinmann, G., 2009. Differential synthetic aperture radar interferometry in monitoring large landslide (La Frasse, Switzerland). Geophys. Res. Abstr. 11, EGU2009−5129.

Mikoš, M., Fazarinc, R., Pulko, B., Petkovšek, A., Majes, B., 2005. Stepwise mitigation of the Macesnik landslide, N Slovenia. Nat. Hazards Earth Syst. Sci. 5, 947−958.

Mitchell, J.K., Soga, K., 2005. Fundamentals of Soil Behaviour, third ed. John Wiley & Sons, Hoboken, New Jersey.

Muir Wood, D., 1990. Soil Behaviour and Critical State Soil Mechanics. Cambridge University Press.

Muir Wood, D., 2003. Geotechnical Modelling. CRC Press.

NIWA, 2007. National Climate Summary—March 2007.

O'Sullivan, A., Davies, T., Okada, W., 2008. Ten walls in ten weeks: case studies in the use of launched soil nail supported GCS walls in the far North District. In: Paper Presented at the 18th NZGS Geotechnical Symposium on Soil−structure Interaction, Auckland, N.Z.

O'Sullivan, C., 2011. Applied Geotechnics. Particulate Discrete Element Modelling: A Geomechanics Perspective, vol. 4. Taylor & Francis.

ODESC, 2007. National Hazardscape Report. Officials' Committee for Domestic and external Security Coordination, Wellington.

Parkman, C.C., Bradbury, T., Peeling, D., Booth, C., 2012. Economic, Environmental and Social Impacts of Changes in Maintenance Spend on Local Roads in Scotland. Transport Research Laboratory, Transport Scotland.

Petley, D.N., Dunning, S.A., Rosser, N.J., 2005. The analysis of global landslide risk through the creation of a database of worldwide landslide fatalities. In: Paper Presented at the Landslide Risk Management, Vancouver, May 31−June 3, 2005.

Pfurtscheller, C., Thieken, A.H., 2013. The price of safety: costs for mitigating and coping with Alpine hazards. Nat. Hazards Earth Syst. Sci. 13, 2619−2637. http://dx.doi.org/10.5194/nhess-13-2619-2013.

Prina, E., Bonnard, C., Vulliet, L., 2004. Vulnerability and risk assessment of a mountain road crossing landslides. Riv. Ital. Geotecnica 38 (2), 67−79.

Pulko, B., Majes, B., Mikoš, M., 2014. Reinforced concrete shafts for the structural mitigation of large deep-seated landslides: an experience from the Macesnik and the Slano blato landslides (Slovenia). Landslides 11, 81−91. http://dx.doi.org/10.1007/s10346-012-0372-2.

Rush, M., 2009. Landslides ~ Gravity Always Wins. http://mightyclutha.blogspot.co.uk/2009/03/landslides-gravity-always-wins.html.

Safeland, 2012. Safeland: Living with Landslide Risk in Europe; Assessment, Effects of Global Change, and Risk Management Strategies: Summary Report.

Schuster, R.L., Highland, L.M., 2001. Socioeconomic and environmental impacts of landslides. In: Paper Presented at the Third Panamerican Symposium on Landslides, Cartagena Colombia, July 29 to August 3, 2001.

Schuster, R.L., Highland, L.M., 2004. Impact of landslides and innovative landslide mitigation measures on the natural environment. In: Paper Presented at the International Conference on Slope Engineering, Hong Kong, China, December 8−10, 2003.

Skempton, A.W., Leadbeater, A.D., Chandler, R.J., 1989. The Mam Tor landslide, North Derbyshire. Philos. Trans. R. Soc. London Ser. A Math. Phys. Sci. 329 (1607), 503−547.

Smart, G., 2007. The Northland Floods 28−29 March 2007: Hydrologic Hazards Investigation. NIWA Project: RIWB055. NIWA, Wellington.

Smith, C., 2014. Further £6M Committed to A83 Landslide Prevention. New Civil Engineer.

Solheim, A., Bhasin, R., De Blasio, F.V., Blikra, L.,H., Boyle, S., Braathen, A., Dehls, J., Elverhøi, A., Etzelmüller, B., Glimsdal, S., Harbitz, C.B., Heyerdahl, H., Høydal, Ø.A., Iwe, H., Karlsrud, K., Lacasse, S., Lecomte, I., Lindholm, C., Longva, O., Løvholt, F., Nadim, F., Nordal, S., Romstad, B., Roed, J.K., Strout, J.M., 2005. International centre for geohazards (ICG): assessment, prevention and mitigation of geohazards. Norw. J. Geol. 85 (1—2), 45—62.

Tacher, L., Bonnard, C., Laloui, L., Parriaux, A., 2005. Modelling the behaviour of a large landslide with respect to hydrogeological and geomechanical parameter heterogeneity. Landslides 2 (1), 3—14. http://dx.doi.org/10.1007/s10346-004-0038-9.

Takahashi, T., 2007. Debris Flow: Mechanics, Prediction and Countermeasures. Taylor & Francis.

Tatarniuk, C., 2014. Deep Soil Mixing as a Slope Stabilization Technique in Northland Allochthon Residual Clay Soil. PhD, University of Canterbury, Christchurch, NZ.

Tatarniuk, C., Bowman, E.T., February 2012. Case study of a road embankment failure mitigated using deep soil mixing. In: 4th International Conference on Grouting and Deep Mixing, New Orleans, USA, pp. 16—18.

Terzaghi, S., Okada, W., Houghton, L.D., 2004. Deep soil mixing in New Zealand—role in slope stabilisation. In: Paper presented at the Amélioration des sols en place, Paris.

Tonkin & Taylor Ltd, 2013. Quantitative Landslide Risk Assessment Matata Escarpment: Final Draft. Whakatane District Council.

Transport Scotland, 2014. Previous Landslides at the A83 Rest and Be Thankful. http://www.transportscotland.gov.uk/road/a83/previous-landslides-a83-rest-and-be-thankful (accessed 09.07.14.).

Trezzini, F., Giannella, G., Guida, T., 2013. Landslide and flood: economic and social impacts in Italy. In: Margottini, C., Canuti, P., Sassa, K. (Eds.), Landslide Science and Practice—Vol. 7: Social and Economic Impact and Policies. Springer-Verlag Berlin Heidelberg, pp. 171—176. http://dx.doi.org/10.1007/978-3-642-31313-4_22.

U.S. Census Bureau, 2012. 2010 Census Special Reports, Patterns of Metropolitan and Micropolitan Population Change: 2000—2010, Vol C2010SR-01. U.S. Government Printing Office, Washington, DC.

UNU News Release, 2006. Asia Has Most; Americas, the Deadliest; Europe, the Costliest; Experts Seek Ways to Mitigate Landslide Losses; Danger Said Growing Due to Climate Change, Other Causes.

Vranken, L., Van Turnhout, P., Van Den Eeckhaut, M., Vandekerckhove, L., Poesen, J., 2013. Economic valuation of landslide damage in hilly regions: a case study from Flanders, Belgium. Sci. Total Environ. 447, 323—336. http://dx.doi.org/10.1016/j.scitotenv.2013.01.025.

Waltham, A.C., Dixon, N., 2000. Movement of the Mam Tor landslide, Derbyshire, UK. Q. J. Eng. Geol. Hydrogeol. 33, 105—123.

Wang, Y., Summers, R.D., Hofmeister, R.J., 2002. Landslide Loss Estimation Pilot Project in Oregon. vol Open-file Report O-02-05. State of Oregon Department of Geology and Mineral Industries.

WDC, 2013. Managing Debris Flow and Landslide Hazards — from the Matata Escarpment. Whakatane District Council.

Winter, M.G., Macgregor, F., Shackman, L., 2008. Scottish Road Network Landslides Study: Implementation. Transport Scotland, Edinburgh.

Zhang, L., Shan, W., 2006. Early warning and prevention of geohazards in China. In: Sassa, K., Fukuoka, H., Wang, F., Wang, G. (Eds.), Risk Analysis and Sustainable Disaster Management. Springer, pp. 285—296.

Chapter 13

Analysis Tools for Mass Movement Assessment

Stefano Utili[1] and Giovanni B. Crosta[2]

[1] *School of Engineering, University of Warwick, Coventry, UK,* [2] *Università degli Studidi Milano-Bicocca, Italy*

ABSTRACT

Slope stability assessment is a fundamental step in evaluating landslide hazard and for the safe design of structures and infrastructures. We analyze the physical principles that underlie both the most frequently used computational methods and some less common but accurate methods. The most important limit equilibrium methods, and the most widely adopted by practitioners, are presented in detail in a unified conceptual framework, and then limit analysis approaches are introduced and discussed. The latter have the advantages of performing parametric analyses of slope stability under slightly simplified conditions to obtain dimensionless stability charts. The basic principles of finite element, finite difference, and distinct element methods are described, and their advantages and disadvantages are listed.

13.1 INTRODUCTION

In this chapter, we outline the physical principles underpinning the onset of landslides. Put into simple terms, the information of interest is how big a potential landslide will be at a chosen site, under what conditions it can occur, and how close the slope is to releasing material through landsliding. To answer the first question, geotechnical engineers and geologists need to identify the location of the underground failure surface, which determines the shape of the failing soil/rock mass and therefore the landslide volume. This task is simple in the ideal case of homogeneous slope materials, but it may become very challenging in the cases of complex stratigraphy and/or the presence of discontinuities since there may be several potential failure surfaces within the investigated slope on which there may be very different mechanical resistances. In this case, a small error in the evaluation of the parameters governing the ground resistance (e.g., the cohesion c and internal friction angle ϕ) along one of the candidate failure surfaces may have a profound effect on the assessment

Landslide Hazards, Risks, and Disasters. http://dx.doi.org/10.1016/B978-0-12-396452-6.00013-6
441

of the landslide volume. For instance, a shallower failure surface may be identified as the most critical one, whereas the most critical surface is in fact a much deeper one whose resistance has been slightly overestimated (this may also occur due to the fact that the level of uncertainty on the determination of the properties of mechanical resistance for the various candidate failure surfaces can be significantly different). Typically, when more than one potential failure surface (or more than one failure mechanism) is considered in a geomechanical analysis, they are compared in terms of a dimensionless number, in most cases the factor of safety (less often, the so-called stability number is used), which according to engineering practice is defined as the ratio between the resistance offered by the ground within the slope (in this case, the resistance developed along the failure surface) and the factors causing instability, the main one being due to the self-weight of the volume involved in the landslide, together with the weight of any installation on the slope, seepage, and/or seismic forces. To answer the second question, "Under what conditions will a landslide occur?", a knowledge of the most critical failure surface and its associated factor of safety is not enough. In fact, it requires also a reasonable estimate of potential future variations of the factors that affect the slope (e.g., hydrogeological conditions and anthropic activities) and of chemophysical factors that may decrease the ground resistance over time (e.g., creep behavior in cohesive soils and weathering in rocks).

The acquisition of reliable information on the geology of the slope is paramount for any geomechanical analysis of the safety of the slope, and for subsequent geotechnical work undertaken to increase the stability of the slope, since any type of analytical or computational calculation and design is based on data from geological maps (especially concerning the potential stratigraphy) and in situ investigations (e.g., borehole stratigraphic and lithologic logs, ground properties, structural features, hydrogeological modeling, and geophysical surveys).

The aim of this chapter is to outline the methods available for analyzing the critical failure mechanisms (hence the potential landslide volume) and the factor of safety of slopes. This is useful to geologists, engineering geologists, and geotechnical engineers and to various other professionals engaged in acquiring data and in slope stability analyses or hazard and risk assessment. For instance, information on the location of layers of different mechanical properties and the determination of these properties through standard laboratory and in situ tests provide more comprehensive information than can be derived simply from the mineralogical composition of the materials.

13.2 THE COMPUTATIONAL TOOLS AVAILABLE

Nowadays, there is a variety of tools available for slope stability analyses and modeling. In this chapter, the ones most used in practice are emphasized. The first method (and nowadays still the one most used in practice for the assessment of the stability of slopes) is the so-called limit equilibrium method (LEM: Fellenius, 1927). Limit analysis, especially the upper bound form, is very important for slope stability since several stability charts generally employed by practitioners (e.g., Michalowski, 2002; Duncan and Mah, 2004; Utili, 2013) have been produced from this method. Unfortunately, the underlying analytical procedure is less well known among practitioners since they can directly use the derived standard charts without necessarily having to know the theory behind it. However, we believe it is important to include limit analysis in any textbook dedicated to the stability of slopes for its importance in deriving stability charts largely used by practitioners. In addition, finite element analyses are becoming increasingly popular (especially for cohesive frictional soils) to run stability analyses of slopes with complex stratigraphy. In slopes where several discontinuities are present (typically rock slopes), an equivalent continuum approach is not feasible because the scale of the discontinuities is too large in comparison with the scale of the slope, so discontinuous analyses are required. To this end, block theory was developed by Goodman and Shi (1985), and the Distinct Element Method (DEM; Cundall, 1971) is becoming increasingly popular. In this chapter, we will outline the basic tenets of LEM, limit analysis, finite element method (FEM), and some elements of discontinuity analysis, for example, Distinct Element Method and block theory.

The methods of analyses illustrated are very different in terms of both the conceptual framework of reference (e.g., continuum versus discontinuum soil mechanics) and the underlying physical assumptions (e.g., the validity of constitutive equations at the level of a representative element volume versus equations governing the mechanical interaction of blocks, and imposition of global equilibrium at the level of a finite wedge of soil/rock versus imposition of local equilibrium at the level of an infinitesimal element of soil/rock). Thus, it is important to start with an overview of the different conceptual frameworks of reference for the various methods illustrated.

Limit equilibrium methods are based on subdividing the mass of potentially unstable ground to be analyzed into (often) vertical slices of a finite size (if the slices are not vertical, they are sometimes called wedges), imposing the equations of (global) equilibrium on each slice, and assuming reaction forces along the boundaries of the slices according to some physical assumptions concerning both the interslice forces (forces exchanged between slices) and the forces at the base of each slide, which stem from the reaction offered by the ground underneath the failing mass and the water pressure. Typically, the water pressure acting at the base of each slice is either assumed on the basis of a hypothetical phreatic line or taken from a seepage analysis performed on the basis of some known hydraulic boundaries (for instance, the position of a horizontal phreatic line at a considerable distance from the slope). In some methods, some of the equations of equilibrium may not be satisfied, and the methods are called non-rigorous, whereas if all the equations are satisfied, the methods are called rigorous. These methods provide no information on the stress state inside the failing mass, nor the deformations, nor the displacements.

In limit analysis, there are two methodologies that have been employed for slope stability: the so-called kinematic (or upper bound) and static (or lower bound) methods. The upper bound method is the more popular among practitioners, and for this reason, we shall present the upper bound method only. In this method, a failure mechanism has to be assumed as in LEM, but with the additional constraint of being kinematically compatible. This means that the failure mechanism has to satisfy equations imposing the constraint that the body can deform but remains a continuum at all times, that is, if we consider two adjacent points located at an infinitesimal distance from each other, neither detachment nor penetration between them is allowed. Then, the energy balance between the rate of external work done by the load applied on the failing mass and the rate of internal energy dissipation, that is, the energy dissipated by the deforming soil, is imposed for all the potential failure mechanisms considered. The critical failure mechanism is identified as the mechanism giving rise to the minimum (lowest) stability number. The energy balance equation translates the well-known principle of virtual work. Both methods assume that the materials constituting the slope behave as an elasto-perfectly plastic body, that is, they assume the validity of the normality rule according to which plastic deformations occur proportionally to the incremental stresses applied according to the so-called associate constitutive law. Considering a linear failure criterion, such as the Mohr–Coulomb criterion, means that the so-called dilation angle is assumed to be equal to the angle of internal friction. However, real frictional-cohesive soils and rocks do not obey the normality rule. In fact, soft rocks, overconsolidated clays, and cemented sands are usually characterized by a dilation angle that is smaller than the friction angle. Unfortunately, the limit theorems are only applicable to materials obeying a nonassociated flow rule in the case of translational failure (Drescher and Detournay, 1993), which is in general far less critical than rotational mechanisms. According to the limit analysis upper bound theorem (Chen, 1975), the collapse load for a material with a nonassociated flow rule is smaller than those obtained for the same material when an associated flow rule is assumed. Manzari and Nour (2000) were the first to examine the effect of soil dilatancy on homogeneous slopes, performing nonlinear finite element analyses of slopes by the strength reduction technique. They showed that the stability numbers obtained from limit analysis are not conservative (i.e., higher than the real value) for soils exhibiting dilation angles smaller than friction angles. Recently, Crosta et al. (2013) ran FEM analyses on straight homogeneous c, ϕ slopes with both the associative flow rule as assumed in limit analysis ($\Psi = \varphi$) and with a dilation angle $\Psi = 1/4\varphi$, typical for materials with little dilatancy, for a range of slope inclinations of $20-30°$, with φ values ranging from 8 to 28°. It emerged that the influence of the dilation angle on the volume of the sliding mass is negligible. This is due to the fact that the soil/rock is little constrained from a kinematic point of view in a slope (or in other words, the degree of confinement on the material is small), whereas dilatancy may have a very important effect in the case of high confinement (e.g., tunneling).

As in LEM, the presence of water pressure acting on the sliding mass is either assumed on the basis of a hypothetical phreatic line or taken from a seepage analysis performed on the basis of some known hydraulic boundaries (e.g., elevation of a horizontal phreatic line at a considerable distance from the slope). In the lower bound method, a stress field satisfying the local equations of equilibrium (i.e., the differential equations of static equilibrium for a deformable body) is imposed together with the already mentioned associated flow rule. This method is dual (i.e., where duality is in the sense of an optimization problem for which both the primal and the dual solutions are possible, and are associated, one with the lower bound and the other with the upper bound of the optimal sought value; Boyd and Vandenberghe, 2004. Their difference is called the duality gap) to the upper bound since the maximum (highest) stability number is sought. The upper bound limit analysis method will be expanded in section 13.4, since it is the most important for slope stability calculations.

The FEM is a very large subject to which several books have been devoted. Some cornerstone textbooks are Zienkiewicz and Taylor (2005), and, specifically for geotechnical engineering, Potts and Zdravkovic (1999, 2001). Here, we provide a brief overview of the main aspects of finite element analyses performed for slope stability. Unlike LEM and limit analysis, any constitutive law can be considered, so there are no restrictions on the type of mechanical behavior that can be considered for the soil/rock of the slope analyzed. This is a continuum mechanics approach since the materials constituting the slope are assumed to be one continuum or several continua separated by known boundaries (e.g., between different strata) along which a mechanical law ruling the interaction has to be specified. The differential equations of classical solid mechanics are applied, that is, equations imposing equilibrium (two in two-dimensional (2D) and three in 3D) on the stress field, equations imposing kinematic compatibility (three in 2D and six in 3D) on the deformation or strain field, and constitutive equations imposing the law of material behavior (three in 2D and six in 3D) linking stresses to strains. The equations of fluid mechanics ruling the behavior of the water (assumed to be in laminar regime, i.e., Darcy's law) and its interaction with the solid phase (seepage forces exchanged between the solid and the fluid phases) are also considered. These differential equations are coupled (in 3D, they amount to a set of 16 independent differential equations), and therefore, no analytical solutions are available. However, they can be solved by discretization via the FEM so that they are imposed at the level of each small element used to discretize the continuum considered (at times, different discretizations may be used for the solid and the fluid phases).

In order to find the potential failure surface, usually the so-called strength reduction technique is employed. First, a solution is found for the whole slope in its current stable state, then the strength parameters of the slope (e.g., c and ϕ or the Hoek–Brown parameters) are decreased by steps, with a new solution being sought after each step of strength decrease has been applied. After each step, the slope suffers extra deformations that typically tend to localize in a narrow band called a shear band that identifies the failure surface in the slope. However, loss of convergence (i.e., a solution satisfying all the differential equations ruling the problem cannot be found) often occurs before a shear band is clearly identifiable within the slope, so that some considerable approximation may be involved in the definition of the failure surface and in the assessment of the stability factor. In the latter case, for instance, if the loss of convergence is due to numerical reasons (i.e., the strength reduction stopped due to a numerical issue rather than because the strength at which failure occurs has been found), the calculated safety factor may be significantly lower than the real one. More recently, analyses where the failure surface is sought by gravity increase (Li et al., 2009; Nishimura et al., 2010) have been presented where the ground strength assigned is kept fixed and the self-weight of the slope is increased until a failure surface is detected. Whether the two methods give rise to similar results is still to be seen, but some studies support this hypothesis (Scholtes and Donzé, 2012). The advantage of using the FEM lies in the accuracy of the constitutive model that can be used. However, a drawback is that very often the failure surface found is poorly defined, and there is no certainty whether the factor of safety found reflects a loss of convergence of the static solution for physical or numerical reasons.

The DEM was introduced by Cundall (1971) for the analysis of rock mechanics problems and then applied to soils by Cundall and Strack (1979). A thorough description of the method is given in the two-part paper of Cundall (1988) and Hart et al. (1988). A distinct element code simulates the mechanical behavior of a system that comprises a collection of arbitrarily-shaped bodies that displace independently from one another and interact only at contacts or interfaces between the particles. In the Distinct Element Method, also called the soft contact approach of the more general discrete element method, the blocks are assumed to be rigid, and the behavior of the contacts is characterized using relative displacement–force laws. The mechanical behavior of such a system is described in terms of the movement of each particle and the interparticle forces acting at each contact point. Finite rotations and displacements of any body of the assembly, including complete detachment between bodies, are allowed. The interaction of the particles is treated as a dynamic process. The calculations alternate between the application of Newton's second law to the particles and a force–displacement law at the contacts. Newton's second law is used to determine the motion of each particle arising from the contact and body forces acting upon it, while the force–displacement law is used to update the contact forces arising from the relative motion at each contact.

The DEM is used in rock slope problems where the presence of single joints and discontinuities in the rock mass plays a dominant role with respect to the overall behavior. For instance, Allison and Kimber (1998) used the UDEC code (Itasca, 2006) to simulate the possible failure mechanisms of a rocky cliff featured by two sets of joints. More recently, Boon et al. (2014) simulated the occurrence of the famous Vaiont slide employing a 3D analysis. The DEM has great potential since unlike the other methods described so far (LEM, limit analysis, and to a certain extent finite element), it has the potential to simulate the whole landslide from its onset throughout the collapse and runout phases until the material comes to a full stop. However, a simulation (especially in 3D) of a full landslide is still computationally very onerous so that very important simplifications have to be undertaken. The most important of these is the upscaling of the block size employed so that potentially un-realistic failure surface and runout distances will be obtained. Although this tool is still mainly confined to academic research, we believe that its description is justified in light of the fact that the increasing computational power available to practitioners makes it likely that this may soon become a mainstream tool for the assessment of slope stability.

13.3 LIMIT EQUILIBRIUM METHODS

13.3.1 Introduction

Limit equilibrium methods have been used in geotechnical engineering for decades to assess the stability of slopes. The idea of discretizing a potential sliding mass in vertical slices was introduced early in the twentieth century. In 1916, Petterson (1955) presented the stability analysis of the Stigberg Quay in Gothenberg, Sweden, where the slip surface was taken to be circular. But the first method of slices is associated with Fellenius (1927, 1936). His method (also known as the Ordinary method, the Swedish circle method, the con-ventional method, and the US Bureau of Reclamation method) assumes no interslice forces, and the factor of safety is achieved by the overall moment equilibrium around the center of a circular slip surface.

In the mid-1950s, Janbu (1954) and Bishop (1955) made advances in the method. Janbu developed his method for generic slip surfaces, whereas Fellenius and Bishop developed their methods for circular surfaces only (later Bishop extended his method to generic surfaces). In the 1960 and 1970s, most other methods were invented, some making the LEM a more powerful and refined tool of slope stability analysis (Spencer, Morgenstern & Price, Sarma methods) and others making it more suitable for hand calculations (force equilibrium methods). Many articles were published in these years on this topic: some of them made a real contribution to the improvement of the method, whereas in others, only slight modifications or different formulations of earlier methods were given.

In the late 1950s, these methods began to be implemented in computer codes. The advent of powerful desktop personal computers in the 1980s made it economically viable to develop commercial software based on LEMs. Nowadays, these methods are routinely used for stability analyses in geotechnical engineering practice, and many programs are available (examples will be supplied further on).

Methods of slices can be classified according to different criteria:

1. suitable only for circular failure surfaces or applicable to any shape of surface.
2. Rigorous or simplified: the former satisfy all the equilibrium equations, whereas the latter satisfy only some of the equilibrium equations. Some authors developed two versions (simplified and rigorous) of the same method: Bishop, Sarma, Janbu, etc. Within simplified methods, a large group is covered by the methods of forces (Lowe & Karafiath, 1960; Corps of Engineers method, Seed and Sultan, 1967).
3. Depending on assumptions made to render the problem statically determinate: three groups of methods can be recognized on the basis of the hypotheses introduced about the interslice forces (Espinoza et al., 1992, 1994).
4. Based on the parameter used to determine the critical surface: the traditional factor of safety Fs or other parameters such as the critical horizontal uniform acceleration (Sarma, 1973, 1979; Spencer, 1978).

All methods approximate the bottom boundary of slices with linear bases. Formulations are based either on differential equations (e.g., Janbu, 1954; Bishop, 1955; Spencer, 1967) or algebraic equations, making it difficult for the inexperienced reader to compare different methods. As the factor of safety is calculated by algebraic equations, and LEMs are based on a slope divided into a discrete number of slices, here the latter formulation is preferred.

13.3.2 Assumptions That Make the Problem Determinate

13.3.2.1 First Group

In the first group, assumptions address the inclination of the resultants of interslice forces with respect to the horizontal direction:

$$T(x) = \lambda_1 f_1(x)E(x), \tag{13.1}$$

where λ_1 is a dimensionless scaling parameter to be evaluated with the factor of safety, and $f_1(x)$ is a chosen scalar function of the abscissa (x) representing the distribution of the inclination of the interslice forces. Morgenstern and Price (1965) were the first to propose this type of assumption. To solve their method, they used the Newton–Raphson numerical technique.

Subsequently, Fredlund (1974) at Saskatchewan University implemented a different numerical procedure (Slope code) based on the so-called "method of best fit regression" (Fredlund and Krahn, 1977). This technique is illustrated since it is common to most methods of slices and it gives the reader a better understanding of the use of equilibrium equations in determining the factor of safety than does the Newton–Raphson technique. Moreover, the structure of the algorithm is almost the same as that implemented, later on, in other computer codes (Slope/w, 2002).

According to Fredlund and Krahn, the sensitivity of F_{ff} and F_{mm} to the distribution of the inclination of the interslice forces $f_1(x)$ is very different (Fredlund and Krahn, 1977). In fact, F_{ff} shows a strong dependence on $f_1(x)$, whereas F_{mm} shows no significant variations with $f_1(x)$ (Figure 13.1). However, in the case of uniform slope, the global factor of safety F shows very little dependence on $f_1(x)$.

Spencer (1967) proposed a simpler expression than the Morgenstern & Price assumption: $T(x) = \tan \theta \, E(x)$. This assumption corresponds to $f_1(x) = 1$

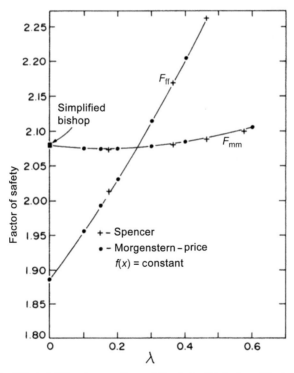

FIGURE 13.1   Effect of different assumptions relative to the distribution of the inclination of the interslice forces (constant, sine, and clipped-sine) on the factors of safety. The analyzed slip surface is circular and crosses a uniform slope (after Fredlund and Krahn, 1977).

and $\lambda_1 = \tan \theta$, where θ is the angle between the interslice resultants and the horizontal direction. Therefore, it is a particular case of the Morgenstern & Price method. In the case of $\lambda_1 = 0$, the factor of safety coincides with F_{Bishop}.

According to Spencer's static assumption, all interslice forces are parallel. However, a variation of the inclination of the interslice resultants along slices must be expected since physics suggests that the soil mass above the slip line is characterized by different stress states: it could be roughly divided into an active region, a transition region, and a passive region. Therefore, the assumed interslice force distribution is not realistic.

Lowe and Karafiath (1960) assumed the direction of the resultants of the interslice forces $\tan \theta$ equaled the average of the slope surface and the slip surface. The U.S. Corps of Engineers method (1970) takes $\tan \theta$ to be equal to either the changing slope of the ground surface or the average slope of the slip surface between the two end slices. Neither method introduces an extra unknown. In fact, they compute only F_{ff}, calculated with the pre-scribed distribution of $\tan \theta$, assuming the factor of safety of forces, as the final factor of safety. Of course, this factor does not satisfy all the equilibrium equations, and it can be very different from the factor of safety calculated by rigorous methods. In fact, F_{ff} is very sensitive to the assumptions made (Figure 13.2).

Chen and Morgenstern (1983) were the first to focus on the physical admissibility of solutions; however, the main interest here concerns the new relationship among interslice forces, which they proposed. They took into consideration the slices located at the edges of slopes (end slices: 1, n). Assuming the slices to be infinitesimal and homogeneous, equilibrium considerations together with the Mohr–Coulomb failure criterion led them to infer that the direction of the interslice resultants of the end slices must be equal to the slope of the ground surface above the slices. They concluded that this condition is necessary to achieve a physically admissible solution.

But, this is not true. In fact, their demonstration is based on the implicit hypothesis of a uniform state of stress at failure throughout the end slices. In the case of finite slices, this hypothesis is not acceptable anymore. Therefore, taking the interslice resultants of the end slices to be equal to the slope of the ground surface cannot be judged a condition of physical admissibility. Nevertheless, Chen and Morgenstern's requirement on end interslice resultants is reasonable since the stress field relative to these slices is, of course, well approximated by a uniform active and passive stress field.

Sarma (1979) suggested assuming a local factor of that is safety constant and equal to the factor of safety along the slip surface. In this case, the same degree of mobilization of the strength parameters along all the vertical slices is assumed. Sarma's method assumes that the whole shear strength along the slip surface is mobilized under the action of a uniform horizontal acceleration K, defined as a fraction of the acceleration of gravity $K = a/g$. Therefore, a critical

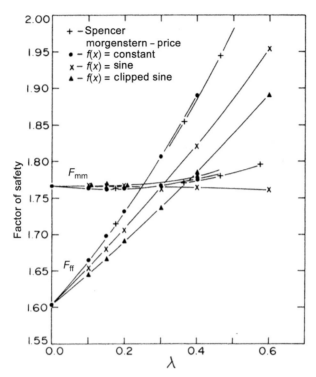

FIGURE 13.2 λ versus factors of safety. The factor of safety F is given by the point of intersection of the two curves (F_{ff} and F_{mm}) obtained by best fitting polynomial regression. The analyzed slip surface is circular and crosses a uniform slope (after Fredlund and Krahn, 1977).

acceleration factor $K_c = a_c/g$ is sought instead of the factor of safety. At the end of an iterative procedure, K_c and the extra unknown λ are determined.

   To obtain the static factor of safety, the value of the factor of safety that gives zero critical acceleration needs to be computed. To this end, the strength parameters of the material along the slip surface have to be reduced by a trial factor of safety F_i, and the critical acceleration K_{Ci} has to be computed. If K_{Ci} is positive, the new trial factor must be $F_{i+1} > F_i$; otherwise, $F_{i+1} < F_i$. A few trials suffice to produce a relationship between K_c and F from which the static factor of safety can be inferred. Note that the surface with the lowest K may not have the lowest F and the surface with the lowest F may not have the lowest K_c. This has been recognized by Sarma as well (Sarma and Bhave, 1974).

13.3.2.2 Second Group

In the second group of methods, assumptions concern the shape of internal shear force distribution $T(x)$:

$$T(x) = \lambda_2 f_2(x), \tag{13.2}$$

where λ_2 is a scaling factor with the dimension of a force and $f_2(x)$ a chosen scalar function. The expression proposed by Sarma belongs to this class of hypotheses:

$$T(x) = \lambda_2 f_2(x)\left[c_{\text{avg}}H(x) + k'\frac{\gamma H^2}{2}\tan\phi_{\text{avg}}\right],$$

where k' depends on the soil strength parameters and the geometric characteristics of the analyzed slice (Sarma, 1973). An assumption of this type was also proposed by Correia (1988).

Note that methods belonging to the first two groups do not use moment equilibrium equations for each slice in determining the factor of safety. The line of thrust can be determined at the end of the iterative process, once Fs and λ have been found, in order to assess the reasonableness of a solution. To do so, working from the first slice to the last, the points of action of the interslice forces h_i are found by taking the equilibrium of moments for each slice in turn. The independent moment equilibrium equations available are $n-1$, since the overall moment equilibrium has been used to determine F_{mm}, and they coincide with the remaining unknowns: $h_2,\ldots, h_n$.

13.3.2.3 Third Group

Unlike the methods previously discussed, the third group of methods of slices makes use of moment equilibrium equations for each slice in determining the factor of safety.

13.3.3 Limit Equilibrium Solutions: Physical Admissibility and Optimal Solutions

Chen and Morgenstern (1983) stated two conditions of physical admissibility. The first condition applies to interslice shear forces, which must not exceed the shear strength that can be mobilized along the slip surface, that is,

$$-1 \le \rho \le 1, \tag{13.3}$$

where ρ is defined as the ratio between the local factor of safety averaged along the vertical face of each slice:

$$F(x) = \frac{\left[c_{\text{avg}}H(x) + E(x)\tan\phi_{\text{avg}}\right]}{T(x)}$$

and the factor of safety along the slip surface. The second condition concerns the line of thrust that must not lie outside the vertical surface of slices:

$$0 \le h(x) \le H(x). \tag{13.4}$$

According to Zhu et al. (2003), two more conditions can be stated: the effective normal forces P_i and the effective interslice forces E_i must be nonnegative. These conditions apply to granular and cohesive soils, but do not apply to rock. In fact, LEMs are mainly used to analyze cohesive and non-cohesive slopes rather than rock slopes. For rock slopes, it is reasonable to accept negative interslice forces limited by the rock tensile strength. If the condition of Eqn (3) is violated, the factor of safety determined refers to a surface that is no longer the slip surface. In fact, slip should occur along the vertical interslice face where Eqn (3) is violated and along the remaining part of the analyzed slip surface.

Eqn (4) is required by elementary principles of mechanics, since the line of action of a force must intersect the body that the force is acting on. Zhu et al. (2003) showed that the condition of Eqn (3) may not be satisfied in simplified as well as rigorous methods (Figure 13.3).

Physical admissibility is a necessary requirement to achieve a reasonable solution from LEMs, but it is not yet sufficient. In fact, a solution may be physical admissible, but not reasonable if the achieved interslice force distribution does not "agree" with the state of stress within the slope mass which, in simple cases, can be guessed by engineering experience or otherwise needs to be determined by a finite element analysis. To perform such an analysis is a simple task as it is required to analyze the slope in its current stable state (far from collapse). No problems due to localization of deformations intervene, and even simple constitutive relations may be used. In fact, the purpose of such an analysis is to determine horizontal and shear stresses (σ_{xx}, τ_{xy}) which, integrated along vertical lines from the slip to the ground surface, give rise to a distribution of "mobilized interslice forces" suitable to be compared from a qualitative point of view with the distribution resulting from the limit equilibrium analysis. If more than one failure surface has been analyzed by LEMs, only the distribution of forces relative to the most critical surface has to be judged.

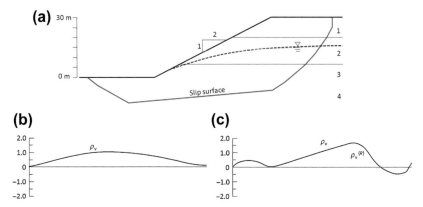

FIGURE 13.3   (a) Composite slip surface. (b) Rho distribution obtained by the Morgenstern & Price method. (c) Rho distribution obtained by Janbu's rigorous method (after Zhu et al., 2003).

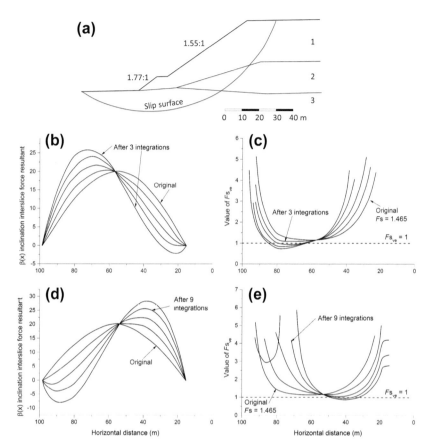

FIGURE 13.4   (a) The slope and the circular failure surface. (b) and (d) The distribution of the inclination of the interslice force resultants, initially sinusoidal, varied toward two opposite directions. (c) and (e) The variation of the local factor of safety relative to (b) and (d) distributions, respectively. The limiting condition is reached at Fs = 1 (after (Chen and Morgenstern, 1983)).

Chen and Morgenstern (1983) aimed to find bounds to $f(x)$ from physical admissibility conditions. They analyzed two example cases, one of which is shown in Figure 13.4(a), by the Morgenstern & Price method. They used the numerical procedure based on the Newton–Raphson technique developed at the University of Alberta in order to find the factor of safety (Morgenstern and Price, 1967). They selected $f_0(x) = \sin x$ as the starting function expressing the variation of the inclination of the interslice resultants and determined the associated Fs. The starting function $f_0(x)$ was then modified by repeatedly adding $\Delta f(x) = K(\eta_1(x) + \eta_2(x))$ where $\eta_1(x)$ and $\eta_2(x)$ were arbitrarily chosen functions (parabolic and elliptic) and K is a constant value that makes $\Delta f(x)$ sufficiently small in comparison to the corresponding value of $f(x)$ in order to not lose convergence. They then determined a new Fs_{i+1} for each iteration

$f_{i+1}(x) = f_i(x) + \Delta f_i(x)$. According to the authors' intentions, the iterative process should have ended when the function $f_{min}(x)$, giving rise to the minimum value of F_{min}, was found. In all cases examined, the iterative procedure was stopped on reaching the limiting condition in one or more slices (Figure 13.4(c,e)). Therefore, the F_{min} determined, in terms of mathematical analysis, is a constrained minimum and not a free minimum.

In general, adopting different η_i functions, any type of distribution $f(x)$ could be tested. But it is not necessary to use complicated functions, since the cubic functions adopted (η_1, η_2) were enough to get a reasonable distribution (Figure 13.4(d)). From this work stems the idea of determining the best interslice force distribution function $f(x)$, defined as the function among all the possible ones giving rise to the minimum factor of safety, as an optimization problem with constraints (physical admissibility conditions). Considering that the determination of the most critical slip surface is an optimization problem too, determining the most critical surface having assumed the best interslice force distribution becomes a double optimization problem.

Therefore, implementing an efficient and time-inexpensive algorithm capable of seeking the optimum interslice force distribution $f(x)$ for each slip surface analyzed appears to be a difficult task, since the variation in the factor of safety for the assumed functions was small and relative minima of Fs were not found, but only constrained minima. Anyway, to assess the influence of interslice force distribution on Fs, an extensive numerical campaign is required, and this task appears prohibitive since Fs depends on many factors such as stratigraphy, soil strength, number of slices, and slip surface assumed.

To obtain physically admissible solutions, simple modifications to the existing algorithms of limit equilibrium computer codes could be introduced. Of course, further constraints may lead to nonconvergence since, in some situations, there are no physically admissible solutions for any interslice force distribution assumed. In these cases, the only way to achieve a safety factor consistent with a physically sound interslice force distribution is by repeating the analysis with another LEM. In general, methods belonging to the third group are less likely to converge to physically admissible solutions.

13.3.4 Some Critical Considerations

In simple cases, that is, when the slip surfaces analyzed are circular, Bishop's simplified method is a good method since F_{mm} is only slightly sensitive to the extra unknown λ_1 (Figure 13.1), and it is lower than the factor of safety found by rigorous methods; therefore, it is on the safe side. This has been found to be an accurate method of analysis for circular slip surfaces in any case (Duncan and Wright, 1980). The Ordinary method, by contrast, gives bad results in the case of flat slopes and deep failure surfaces.

However, the stratigraphy of many slopes features the presence of thin weak layers (e.g., lenses of sand) or bedrock at shallow depths. Hence, the slip

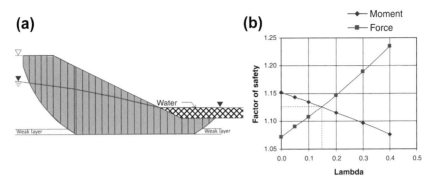

FIGURE 13.5   (a) Composite slip surface. (b) F_{mm} decreases at increasing λ; therefore, $F < F_{mm}$. Results obtained by Slope/w, 2001 code (after Krahn, 2003).

surface becomes composite (partly planar and partly circular). In this case, F_{mm} may be strongly sensitive to λ_1, and above all, it is no longer lower than F (Figure 13.5). Therefore, in all cases of noncircular slip surfaces (composite, logspiral, etc.), rigorous methods have to be preferred since it is not guaranteed that F_{Bishop} is an approximation on the safe side of the factor of safety found by rigorous methods.

Force equilibrium methods determine a safety factor F_{ff} that already shows strong sensitivity to λ_1 for simple cases. In the past, when computers were not readily available, these methods were advantageous because they made it possible to calculate the factor of safety only by drawing force equilibrium polygons, and they had some success. But, nowadays, they should be avoided as they are unnecessary in any case.

Among rigorous methods, there is no physical evidence of the superiority of one method over another. Moreover, all methods can be reformulated in a unified framework (Espinoza et al., 1992, 1994), and there are computer codes where the main methods are implemented so that comparative analyses may be easily run. In any case, methods that make use of the moment equilibrium equation for each slice (third group) require the analyst to guess the line of thrust. This is a nonintuitive task especially for complex slopes. In these cases, methods belonging to the first group should be preferred as an engineer is usually better able to make a mental picture of the inclination of the interslice resultants than of their location. In other words, it is easier to guess forces than points of action as dealing with equilibrium of forces is usually simpler than equilibrium of moments.

Sarma's method (based on the search for the critical acceleration factor) may be useful for slope stability analysis when the main cause of collapse is earthquake shaking. Of course, an analysis based on LEMs that considers the seismic effect only by means of a uniform static force appears very poor in comparison with other methods of analyses (FEM, spectral element method, boundary element method, coupled finite and spectral element method, etc.)

and cannot take into account fundamental phenomena such as soil liquefaction, which intervene during the seismic event. But if a first rough estimation by a pseudostatic analysis is sought, Sarma's method appears to be the best among all the LEMs. In fact, it makes it possible to determine the acceleration level causing the slope to collapse and the failure surface as result of the analysis. Then, a safety factor may be defined as the ratio between the critical acceleration $g \cdot K_c$ and the design seismic acceleration PGA (peak ground acceleration) of the chosen return period. This safety factor has the advantage of being based on the main cause of collapse. By contrast, all the other LEMs determine the factor of safety as the reduction of soil strength causing slope collapse. The seismic effect is represented by a uniform inertial force assumed equal to the PGA acceleration multiplied by the soil mass. The factor of safety found is not based on the cause of collapse and the slip surface found as critical is likely to not coincide with that determined by Sarma's method.

The concept of physically admissible solutions, first applied to LEMs by Chen and Morgenstern, is useful but not sufficient to obtain physically meaningful solutions in terms of interslice forces. This is due to a limitation intrinsic to all the LEM methods: they are based on satisfying the equations of global static equilibrium written for finite soil slices, but they do not provide any information on the actual state of stress within the soil mass. To obtain such information, finite element analyses of the soil wedge (or more recently DEM analyses) are needed.

A final observation concerns the very large literature available on LEMs. Unfortunately, as already mentioned, there is no uniform way of presenting the methods, for example, variational formulation versus discrete formulation. Moreover, a great deal of difficulty is due to the fact that, in many papers, people who publish results referring to algorithms relative to methods of slices, do not clearly and exhaustively report which equations they implemented in their codes, but only quote the method that they intended to implement or to generalize. Unfortunately, most times, their implementation does not correspond to the original method and therefore it is very difficult to understand to what extent the method used is different from the original one and to use the results published to compare the performances of the methods they refer to.

13.4 LIMIT ANALYSIS

13.4.1 The Limit Analysis Upper Bound Theorem

Considering a 3D solid, a virtual rate of displacement that satisfies the following relations:

$$\dot{W}_{\text{ext}} = \int_V F_j \cdot \dot{u}_j \mathrm{d}V + \int_{S_F} f_j \cdot \dot{u}_j \mathrm{d}S > 0 \text{ and } \dot{u}_i = 0 \text{ on } S_u, \qquad (13.5)$$

$$\dot{\varepsilon}_{ij} = \frac{1}{2}\left(\dot{u}_{i,j} + \dot{u}_{j,i}\right),$$ (13.6)

$$\dot{\varepsilon}_{ij} = \frac{\partial g}{\partial \sigma_{ij}}\dot{\lambda} \text{ and } \dot{\lambda} \geq 0$$ (13.7)

with **g** vector of plastic modes making a convex domain in the stress space, gives rise to a kinematically admissible act of motion. Assuming such an act of motion, the upper bound limit analysis theorem states that "the loads determined by equating the rate at which the external forces do work:

$$\dot{W}_{ext} = \int_V F_j \cdot \dot{u}_j dV + \int_{S_F} f_j \cdot \dot{u}_j dS$$ (13.8)

to the rate of internal dissipation:

$$\dot{W}_d = \int_V \sigma_{ij}\dot{\varepsilon}_{ij}dV$$ (13.9)

will be either higher than or equal to the actual limit load." Therefore, it can be inferred that the lowest load among all the loads relative to admissible failure mechanisms, determined by equating the rate of external work to the rate of energy dissipation, is the best approximation to the limit load. This load is an upper bound on the limit load.

In this case, the only force present is the weight force (a body force) and no tractions are present on the solid boundary. Eqn (13.7) are satisfied since a $c - \varphi$ soil type has been assumed. Further, the problem is 2D.

A further assumption about kinematics has been made: rigid body motions are considered. This means that strains only develop along a narrow separation layer (discontinuity surface) between a sliding rigid body and a fixed one (Figure 13.6(a)) where all energy dissipation occurs. According to the assumptions made, the rate of energy dissipation is given by

$$\dot{W}_d = \int_\Gamma (\sigma\dot{\varepsilon} + \tau\dot{\gamma})dl.$$ (13.10)

Strains develop according to an associated flow rule (Figure 13.6(b)).

The slope self-weight is given by $F = mg = \rho gA$. Since the area A is proportional to the slope height H, the load is proportional to H as well. Finally, the energy balance equation,

$$\dot{W}_{ext} = \dot{W}_d$$ (13.11)

makes it possible to determine a function $c = c$ (considered mechanisms) by which the most critical mechanism can be determined. The maximum of this function gives an upper bound on the limit value.

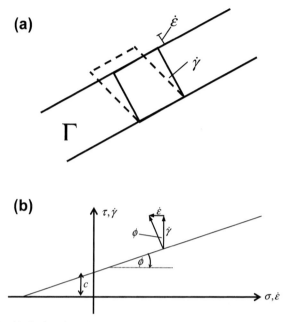

FIGURE 13.6   (a) Strains along the separation layer. (b) Stresses and strains according to the associated flow rule.

13.5 CONTINUUM NUMERICAL METHODS

For most of the problems of practical interest in engineering geology and geotechnical engineering, the integration of the governing system of partial differential equations is possible only by means of approximate numerical methods. As mentioned above, among the various numerical techniques adopted to solve geomechanical problems, two methods are commonly used in practice for geomaterials characterized by nonlinear and irreversible behavior. These are the finite difference (FDM) and the finite element (FEM) methods that are based on the discretization of the governing partial differential equations. The derivative operator becomes the difference quotient in the FDM and the solution improves with reduction in the discretization step. Unfortunately, for complex material behaviors, the solution to the problem does not always exist, nor is it always unique, as is the case for linear governing equations. In the case of nonlinear equations (e.g., for coupled seepage and static problems), the efficiency and accuracy of FDM reduce and FEM is preferred, unless very fine discretization steps (in both space and time) are adopted.

The FEM approach is based on a local formulation of the governing partial differential equations and the transformation from a differential to an algebraic problem. FEM is based again on a spatial discretization, but the unknown functions are approximated through some functions with a constrained number of unknown scalar coefficients that represent the values of the unknown functions at specific points. A detailed presentation of the FEM method for general problems in solid mechanics is given in Zienkiewicz and Taylor (2005). A specific textbook dedicated to the application of finite elements into geotechnical engineering with considerable emphasis on slope stability is that of Potts and Zdravkovic (1999).

The main advantage of FEM is the possibility of including complex constitutive laws (e.g., creep via viscosity) and inclusion of progressive failure (e.g., gradual development of the failure line) implying the redistribution of stresses in the continuum and along the failure surface. The disadvantage of FEM is the onerous computational time when parametric analyses or back analyses are sought, and approximations in the computation of the landslide volume because of the mesh size. The latter can be solved by the used of more advanced meshing techniques (e.g., adaptive remeshing). Recently, FEM analyses based on an arbitrary lagrangian eulerian framework have been adopted to simulate complex landsliding inclusive of the initial instability, the collapse, and runout and the associated erosion of basal material (Crosta et al., 2009a,b, 2013; Roddeman, 2008).

Finally, during 1994—2004, the combined finite-discrete element method has been developed and applied to the solution of complex solid mechanics problems including failure, fracture, fragmentation, collapse, and extensive material damage mechanisms (Munjiza, 2004).

13.6 DISTINCT ELEMENT METHOD

For jointed rock masses intersected by discontinuities, the design philosophy is different between sparsely and moderately jointed rock masses (Hoek et al., 1995). For sparsely jointed rock masses with large joint spacing, stability is governed by key blocks whose shapes permit free kinematic movement. Failure involves either sliding or falling of key blocks. The study of key blocks has led to established analytical design procedures using stereographic projection techniques and block theory (Londe, 1970; Goodman and Shi, 1985; Goodman, 1995). However, slope stability assessments based on key blocks are not suitable for slopes with moderately to heavily jointed rock masses. Sliding of large wedge structures is not normally encountered in such materials, as maintained by Hoek et al. (1995). In contrast, failure usually involves raveling or loosening of rock mass material (Utili and Crosta, 2011a,b). The failed material usually consists of numerous rock blocks. The complexity of the problem has led rock engineers to develop, for the purpose of routine design, rock mass classifications based on past field data, such as the Rock Mass Rating (Bieniawski, 1983) and the Q-system (Barton et al., 1974).

Jointed rock masses are typically made up from numerous polyhedral rock blocks, whose faces are determined by discontinuities in the rock field. The spatial distribution, size, and orientation of these discontinuities are rarely

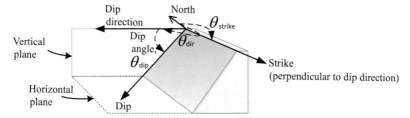

FIGURE 13.7 Definition of strike, dip, and dip direction according to Hoek et al. (1995).

regular and are usually assumed to follow probabilistic distributions. As a result, the sizes and shapes of each block in the jointed rock mass are different. For the purpose of distinct element modeling (DEM) or discontinuous deformation analysis (DDA), one has to invest significant effort to identify polyhedral blocks from the discontinuities, whose orientations are typically defined using their dip directions and dip angles (Figure 13.7).

Broadly, there exist two approaches in block generation algorithms, a summary of which is given in Boon (2013). The first approach is based on subdivision, in which discontinuities are introduced sequentially (Warburton, 1985; Heliot, 1988; Yu et al., 2009; Zhang and Lei, 2013). Each discontinuity is introduced one at a time. If the discontinuity intersects a block, the parent block is subdivided into a pair of so-called child blocks. This process is repeated until all the discontinuities have been introduced. The number of blocks increases as more "slices" are introduced, and a data structure of every block is maintained throughout the slicing process. The blocks generated through sequential subdivision are convex because a discontinuity has to terminate at the face of a neighboring block. Concave blocks can, nonetheless, be generated through the use of clustering, which can be automated (Yu et al., 2009) or guided by specifying fictitious construction joints (Warburton, 1985). Blocks subdivided by a construction joint are clustered together by imposing a kinematic constraint that prevents any relative movement between the two sides of the joint. Likewise, nonpersistent joints, that is, joints of finite sizes (Einstein et al., 1983; Zhang and Einstein, 2010), can be modeled through clustering or specifying fictitious construction joints. This is discussed in more detail in the next paragraph.

On the other hand, in the second approach ("face-tracing" based on homology theory), discontinuities are introduced all at once. All the vertices and edges in the domain are first calculated from the intersections between the discontinuities. From these vertices and edges, there are ways by which the faces and polyhedra in the rock mass can be identified (Lin et al., 1987; Ikegawa and Hudson, 1992; Jing, 2000; Lu, 2002). The necessary algorithms are, however, rather complex. The advantage of this approach is that convex and concave blocks are identified in the same manner. Nonpersistent joints and dangling joints, that is, joints that terminate inside intact rock without contributing to block formation (Jing, 2000), are also treated in the same manner as persistent joints, that is, joints of infinite size. Depending on the type of mechanical analysis that is to be performed on the generated rock mass, these dangling joints may have to be removed; for instance, they have to be removed if either the DEM (Cundall, 1988; Itasca, 2006, 2007) or DDA is used later on for analysis, but they do not need to be removed if fracturing has to be modeled, for instance, by employing the discrete FEM (Pine et al., 2006).

13.7 CONCLUSIONS

Assessing the degree of stability of natural or artificial slopes requires a series of steps: first, geological and geomorphological characterization; second, assessment of mechanical properties; third, choice of the boundary conditions; and finally, choice of a calculation method for assessing slope stability. This chapter addresses the last point, outlining limit equilibrium and limit analysis methods, and the basic principles and features of finite element, finite difference, and discrete element methods. Limit equilibrium methods are still the most widely used for practical applications, and for obtaining a rapid assessment of slope stability under complex geological conditions as well. Recently, 3D LEMs and codes have become increasingly available, enlarging the possible applications, and details can be found in the specialized literature. In this chapter, we have not discussed this subject, but the main principles of 2D methods are straightforwardly extendible to them. Limit analysis is a more robust method that has the great advantage of supporting parametric slope stability studies and for which the initial solutions have been extended to more complex conditions (e.g., slope geometry, tension cracks, and reinforcements). Numerical methods, for both continuum and discontinuum materials, are extremely powerful and versatile. The former are employed for frictional-cohesive slopes; the latter are used for rock slopes where the stability of the slope is ruled by the presence of discontinuities. Nevertheless, they require specific expertise in preparing data, characterizing the material behavior, and setting up the numerical model (e.g., discretization and meshing, boundary conditions), and their presentation requires a dedicated chapter or even a volume. Finally, the choice of the best method to be used depends on the quality of the input data and the target level of accuracy. The use of more than one method is recommended for providing a more complete understanding of the significant variables and parameters controlling the stability of the considered slope.

REFERENCES

Allison, R.J., Kimber, O.G., 1998. Modelling failure mechanisms to explain slope change along the isle of Purbeck coast, UK. Earth Surf. Processes Landforms 23, 731−750.

Barton, N., Lien, R., Lunde, J., 1974. Engineering classification of rock masses for the design of tunnel support. Rock Mechanics Felsmechanik Mécanique des Roches 6 (4), 189−236.

Bieniawski, Z.T., 1983. Geomechanics classification (RMR system) in design applications to underground excavations. In: Paper Presented at the Proceedings − International Symposium on Engineering Geology and Underground Construction. Lisbon.

Bishop, A.W., 1955. The use of the slip in the stability analysis of earth slopes. Geotechnique 5 (1), 7−17.

Boon, C.W., 2013. Distinct Element Modelling of jointed rock masses: algorithms and their verification. DPhil Thesis. University of Oxford, Oxford, UK.

Boon, C.W., Houlsby, G.T., Utili, S., 2014. New insights in the 1963 Vajont slide using 2D and 3D distinct element method analyses. Geotechnique (in print). http://dx.doi.org/10.1680/geot.14.P.041.

Boyd, S.P., Vandenberghe, L., 2004. Convex Optimization. Cambridge University Press. ISBN 978-0-521-83378-3, 716 pp.

Chen, W.F., 1975. Limit Analysis and Soil Plasticity. Elsevier, Amsterdam.

Chen, Z., Morgenstern, N.R., 1983. Extensions to the generalized method of slices for stability analysis. Can. Geotech. J. 20 (1), 104−119.

Correia, R.M., 1988. A limit equilibrium method of slope stability analysis. In: Proc. 5th Int. Symp. Landslides. Switzerland, Lausanne, pp. 595−598.

Crosta, G.B., Imposimato, S., Roddeman, D., 2009a. Numerical modeling of 2-D granular step collapse on erodible and non erodible surface. J. Geophys. Res. 114, F03020.

Crosta, G.B., Imposimato, S., Roddeman, D., 2009b. Numerical modelling of entrainment/deposition in rock and debris-avalanches. Eng. Geol. 109 (1−2), 135−145, 10.1016/j.enggeo.2008.10.004.

Crosta, G.B., Imposimato, S., Roddeman, D., Frattini, P., 2013. On controls of flow-like landslide evolution by an erodible layer. In: Margottini, C., Canuti, P., Sassa, K. (Eds.), Landslide Science and Practice, Spatial Analysis and Modelling, vol. 3. Springer Heidelberg New York Dordrecht London, pp. 263−270. http://dx.doi.org/10.1007/978-3-642-31427-8. ISBN 978-3-642-31426-1,28.

Cundall, P.A., 1971. A computer model for simulating progressive large scale movements in blocky rock systems. In: Proceedings of the Symposium of the International Society for Rock Mechanics (Nancy, France, 1971), vol. 1, pp. II−8.

Cundall, P.A., 1988. Computer simulations of dense sphere assemblies. In: Satake, M., Jenkins, J.T. (Eds.), Micromechanics of Granular Materials. Elsevier Science, Amsterdam, pp. 113−123.

Cundall, P.A., Strack, O.D.L., 1979. A discrete element model for granular assemblies. Geotechnique 29 (1), 47−65.

Drescher, A., Detournay, E., 1993. Limit load in translational failure mechanisms for associative and non-associative materials. Géotechnique 43 (3), 443−456.

Duncan, J.M., Wright, S.G., 1980. The accuracy of equilibrium methods of slope stability analysis. Eng. Geol. 16 (1), 5−17.

Duncan, D.C., Mah, C.W., 2004. Rock Slope Engineering, Civil and Mining. Taylor and Francis, 431 pp.

Einstein, H.H., Veneziano, D., Baecher, G.B., O'Reilly, K.J., 1983. The effect of discontinuity persistence on rock slope stability. Int. J. Rock Mech. Min. Sci. Geomech. Abstr. 20 (5), 227−236.

Espinoza, R.D., Repetto, P.C., Muhunthan, B., 1992. A general framework for slope stability analysis. Géotechnique 42 (4), 603−615.

Espinoza, R.D., Bourdeau, P.L., Muhunthan, B., 1994. Unified formulation for analysis of slopes with general slip surface. J. Geotech. Engrg. ASCE 120 (7), 1185−1204.

Fellenius, W., 1927. Erdstatische Berechnungen mit Reibung und Kohasion. Ernst, Berlin (in German).

Fellenius, W., 1936. Calculation of the stability of earth dams. In: Proc. 2nd Congr, 4. Large Dams, Washington, USA, pp. 445–462.

Fredlund, D.G., 1974. Slope Stability Analysis. University of Saskatchewan, Saskatoon, Canada. User' manual CD-4.

Fredlund, D.G., Krahn, J., 1977. Comparison of slope stability methods of analysis. Can. Geotech. J. 14 (3), 429–439.

Goodman, R.E., 1995. Block theory and its application. Geotechnique 45 (3), 383–423.

Goodman, R.E., Shi, G.H., 1985. Block theory and its application to rock engineering. Prentice-Hall International Series in Civil Engineering and Engineering Mechanics, p. 338.

Hart, R., Cundall, P.A., Lemos, J., 1988. Formulation of a three dimensional distinct element model—Part II. Mechanical calculations for motion and interaction of a system composed of many polyhedral blocks. Int. J. Rock Mech. Min. Sci. Geomech. Abstr. 25 (3), 117–125.

Hoek, E., Kaiser, P.K., Bawden, W.F., 1995. Support of Underground Excavations in Hard Rock. Taylor & Francis, pp. 1–215.

Ikegawa, Y., Hudson, J.A., 1992. Novel automatic identification system for three-dimensional multi-block systems. Eng. Comput. 9 (2), 169–179.

Itasca, 2006. UDEC Universal Distinct Element Code. User Manual. 2006. Itasca Consulting Group Inc, Minneapolis, MN.

Itasca, 2007. 3DEC 3-Dimensional Distinct Element Code, Ver 4.1. 2007. Itasca Consulting Group Inc, Minneapolis, MN.

Janbu, N., 1954. Application of composite slip surfaces for stability analyses. In: Proc., European Conf. on Stability of Earth Slopes, Stockholm, Sweden, 3, pp. 43–49.

Jing, L., 2000. Block system construction for three-dimensional discrete element models of fractured rocks. Int. J. Rock Mech. Min. Sci. 37 (4), 645–659.

Li, C., Tang, C.A., Zhu, H.H., Liang, Z.Z., 2009. Numerical analysis of slope stability based on the gravity increase method. Comput. Geotech. 36 (7), 1246–1258.

Lin, D., Fairhurst, C., Starfield, A.M., 1987. Geometrical identification of three-dimensional rock block systems using topological techniques. Int. J. Rock Mech. Min. Sci. Geomech. Abstr. 24 (6), 331–338.

Londe, P., 1970. Stability of rock slopes—graphical methods. Journal of the Soil Mechanics and Foundations Divisions 96 (4), 1411–1434.

Lowe, J., Karafiath, L., 1960. Stability of earth dams upon drawdown. In: Proc., 1st Pan-Am. Conf. On Soil Mech. and Found, 2. Engrg., Mexico City, Mexico, pp. 537–552.

Lu, J., 2002. Systematic identification of polyhedral rock blocks with arbitrary joints and faults. Comput. Geotech. 29 (1), 49–72.

Manzari, M.T., Nour, M.A., 2000. Significance of soil dilatancy in slope stability analysis. J. Geotech. Geoenv. Engrg. ASCE 126 (1), 75–80.

Michalowski, R.L., 2002. Stability charts for uniform slopes. J. Geotech. Geoenv. Engrg. ASCE 128 (4), 351–355.

Morgenstern, N.R., Price, V.E., 1965. The analysis of the stability of general slip surfaces. Géotechnique 15 (1), 79–93.

Morgenstern, N.R., Price, V.E., 1967. A numerical method for solving the equations of stability of general slip surfaces. Comput. J. 9, 388–393.

Munjiza, A., 2004. The Combined Finite-Discrete Element Method. John Wiley & Sons Ltd, The Atrium, Southern Gate, Chichester, England, 350 pp.

Nishimura, T., Fukuda, T., Tsujino, K., 2010. Distinct element analysis for progressive failure in rock slope. Soils and Foundations 50 (4), 505–513.

Petterson, K.E., 1955. The early history of circular sliding surfaces. Géotechnique 5, 275–296.

Pine, R.J., Coggan, J.S., Flynn, Z.N., Elmo, D., 2006. The development of a new numerical modelling approach for naturally fractured rock masses. Rock Mech. Rock Eng. 39 (5), 395–419.

Potts, D.M., Zdravkovic, L., 1999. Finite Elements Analysis in Geotechnical Engineering: Theory. Thomas Telford, 458 pp.

Potts, D.M., Zdravkovic, L., 2001. Finite Element Analysis in Geotechnical Engineering: Application. Thomas Telford, London.

Roddeman, D.G., 2008. TOCHNOG User' Manual. FEAT, 255 pp. www.feat.nl/manuals/user/user.html.

Sarma, S.K., 1973. Stability analysis of embankments and slopes. Géotechnique 23 (3), 423–433.

Sarma, S.K., 1979. Stability analysis of embankments and slopes. J. Geotech. Engrg. Div. ASCE 105 (12), 1511–1524.

Sarma, S.K., Bhave, M.V., 1974. Critical acceleration versus static factor of safety in stability analysis of earth dams and embankments. Géotechnique 24 (4), 661–665.

Scholtes, L., Donze, F.V., 2012. Modelling progressive failure in fractured rock masses using a 3D discrete element method. Int. J. Rock Mech. Min. Sci. 52, 18–30.

Seed, H.B., Sultan, A., 1967. Stability analyses for a sloping core embankment. J. Soil Mech. Found. Div. ASCE 93 (SM4), 69–83.

Slope/w, 2002. Slope/w for Slope Stability Analysis. Version 5. User's Guide. Geo-Slope Office. Geo-slope Int. Ltd., Calgary, Alberta, Canada.

Spencer, E., 1967. A method of analysis of the stability of embankments assuming parallel interslice forces. Géotechnique 17 (1), 11–26.

Spencer, E., 1978. Earth slope subjected to lateral acceleration. J. Geotech. Engng. Div. ASCE 104 (GT12), 1489–1500.

U.S. Army Corps of Engineers, 1970. Engineering and design-stability of earth and rock fill dams. In: Engrg. Manual EM 1110-2-1902. Dept. of the Army, Corp of Engrs., Ofc. of the Chf. of Engrs.

Utili, S., 2013. Investigation by limit analysis on the stability of slopes with cracks. Geotechnique 63, 140–154.

Utili, S., Crosta, G.B., 2011a. Modelling the evolution of natural slopes subject to weathering: part I. Limit analysis approach. J. Geophys. Res. – Earth Surf. 116, F01016.

Utili, S., Crosta, G.B., 2011b. Modelling the evolution of natural slopes subject to weathering: part II. Discrete element approach. J. Geophys. Res. – Earth Surf. 116, F01017.

Warburton, P.M., 1985. A computer program for reconstructing blocky rock geometry and analyzing single block stability. Comput. Geosci. 11 (6), 707–712.

Yu, Q., Ohnishi, Y., Xue, G., Chen, D., 2009. A generalized procedure to identify three-dimensional rock blocks around complex excavations. Int. J. Numer. Analytical Methods in Geomech. 33, 355–375.

Zhang, L., Einstein, H., 2010. The planar shape of rock joints. Rock Mech. Rock Eng. 43 (1), 55–68.

Zhang, Z.X., Lei, Q.H., 2013. Object-oriented modeling for three-dimensional multi-block systems. Comput. Geotech. 48, 208–227.

Zhu, D.Y., Lee, C.F., Jiang, H.D., 2003. Generalised framework of limit equilibrium methods for slope stability analysis. Géotechnique 53 (4), 377–395.

Zienkiewicz, O.C., Taylor, R.L., 2005. The Finite Element Method, sixth ed. Butterworth-Heinemann, Oxford. vol. 1 and 2.

Note: Page numbers followed by "b", "f" and "t" indicate boxes, figures and tables respectively